IMPACT OF CLIMATE CHANGE ON HYDROLOGICAL CYCLE, ECOSYSTEM, FISHERIES AND FOOD SECURITY

Climate change has emerged as the most pressing global challenge of the 21st century. Sea level has been reported to be increasing at a rate of 0.1 to 0.2 mm/year while Sea Surface Temperature (SST) showed an increase of 0.6 to 1°C, over the last century. Climate change has a dramatic effect on natural ecosystems and environment often leading to acceleration in biodiversity loss. These impacts will have a knock-on effect on many sectors that depend on natural resources, such as agriculture, fisheries, aquaculture, energy and tourism. Thousands of peer-reviewed studies from around the world have documented evidence for extensive climate effects on environment that are already occurring and many of which are irreversible. Climate change is altering aquatic ecosystems, driving shifts in physical and chemical processes, ecological communities and the distribution and abundance of commercially important species. Ocean warming and acidification also affect growth and reproduction processes of many marine organisms, which may reduce stock availability of several commercial species and food and nutritional security of the masses. Intelligent mitigation strategies to minimise climate change impacts can result in advanced, novel technologies; healthier aquatic ecosystems and higher food security and well-being for humans.

Prof. (Dr.) B. Madhusoodana Kurup, an internationally renowned fisheries researcher, academician, visionary and administrator, is the Founder Vice-Chancellor of Kerala University of Fisheries and Ocean Studies (KUFOS), Cochin and also served as the third Vice Chancellor of Shri Venkateshwara University, Gajraula, Uttar Pradesh. He has established and successfully developed the first Fisheries and Ocean Sciences University in the country.

Dr. M.R. Boopendranath retired as Principal Scientist from Fishing Technology Division, Central Institute of Fisheries Technology, Indian Council of Agricultural Research. He has 35 years of professional experience in fisheries and harvest technology of fish, with focus on responsible fishing techniques, fishing gear selectivity, bycatch reduction devices and turtle excluder devices, Antarctic krill and Myctophid harvesting techniques and energy optimization in fish harvesting systems.

Prof. (Dr.) M. Harikrishnan was educated in Cochin University of Science and Technology and has been teaching post graduate students for the past 20 years besides undertaking serious research in freshwater and deep sea crustaceans and fish bionomics.

Dr. A.V. Shibu, is an alumnus of School of Industrial Fisheries and has been involving in fisheries academics for the past 12 years. He started research career in fishing technology and has been actively contributing in developing, implementing and evaluating various academic programmes of the school.

IMPACT OF CLIMATE CHANGE ON HYDROLOGICAL CYCLE, ECOSYSTEM, FISHERIES AND FOOD SECURITY

Edited by:

B. Madhusoodana Kurup
M.R. Boopendranath
M. Harikrishnan
and
A.V. Shibu

NARENDRA PUBLISHING HOUSE
DELHI (INDIA)

First published 2022
by CRC Press
4 Park Square, Milton Park, Abingdon, Oxon, OX14 4RN

and by CRC Press
6000 Broken Sound Parkway NW, Suite 300, Boca Raton, FL 33487-2742

CRC Press is an imprint of Informa UK Limited

British Library Cataloguing-in-Publication Data
A catalogue record for this book is available from the British Library

Library of Congress Cataloging-in-Publication Data
A catalog record has been requested

ISBN: 978-1-032-29043-0 (hbk)
ISBN: 978-1-003-29976-9 (ebk)

DOI: 10.1201/9781003299769

Contents

JUSTICE DEVAN RAMACHANDRAN

सत्यमेव जयते

Foreword

7th July, 2020

This Book has its parturition in the International Conference on '*Impact of Climate Change on Hydrological Cycle, Ecosystem, Fisheries and Food Security*', organized and held by the Cochin University of Science and Technology along with the Departments of the Government of Kerala and the Government of India, between the 12th and 14th February 2020, in Kochi.

It was my honour to have been invited to inaugurate the Conference; and the couple of hours of the Inaugural Ceremony was indeed a treat, with the foremost experts and analytical minds offering a glimpse - in the nature of a virtual curtain raiser - into the depth and thoroughness of the deliberations to take place in the next three days. The carefully curated assemblage of foremost scientists, academicians, researchers, aqua farmers, administrators and policy makers on a global platform made the Conference unique and its outcome never to be ignored.

The unmistakable changes in climate all around the world, with its attendant inevitable consequences of altered water cycle, natural disasters, deluges and famines, compels us all - from the prince to the pariah - to be partners in adaptation and mitigation of global strategies since, on this, depends the survival of humanity. This is where the conference really had its acme worth.

Being the compilation of the various papers presented in the Conference, this Book is a treasure-trove for all involved seriously in the study of climate change and presents an invaluable opportunity to others to read and assess the future of our existence, juxtaposed by the imperatives of conservation and protection of the natural resources and elements of nature, we have always taken for granted.

I commend all those who have contributed to this book; as also the editors, for the effort taken to choose and arrange the papers in a manner to make the experience of the reader richer to the topics chosen.

The contextual analysis in each of them is done so lucidly that it can relate to the expert and the dilettante alike and this is its success, because the topics, otherwise, are specialized and generally esoteric to the layman.

In all, this book has abundance without profusion; elegance without affectation and profundity without superficiality.

Justice Devan Ramachandran

HIGH COURT OF KERALA

ERNAKULAM, KOCHI-682031, PHONE: 0484-2393901 to 2393906/0484-2562502 (Direct)

Preface

Climate change has emerged as the most pressing global challenge of the 21st century. Sea level has been reported to be increasing at a rate of 0.1 to 0.2 mm/year while Sea Surface Temperature (SST) showed an increase of 0.6 to 1°C, over the last century. Climate change has a dramatic effect on natural ecosystems and environment often leading to acceleration in biodiversity loss. These impacts will have a knock-on effect on many sectors that depend on natural resources, such as agriculture, fisheries, aquaculture, energy and tourism. Thousands of peer-reviewed studies from around the world have documented evidence for extensive climate effects on environment that are already occurring and many of which are irreversible. Climate change is altering aquatic ecosystems, driving shifts in physical and chemical processes, ecological communities and the distribution and abundance of commercially important species. Ocean warming and acidification also affect growth and reproduction processes of many marine organisms, which may reduce stock availability of several commercial species and food and nutritional security of the masses. Intelligent mitigation strategies to minimise climate change impacts can result in advanced, novel technologies; healthier aquatic ecosystems and higher food security and well-being for humans.

This volume titled *Impact of Climate Change on Hydrological Cycle, Ecosystem, Fisheries and Food Security,* published by Narendra Publishing House, Delhi, includes papers presented in the *International Conference on Impact of Climate Change on Hydrological Cycle, Ecosystems, Fisheries and Food Security* (ClimFishCon 2020), 11-14 February 2020, Kochi, jointly organised by the School of Industrial Fisheries, Kochi University of Science and Technology, Kochi and Department of Fisheries, Government of Kerala, in association with various ministries of the State of Kerala and Union Government of India. The volume includes 45 Chapters by expert authors, covering (i) Hydrometeorology and hydrology, (ii) Natural hazards and disaster risk management, (iii) Aquaculture, (iv) Changing biodiversity scenarios, (v) Capture fisheries, (vi) Food and nutritional insecurity, (vii) Climate change and socio-economic scenarios, and allied areas. It is hoped that this volume will further our understanding and research achievements in the field of climate change and its consequences and facilitate the synthesis of information on how climate-related changes will influence oceans, marine and inland ecosystems, hydrological cycles, fisheries and aquaculture and coastal communities.

The cooperation of the Cochin University of Science and Technology, R&D Institutes, State and Union Governments and agencies, and M/s. Narendra Publishing House, Delhi, in facilitating the publication of this volume is highly appreciated. It is hoped that this volume will be useful to planners and resource managers, scientists, conservationists, environmentalists, academicians, students and all those who are directly or indirectly involved in the impact of climate change and mitigation measures.

Kochi **Editors**

16 July 2020

Acknowledgments

The International Conference on ***Impact of Climate Change on Hydrological Cycle, Ecosystem, Fisheries and Food Security*** **(ClimFishCon 2020)** was jointly organised by Cochin University of Science and Technology, School of Industrial Fisheries and Dept. of Fisheries, Govt. of Kerala, in association with Govt. of India and Govt. of Kerala and was held at Le Meridian International Convention Centre, Kochi, during 12-14 February 2020. The Conference was well attended by more than 300 delegates including Scientists from over 12 countries. ClimFishCon2020 encompassed 11 Technical Sessions, one Special Workshop on Climate Change and Wetlands, Stakeholders' Conclave and ClimFishCon Expo 2020. Twenty-two keynote papers, 41 invited and 150 contributed papers were presented. A Stakeholders' Conclave conducted on 14 February 2020 has provided a unique opportunity for stakeholders belonging to fishing, aqua farming, processing, marketing and exporting of seafood sectors for expressing their concerns on the likely impacts of global warming on livelihood security and employment opportunities in the respective sectors. From the presentations made in the Conference, 45 manuscripts were selected, peer reviewed and included in this text book.

We are immensely thankful to Prof. (Dr.) K.N. Madhusoodanan,Vice-Chancellor, Cochin University of Science and Technology (CUSAT) for his whole hearted support, cooperation and active involvement in the conduct of the Conference. We also thank the CUSAT administration and staff and students of School of Industrial Fisheries for the help rendered in the conduct of the Conference. The organisers of the Conference are greatly thankful to Indian Council of Agricultural Research (ICAR) Govt. of India; Ministry of Jal Shakthi, Dept. of Water Resources, River Development and Ganga Rejuvenation, (Govt. of India); Ministry of Earth Sciences (Govt. of India); National Bank for Agriculture and Rural Development (NABARD); Science and Engineering Research Board (SERB); Seafood Exporters Association of India (SEAI), Kerala State Disaster Management Authority (KSDMA); Indian National Science Academy (INSA), Council for Scientific and Industrial Research (CSIR), Govt of India; and Kerala State Coastal Area Management Authority (KSDMA) for their funding support in organising the Conference and printing of Book of Abstracts, Souvenir, Programme Guide and proceedings.

The participation of sixteen national institutions such as Centre for Marine Living Resources and Ecology (CMLRE), ICAR Central Institute of Fisheries Technology (ICAR-CIFT), the Marine Products Export Development Authority (MPEDA), Central Institute of Fisheries Education (CIFE), National Bureau of Fish Genetic Resources (ICAR-NBFGR), Central Inland Fisheries Research Institute (ICAR-CIFRI), National Centre for Earth Science Studies (NCESS), ICAR Central Institute of Brackish water Aquaculture (ICAR-CIBA), ICAR Central Marine Fisheries Research Institute (ICAR-CMFRI), Indian Space Research Organization (ISRO), National Institute of Fisheries Post Harvest Technology and Training (NIFPHAT), Central Institute of Fisheries Nautical Engineering and Training (CIFNET), Fishery Survey of India (FSI), Kerala University of Fisheries and Ocean Studies (KUFOS), Indian National Centre for Ocean Information Services (INCOIS) and The Nansen Environmental and Remote Sensing Centre (NERSC), in the CimfishCon Expo 2020 are thankfully acknowledged.

We are immensely thankful to the Chairman and Rapporteurs of various Technical Sessions of the Conference, for their benign presence and cooperation rendered for the successful conduct of various sessions and framing valuable recommendations. We also thank Shri. Sileesh Mullasseri, Researcher, KUFOS, for his untiring efforts in compilation of the manuscripts from the authors and helping in the preliminary editing.

Kochi **Editors**

16 July, 2020

I

Hydrometeorology and Hydrology

Impact of Climate Change on Hydrological Cycle, Ecosystem, Fisheries and Food Security, Pages: 3–12
Edited by: B. Madhusoodana Kurup, M.R. Boopendranath, M. Harikrishnan and A.V. Shibu

CHAPTER - 1

The Climate Variability in the Indo-Pacific Sector

Swadhin Behera

Application Laboratory, JAMSTEC, Yokohama, Japan
E-mail: behera@jamstec.go.jp

Global warming is a serious concern for the present society as studies have shown a rise of 1 to 1.5°C in global surface temperature based on the available instrumental observations (~1860). The oceans have absorbed more than 80% of the total heat over the past 50 years (e.g. Levitus *et al.*, 2005). While it is clear that oceans are absorbing the heat and warming rapidly, it is unclear how that is affecting the ocean processes and subsequently the air-sea interactions related to the natural modes of climate variations. This to a large extent dependent on the regional distribution of the warming, which is not necessarily uniform over the globe as seen in the trends of the sea surface temperature anomaly (SSTA). The SSTA trends, after 1982 since when the satellite observations became available, show clear asymmetries in regional distribution of temperature trends particularly in the tropical Indo-Pacific domain (Fig. 1). The tropical Indo-Pacific domain is important for global climate as the warm pool region rooted in both basins plays an important role in the modulation of ocean and atmosphere variability on multiple scales of space and time.

The tropical parts of the Indian and Pacific Oceans share the extended warm pool and both of have their own modes of ocean and climate variations. The El Niño/Southern Oscillation (ENSO) phenomenon with warmer SSTA in the tropical eastern Pacific is the dominant mode of ocean-atmosphere variability in the tropical Pacific Ocean. In its opposite phase of La Niña, cooler SSTA prevails in the tropical eastern Pacific region. Both phases have large impacts on the weather and climate of the world. Therefore, it is important to understand how the trends and regional distribution of ocean warming going to affect the intensity and behavior of ENSO in the future climate. An El Niño phase is what one would imagine in a warmer world. However, the cooler trends in the eastern Pacific (Fig. 1) suggest a La Niña-like state is not unexpected in the future climate. A La Niña like state is desirable as it helps the oceans to absorb the heat. This has contributed to the global warming hiatus recently (Kosaka and Xie, 2013). This article reviews the present status of the climate variation studies under the stress of global warming.

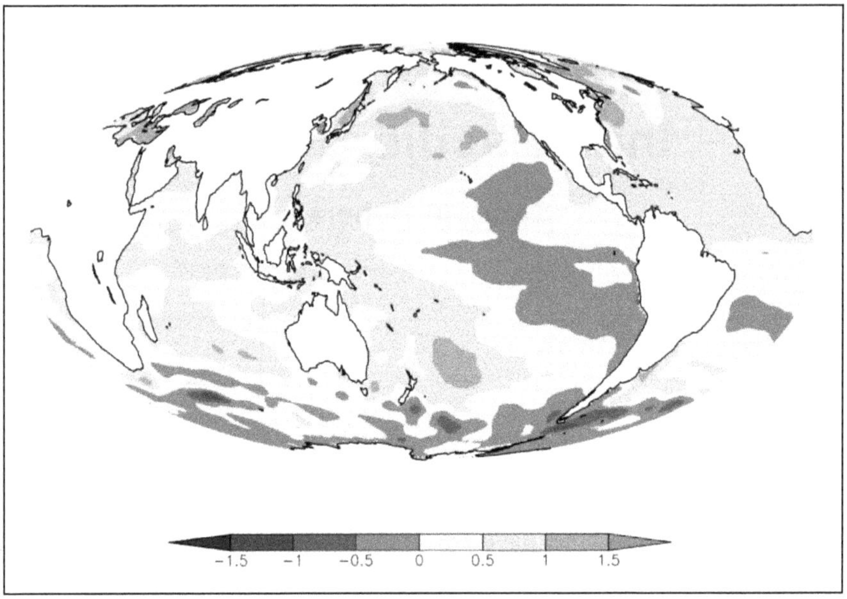

Fig. 1. The spatial pattern of SSTA trends (°C) between 1982 and 2018

ENSO flavors

In recent years, it is identified that the tropical Pacific Ocean also has another mode of variability called the ENSO Modoki (Ashok *et al.*, 2007). The ENSO Modoki events are similar to that of ENSO but the action center is located in the central Pacific. For this reason, the phenomenon is also referred to as the central Pacific ENSO, the dateline ENSO and the warm pool ENSO. During El Niño Modoki (La Niña Modoki) years warmer (cooler) central Pacific SSTA is flanked by cooler (warmer) SSTA on both sides of the tropical Pacific. Unlike the strong seasonal phase-locking of ENSO events, that typically develop in boreal summer and mature in boreal winter, ENSO Modoki events not necessarily always continue into boreal winter after appearing in boreal summer. Therefore, the ENSO Modokis are studied separately for boreal summer and winter seasons.

The ENSO Modoki is not only different in its characteristic of the sea surface temperature but also in terms of global teleconnections (Behera and Yamagata 2018). For example, the boomerang pattern of warm SST anomalies, extending from central Pacific to the coasts of America, associated with El Niño Modoki favors warmer than normal surface temperature in the western states of USA (Weng *et al.*, 2007). In addition, the cooler SST anomalies in the eastern Pacific Ocean associated with the event leads to below normal rainfall. Therefore, the combined effects of rainfall and temperature anomalies during the El Niño Modoki-related summers exacerbate the drought conditions in the Western USA. In the western North Pacific, El Niño Modoki is shown to be associated with a positive Pacific-Japan pattern, enhanced western-north Pacific summer monsoon but weakened East Asian summer monsoon. This dipole in the response causes drier conditions in most parts of Japan and the central eastern China, while southern China is flooded. Opposite conditions are expected during La Niña Modoki (Weng *et al.*, 2007). Some of the El Niño Modoki events, especially if blended with canonical El Niño like that of 2009, cause drier conditions over India by suppressing summer monsoon rainfall (Ratnam *et al.*, 2010).

The oceanic teleconnection of ENSO Modokis are also different from that of ENSOs. Particularly, the sea level variations associated with both type of phenomena are different (Fig. 2). The rise in sea level associated with the global warming is a major concern and it is mainly attributed to two processes; thermal expansion and melting of ice. But some of the regional variations (decadal) could be related to regional processes and ocean-atmosphere dynamics; e.g. El Niño Modoki related sea level variation (comparable to climate change scales) in the central Pacific. During El Niño Modoki, higher than normal sea level prevails in the central Pacific flanked by lower than normal sea level on either side (Fig. 2). On decadal scale, such a pattern dominated the basin during the early part of this century. The abnormal condition is evidently aided by frequent occurrences of El Niño Modoki events and associated wind convergence to the dateline during 2000–2004 (Behera and Yamagata, 2010).

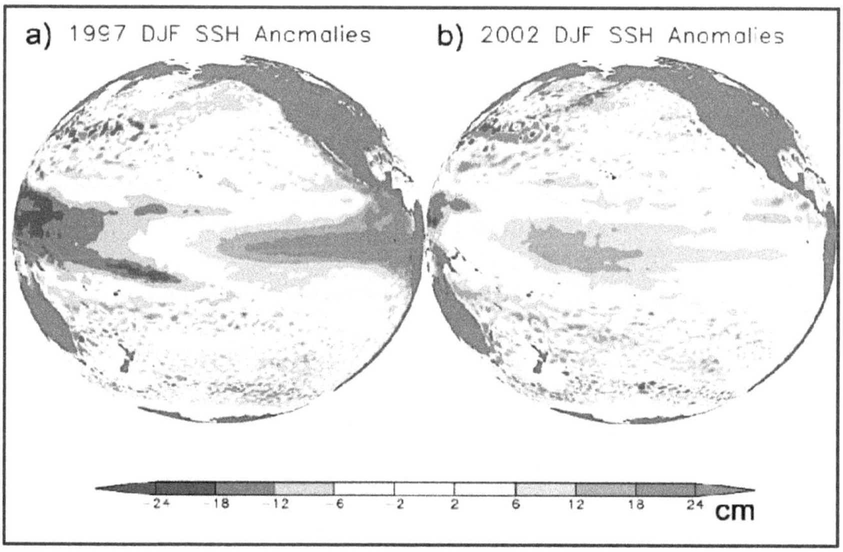

Fig. 2. Spatial pattern of sea level anomalies associated with typical (a) 1997 El Niño and (b) 2002 El Niño Modoki events.

The Indian Ocean Dipole and Monsoon

The Indian Ocean Dipole (IOD) is a prominent mode of climate variability (Saji *et al.*, 1999) with large impacts on the variability of ocean and atmosphere not only within the basin but also outside of it. A positive IOD event is characterized by cooler than normal sea surface temperature in the eastern tropical Indian Ocean off Sumatra while the sea surface temperature in the western Indian Ocean off Somali is warmer at this time. Stronger than normal seasonal southeasterly winds along the Sumatra coast during positive IOD events cause SST cooling by coastal upwelling and evaporation. The most prominent positive IOD occurred just in 2019 without any accompanying El Niño in the Pacific.

Under the changing climate/anthropogenic global warming, the upper Indian Ocean is seen to warm continually for last several decades. The warming, which is particularly pronounced on the western side is linked to the devastating drought along the southern edge of the Sahara Desert in the 1970s and '80s. This warming in the Indian Ocean has also changed the IOD characteristics. Coral records taken from Kenya and Sumatra reveal a mode shift in the IOD behavior (Nakamura *et al.*, 2009; Abram *et al.*, 2008). More frequent positive IODs are seen in recent decades compared to the early part of the 20th century. This is further confirmed by the analysis of CMIP5 model results. Extreme positive IOD affects weather, agriculture, ecosystems, and public health worldwide.

The ENSO is historically linked to the Indian summer monsoon rainfall variability (e.g. Rasmusson and Carpenter, 1983). Some of the mega droughts/famines of the 19th century are shown to be associated with the El Niños. The ENSO connection to Indian summer monsoon rainfall, however, is shown to have weakened in recent decades (Krishna Kumar *et al.*, 1999). It has been revealed that the influence of the ENSO on the monsoon is complemented by the Indian Ocean Dipole (IOD), discussed in the next section, (Behera *et al.*, 1999) and that the frequent occurrence of the IOD in the last couple of decades has weakened the ENSO- monsoon relationship (Ashok *et al.*, 2001). While investigating the unusual monsoon of 1994, Behera *et al.* (1999) observed that the moisture divergence in the southeast Indian Ocean, associated with the IOD of that year, helped maintain the monsoon rainfall over India and several parts of Southeast Asia. The role of the eastern pole of the IOD as an important source of moisture was depicted by the anomalies of moisture divergence from that region and convergence over Indian sub-continent.

Ashok *et al.* (2001) further confirmed this relationship with a long record of data and found that the emergence of the IOD influence in the Indian Ocean has reduced the ENSO impacts on monsoon. This explains the observed normal monsoon rainfalls over India during some of the strong El Niño events such as that of 1997 and 2006, which co-occurred with positive IODs. The Indian summer monsoon rainfall (ISMR) usually suffers from an ensuing El Niño in the absence of positive IOD. The past variation of this trilateral relationship is obscure due to the lack of reliable instrumental observations though at present it appears quite robust. Nevertheless, based on available observations, it is noted that the relationship between the SOI and ISMR has dramatically changed over the previous decades. The correlation between them was close to 1 during the first decade of the 20th century as compared to a negligible relationship during the last decade (Fig. 3). The evolution of this relationship in the future climate remains to be seen as the trilateral relationship evolves in the global warming scenario. Some studies indicate a greater number of intense and positive IODs in the future climate (e.g. Cai *et al.*, 2013). This would certainly help for better ISMR seasons in future.

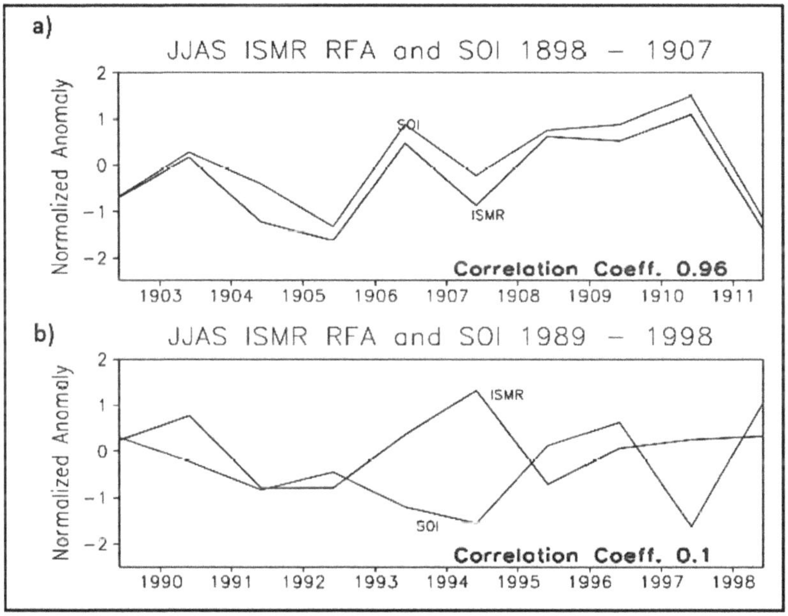

Fig. 3. The time series of Southern Oscillation Index (SOI) and ISMR for (a) first and (b) last decades of 20th Century (after Behera 2019).

Discussion

It is generally believed that the ENSO frequency and intensity are going to increase under the global warming. The projected surface warming over the eastern equatorial Pacific is said to cause more occurrences of atmospheric convection in the eastern equatorial region, giving rise to a greater number of El Niño events in the future climate. In addition, we have started observing frequent occurrences of consecutive positive IODs (e.g. the 2017-19 cases recently). The effect of the ocean warming and changes in IOD and ENSO characteristic are going to affect their impacts; e.g. intense heat waves over the Indian subcontinent. Those will also affect the predictability of ENSO, ENSO Modoki, IOD, the Indian summer monsoon rainfall in addition to other climate modes over the coming decades.

The climate variation impact on the society is recognized through some of the societal implications. Several social sectors are affected by year to year variability of El Niño/La Niña. Agriculture in many parts of the world could be severely affected by the El Niño, El Niño Modoki and IOD phenomena. While drought is very likely in some regions associated with those phenomena (such as the one on-going in Australia) some other regions experience heavy rains, flooding (like the one caused by IOD in East Africa during spring of 2006) or extremely hot or cold weather. The deficit of rain leads to droughts, severely impacting the ground hydrology and catchments of dams. The lack of water in the dams significantly affects their capacity to generate hydroelectricity in many parts of the world. On the other hand, excess rains by extreme events linked to climate variation, pose a risk for the physical vulnerability of the infrastructures such as housing, highways and even the electrical transmission system. In addition, over saturated ground (above the maximum capacity of the ground to absorb water) may lead to landslides, mud avalanches, destruction of properties, loss of electric transmission towers, distribution posts, and clogging of dams.

The climate phenomena are also associated with a heightened risk of certain vector-borne diseases in specific geographical areas where weather patterns are linked to modes of climate variations. This is particularly seen for malaria. Such associations are also suggested for some other epidemics such as diarrhea and cholera (for example in India and Bangladesh). There is evidence that ENSO is associated with a heightened risk of certain vector-borne diseases in specific geographical areas where weather patterns are linked with the ENSO cycle and disease control is limited. This is particularly true for malaria, but associations are also suggested for other vector-borne diseases that can be triggered by extreme weather condition associated with not only ENSO but ENSO Modoki, IOD and some other regional climate modes.

Besides the tropical climate variations, the subtropical climate variations such as the Indian Ocean subtropical dipole (IOSD) and South Atlantic Ocean subtropical dipole (SASD) affect the rainfall and temperature of the continental regions of the Southern Hemisphere. Especially, the IOSD (Behera and Yamagata 2001) is shown to influence the rainfall and temperature of southern Africa thereby influencing certain types of infectious diseases like malaria (Ikeda *et al.*, 2017). Also, several new modes of coastal air-sea interactions have been reported recently (e.g. Yamagata *et al.,* 2015). For example, the local Bjerknes feedback, in addition to the ENSO variability, gives rise to coastal Niños and Niñas off the coast of Weste A Australia. Named as Ningaloo Niños/Niñas (Kataoka *et al.,* 2014), these coastal coupled modes could profusely affect the coastal marine ecosystems besides the land processes in the adjacent regions.

Considering the huge impacts of climate variations on the society, it is important to develop climate prediction based early warning system. Sophisticated ocean-atmosphere coupled general circulation models are now able to predict climate variations and their global teleconnection. For example, the ENSO predictability has improved dramatically over the years. The chaotic spring-time variability was often cited for low ENSO predictability during boreal spring. However, new prediction systems are able to overcome these seasonal barriers. For example, the SINTEX-F1 model can successfully predict ENSO across the first spring barrier. The correct prediction of subsurface signals in the equatorial Pacific (Luo *et al.*, 2005) is shown to be the reason for the improvement. On an average, ENSO prediction skill for the period 1982–2012 reaches about 0.6 (0.4) at 14 (24) months lead with root mean square errors smaller than 0.8°C (e.g. Luo *et al.*, 2008; Luo *et al.*, 2015). Unlike the ENSO, the IOD has limited predictability (Luo *et al.*, 2007) as air-sea coupling related to IOD is usually weak and more localized compared to ENSO. Also, other phenomena, most prominently the intra-seasonal oscillations, affect the predictability in the basin through complicated scale-interactions.

It is also suggested that increasing the ensemble size is beneficial for improving seasonal prediction skills. Generally, an ensemble size of 10 to 50 members is adopted for the retrospective forecasts with operational dynamical seasonal prediction systems in the world, although more ensemble members are used in the multi-model forecast systems. It is generally expected that the prediction performance based on a 10 member-ensemble system will be less efficient as compared to that of a 100-member ensemble system, when predicting the occurrence of rare and extreme climate events. This is because a smaller number of members in the 10-member system means there are fewer chances of capturing extreme events at the ends in the observed probability distributions. In an ambitious project, Doi *et al.* (2019a) have increased the original ~10-member ensemble system to ~100-member ensemble system based on the SINTEX-F2 prediction system. This was the first such attempt in the world with a single model prediction system. It has turned out that about 10 member-ensemble could not perform as well as about 100-member ensemble when predicting the occurrence of 15% tails extreme climate events of the ENSO and the IOD. Also, the predictions of extreme temperature and precipitation events are improved over some regions. Those improvements may be due to the fact that large ensemble members capture tails of the PDFs and effectively retain the probability of the extremely strong, but rare events.

The climate predictions on interannual scales still suffer from occasional prediction lapses despite the improvements in the modeling strategies and predictability. Some of those lapses could be addressed if we have a reliable, sustained and dense observational network. While it is desirable to develop such a network, it is often constrained by research support and committed funding. Therefore, we explore other possibilities specially to assimilate subsurface data. In a collaborative effort, Doi *et al.* (2019b) have succeeded in increasing the accuracy of a new system that predicts SSTs a few months in advance by applying subsurface ocean observation data obtained from biologging of sea turtles.

Besides IOD and ENSO, predictability studies must include the subtropical dipole modes, coastal Niños/Niñas and the decadal mode in the southern Indian Ocean linked to the climate variability in the southern Atlantic Ocean and the Weddle Sea. Some studies suggest intensification of trade winds in the Indian Ocean would increase the transport of warm waters to the southern Indian Ocean. This could affect those climate modes and the long-term variations in the southern Indian Ocean. Further studies are needed to understand the changes in their characteristics.

References

Abram, N.J., Gagan, M.K., Cole, J.E., Hantoro, W.S. and Mudelsee, M.,2008. Recent intensification of tropical climate variability in the Indian Ocean. Nature Geoscience. 1: 849-853. 10.1038/ngeo357.

Ashok, K., Behera, S.K., Rao, S.A., Weng, H. and Yamagata, T., 2007: El Niño Modoki and its possible teleconnection. J. Geophys. Res., 112: C11007. doi:10.1029/2006JC003798.

Ashok, K., Guan, Z. and Yamagata, T., 2001. Impact of the Indian Ocean Dipole on the relationship between the Indian Monsoon rainfall and ENSO. Geophys. Res. Lett., 26: 4499–4502.

Behera, S. and Yamagata, T., 2010. Imprint of the El Niño Modoki on decadal sea level changes, Geophysical Research Letters 37: L23702. doi:10.1029/2010GL045936.

Behera, S.K. and Yamagata, T., 2001. Subtropical SST dipole events in the southern Indian Ocean. Geophysical Research Letters 28: 327–330.

Behera, S.K. and Yamagata, T., 2018: Climate Dynamics of ENSO Modoki Phenomenon, Oxford Research Encyclopedia of Climate Science, DOI: 10.1093/acrefore/9780190228620.013.612

Behera, S.K., 2019: The Indo-Pacific climate dynamics and teleconnections with a special emphasis on the Indian summer monsoon rainfall, Mausam70(1): 87-110.

Behera, S.K., Krishnan, R. and Yamagata, T., 1999: Unusual ocean-atmosphere conditions in the tropical Indian Ocean during 1994. Geophysical Research Letters 26: 3001-3004.

Cai, W., Zheng, X.T., Weller, E., Collins, M., Cowan, T., Lengaigne, M., Yu, W. and Yamagata, T., 2013. Projected response of the Indian Ocean Dipole to greenhouse warming. Nature Geoscience 6(12): 999-1007. https://doi.org/10.1038/NGEO2009

Doi, T., Storto, A., Behera, S. and Yamagata, T., 2019a. Merits of a 108-member ensemble system in ENSO and IOD predictions, J. Climate 32: 957-972, doi: 10.1175/JCLI-D-18-0193.1

Doi, T., Storto, A., Fukuoka, T., Suganuma, H. and Sato, K., 2019b. Impacts of temperature measurements from sea turtles on seasonal prediction around the Arafura Sea. Front. Mar. Sci. 6: 719. doi: 10.3389/fmars.2019.00719

Ikeda, T., Behera, S., Morioka, Y., Minakawa, N., Hashizume, M., Tsuzuki, A., Maharaj, R. and Kruger, P. 2017: Seasonally lagged effects of climatic factors on malaria incidence in South Africa, Scientific Reports 7(1): 2458, doi:10.1038/s41598-017-02680-6

Kataoka, T., Tozuka, T., Behera, S.K. and Yamagata, T., 2014: On the Ningaloo Nino/Nina, Climate Dynamics, 10.1007/s00382-013-1961-z

Kosaka, Y., and Xie, S.P., 2013. Recent global-warming hiatus tied to equatorial Pacific surface cooling. Nature 501: 403-407.

Krishna Kumar, K., Rajagopalan, B. and Cane, M., 1999. On the weakening relationship between the Indian monsoon and ENSO. Science 284: 2156– 2159.

Levitus, S., Antonov, J. and Boyer, T.T., 2005. Warming of the world ocean, 1955–2003. Geophys. Res. Lett. 32: L02604 10.1029/2004GL021592 (2005).

Luo, J.J., Behera, S., Masumoto, Y., Sakuma, H. and Yamagata, T., 2008. Successful prediction of the consecutive IOD in 2006 and 2007. Geophys. Res. Lett., 35: L14S02.

Luo, J.J., Lee, J.Y., Yuan, C., Sasaki, W., Masson, S., Behera, S., Masumoto, Y. and Yamagata, T., 2015. Current status of intraseasonal-seasonal-to-interannual prediction of the Indo-Pacific climate. Chapter 3, World Scientific Series on Asia-Pacific Weather and Climate, Vol. 7, Behera, S. and Yamagata, T. (Eds.).

Luo, J.J., Masson, S., Behera, S. and Yamagata, T., 2007. Experimental forecasts of Indian Ocean Dipole using a coupled OAGCM. J. Clim., 20: 2178-2190.

Luo, J.J., Masson, S., Roeckner, E., Madec, G. and Yamagata, T., 2005. Reducing climatology bias in an ocean-atmosphere OGCM with improved coupling physics. J. Clim., 18: 2344-2360.

Nakamura, N., Kayanne, H., Iijima, H., McClanahan, T.R., Behera, S.K. and Yamagata, T., 2009. Mode shift in the Indian Ocean climate under global warming stress. Geophys. Res. Lett., 36(23). doi:10.1029/2009GL040590

Rasmussen, E.M. and Carpenter, T.H. 1983. The relationship between eastern equatorial Pacific sea surface temperature and rainfall over India and Sri Lanka. Monthly Weather Review 110: 354–384.

Ratnam, J.V., Behera, S.K., Masumoto, Y., Takahashi, K. and Yamagata, T., 2010. Pacific Ocean origin for the 2009 Indian summer monsoon failure, Geophys. Res. Lett., 37: L07807. doi:10.1029/2010GL042798.

Saji, N.H., Goswami, B.N., Vinayachandran, P.N. and Yamagata, T., 1999. A dipole mode in the tropical Indian Ocean. Nature 401: 360-363.

Weng, H., Ashok, K. Behera, S.K., Rao, S.A. and Yamagata, T., 2007: Impacts of recent El Niño Modoki on dry/wet conditions in the Pacific Rim during boreal summer. Climate Dynamics, 29: 113-129.

Yamagata, T., Morioka, Y. and Behera, S., 2015. Old and new faces of climate variations. Chapter 1, World Scientific Series on Asia-Pacific Weather and Climate, Vol. 7 (Behera, S. and Yamagata, T., Eds.).

Impact of Climate Change on Hydrological Cycle, Ecosystem, Fisheries and Food Security, Pages: 13–26
Edited by: B. Madhusoodana Kurup, M.R. Boopendranath, M. Harikrishnan and A.V. Shibu

CHAPTER - 2

Embracing the Water: Design-led Ways to deal with Change

Rob Roggema

Professor of Spatial Transformations
Research Centre for the Built Environment NoorderRuimte
Hanze University of Applied Sciences Groningen, The Netherlands
E-mail: r.e.roggema@pl.hanze.nl

The Kuttanad wetland is located at the southern portion of India's largest Ramsar site the Vembanad-Kole wetland (Sreejith, 2013). It is a delta area consisting of five river systems: Meenachil, Pamba, Manimala, Muvattupuzha and Achencovil, located in the fertile low-lying areas of the Vembanad wetland system. The region encompasses vast stretches of backwaters, bordering mangrove formations, and rice fields, the latter mostly reclaimed from the shallow stretches of the lake during the recent past. The garden lands, or the reclaimed purayidams or homesteads with coconut groves, fringed by canals and channels make this a land of richness and beauty (Thambatti and Padmakumar, 1999). Kuttanad lies 0.6 to 2.2 m below mean sea level and has a warm climate and a very high humidity (Shari and Chitra, 2005). The average annual rainfall in Kuttanad is about 3,000 mm. Although distributed throughout the year, 60% rain is received between May and August, and 30% from October to November. During the months of June to November, the inflow into Kuttanad exceeds the inflow into the Vembanad Lake. During the months of December to almost mid-May, which coincide with summer, the reverse happens. There is a net decrease in the total inflow of the rivers into Kuttanad. This brings down the water level in the Vembanad Lake, which creates a reverse flow from the sea to the inland water body bringing the salinity all over the Lake and low-lying downstream parts of rivers (Koonan, 2007).

Key Problem

In recent years the Kuttanad area faced a yearly deluge. In August 2018 (Jayaraj, 2018) and 2019 (Mathrubhumi News, 2019) large parts of the area severely flooded (Fig. 1). Due to the elevation of the area, of which major parts are up to 2.2 meter below current mean sea level, it is extremely susceptible to flooding, at the same time to salt intrusion. This poses a risk to livelihoods, but also to the main sources of income, agriculture and tourism. What the area can expect in the future is not very hopeful. Taking into account that the current situation, the recent deluges have occurred while sea level rose quite slowly at a pace of about 11 cm/century (Baart *et al.*, 2014) and this is expected to accelerate in the next century to probably 2-3 m (Le Bars *et al.*, 2017). In addition, recent projections for 2050 show that large parts of major cities in the world are threatened by the water (Kulp and Straus, 2019). The maps of Mumbai are illustrative (Fig. 2).

Fig. 1. An aerial view of flooded Kuttanad in the Alappuzha district (19 August 2018) (Source: https://www.hindustantimes.com/india-news/around-70-000-participate-in-clean-up-drive-in-kerala-s-kuttanad/story-UZ22s3Q9NtoyyDPPJlwwWO.html)

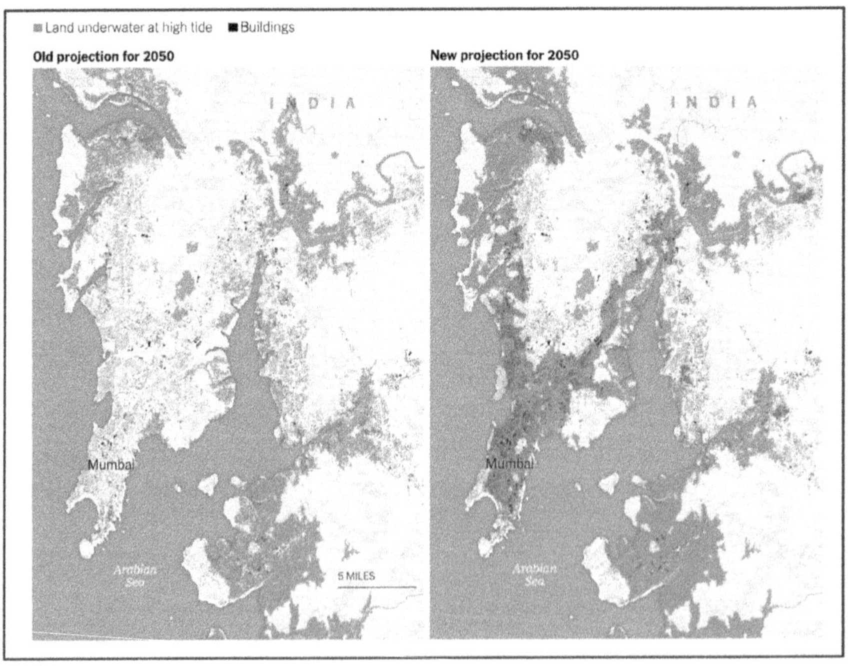

Fig. 2. Potential flood projections for Mumbai in 2050 (Source: Kulp and Straus, 2019)

With the area of Kuttanad being up to 2.2 m below mean sea level at the moment and its soil subsiding, an additional rise of 2 m implies the total relative sea level might rise up to 4.2 m; maybe even 5 m or more. In this situation no dam, seawall or dike will offer solace and it is urgent to envision an alternative future perspective. Past approaches to deal with these threats mainly consist of engineering constructions that prevent the change. For instance, the Thottappally Spillway, Thanneer Mukkam Bund and Alappuzha-Changanassery Road are historic relics that increased human interference, not necessarily creating an adaptive hence safe future for the area (Sreejith, 2013). On the contrary, these rigid structures only increase the risk, the negative impact of salinity, flood-risk, and loss of biodiversity, indirectly decreasing the attractiveness for tourism, and the potential for economic development.

Problems in Detail

The current problems in Kuttanad are described in detail by Sreejith (2013). A total of 15 different topics are distinguished. When these problems are categorized and plotted in a square it turns out that the majority of the problems are related to water (eg. Balchand, 1983; Gopakumar and Takara, 2009; Sulekha and Mercy, 2009) and nature (eg. Remani *et al.*, 2013; Menon *et al.*, 2000; Narayanan *et al.*, 2011) (Fig. 3). However, the main attention seems to be given to economic factors, income for farmers and fishermen, and their living standards (eg. Krishnan and Sharma, 1996; Ranjeet and Kurup, 2013; Padmanabhan *et al.*, 2001). Creating a safer future is mostly proposed through sectoral investigations and technical solutions in the form of engineered protective structures. Research carried out in the past, such as the Kuttanad Water Balance Study (KWBS, 1989), which mainly focuses on understanding the ecological and water systems, or the Kuttanad Package (Government of Kerala, 2012), mainly focusing on improving conditions for agriculture. However, the continuation of sectoral research approaches will only lead to slight improvements for one subject, leading to degradation of others. This has been the case over recent decades. Water and ecological system are under pressure due to increasing salinity and floods and further subsidence of the soil. This subsequently makes it more and more difficult to keep living standards up, with higher income and better lives for the inhabitants of the area.

Spatial strategy

The risk of current initiatives and action plans is that it will, despite all ambitions only add to the downward cycle. The more investments in sectoral policies and programs, the more the ecosystem of Kuttanad comes under pressure eventually leading to a complete subsidence of the area, with no prospective for the people than to flee. There is an urgent need to start planning in an integrative way for sustainable development (Kumar and Devadas, 2016), paying attention to complex systems, participatory processes and integrated planning. However, even integrated planning often lacks the imagination or attractive vision to coherently and consistently work towards a resilient future for the entire living eco-system. This requires a spatial strategy, in which all factors fuse together.

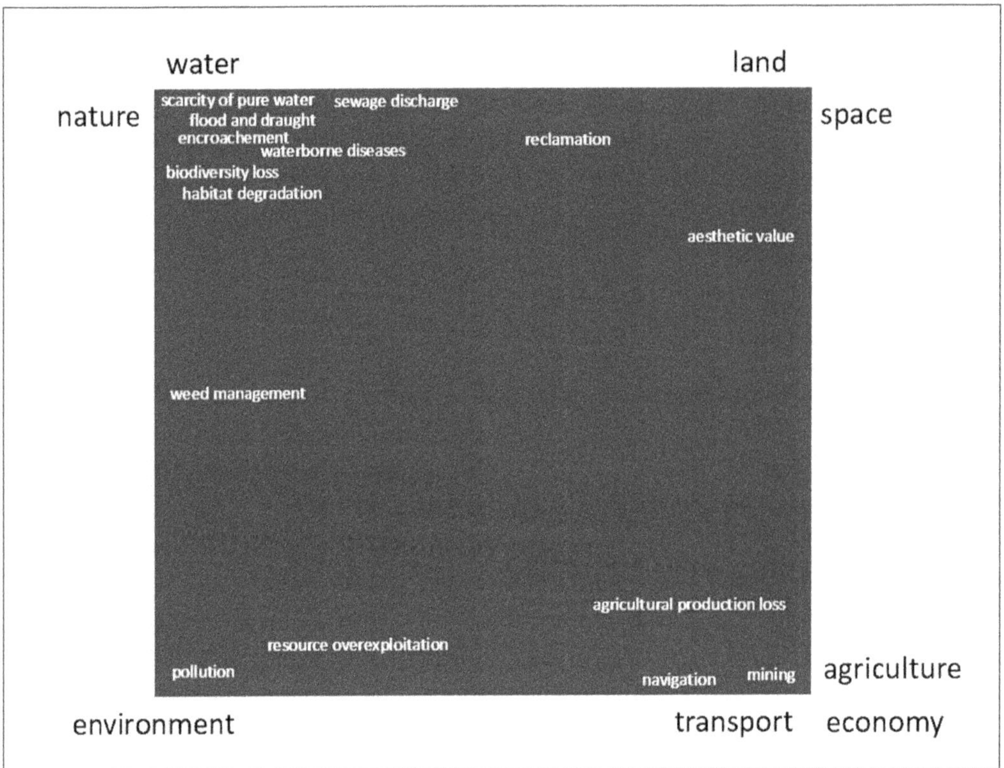

Fig. 3. Current problems in the Kuttanad area, organized by theme (adapted from Sreejith, 2013)

This implies that, instead of developing an engineered construction of a brave new world, in which mankind is leading and intervenes in the delta system, this system should firstly be understood as a natural living organism and investigate what a natural functioning system for the future would contain. Once this is envisioned, the current landscape can adapt to this future and, within this context, a sound economic development will emerge. This offers a long-term solution which increases the living circumstances and a higher income in the end.

This requires traditional thinking in silos and sectors is replaced by thinking more holistically. Instead of creating divides and barriers, such as illustrated by name-giving of natural landscapes such as the Great Dividing Range or the Great Barrier Reef in Australia, in fact these systems are connecting land and water and creating living systems (Yunkaporta, 2019). In a context where a delta area is degrading and under threat, what would it mean to allow to transform it in one living organism? Would nature shape the landscape and would it, by itself, develop a resilient coastal system? It would regenerate the delta, in which mangrove forests and wetlands would emerge, a natural coastal protection system would grow, and the coast would operate as one systemic entity. If only the fixed engineered structures and hard boundaries could be replaced by a flexible adaptive whole, a yin-yang type of thinking (Fig. 4) allowing the sea to enter the land, and land extends into the sea.

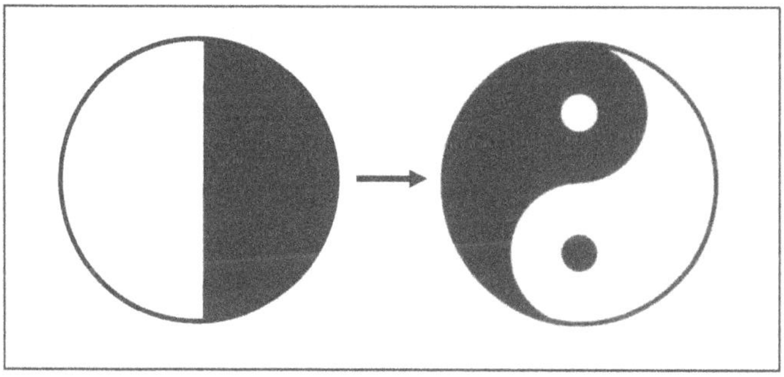

Fig. 4. From division to holistic thinking

As soon as the systemic properties of this coastal system are understood, strategic spatial interventions can be designed to speed up this process of natural emergence (Roggema, 2012). Within this emerging system, human uses can be implemented, such as floating houses, saline agriculture, floating agriculture and fisheries.

A design-led approach to change

A design-led approach to allow for a structured development of a resilient natural system consists of the following steps:

i. Understand the system: identify the natural principles that shape the system. Rivers, mangrove forests, landscape forming.
ii. Design a natural living coastal organism: use the principles to design the entire system and how it would function as an organism.
iii. Define the points of intervention, to ignite the emergence of natural processes, and identify the points where change can be enforced.
iv. Imagine the natural delta: the entirety of the natural delta can be envisioned and designed.
v. Insert human uses within that system: on the basis of the vision human use, where suitable and applicable, can be proposed.

Implementable examples

Without extensively designing this future system, a range of examples can be used to inspire such an emerging natural system.

The Double Defense Project

The Double Defence project (Fig. 5) is located in the northern part of the Netherlands. The north shores and the Wadden Sea are currently under threat of sea level rise and storm surges. This area is very vulnerable because it is low lying and the high ecological values of its tidal flats, which, under accelerated sea level rise, might drown. Storm surges could attack the coastal protection, which has, compared to other parts of The Netherlands, a lower safety level. The proposed response to these potential disruptions is to introduce a second row of barrier islands in front of the existing ones. Because of the new islands, a storm surge is not only prevented from harming the coastline, it also adds a huge amount of extra sediment, clay and sand particles, to the area behind and in between the existing and new islands. Because the islands create a more tranquil maritime environment, these particles get the chance to sink down and build up existing or new sandbanks. This enhances the wetland ecosystem of the Wadden Sea, not only in quantity but also in quality. In contrast to the existing islands and sandbanks being threatened of washing away during a storm surge, the new islands provide the incentive for more and new ecological area to emerge.

Fig. 6. Plan for Double Defence (Roggema *et al.*, 2006)

The introduction of a second line of barrier islands starts a new development of land-genesis. The main, long-term objective is clear, whilst the specific shape of the wetlands is dependent on the interaction of the storm surge, amount of sediment, existing sandbanks and wave strength and direction. The shape of the new Wadden will emerge under influence of the currents and available sediments, and constantly change shape. This adaptivity is possible because the area is effectively an extensive absorption zone where water, wind and sand self-organise the build-up of new sandbanks. The result is that a rich nature emerges, which by its land-forming increases the safety of the entire northern part of the Netherlands. On the new islands, opportunities for new uses, such as recreation, urban living, agriculture and renewable energy generation arise. At the same time, the existing hinterland, due to its better protection, could explore new developments and become a place for food supply, urban living, water retention and high-tech industrial activities, which would be under serious scrutiny if the design intervention would not be effectuated.

Marker Wadden

At the initiative of Natuurmonumenten, the Marker Wadden (Fig. 6) are constructed, a group of five islands located four kilometers to the south of the Houtribdijk causeway between Enkhuizen and Lelystad. The islands are formed from sand, clay and silt from the Markermeer. The construction of a new paradise for birdlife started in 2016. This project will turn the tide and reintroduce nutrients to the food chain in the lake. The swampy areas between the islands will become new spawning and nesting grounds for fish and birdlife. The interaction between the natural processes and soil fauna and fish populations, and the effect of the new nature on the fisheries and the ecological targets, are the two main objectives of the project. One of the islands will be made accessible to visitors, with a small jetty for boats. The island will have a visitor's center, walking

Fig. 6. The Marker Wadden bird's eye view (Source: Vista/Natuurmonumenten)

paths and a bird hide on its farthest corner. This will ensure that, alongside the fish and birds, people too can enjoy all that the new nature has to offer. The project is also expected to generate more income for Lelystad through the increase in water recreation and hotel stays. The project is an example of how new opportunities for nature can be found even in a highly densely populated delta region, making use of natural ecosystems and the region's natural resilience.

Sand engine

An alternative design approach is to rely on the natural coastal processes of land forming, such as reefs, islands, mangroves, beaches and dunes. Humans can help the formation of these natural protectors by providing the triggers for them to emerge. A specific way of allowing ecological principles to form the landscape is Building with Nature (Waterman, 2010). Natural forces, such as wind, waves and currents are used to understand how the landscape would form itself, and how it can become most resilient. This knowledge is then used and applied to intervene in the system to enhance or start these processes. The purpose is to start a process of natural strengthening of the landscape without largescale human or technical interventions.

For example, these principles are applied in the design for the Zandmotor (Sand-engine, Fig. 7), and results show that the sea manages to increase the size of the beaches and the dunes due to dropping sand in a crucial place in the sea in front of the Dutch coast. The sand-engine (Stive *et al.*, 2013) is a project in front of the Western coast in The Netherlands which saw a one-off deposit of 21.5 million cubic metres of sand which will stay in the system for about 20 years. It will allow the development of a permanent ecosystem. The sand is redistributed along the coast by currents and wind, natural forces that deposit sand in the right places to build up coastal protection, ecological systems and provides the environment for recreational use (De Vlieger, 2017). In this process, nature builds new beaches and dunes to form a much stronger system than humans ever could, preventing coasts, beaches and real estate from being washed away. This requires design thinking, insights into the resilience of the coastal system and an understanding of the natural forces at play.

Floodable Eemsdelta

The Eemsdelta region is located in the northeast of The Netherlands. It is a historically valuable area where until 1000AC people used to live on artificial hills, created by piling up their own waste, amidst a tidal flat. Once the residents started diking the land a coherent system of coastal protection emerged up to the point they started to make their own land and reclaimed tis from the sea by introducing a wooden structure system in the sea that was capable of capturing sediment and this way grew above sea levels and became new land. This region is now under threat of accelerated sea level rise and occasional storm surges.

Fig. 8. Bird's eye view of the Sand-engine in 2013 (Source: https://climate-adapt.eea.europa.eu/metadata/case-studies/sand-motor-2013-building-with-nature-solution-to-improve-coastal-protection-along-delfland-coast-the-netherlands/delfland-coast_figure-2.jpg/view)

The proposed intervention in this landscape is to create consciously a hole in the coastal protection system. This way a Floodable Eemsdelta (Fig. 8) (Roggema, 2009; Roggema and Van den Dobbelsteen, 2012) will emerge over time. The water will enter the hinterland and fill up the landscape as high as the sea level will rise. This implies a twofold mechanism. The first element is that it doesn't matter how fast sea level rise happens. At every stage of a sea level rise the landscape is ready and used to the water in the local environment. Secondly, an eventual disaster, e.g. a dike breech, will not happen due to the hole that has been there forever. This way, people in the Eemsdelta region can predict the future circumstances they will live in. As a matter of fact, the water increases the quality of living in the area, and provides an abundant resource for cooling, against drought and other uses. A second intervention optimally profits from the seawater. Applying the historic system of land reclamation, also in this landscape wooden structures are proposed that will capture sand and clay sediments hence grow the ground level of the landscape. Once the wooden structures are in place, poles and to them floating houses can be attached. This way the houses are prepared for coming seawater, they won't float away, and the residents are certain of the most beautiful and safe landscape to live in.

In normal situations, sea level rise will be dealt with by increasing the protective seawall and increase its height and strength. To purposely make a gap in the dike is, especially in the Netherlands, counterintuitive. Using a similar principle as applied in the Sand Engine, the incoming water enforces the landscape behind the dike to adjust and adapt. This delta landscape self-organises its components to make sure the typology of housing and their location is chosen in safe places. Because the water will flow to the lowest point, after the inlet the main direction of how the landscape will evolve is clear.

In detail natural processes such as wind speed and direction, the amount and type of sediment, the shape of contour lines are the few guiding principles that determine the exact land forming, the build-up of new soil and where new housing can be developed. The entire zone under influence of the water is redundant and functions as an absorption zone, in which many different developments can be accommodated. The pace of sea level rise is not relevant, as the area is always prepared for any change in sea level.

Besides accommodating sea level rise in a secure way, this proposition also creates the absorption zone for collecting and storing seawater during and after a storm surge, as well as the impact of heavy rainfall in the landscape itself. The Floodable Eemsdelta project predominantly focuses on disruptive sea level rise and to a lesser extent storm surges and heavy rainfall, using an unorthodox and counterintuitive approach in which an entire landscape provides the space to adapt to constantly changing circumstances.

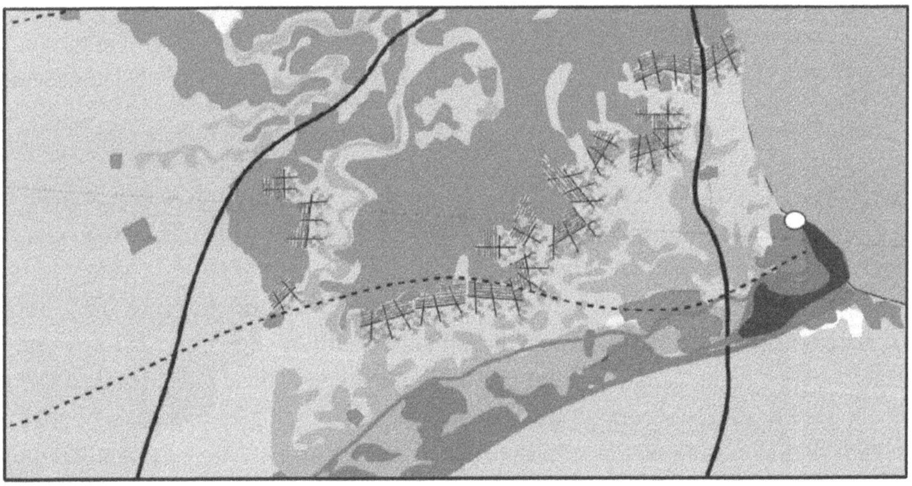

Fig. 8. Floodable Eemsdelta (Roggema, 2009; Roggema and Van den Dobbelsteen, 2012)

Floating agriculture

A farmer, 40 cows, and robots to collect manure and distribute fodder – all floating on the water of the port of Rotterdam. This Floating Farm, the first self-sufficient farm floating on water, which is completed and will begin operation in the Merwe4 Haven of Rotterdam. Important factors behind the Floating Farm project are land scarcity, climate change, rapid population growth concentrated in and around large cities, long distances between the sites of dairy production and consumers leading to increased transport and pollution. These are the reasons to bring dairy production as close to consumers as possible. More and more people live in large cities, far away from dairy production. A massive amount of transport is required to deliver dairy products to consumers, and this adds pollution and puts a heavy burden on infrastructure.

The cows can rest on the upper level of the structure (Fig. 9) or head to a feeding station for a mix of hay, grass clippings and beer byproducts. Milk and manure processing facilities are located on the lower deck, as well as the visitors' entrance and store. A quay side pasture gives the animals the chance to be on dry land. The structure's roof is used to collect rainwater. Power for the farm comes from a solar panel array floating nearby. The whole farm site is kitted out with the latest in dairy tech such as an automated feeding system, manure-scooping robots, self-serve cleaning stations, and the smartphone app is used on-site by the farmer to monitor the cows. As part of its aim to be self-sustainable, the farm already has a manure separator used to separate dry material from urine, with the dry part used as bedding for the cows and the urine turned into an organic fertilizer. It sells some bottles of raw milk on-site to visitors, while the rest is pasteurized and turned into milk or yoghurt. It is sold to customers who live nearby via an online grocer, Picnic, known for its small, electric delivery vans.

Fig. 9. The Floating Farm in Rotterdam (Source: https://www.dezeen.com/2019/05/24/floating-farm-rotterdam-climate-change-cows-dairy/)

Design-led Process

To propose projects that can help to increase the resilience of the coastal area of Kuttanad is one thing, but to realize these is a different task. The projects presented in this chapter are designed in a different context with other parameters and, hence, cannot be implemented as such in the Kuttanad context. However, these ingredients can be used to start a design-led, inclusive research-by-design project, in which all involved, governments, businesses, farmers and fishermen, residents and academics collaborate in a series of design charrettes to find common ground, and come up with a supported plan which can be implemented and solves long-term problems. This design-led approach has been proven successful to bind people in a highly complex situation. As an illustration, this method has been used to develop resilient coastal planning in the tsunami-hit area of Fukushima, Japan (Roggema and Yan, 2017). In the towns of Kessennuma and Minamisoma, a series of charrettes have led to integrated coastal plans that were designed by local authorities, businesses and inhabitants together. The support for collaborative design, the content and realization is enhanced due to the collective design process.

Proposition

This article does not provide a concrete solution for Kuttanad area. It proposes a design-led approach in which the water is embraced as a primer for resilient coastal development. This should be read as a research agenda, using a research by design methodology to create the outcomes that are tactile, can be realized instantly and are measurable. This could look like a two-year trajectory with twice every year a design charrette, resulting in an overall plan for Kuttanad enabling sustainable development on the long term, instigate regenerative coastal processes and increase the wellbeing and quality of living of its inhabitants, brought together in a multi-year program, financing individual projects.

References

Baart, F., Leander, R., De Ronde, J., De Vries, H., Vuik, V. en Nicolai, R., 2014. Zeespiegelmonitor 2014. Rekenmethode voor huidige en toekomstige zeespiegelstijging. Delft: Deltares

Balchand, A.N., 1983. Kuttanad: A case study on environmental consequences of water resources mismanagement. Water International 8(1): 35-41.

De Vlieger, B., 2017. The New Delta; The Rhine-Meuse-Scheldt delta in transition. Heijningen: Japsam Books.

Gopakumar, R. and Takara, K., 2009. Water balance and mean water residence time of the Vembanad Wetland of Kerala State, India. IAHS Publ. 328: 223-232.

Government of Kerala, 2012. Kuttanad Package. A government of Kerala initiative: http://www.kuttanadpackage.in/index.php?option=com_contentandview=articleandid=77andItemid=71

Jayaraj, V.R., 2018. Deluge in God's Own Country. The Pioneer, published 29 July 2018.

Koonan, S., 2007. Legal Implications of Plachimada - A Case Study. IELRC Working Paper 05.URL: http://www.ielrc.org/content/w0705.pdf

Krishnan, M. and Sharma, B.N., 1996. Multi-objective approach for evaluation of farming systems in Kuttanad region of Kerala State, India: A model for decision making. Journal of Aquaculture in the Tropics 11: 205-214.

Kulp, S.A. and Strauss, B.H., 2019. New elevation data triple estimates of global vulnerability to sea-level rise and coastal flooding. Nat. Commun. 10(4844). https://doi.org/10.1038/s41467-019-12808-z

Kumar, S. and Devadas, V., 2016. Integrated Planning for Sustainable Development of Kuttanad Wetland Region, Kerala State. Procedia Technology 24: 1660-1667. 10.1016/j.protcy.2016.05.185.

KWBS, 1989. Kuttanad Water Balance Study, Vol. 1. Main Report. BKH Consulting Engineers, Kingdom of the Netherlands and Ministry of Foreign Affairs, Republic of India

Le Bars, D., Drijfhout, S. and De Vries, H., 2017. A high-end sea level rise probabilistic projection including rapid Antarctic ice sheet mass loss. Environ. Res. Lett. 12(044013). https://doi.org/10.1088/1748–9326/aa6512.

Mathrubhumi News, 2019. Deluge batters Kuttanad on Kerala Flood's First Anniversary. YouTube, 18 August 2019. https://www.youtube.com/watch?v=SXuWgEr0dSk&vl=en.

Menon, N.N., Balchand, A.N. and Menon, N.R., 2000. Hydrobiology of the Cochin backwater system–a review. Hydrobiologia 430(1-3): 149-183.

Narayanan, S.P., Thomas, A.P. and Sreekumar, B., 2011. Ornithofauna and its conservation in the Kuttanad wetlands, southern portion of Vembanad-Kole Ramsar site, India. Journal of Threatened Taxa 3(4): 1663-1676.

Padmanabhan, P.G., Narayanan, N.C. and Padmakumar, K.G., 2001. Economic Viability of an Integrated and Sustainable Resource Use Model for Kuttanad. Discussion Paper No. 33, Kerala Research Programme on Local Level Development, Centre for Development Studies, Thiruvananthapuram.

Ranjeet, K. and Kurup, B.M., 2013. Economic analysis of polder based freshwater prawn farming systems in Kuttanad, India., Int. J. Fish. Aquacult. 5: 110-121.

Remani, K.N., Jayakumar, P. and Jalaja, T.K., 2010. Environmental problems and management aspects of Vembanad kol wetlands in South West coast of India. Nature, Environment and Pollution Technology 9(2): 247-254.

Roggema, R., 2009. Adaptation to climate change, does spatial planning help? Swarm planning does! In: Brebbia, C.A., Jovanovic, N. and Tiezzi, E. (Eds.) Management of natural resources, sustainable development and ecological hazards. Southampton: WIT Press, pp. 161–172.

Roggema, R., 2012. Swarm Planning: The development of a Methodology to Deal with Climate Adaptation. Ph.D. thesis, Delft University of Technology and Wageningen University and Research Centre, Delft, Wageningen.

Roggema, R. and Van den Dobbelsteen, A., 2012. Swarm Planning for Climate Change: an alternative pathway for resilience. Journal for Building Research and Information 40(5): 606-624.

Roggema, R. and Wanglin Yan (Eds.), 2017. Tsunami and Fukushima Disaster: Design for Reconstruction. Dordrecht, Heidelberg, London: Springer, 196 p.

Roggema, R., Van den Dobbelsteen, A. and Stegenga, K., 2006. Pallet of Possibilities, Report Spatial team, Groningen: Grounds for Change.

Shari, P.V. and Chithra, K.P., 2005. Kuttanadinte charithravum bhoomisastravum (History and geography of Kuttanad), pp. 9–18. In: Joseph, M.J. (Ed.) Kuttanattile jala punarujjeevana prayoga sadhyathakal (Water reclaiming in Kuttanad; Possible chances). Planet Kerala: Trivandrum, 131 p. (In Malayalam).

Sreejith, K.A., 2013. Human impact on Kuttanad wetland ecosystem - an overview. International Journal of Science, Environment and Technology 2(4): 679 - 690.

Stive, M.J.F., De Schipper, M.A., Luijendijk, A.P., Aarninkhof, S.G.J., Van Gelder-Maas, C., Van Thiel de Vries, J.S.M., De Vries S., Henriquez, M., Marx, S. and Ranasinghe, R., 2013. A New Alternative to saving our beaches from sea-level rise: The Sand Engine. Journal of Coastal Research 29(5): 1001-1008. https://doi.org/10.2112/JCOASTRES-D-13-00070.1

Sulekha, B.T. and Mercy, T.V.A., 2009. Pesticide-induced histopathological changes in the freshwater fishes of Kuttanand, Kerala - A tool to assess water quality and the health status of fishes. Asian Fish. Science 22(2): 729-749.

Thampatti, K.C.M. and Padmakumar, K.G., 1999. Rice bowl in turmoil: The Kuttanad Wetland Ecosystem. Resonance 4(3): 62–70.

Waterman, R.E., 2010. Integrated coastal policy via building with nature, Ph.D. thesis. Delft University of Technology, Delft.

Yunkaporta, T., 2019. Sandtalk. How indigenous thinking can save the world. Melbourne: The Text Publishing Company, Melbourne, 224 p. ISBN: 9781925773996

Impact of Climate Change on Hydrological Cycle, Ecosystem, Fisheries and Food Security, Pages: 27–37
Edited by: B. Madhusoodana Kurup, M.R. Boopendranath, M. Harikrishnan and A.V. Shibu

CHAPTER - 3

Accelerated Hydrological Cycle under Warming Climate and Role of Extended Range Hydro-Meteorological Prediction System

S. Abhilash*, A.K. Sahai, R. Mandal, A. Dey, S. Joseph, R. Chattopadhyay, R. Phani and G. Saranya

Indian Institute of Tropical Meteorology (IITM)
Dr. Homi Bhabha Road, Pashan, Pune-411008, India
**E-mail: abhimets@gmail.com*

It is evident from instrument (thermometer) based observations that, the planet's average surface temperature has risen about 1.1 degree Celsius since the late 19th century, a change driven largely by increased carbon dioxide and other man-made emissions into the atmosphere driven by fossil fuel burning and deforestation. A potential consequence of global warming is the acceleration of the hydrological cycle, which may manifest itself in the form of increased precipitation and land evaporation. Due to this, both the extreme hydrological events such as floods and droughts are increasing especially over wet regions of the Earth. In a warming world, differential heating between land and oceans and between polar and tropical regions will affect global circulation patterns and thereby change the intensity, frequency and seasonality of climate patterns and distribution of extreme weather events.

In recent decades, stability of the monsoon system is found to be decreasing and monsoon will become more chaotic in nature with increased year-to-year (inter-annual) variability along with increased intra-seasonal (within season) variability. Though, there is not much changes expected for seasonal mean monsoon rainfall, the variability on shorter time scale manifested as short duration floods or prolonged break spells. The increased variability and uncertainty driven by global warming had made monsoon climate more unpredictable in future. Immediate attention is needed in understanding and predicting these high impact weather events with sufficient lead time. It is therefore essential to review the capability and limitations of current observational techniques like satellite and radar in understanding the underlying physical processes responsible for the extremes as well as monitoring them in near-real time. Large uncertainty in different models and inability of reliable metrics in verifying different measures of the extremes on all time scales is also challenging. In this context, identifying relevant physical processes in controlling and modulating these extremes are difficult and essential to be incorporated into the model formulation to capture the large scale favourable conditions at sufficiently longer lead time for effectively managing the hazards. In this work we will discuss the importance of satellite and re-analysis based monitoring and model based prediction system in real-time monitoring and disaster management.

Extended range forecast (ERF) covering the time scale from one week to about a month in the tropics are one of the most challenging tasks in atmospheric sciences. It fills the gap between medium-range weather forecasting and seasonal forecasting. It is not only the agriculture sector which is benefited from the proper outlook of extended range forecast, a skilful extended range forecast can also be very useful for reservoir operation in managing floods. Under the rapid Indian Ocean warming, Arabian Sea is expected produce more intense storms in the future and that will affect the life and livelihood of coastal communities along the West coast of India. Hence, forecasting of genesis of tropical cyclone and associated rainfall in the extended range time scale (about 2 weeks in advance) is also very useful for applications in disaster risk reduction. The usefulness of extended range forecast system developed indigenously at Indian Institute of Tropical Meteorology for applications in agriculture, water resource management, disaster management. The application of ERP system for the prediction of monsoon heavy rainfall during south west monsoon, extended breaks, tropical cyclones and extreme rainfall events during north east monsoon will be discussed in detail.

The importance of sub-seasonal extended range forecasts was primarily appreciated by the hydro-climatic and agro-meteorological stakeholders when the experimental forecasts were released earlier in pseudo-operational mode. Skillful forecasts of onset date and the active/break spells at multiple spatial scales could provide essential information to the agrarian community and water managers about the pattern of decision support system to be implemented in case of likely extremes forecasted over a small region. An example with regard to onset date may be helpful. The onset is awaited eagerly by the farming community, whose arrival and intensity over Kerala gives a confidence boost regarding the sowing pattern, crop selection, watering pattern, pesticide spraying pattern, etc. The lead-time of 2-3 weeks provides a sufficient lead-time for alternative planning and solutions in agricultural and health sector (for example selection of alternate crops).

Extended Range Prediction System

As part of India's National Monsoon Mission (NMM) program towards improving Indian monsoon region prediction capabilities, a multi-model ensemble (MME) prediction system has been developed at Indian Institute of Tropical Meteorology (IITM), and is now operational at IMD. The dynamical model is derived from an operational climate model developed at the U.S. National Centers for Environmental Prediction (NCEP) for the extended range (2-3 weeks). For generating the MME, we have used a version of NCEP's Coupled Forecast System version 2 or CFSv2 (Saha *et al.*, 2006; 2014) run at two horizontal resolution: T126 (110km) and T382 (38km). The MME also consist of the standalone atmospheric component of CFSv2, the Global Forecast System (GFS) model also run at T126 and T382 horizontal resolutions. GFS is coupled to an ocean model, sea-ice model, and land surface model. For its ocean component, the CFSv2 uses the GFDL Modular Ocean Model version 4p0d (MOM4; Griffies *et al.*, 2004). The standalone GFS with slightly different physics options than CFSv2 is forced with daily bias corrected forecasted SST from CFSv2. Bias correction involves subtracting the mean bias as a function of a calendar day and lead time, with Optimum Interpolation Sea Surface Temperature (OISST) observations (Reynolds *et al.*, 2007) as the reference. We denote this two-tier standalone GFS forecast as GFSbc, with 'bc' indicating bias corrected boundary conditions (Abhilash *et al.*, 2015).

Choosing a proper method to generate initial conditions is a pre-requisite for MME predictions. Although there are several approaches to generate ensembles of different initial conditions; we use a method which is similar to the 'complex-and-same-model environment group' as classified in Buizza (2008). An ensemble of perturbed atmospheric initial conditions was developed in addition to one actual initial condition. The actual (i.e., unperturbed) initial condition is prepared from NCEP's coupled data assimilation system (CDAS) with T574 L64 (~23km horizontal resolution with 64 vertical levels) resolution atmospheric assimilation and MOM4 based oceanic assimilation, a real-time extension of the CFSR (Saha *et al.*, 2010). These are generated in real-time and made available for operational implementation by the National Centre for Medium Range Weather Forecasting (NCMRWF), New Delhi and the Indian National Centre for Ocean Information Services (INCOIS), Hyderabad through an ongoing collaboration. Each ensemble member forecast was generated by perturbing these actual initial climatic conditions. We perturb the wind, temperature and moisture fields and the amplitude of perturbation for all variables are consistent with the magnitude of the variance of each variable at a given vertical level. More details of the ensemble generation techniques can be found from Abhilash *et al.* (2014). More model and experimental details and skill assessment of GFSbc, CFST126, and CFST382 can be found in Abhilash *et al.* (2013; 2014) and Sahai *et al.* (2015).

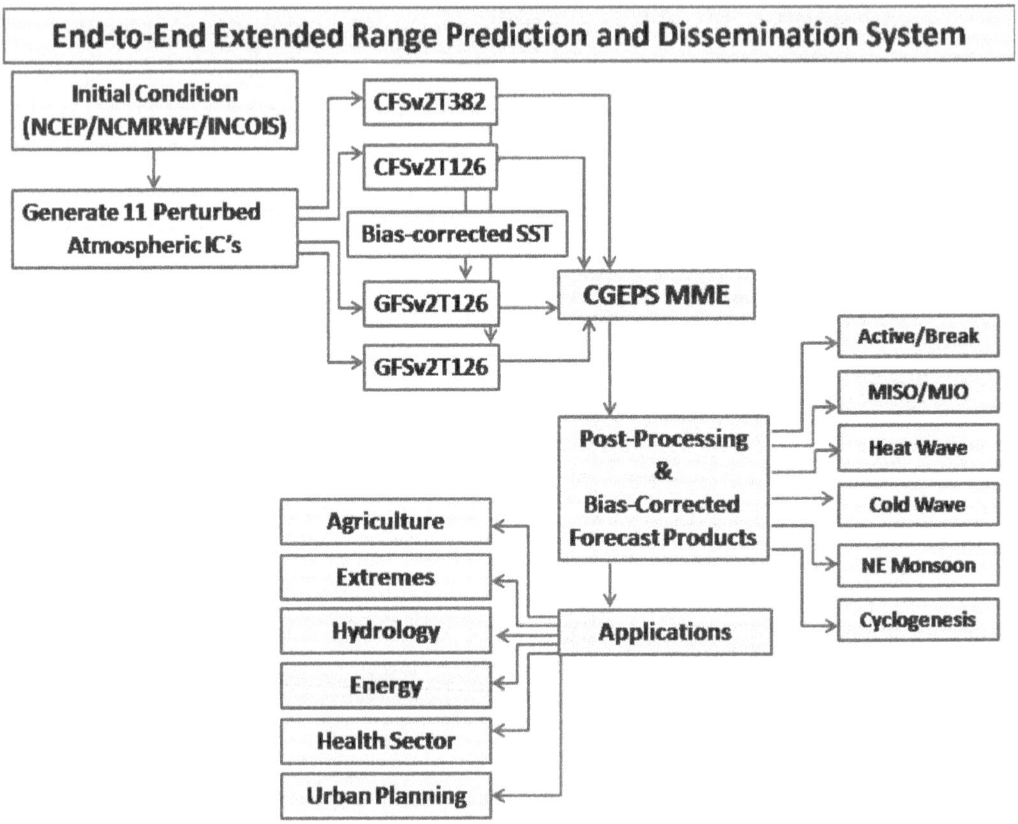

Fig. 1. Schematic of the end-to-end forecast and dissemination system implemented for ERP.

The MME, the initialization, and the modeling strategy are created in this way to increase model diversity within available computing resources. This MME uses different variants of the CFSv2 and its atmosphere component, GFS with different resolutions, parameters, and coupling configurations (to address coupled SST biases), motivated by the various physical mechanisms thought to influence monsoon forecast errors in the extended range (10-20 day lead times) using the same dynamical core. Based on performance tests, and aiming to maximize the operational skill for our available computer resource, a 16 member forecast ensemble was chosen for operational purposes. Using CFS, we run 4 members of CFST126 (~100km) and 4 members of CFST382 (~38km). Similarly, for the GFS, we run 4 members each from GFSbc at two resolutions forced with bias-corrected forecasted SST from CFS. IMD adopted this system for operational extended range prediction (ERP) over Indian region from the 2017 monsoon season (refer Fig. 2 for a flowchart of the model and prediction strategy). The 4 week lead-time forecast is operationally made once every week (every Wednesday) at present. Fig. 1 shows the schematic of the end-to-end forecast and dissemination system developed and implemented for operational ERP.

Applications of ERP System

Various customized forecast products are being disseminated from this ERP which include, extended range prediction of active-break spells of Indian summer monsoon, monsoon onset, progression, withdrawal, heat-cold wave, monitoring of Monsoon intraseasonal Oscillations (MISO) and Madden Julian Oscillation (MJO), Cyclogenesis and many other. Now this ERP system is capable of generating extended outlook for various sector-specific applications such as agriculture, hydrology, energy, insurance, re-insurance, urban planning, health, etc. Since these sectors rely on forecasts on different spatial and temporal scales, a seamless sub-seasonal forecast scheme spanning multiple spatial and temporal scales is required for operational reasons.

The onset of ISM happens over the southern peninsular state of Kerala, and hence Monsoon Onset over Kerala (MOK) has tremendous operational importance. Deterministic forecasts have been generated in real-time for the last few years. Criterion for MOK to have been defined from model hindcasts (Joseph *et al.,* 2015) and forecasted MOK for the period 2001-2014, as shown in Table 1. The main characteristics that differentiate the monsoon rains from the pre-monsoon thundershowers (the seasonal reversal of the low-level winds, the persistence of both rainfall and low-level wind after the MOK date) have been considered while formulating the criterion to avoid the occurrence of "bogus onsets" that are unrelated to the large-scale monsoon system. The onset date (MOK) based on rainfall over Kerala (ROK) and strength of zonal wind over the Arabian Sea (uARAB) are plotted in the left panel of Fig. 3 based on forecast run from 16th May initial condition

Table1: Forecasted onset date based on ERP and observed onset dates.

Year	Actual MOK	Forecasted MOK	SD among ensemble members	Difference between actual and forecasted MOK
2001	23 May	25 May	2	2
2002	29 May	21 May	5	8
2003	08 Jun	30 May	5	9
2004	18 May	18 May	1	0
2005	05 Jun	05 Jun	3	0
2006	26 May	25 May	2	1
2007	28 May	02 Jun	8	5
2008	31 May	01 Jun	7	1
2009	23 May	24 May	2	1
2010	31 May	30 May	5	1
2011	29 May	01 Jun	2	3
2012	05 Jun	04 Jun	4	1
2013	01 Jun	29 May	2	3
2014	06 Jun	05 Jun	6	1

for two selected years (2003 and 2005). Also, composites of wind and rainfall pattern based on the multi-model ensemble (MME) forecast ensemble means are shown in the right panel of Fig. 2, and are compared with observation (OBS) during MOK for the 14-year period. The plot indicates that the MME has good fidelity in matching the observed spatial pattern during the onset phase.

Any unexpected, unusual and unseasonal weather event such as heavy rainfall events, heat/cold waves, cyclones, etc. could be defined as an extreme weather event. If such events are modulated by large-scale forcing, their probabilities ought to be predictable on the extended range. The application of the MME in predicting the heavy rainfall event that occurred in June 2013 over the north Indian state of Uttarakhand is shown in Fig. 3. The solid black line with dots represent the observed rainfall amount over Uttarakhand region (78-80°E; 29-31°N), while solid red line indicates the ensemble mean forecast, and the dashed brown lines indicate individual ensemble members. It is clear from the figure that the MME could predict the event at least 10 to 15 days in advance.

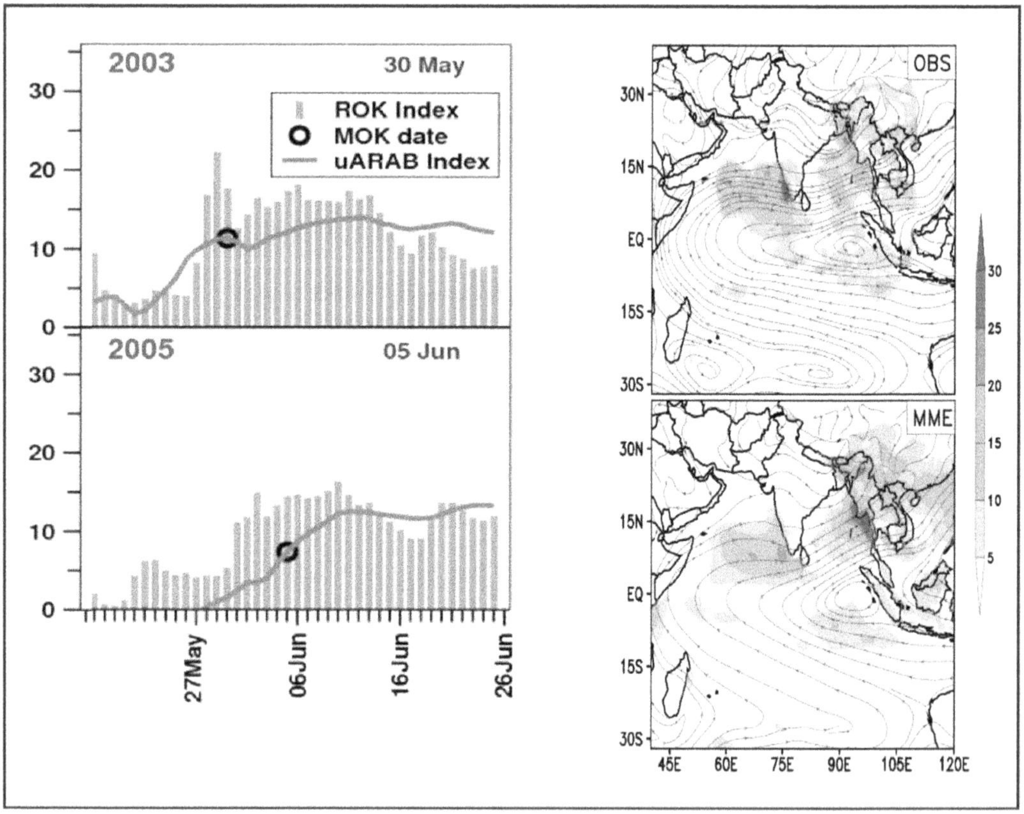

Fig. 2. Example of extended range prediction of monsoon onset over Kerala (MOK) for selected years (left panel). The composite of rainfall (in mm/day) and 850 hPa wind (in m/s) during MOK day of 2001-14 in observations and MME is shown in the right panel.

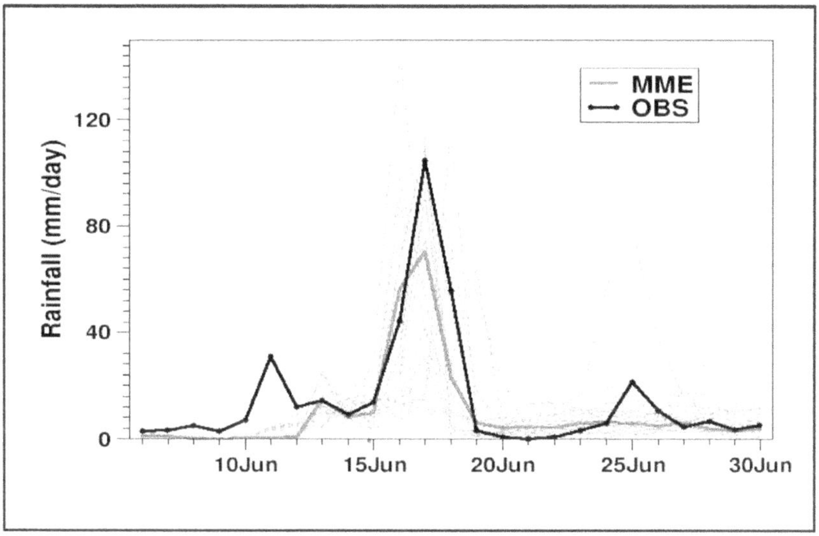

Fig. 3. Example of extreme weather forecast for the Uttarakhand heavy rainfall event of 2013.

Further, implementation of a Bias-Correction and Signal Amplification (BCSA) Technique is applied to the raw model output for improving track predictions Cyclonic Storms (CS) over North Indian Ocean (NIO). Bias-Correction method involves the removal of lead-dependent climatological-bias from MME forecasts by using European Centre for Medium-Range Weather Forecasts Re-analysis (ERA-Interim) daily-averaged datasets as observations. The corrected data is then subjected to Signal Amplification procedure involving a two-point space and time correction of ensembles based on the leading signal (ensemble mean), whereby large uncertainties and disagreements between different model outputs are reduced. Figure 4 shows the flowchart of BCSA technique. Figure.5 presents the improvement of track forecast by applying BCSA technique as compared to raw MME forecast. Results show that BCSA Technique is, indeed, improving the track forecasts of selected CS cases with significant reduction in track errors even at longer lead times. The pre-genesis track and intensity predictions were promising and the MME system is capable of capturing almost all desirable features of the cyclone forecasting system such as cyclogenesis, pre-genesis track and intensity even at longer lead times of more than one week.

Summary

Extended Range forecast of sub-seasonal variability beyond weather scale is a critical component in climate forecast applications over Indian region. The sub-seasonal to seasonal (s2s) project, undertaken by WCRP, started in 2013 to improve the forecast beyond weather scale which is a challenging gap area both in research and operational forecast domain. The primary objective of this s2s project is to provide the sub-seasonal to

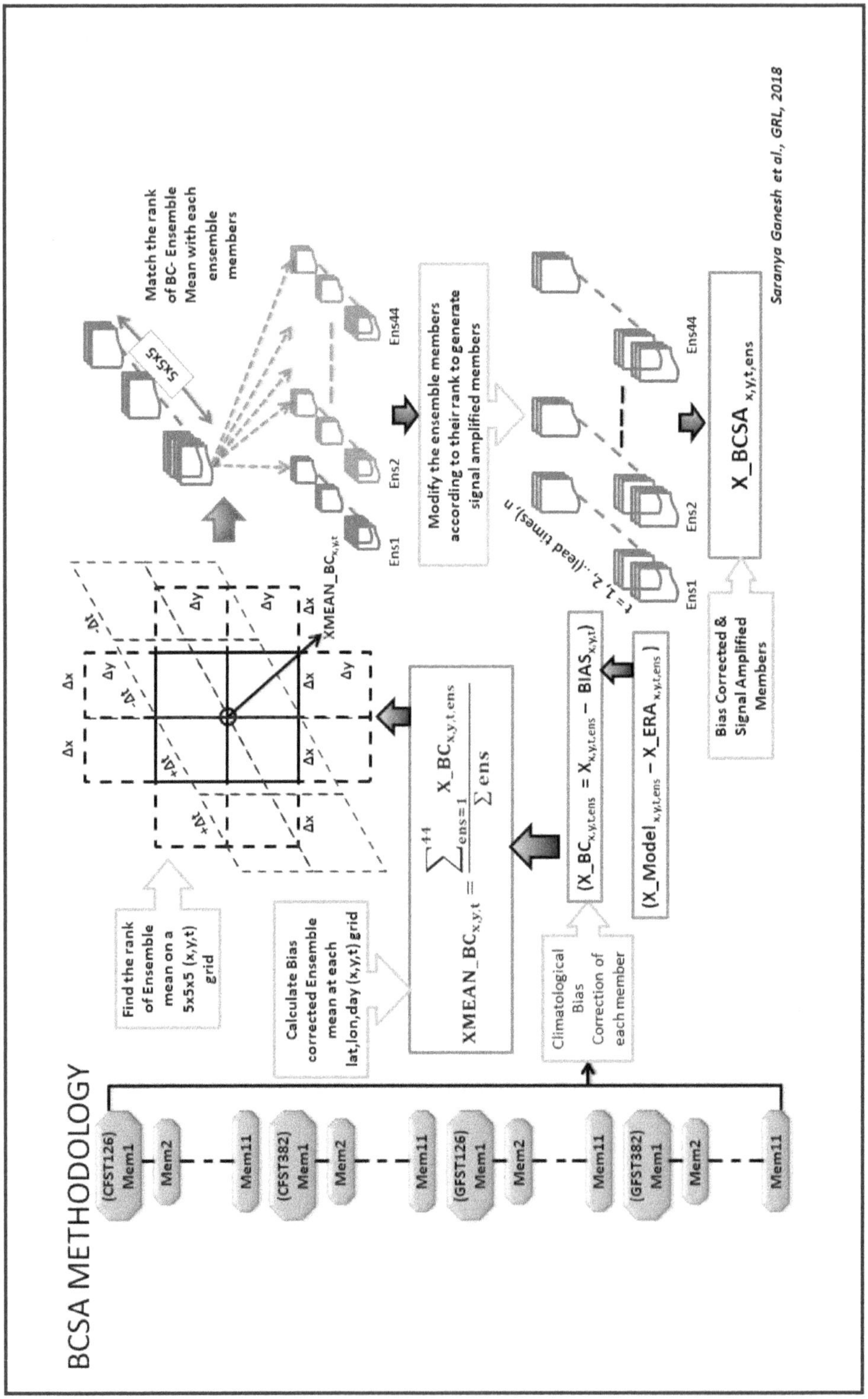

Fig. 4. Flow chart of BCSA technique

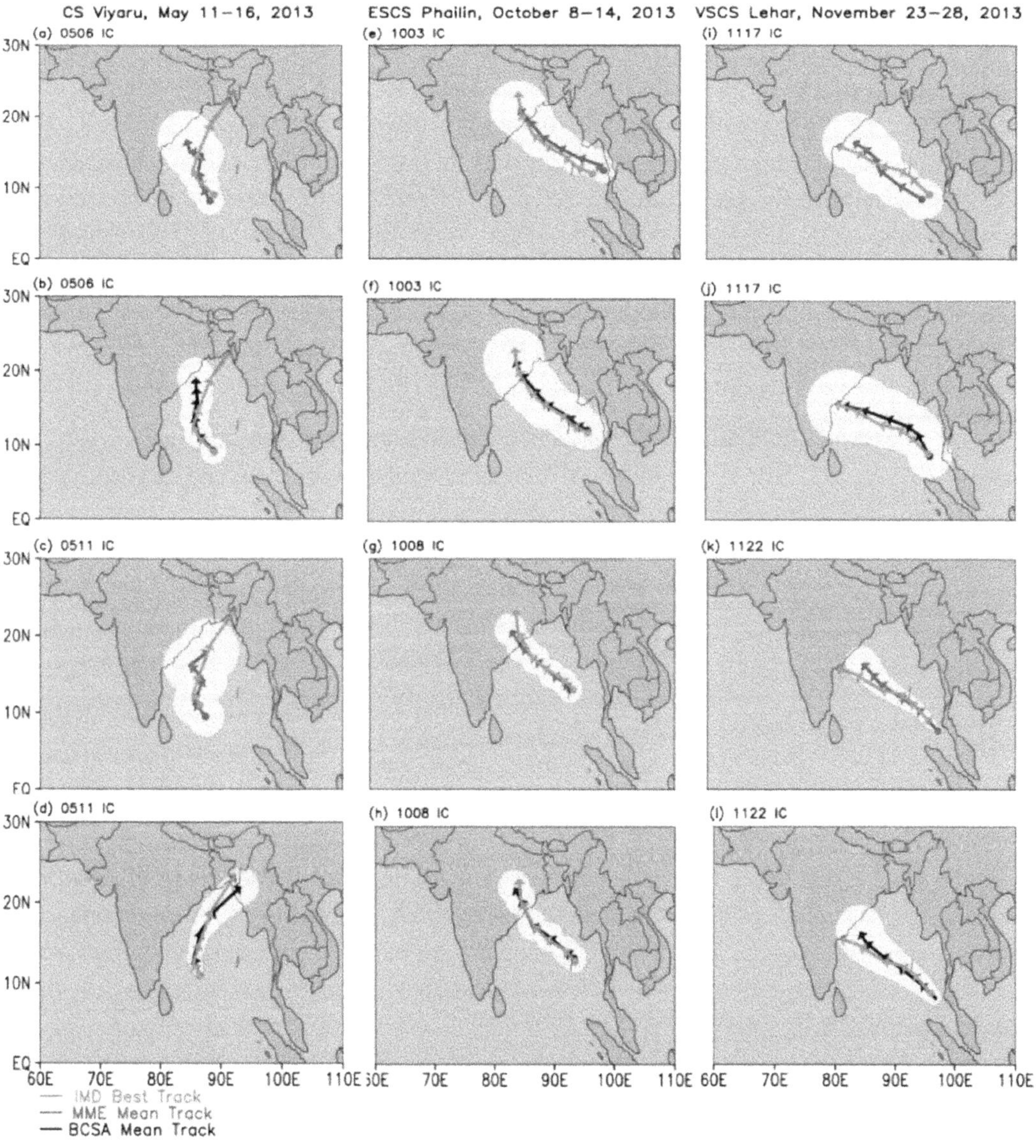

Fig. 5. Track forecasts for all storm cases. (Viyaru: a-d; Phailin: e-h; Lehar: i-l. Plots (a) and (b) show IMD Best Track Observations vs. MME mean Track (a) and BCSA mean Track (b) predictions, respectively, for 0506 IC, (c) and (d) the same for 0511 IC, (e) and (f) for 1003 IC, (g) and (h) for 1008 IC, (i) and (j) for 1117 IC, (k) and (l) for 1122 IC (c), respectively).

seasonal forecast in various lead times. Extended Range forecast project at Indian Institute of Tropical Meteorology (IITM) was envisioned more than a decade ago in this way, keeping in view of the demands from the stakeholders to create a robust research background required for improved delivery of operational forecast in this scale. Extended range forecast implies an outlook with a lead-time of 2-3 weeks. Several efforts have been undertaken to improve the extended range forecast in the past decade or so at IITM. The current study summarizes the development of extended range prediction system for operational implementation at IMD. The research and development that lead to the development of extended range prediction are summarized at first. It describes the efforts that were undertaken to establish the statistical properties of monsoon intraseasonal oscillations, establish the predictability horizon, represent the oscillations in state-of-art climate models and finally develop the latest state-of-art forecast technique. This technique is now operationally implemented at India Meteorological Department. The primary use of this forecast system is the development of an operational country-wide forecast product in the extended range time scale. The application of this forecast can be made in several allied fields like agro-meteorology, hydrometeorology, health sector, etc.

In the backdrop of climate change effects on the seasonal and sub-seasonal variability (Sharmila *et al.,* 2015b), applications of extended range prediction can be useful for farmers, dam managers, and several other stakeholders. With erratic rainfall pattern and unpredictable trends, new stakeholders are showing interest, and the number of potential users of this forecast is increasing every day. Several new climate sensitive stakeholders related to artificial groundwater recharging, rainwater harvesting, crop insurance, crop planning, urban planning are showing interest in this forecast as this would help them to mitigate the uncertainties arising due to projected erratic behaviour of rainfall in the near future.

Acknowledgments

Most of this report is based on past and ongoing work on extended range prediction. We acknowledge the support of IITM, an autonomous institute under Ministry of Earth Sciences (MoES), scientists, researchers, technical support staff and operational forecasters who are directly or indirectly involved in National Monsoon Mission project and have provided valuable feedback from time to time. The CFS model-based forecast system is implemented as a part of MoU between MoES, Govt. of India and NOAA, US.

References

Abhilash S., Sahai, A.K., Borah, N., Chattopadhyay, R., Joseph, S., Sharmila, S., De, S., Goswami, B.N. and Arun Kumar, 2014. Prediction and monitoring of monsoon intraseasonal oscillations over Indian monsoon region in an ensemble prediction system using CFSv2. Climate Dynamics 42: 2801-2815 . doi: 10.1007/s00382-013-2045-9.

Abhilash, S., Sahai, A.K., Borah, N., Joseph, S., Chattopadhyay, R. and Sharmila, S., Rajeevan, M., Mapes, B.E. and Kumar, A., 2015. Improved spread-error relationship and probabilistic prediction from CFS based Grand Ensemble Prediction System. J. Appl. Meteor. Clim., 54: 1569-1578. doi: 10.1175/JAMC-D-14-0200.1.

Abhilash, S., Sahai, A.K., Pattnaik, S., De, S., 2013. Predictability during active break phases of Indian summer monsoon in an ensemble prediction system using climate forecast system. J. Atmospheric Sol-Terr. Phys., 100-101: 13-23. doi: 10.1016/j.jastp.2013.03.017.

Buizza, R., Leutbecher, M., Isaksen, L., 2008. Potential use of an ensemble of analyses in the ECMWF Ensemble Prediction System. QJR Meteorol. Soc., 134: 2051-2066. doi: 10.1002/qj.346.

Griffies, S., Harrison, M., Pacanowski, R. and Rosati, A., 2004. A Technical Guide to MOM4. GFDL Ocean Group, NOAA GFDL.

Joseph, S., Sahai, A.K., Abhilash, S., Chattopadhyay, R., Borah, N.; Mapes, B.E., Rajeevan, M. and Kumar, A., 2015. Development and evaluation of an objective criterion for the real-time prediction of Indian summer monsoon onset in a coupled model framework. J. Clim., 28(15): 6234-6248 . doi: 10.1175/JCLI-D-14-00842.1.

Reynolds, R.W., Smith, T.M., Liu, C., Chelton, D.B., , Casey, K.S. and Schlax, M.G., 2007. Daily high-resolution-blended analyses for sea surface temperature. J. Climate 20(22): 5473-5496 . doi: 10.1175/2007JCLI1824.1

Saha, S., Moorthi, S., Wu, X., Wang, J., Nadiga, S., Tripp, P., Behringer, D., Hou, Y.T., Chuang, H.Y., Michael Ek, M.I., Meng, J., Yang, R.; Mendez, M.P., Van den Dool, H., Zhang, Q., Wang, W.; Chen, M. and Becker, E., 2014. The NCEP Climate Forecast System Version 2. J Clim 27:2185–2208 . doi: 10.1175/JCLI-D-12-00823.1

Saha, S., Nadiga, S., Thiaw, C., Wang, J., Wang, W., Zhang, Q., Van den Dool, H.M., Pan, H.-L., Moorthi, S., Behringer, D., Stokes, D., Peña, M., Lord, S., White, G., Ebisuzaki, W., Peng, P. and Xie, P., 2006. The NCEP Climate Forecast System. J. Climate 19(15): 3483–3517 . doi: 10.1175/JCLI3812.1.

Saha, S., Shrinivas Moorthi, Hua-Lu Pan, Xingren Wu, Jiande Wang, Sudhir Nadiga, Patrick Tripp, Robert Kistler, John Woollen, David Behringer Haixia Liu, Diane Stokes, Robert Grumbine, George Gayno, Jun Wang, Yu-Tai Hou, Hui-ya Chuang, Hann-Ming H. Juang, Joe Sela, Mark Iredell, Russ Treadon, Daryl Kleist, Paul Van Delst, Dennis Keyser, John Derber, Michael Ek, Jesse Meng, Helin Wei, Rongqian Yang, Stephen Lord, Huug van den Dool, Arun Kumar, Wanqiu Wang, Craig Long, Muthuvel Chelliah, Yan Xue, Boyin Huang, Jae-Kyung Schemm, Wesley Ebisuzaki, Roger Lin, Pingping Xie, Mingyue Chen, Shuntai Zhou, Wayne Higgins, Cheng-Zhi Zou, Quanhua Liu, Yong Chen, Yong Han, Lidia Cucurull, Richard W. Reynolds, Glenn Rutledge and Mitch Goldberg, 2010. The NCEP Climate Forecast System Reanalysis. Bull. Amer. Meteor. Soc., 91(8): 1015-1058.

Sahai, A.K., Chattopadhyay, R., Joseph, S., Mandal, R., Dey, A., Abhilash, S., Krishna R.P.M. and Borah, N., 2015. Real-time performance of a multi-model ensemble-based extended range forecast system in predicting the 2014 monsoon season based on NCEP-CFSv2. Curr. Sci., 109: 1802-1813.

Impact of Climate Change on Hydrological Cycle, Ecosystem, Fisheries and Food Security, Pages: 39–56
Edited by: B. Madhusoodana Kurup, M.R. Boopendranath, M. Harikrishnan and A.V. Shibu

CHAPTER - 4

Extreme Significant Wave Heights and Associated Peak Periods off Kozhikode, Southwest Coast of India

Ramu Pachanattu*, K. Ajith Joseph[1], G. Muraleedharan[1], S. Sureshkumar[2], T.M. Balakrishnan Nair[3] and R. Harikumar[3]

[1]*Nansen Environmental Research Centre, Kochi-682506, India*
[2]*School of Ocean Science and Technology, Kerala University of Fisheries and Ocean Studies, Kochi-682506, India*
[3]*Ocean Information and Forecast Services Group, Indian National Centre for Ocean Information Services, Hyderabad-500090, India*
**E-mail: ramupachu@gmail.com*

Floating structures require bivariate distributions of significant wave heights (H_3) and characteristic (mean or peak) periods as design and operational criteria (Haver, 1985; Bitner-Gregersen *et al.*, 1998; Ferreira and Guedes Soares, 2002). Albeit the survivability of floating energy converters that operate in shallow waters depend on extreme sea states, their operability mainly depends only on the wave period (Silva *et al.*, 2013).

In this case study, the significant wave heights (H_s) and associated peak periods (T_p), mean energy periods (T_e) and zero up-crossing periods (T_z) off Kozhicode, southwest cost of India recorded by WAMAN (Wave Monitoring Along Nears-shore) buoy deployed by INCOIS at this location for a continuous span of April to May 2013 are considered. The peak wave period or dominant wave period is defined as the wave period associated with the most energetic waves in the total wave spectrum at a specific point. It is the reciprocal of the peak frequency f_p. Wave regimes that are dominated by wind waves tend to have smaller peak wave periods, and regimes that are dominated by swell tend to have larger peak wave periods. The energy period T_e, is the mean wave period with respect to the spectral distribution of energy (Mollison, 1994). Typical values of energy periods are in the range 5-15 s (Mollison *et al.*, 1976). The energy period is also used in the estimation of wave power. The zero-crossing period is approximately equal to the average time between waves crossing the mean sea level in an upward direction. The joint behaviour of significant wave heights and associated periods are characterised by constructing joint frequency distribution tables. The frequency distribution of the wave period statistics against wave height intervals are also plotted for further clarification on their joint behaviour. The dependency of the wave statistics to each other was examined.

Coastal engineering projects are badly in need of reliable wave data for design and implementation. Since measurement of waves at each and every coastal location is expensive, predictive models can estimate the required parameters. Verification of such models against real world situation is a pre-requisite for their use in ocean engineering. Ochi (1998) emphasised the importance of probabilistic approach as an integrated part of modern marine design technology. In order to estimate the wave statistics, it is essential to fit probability models to observed frequency distribution of the respective wave parameters. Also it is to be noted at this juncture that when the data are generated from a physical process that can produce anomalies (outliers), then a probability model that is a close fit to the current observed data will not guarantee the same to the future data. So, it is preferable to use a robust approach that is based on a model that can yield sufficiently precise estimates of the wave statistics even if the true frequency distribution deviates from the fitted frequency distribution.

The computed conditional mean functions (*CMF*) of the H_s and associated T_p, T_e and T_z are attempted to simulate by *CMF* of three-parameter *Weibull* and generalised *Pareto* distributions. Since *CMF* determines the distribution uniquely, it is sufficient to find the functional form of *CMF* consistent with the data. The models are estimated by the method of *L-moments*. *L-moments* (Linear combination of probability weighted moments) are relatively a recent development in statistics and it possess an elegant mathematical theory and facilitates in the estimation process. The significance of *L-moments* over the conventional moments is that it can characterise a large number of distributions and when estimated from a sample, it is more robust to the presence of outliers in the data. L-moments are less subjected to bias in estimation Analogous to the method of moments, the method of *L-moments* obtain parameter estimates by equating the first n sample *L-moments* to the population quantities. The method of *L-moments* yields efficient and computationally convenient estimates of parameters of quantile functions.

The efficacy of the models to estimate certain important wave statistics such as average of one-third the highest significant wave heights and associated peak periods, mean energy periods and zero up-crossing periods are ascertained by comparing with the respective computed values from the data. The wave statistics estimates were deduced from the *CMF* of *Weibull* distribution. *GP*3 estimates of the wave statistics are obtained by parametric relations derived from *GP*3 based on order statistics.

Frequency of extreme events (H_{ex}) of designated return periods are of great importance for the design of marine structures and for the planning of maritime operations. An extreme height H_{ex} in the upper tail of the distribution is given by (Muraleedharan *et al.*, 2007) as

$H_{ex} = Q\,[(1-1/T)^{1/n}]$, $Q(.)$ is the quantile function; n-sample size. Since extreme wave height constitutes the peak energy in a wave energy spectrum, the period associated with the extreme event can be considered as the peak period of the spectrum.

The maximum steepness (ultimate) that wave can be attained before wave breaking is considered in this study to attain the error bounds of the peak period (T_p) associated with the extreme significant wave height (H_{ex}), period of the modal value of the spectrum, by the dispersion relation for deep water waves and by the investigations on maximum steepness (0.55 as cut-off for front-face steepness and 0.44 as cut-off for rear-face steepness) of wave profile (Toffoli *et al.*, 2010). These two limits of the steepness are considered to estimate the extreme significant wave height associated peak period bounds.

Methodology

The empirical average conditional exceedances are computed using the expression (Muraleedharan *et al.*, 2015; 2016a; 2016b)

$$m\left(x_{(r)}\right) = \frac{1}{n-r+1}\sum_{k=r}^{N} x_{(k)} \qquad \dots (1)$$

$x_{(r)} - r^{th}$ order wave statistic of a sample size N arranged in ascending order. Empirical average conditional exceedances of H_s, T_p, T_e and T_z are simulated using generalised *Pareto* (*GP*3) and three-parameter *Weibull* distributions by defining $m(x)$ (Muraleedharan *et al.*, 2009, 2015, 2016a, 2016b) as:

$$m(x) = E(X/X>x) = \frac{1}{\bar{F}(x)}\int_x^{\infty} y\left[-\bar{F}'(y)\right]dy \qquad \dots (2)$$

$\bar{F}'(x)$ is the derivative of $\bar{F}(x)$

$$\bar{F}(x) = 1 - F(x)$$

Accordingly the expression for $m(x)$ derived from generalised *Pareto* distribution is given as:

$$m(x) = x + \frac{\lambda}{\alpha+1}\left[1 - \frac{\alpha}{\lambda}(x-\mu)\right];\ \alpha > 0 \text{ or } \alpha < 0$$

$$F(x) = 1 - \left[1 - \frac{\alpha}{\lambda}(x-\mu)\right]^{\frac{1}{\alpha}} \qquad \dots (3)$$

$$\mu \le x < \infty \text{ if } \alpha < 0; \ \mu \le x \le \mu + \frac{\lambda}{\alpha} \text{ if } \alpha > 0$$

And from three-parameter *Weibull* distribution as:

$$m(x) = x + \frac{\lambda^{\Gamma}}{\alpha} \frac{\left(\frac{1}{\alpha'}\left(\frac{x-\mu}{\lambda}\right)^{\alpha}\right)}{\bar{F}(x)} \quad \text{... (4)}$$

$$F(x) = 10 - \exp\left[-\left(\frac{x-\mu}{\lambda}\right)^{\alpha}\right] \quad \text{... (5)}$$

$$x \ge \mu; -\infty < \mu < \infty; \ \lambda, \alpha > 0$$

μ, λ and α are respectively the location, scale and shape parameters.

$\Gamma(s, x)$ is the upper incomplete gamma function defined as:

$$\Gamma(s, x) = \int_{x}^{\infty} \exp[-(t)]\, t^{s-1} dt \quad \text{... (6)}$$

Thus modelling the wave statistic can be accomplished through the function m (x) provided sufficiently accurate information on the functional form of m (x) coherent with the data is known. It is known that the information on the functional form of m (x) will enable to define the distribution of X through the relationship

$$\bar{F}(x) = \exp\left[-\int_{0}^{y} \frac{m'(y)}{m(y) - y} dy\right] \quad \text{... (7)}$$

m' (y) is the derivative of m (y)

The goodness of fit is ascertained by Z-test statistic defined as (Hosking an Wallis, 1997):

$$Z = \frac{(t_4 - \tau_4)}{\sigma_4}$$

Z-Goodness-of-fit measure for the frequency distribution.

t_4-*L-kurtosis* computed from the sample.

τ_4-*L-kurtosis* of the fitted frequency distribution.

σ_4-The standard deviation of t_4, which is obtained from the 100 reiterated simulation of the original data that have the fitted frequency distribution and sample size the same as those of the observed data. The Z statistic has approximately a standard normal distribution. The criterion $|Z|\leq 1.64$ then corresponds to acceptance of the postulated distribution at a confidence level of 90%.

If $H_{(1)}$, $H_{(2)}$ ...., $H_{(n)}$ are for example n wave heights from a sample that are arranged in acending order of magnitude, denoted by $H_{(r)}$ the rth largest among them so that $H_{(1)} <$, $H_{(2)} <$ ...., $< H_{(n)}$. Then

the distribution of $H_{(r)}$ is specified by the density

$$h(h_r) = \frac{n!}{(r-1)!(n-r)!}\left[F(h_r)\right]^{r-1}\left[1-F(g=h_r)\right]^{n-r} f(h_r) \qquad \text{... (8)}$$

$r = 1, 2, ..., n$

F (.) and f (.) are the distribution and density functions of H. The average of the one-third the highest wave height is given by

$$H_{\left(\frac{1}{3}\right)} = \frac{1}{n-\frac{2n}{3}+1} E\left(H_{\left(\frac{2n}{3}\right)} + H_{\left(\frac{2n}{3}+1\right)} + ... + H_{(n)}\right)$$

If the distribution and density function of H are identified as $GP3$, then (Muraleedharan *et al.*, 2015; 2016a),

$$E\left(H_{(r)}\right) = \mu + \frac{\lambda}{\alpha}\left(1 - \frac{\Gamma(n+1)}{\Gamma(n-r+1)}\frac{\Gamma(n-r+1+\alpha)}{\Gamma(n+1+\alpha)}\right) \qquad \text{... (9)}$$

Since derivation of $E\left(H_{(r)}\right)$ is not feasible from *Weibull* distribution $H_{\left(\frac{1}{3}\right)}$, is estimated from CMF (4) of three-parameter *Weibull* distribution. Then the value of x in the CMF is obtained from $\bar{F}(x) = 1/3$.

From (8) the distribution of $H_{(n)}$ or maximum wave height $H_{(\max)}$ is given by the density:

$$h(h_{\max}) = n\left[F(h_{\max})\right]^{n-1} f(h_{\max}) \qquad \text{... (10)}$$

And distribution function is given as:

$$G(h_{\max}) = [F(h_{\max})]^n \qquad \text{... (11)}$$

Muraleedharan et al (2007) obtained the expression for the return period T of an extreme wave height H_{ex} as

$$T = (1-G)^{-1} \text{ ...} \qquad (12)$$

or

$$P = \left(1 - \frac{1}{T}\right)^{\frac{1}{n}} \qquad \text{... (13)}$$

$H_{ex} = Q(P)$ is the extreme wave height for designated return period T. the quantile functions $Q(P)$ of three-parameter *Weibull* distribution is given as:

$$Q(P) = \mu + \lambda\left[-\ln(1-p)\right]^{\frac{1}{\alpha}} \qquad \text{... (14)}$$

And $GP3$ as:

$$Q(P) = \mu + \frac{\lambda}{\alpha}\left[1-(1-p)^{\alpha}\right] \qquad \text{... (15)}$$

Results and Discussion

The joint behaviour of the significant wave heights (H_s) and the associated periods: mean energy periods (T_e), peak periods (T_p) and zero up-crossing periods (T_z) off Kozhicode, southwest coast of India in the months of April and May 2013 are examined (Tables 1-3). The periods have modal value range of 3-6 s and H_s concentrates in the range 0.5 to 0.8m.

Table 1: Recorded joint frequency distribution of significant wave heights (Hs) and associated mean energy periods (Te) off Kozhicode, southwest coast of India in April-May, 2013.

OF	Te(s)								
Hs(m)	**4-5**	**5-6**	**6-7**	**7-8**	**8-9**	**9-10**	**10-11**	**11-12**	***OF***
0.4-0.5				1	3	15	1		20
0.5-0.6		1	3	4	14	21	10	1	54
0.6-0.7		17	27	9	7	8	4		72
0.7-0.8	1	26	8	7	2	2			46
0.8-0.9		7	1	2					10
0.9-1	5	12	1						18
1-1.1	6	3							9
OF	**12**	**66**	**40**	**23**	**26**	**46**	**15**	**1**	**229**

Table 2: Recorded joint frequency distribution of significant wave heights (Hs) and associated peak periods (Te) off Kozhicode, southwest coast of India in April-May, 2013.

OF	Tp(s)												
Hs(m)	3-4	4-5	5-6	6-7	7-8	8-9	9-10	10-11	11-12	12-13	13-14	14-15	Grand Total
0.4-0.5					1	4	10	2	3				20
0.5-0.6		1	1	1	7	9	9	13	6	5		2	54
0.6-0.7		5	24	18	4	5	3	4	2	4	2	1	72
0.7-0.8	1	16	12	8	2	5	1	1					46
0.8-0.9			6	3	1								10
0.9-1		8	9	1									18
1-1.1		6	2	1									9
Grand Total	**1**	**36**	**54**	**32**	**15**	**23**	**23**	**20**	**11**	**9**	**2**	**3**	**229**

Table 3: Recorded joint frequency distribution of significant wave heights (Hs) and associated zero-crossing periods (Tz) off Kozhicode, southwest coast of India in April-May, 2013.

OF	Tz(s)					
Hs(m)	3-4	4-5	5-6	6-7	7-8	Total
0.4-0.5	1	5	12	2		20
0.5-0.6	4	26	15	5	4	54
0.6-0.7	37	29	5	1		72
0.7-0.8	34	11	1			46
0.8-0.9	9	1				10
0.9-1	15	3				18
1-1.1	7	2				9
Grand Total	**107**	**77**	**33**	**8**	**4**	**229**

The frequency distributions of T_e, T_p and T_z associated with H_s are illustrated in Fig. 1-3. T_e in the range 5-7 s and 9-10 s are frequent and are associated with H_s range 0.5-0.8 m (Figure 1). H_s also have wider range of T_e for this height range. Higher frequencies of T_p (4-7 s and 10-11 s)and T_z (3-5 s) are also lie in this H_s range (Fig. 2 and 3).

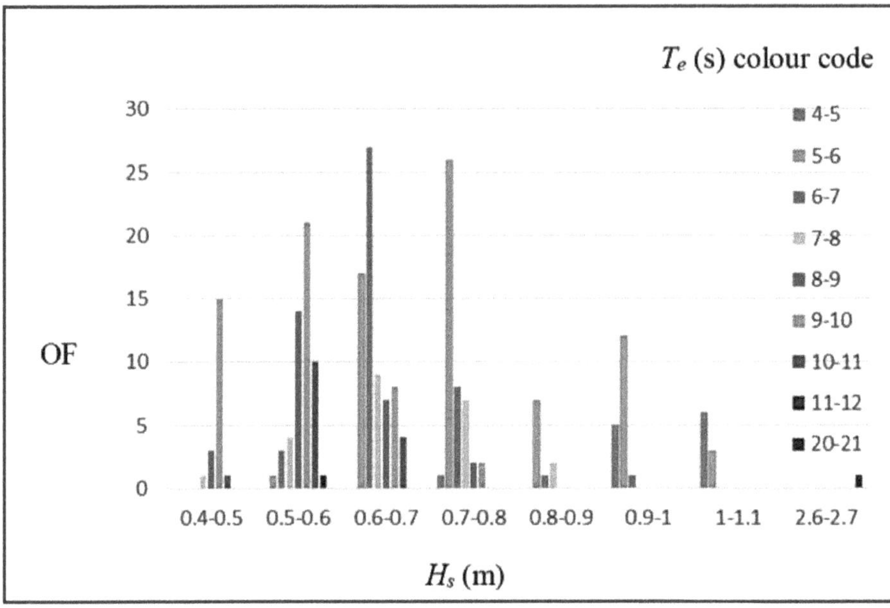

Fig. 1. Recorded frequency distribution of mean energy periods (T_e) associated with significant wave heights (H_s) off Kozhicode, southwest coast of India in April-May, 2013.

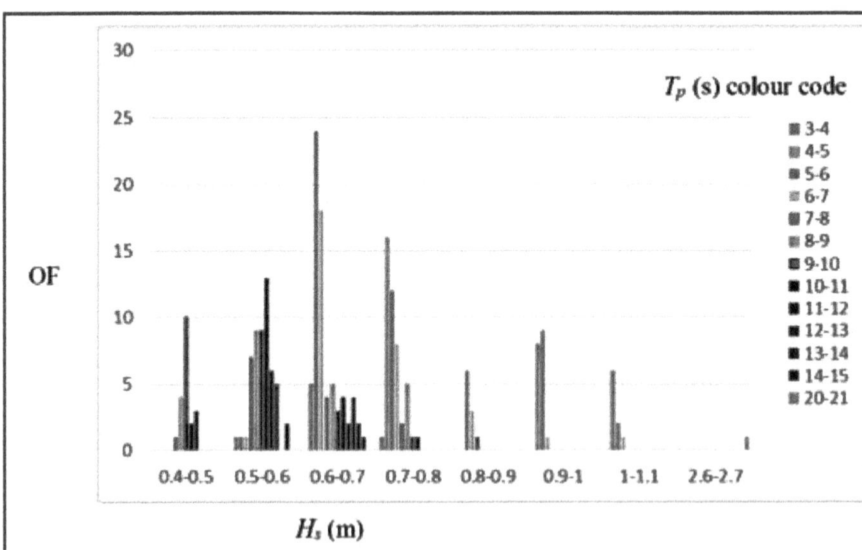

Fig. 2. Recorded frequency distribution of peak periods (T_p) associated with significant wave heights (H_s) off Kozhicode, southwest coast of India in April-May, 2013.

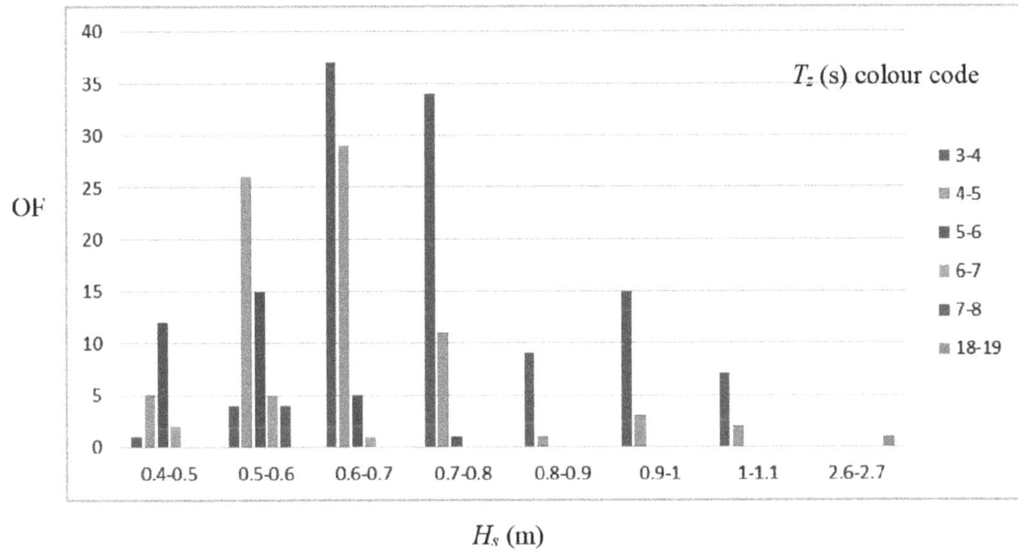

Fig. 3. Recorded frequency distribution of Zero-crossing periods (T_Z) associated with significant wave heights (H_S) off Kozhicode, southwest coast of India in April-May, 2013.

H_S and associated periods exhibit feeble dependency and hence they can be considered as independent variables (Tables 4 to 6) according to the criteria for independence of two random variables. The maximum difference between $P\ (h_S \leq H_S \leq h_S{+}0.1,\ t_x \leq T_x \leq t_x{+}1)$; ($x = e$, p or z) and product of $P\ (h_S \leq H_S \leq h_S{+}0.1)$ and $P\ (t_x \leq T_x \leq t_x{+}1)$ is only 0.07 (Table 4).

A three parameter *Weibull* and generalised *Pareto* distributions are suggested as the marginal distributions of significant wave heights and associated periods respectively by conditional mean functions (*CMF*). *Z*-statistic criteria is also applied as a goodness-of-fit measure at 90% confidence level (Table 7) to measure the compatibility of the *CMFs* with the computed average conditional exceedances of H_S, T_e, T_p and $T\text{-}_z$ (Fig. 4 to 7). *Weibull CMF* provide a nonlinear fit whereas *GP3 CMF* offer a linear fit. Both the postulated models are acceptable by *Z*-test statistic at 90% confidence level.

Average of one-third the highest significant wave heights, mean energy periods, peak periods and zero up-crossing periods were also computed from the data and estimated from the models (Table 8). The estimates are absolutely in consistent with the computed values approximated to one decimal. Average of one-third the highest wave periods are important as there exists substantial linear functional relationship with significant wave periods (Muraleedharan *et al.*, 2015)

Table 4: The joint probability distribution of *Hs* and *Te* from April-May 2013 buoy data off Kozhikode, southwest coast of India are approximately equal to the product of the marginal probability distributions of *Hs* and *Te* indicating that *Hs* and *Te* are independent.

P(0.4dHHs<0.5,7.0dHTe <8.0)	= 0	0.01	=	P(0.4dH Hs <0.5)XP(7.0dH T e <8.0)
P(0.4 ≤ Hs <0.5,8.0 ≤ Te <9.0)	= 0.01	0.01	=	P(0.4 ≤ Hs <0.5)XP(8.0 ≤ T e <9.0)
P(0.4 ≤ Hs <0.5,9.0 ≤ Te <10.0)	= 0.07	0.02	=	P(0.4 ≤ Hs <0.5)XP(9.0 ≤ T e <10.0)
P(0.4 ≤ Hs <0.5,10.0 ≤ Te <11.0)	= 0	0.01	=	P(0.4 ≤ Hs <0.5)XP(10.0 ≤ T e <11.0)
P(0.5 ≤ Hs <0.6,5.0 ≤ Te <6.0)	= 0	0.07	=	P(0.5 ≤ Hs <0.6)XP(5.0 ≤ T e <6.0)
P(0.5 ≤ Hs <0.6,6.0 ≤ Te <7.0)	= 0.01	0.04	=	P(0.5 ≤ Hs <0.6)XP(6.0 ≤ T e <7.0)
P(0.5 ≤ Hs <0.6,7.0 ≤ Te <8.0)	= 0.02	0.02	=	P(0.5 ≤ Hs <0.6)XP(7.0 ≤ T e <8.0)
P(0.5 ≤ Hs <0.6,8.0 ≤ Te <9.0)	= 0.06	0.03	=	P(0.5 ≤ Hs <0.6)XP(8.0 ≤ T e <9.0)
P(0.5 ≤ Hs <0.6,9.0 ≤ Te <10.0)	= 0.09	0.05	=	P(0.5 ≤ Hs <0.6)XP(9.0 ≤ T e <10.0)
P(0.5dH Hs <0.6,10.0dH T e <11.0)	= 0.04	0.02	=	P(0.5dH Hs <0.6)XP(10.0dH T e <11.0)
P(0.5 ≤ Hs <0.6,11.0 ≤ Te <12.0)	= 0	0	=	P(0.5 ≤ Hs <0.6)XP(11.0 ≤ T e <12.0)
P(0.6 ≤ Hs <0.7,5.0 ≤ Te <6.0)	= 0.07	0.09	=	P(0.6 ≤ Hs <0.7)XP(5.0 ≤ T e <6.0)
P(0.6 ≤ Hs <0.7,6.0 ≤ T e <7.0)	= 0.12	0.05	=	P(0.6 ≤ Hs <0.7)XP(6.0 ≤ T e <7.0)
P(0.6 ≤ Hs <0.7,7.0 ≤ T e <8.0)	= 0.04	0.03	=	P(0.6 ≤ Hs <0.7)XP(7.0 ≤ T e <8.0)
P(0.6 ≤ Hs <0.7,8.0 ≤ T e <9.0)	= 0.03	0.04	=	P(0.6 ≤ Hs <0.7)XP(8.0 ≤ T e <9.0)
P(0.6 ≤ Hs <0.7,9.0 ≤ T e <10.0)	= 0.03	0.06	=	P(0.6 ≤ Hs <0.7)XP(9.0 ≤ T e <10.0)
P(0.6 ≤ Hs <0.7,10.0 ≤ T e <11.0)	= 0.02	0.02	=	P(0.6 ≤ Hs <0.7)XP(10.0 ≤ T e <11.0)
P(0.7 ≤ Hs <0.8,4.0 ≤ T e <5.0)	= 0	0.01	=	P(0.7 ≤ Hs <0.8)XP(4.0 ≤ T e <5.0)
P(0.7 ≤ Hs <0.8,5.0 ≤ T e <6.0)	= 0.11	0.06	=	P(0.7 ≤ Hs <0.8)XP(5.0 ≤ T e <6.0)
P(0.7dH Hs <0.8,6.0dH T e <7.0)	= 0.03	0.03	=	P(0.7dH Hs <0.8)XP(6.0dH T e <7.0)
P(0.7 ≤ Hs <0.8,7.0 ≤ T e <8.0)	= 0.03	0.02	=	P(0.7 ≤ Hs <0.8)XP(7.0 ≤ T e <8.0)
P(0.7 ≤ Hs <0.8,8.0 ≤ T e <9.0)	= 0.01	0.02	=	P(0.7 ≤ Hs <0.8)XP(8.0 ≤ T e <9.0)
P(0.7 ≤ Hs <0.8,9.0 ≤ T e <10.0)	= 0.01	0.04	=	P(0.7 ≤ Hs <0.8)XP(9.0 ≤ T e <10.0)
P(0.8 ≤ Hs <0.9,5.0 ≤ T e <6.0)	= 0.03	0.01	=	P(0.8 ≤ Hs <0.9)XP(5.0 ≤ T e <6.0)
P(0.8 ≤ Hs <0.9,6.0 ≤ T e <7.0)	= 0	0.01	=	P(0.8 ≤ Hs <0.9)XP(6.0 ≤ T e <7.0)
P(0.8dH Hs <0.9,7.0dH T e <8.0)	= 0	0	=	P(0.8dH Hs <0.9)XP(7.0dH T e <8.0)
P(0.9 ≤ Hs <1.0,4.0 ≤ T e <5.0)	= 0.02	0	=	P(0.9 ≤ Hs <1.0)XP(4.0 ≤ T e <5.0)
P(0.9 ≤ Hs <1.0,5.0 ≤ T e <6.0)	= 0.05	0.02	=	P(0.9 ≤ Hs <1.0)XP(5.0 ≤ T e <6.0)
P(0.9dH Hs <1.0,6.0dH T e <7.0)	= 0	0.01	=	P(0.9dH Hs <1.0)XP(6.0dHT e <7.0)
P(1.0 ≤ Hs <1.1, 4.0 ≤ T e <5.0)	= 0.03	0	=	P(1.0 ≤ Hs <1.1)XP(4.0 ≤ T e <5.0)
P(1.0dH Hs <1.1, 5.0dH T e <6.0)	= 0.01	0.01	=	P(1.0dH Hs <1.1)XP(5.0dHT e <6.0)

Table 5: The joint probability distribution of H_s and T_p from April to May 2013 buoy data off Kozhicode, southwest coast of India are approximately equal to the product of the marginal probability distributions of H_s and T_p indicating that H_s and T_p are independent.

P(0.4dH Hs <0.5,7.0dHT p <8.0)	= 0	0.01 =	P(0.4dH Hs <0.5)XP(7.0dH T p <8.0)
P(0.4 ≤ Hs <0.5,8.0 ≤ Tp <9.0)	= 0.02	0.01 =	P(0.4 ≤ Hs <0.5)XP(8.0 ≤ T p <9.0)
P(0.4 ≤ Hs <0.5,9.0 ≤ T p <10.0)	= 0.04	0.01 =	P(0.4 ≤ Hs <0.5)XP(9.0 ≤ T p <10.0)
P(0.4dH Hs <0.5,10.0dH T p <11.0)	= 0.01	0.01 =	P(0.4dH Hs <0.5)XP(10.0dH T p <11.0)
P(0.4 ≤ Hs <0.5,11.0 ≤ T p <12.0)	= 0.01	0 =	P(0.4 ≤ Hs <0.5)XP(11.0 ≤ T p <12.0)
P(0.5 ≤ Hs <0.6,4.0 ≤ T p <5.0)	= 0	0.04 =	P(0.5 ≤ Hs <0.6)XP(4.0 ≤ T p <5.0)
P(0.5dH Hs <0.6,5.0dH T p <6.0)	= 0	0.06 =	P(0.5dH Hs <0.6)XP(5.0dH T p <6.0)
P(0.5 ≤ Hs <0.6,6.0 ≤ T p <7.0)	= 0	0.03 =	P(0.5 ≤ Hs <0.6)XP(6.0 ≤ T p <7.0)
P(0.5 ≤ Hs <0.6,7.0 ≤ T p <8.0)	= 0.03	0.02 =	P(0.5 ≤ Hs <0.6)XP(7.0 ≤ T p <8.0)
P(0.5dH Hs <0.6,8.0dH T p <9.0)	= 0.04	0.02 =	P(0.5dH Hs <0.6)XP(8.0dH T p <9.0)
P(0.5 ≤ Hs <0.6,9.0 ≤ T p <10.0)	= 0.04	0.02 =	P(0.5 ≤ Hs <0.6)XP(9.0 ≤ T p <10.0)
P(0.5dH Hs <0.6,10.0dH T p <11.0)	= 0.06	0.02 =	P(0.5dH Hs <0.6)XP(10.0dH T p <11.0)
P(0.5 ≤ Hs <0.6,11.0 ≤ T p <12.0)	= 0.03	0.01 =	P(0.5 ≤ Hs <0.6)XP(11.0 ≤ T p <12.0)
P(0.5dH Hs <0.6,12.0dH T p <13.0)	= 0.02	0.01 =	P(0.5dH Hs <0.6)XP(12.0dH T p <13.0)
P(0.5 ≤ Hs <0.6,14.0 ≤ T p <15.0)	= 0.01	0 =	P(0.5 ≤ Hs <0.6)XP(14.0 ≤ T p <15.0)
P(0.6dH Hs <0.7,4.0dH T p <5.0)	= 0.02	0.05 =	P(0.6dH Hs <0.7)XP(4.0dH T p <5.0)
P(0.6 ≤ Hs <0.7,5.0 ≤ T p <6.0)	= 0.10	0.07 =	P(0.6 ≤ Hs <0.7)XP(5.0 ≤ T p <6.0)
P(0.6dH Hs <0.7,6.0dH T p <7.0)	= 0.08	0.04 =	P(0.6dH Hs <0.7)XP(6.0dH T p <7.0)
P(0.6 ≤ Hs <0.7,7.0 ≤ T p <8.0)	= 0.02	0.02 =	P(0.6 ≤ Hs <0.7)XP(7.0 ≤ T p <8.0)
P(0.6dH Hs <0.7,8.0dH T p <9.0)	= 0.02	0.03 =	P(0.6dH Hs <0.7)XP(8.0dH T p <9.0)
P(0.6 ≤ Hs <0.7,9.0 ≤ T p <10.0)	= 0.01	0.03 =	P(0.6 ≤ Hs <0.7)XP(9.0 ≤ T p <10.0)
P(0.6dH Hs <0.7,10.0dH T p <11.0)	= 0.02	0.03 =	P(0.6dH Hs <0.7)XP(10.0dH T p <11.0)
P(0.6 ≤ Hs <0.7,11.0 ≤ T p <12.0)	= 0.01	0.01 =	P(0.6 ≤ Hs <0.7)XP(11.0 ≤ T p <12.0)
P(0.6 ≤ Hs <0.7,12.0 ≤ T p <13.0)	= 0.02	0.01 =	P(0.6 ≤ Hs <0.7)XP(12.0 ≤ T p <13.0)
P(0.6dH Hs <0.7,13.0dH T p <14.0)	= 0.01	0 =	P(0.6dH Hs <0.7)XP(13.0dH T p <14.0)
P(0.6 ≤ Hs <0.7,14.0 ≤ T p <15.0)	= 0	0 =	P(0.6 ≤ Hs <0.7)XP(14.0 ≤ T p <15.0)
P(0.7 ≤ Hs <0.8,3.0 ≤ T p <4.0)	= 0	0 =	P(0.7 ≤ Hs <0.8)XP(3.0 ≤ T p <4.0)
P(0.7dH Hs <0.8,4.0dH T p <5.0)	= 0.07	0.03 =	P(0.7dH Hs <0.8)XP(4.0dH T p <5.0)
P(0.7 ≤ Hs <0.8,5.0 ≤ T p <6.0)	= 0.05	0.05 =	P(0.7 ≤ Hs <0.8)XP(5.0 ≤ T p <6.0)
P(0.7dH Hs <0.8,6.0dH T p <7.0)	= 0.03	0.03 =	P(0.7dH Hs <0.8)XP(6.0dH T p <7.0)
P(0.7dH Hs <0.8,7.0dH T p <8.0)	= 0.01	0.01 =	P(0.7dH Hs <0.8)XP(7.0dH T p <8.0)
P(0.7 ≤ Hs <0.8,8.0 ≤ T p <9.0)	= 0.02	0.02 =	P(0.7 ≤ Hs <0.8)XP(8.0 ≤ T p <9.0)
P(0.7 ≤ Hs <0.8,9.0 ≤ T p <10.0)	= 0	0.02 =	P(0.7 ≤ Hs <0.8)XP(9.0 ≤ T p <10.0)
P(0.7dH Hs <0.8,10.0dH T p <11.0)	= 0	0.02 =	P(0.7dH Hs <0.8)XP(10.0dH T p <11.0)
P(0.8dH Hs <0.9,5.0dH T p <6.0)	= 0.03	0.01 =	P(0.8dH Hs <0.9)XP(5.0dH T p <6.0)
P(0.8dH Hs <0.9,6.0dH T p <7.0)	= 0.01	0.01 =	P(0.8dH Hs <0.9)XP(6.0dH T p <7.0)
P(0.8dH Hs <0.9,7.0dH T p <8.0)	= 0	0 =	P(0.8dH Hs <0.9)XP(7.0dH T p <8.0)
P(0.9dH Hs <1.0,4.0dH T p <5.0)	= 0.03	0.01 =	P(0.9dH Hs <1.0)XP(4.0dH T p <5.0)
P(0.9 ≤ Hs <1.0,5.0 ≤ T p <6.0)	= 0.04	0.02 =	P(0.9 ≤ Hs <1.0)XP(5.0 ≤ T p <6.0)
P(0.9 ≤ Hs <1.0,6.0 ≤ T p <7.0)	= 0	0.01 =	P(0.9 ≤ Hs <1.0)XP(6.0 ≤ T p <7.0)
P(1.0 ≤ Hs <1.1, 4.0 ≤ T p <5.0)	= 0.03	0.01 =	P(1.0 ≤ Hs <1.1)XP(4.0 ≤ T p <5.0)
P(1.0dH Hs <1.1, 5.0dH T p <6.0)	= 0.01	0.01 =	P(1.0dH Hs <1.1)XP(5.0dHT p <6.0)
P(1.0 ≤ Hs <1.1, 6.0 ≤ T p <7.0)	= 0	0.01 =	P(1.0 ≤ Hs <1.1)XP(6.0 ≤ T p <7.0)

Table 6: The joint probability distribution of H_s and $T\text{-}_z$ from April to May 2013 buoy data off Kozhicode, southwest coast of India are approximately equal to the product of the marginal probability distributions of H_s and T_z indicating that H_s and T_z are independent

P(0.4 ≤ Hs <0.5, 3.0 ≤ Tz <4.0)	= 0.00	0.04 =	P(0.4 ≤ Hs <0.5)XP(3.0 ≤ Tz <4.0)
P(0.4 ≤ Hs <0.5, 4.0 ≤ Tz <5.0)	= 0.02	0.03 =	P(0.4 ≤ Hs <0.5)XP(4.0 ≤ Tz <5.0)
P(0.4dH Hs <0.5, 5.0dH Tz <6.0)	= 0.05	0.01 =	P(0.4dH Hs <0.5)XP(5.0dH Tz <6.0)
P(0.4 ≤ Hs <0.5, 6.0 ≤ T z <7.0)	= 0.01	0.00 =	P(0.4 ≤ Hs <0.5)XP(6.0 ≤ Tz <7.0)
P(0.5dH Hs <0.6, 3.0dH Tz <4.0)	= 0.02	0.11 =	P(0.5dH Hs <0.6)XP(3.0dH Tz <4.0)
P(0.5 ≤ Hs <0.6,4.0 ≤ Tz <5.0)	= 0.11	0.08 =	P(0.5 ≤ Hs <0.6)XP(4.0 ≤ Tz <5.0)
P(0.5dH Hs <0.6,5.0dH Tz <6.0)	= 0.07	0.03 =	P(0.5dH Hs <0.6)XP(5.0dH Tz <6.0)
P(0.5 ≤ Hs <0.6,6.0 ≤ Tz <7.0)	= 0.02	0.01 =	P(0.5 ≤ Hs <0.6)XP(6.0 ≤ Tz <7.0)
P(0.5 ≤ Hs <0.6,7.0 ≤ T z <8.0)	= 0.02	0.00 =	P(0.5 ≤ Hs <0.6)XP(7.0 ≤ Tz <8.0)
P(0.6dH Hs <0.7, 3.0dH Tz <4.0)	= 0.16	0.15 =	P(0.6dH Hs <0.7)XP(3.0dH Tz <4.0)
P(0.6dH Hs <0.7,4.0dH Tz <5.0)	= 0.13	0.10 =	P(0.6dH Hs <0.7)XP(4.0dH Tz <5.0)
P(0.6 ≤ Hs <0.7,5.0 ≤ Tz <6.0)	= 0.02	0.04 =	P(0.6 ≤ Hs <0.7)XP(5.0 ≤ Tz <6.0)
P(0.6 ≤ Hs <0.7,6.0 ≤ Tz <7.0)	= 0.00	0.01 =	P(0.6 ≤ Hs <0.7)XP(6.0 ≤ Tz <7.0)
P(0.7dH Hs <0.8,3.0dH T z <4.0)	= 0.15	0.09 =	P(0.7dH Hs <0.8)XP(3.0dH Tz <4.0)
P(0.7 ≤ Hs <0.8,4.0 ≤ Tz <5.0)	= 0.05	0.07 =	P(0.7 ≤ Hs <0.8)XP(4.0 ≤ Tz <5.0)
P(0.7 ≤ Hs <0.8,5.0 ≤ Tz <6.0)	= 0.00	0.03 =	P(0.7 ≤ Hs <0.8)XP(5.0 ≤ Tz <6.0)
P(0.8dH Hs <0.9,3.0dH Tz <4.0)	= 0.04	0.02 =	P(0.8dH Hs <0.9)XP(3.0dH Tz <4.0)
P(0.8 ≤ Hs <0.9,4.0 ≤ Tz <5.0)	= 0.00	0.01 =	P(0.8 ≤ Hs <0.9)XP(4.0 ≤ Tz <5.0)
P(0.9dH Hs <1.0,3.0dH Tz <4.0)	= 0.07	0.04 =	P(0.9dH Hs <1.0)XP(3.0dH Tz <4.0)
P(0.9 ≤ Hs <1.0,4.0 ≤ Tz <5.0)	= 0.01	0.03 =	P(0.9 ≤ Hs <1.0)XP(4.0 ≤ Tz <5.0)
P(1.0 ≤ Hs <1.1, 3.0 ≤ Tz <4.0)	= 0.03	0.02 =	P(1.0 ≤ Hs <1.1)XP(3.0 ≤ Tz <4.0)
P(1.0dH Hs <1.1, 4.0dH Tz <5.0)	= 0.01	0.01 =	P(1.0dH Hs <1.1)XP(4.0dHTz <5.0)

Table 7: Z-test statistic correspond to postulated distributions *Weibull* and *GP3* of the wave statistics

Wave statistics	*Weibull* dist.	*GP3* dist.
	Z-Test Statistic	
H_s	0.2580	0.4246
T_e	-0.3051	0.1293
T_p	-0.4013	-0.3300
T_z	0.0452	0.0577

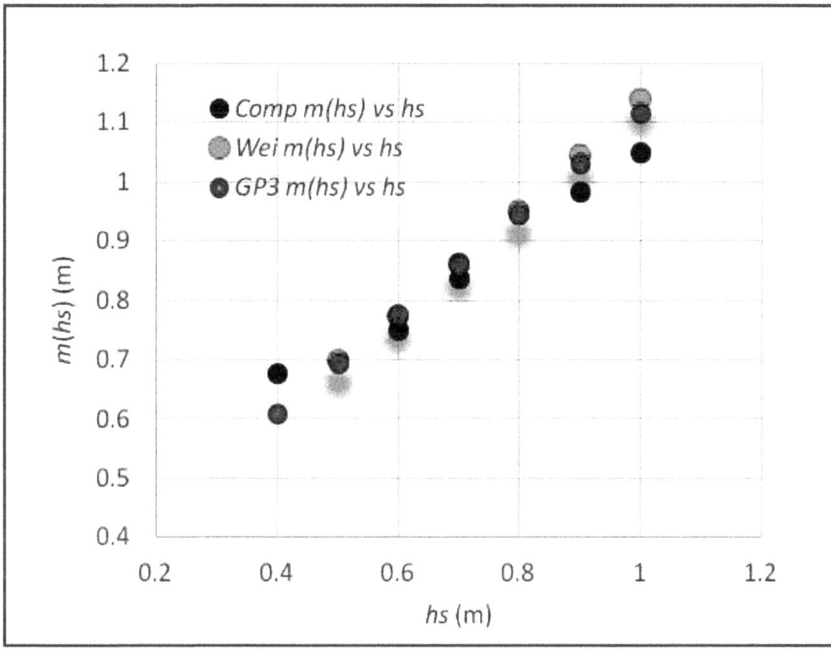

Fig. 4. Computed average conditional exceedances (black dots) versus *CMF*s of *Weibull* (red dots) and *GP*3 (green dots) of H_s

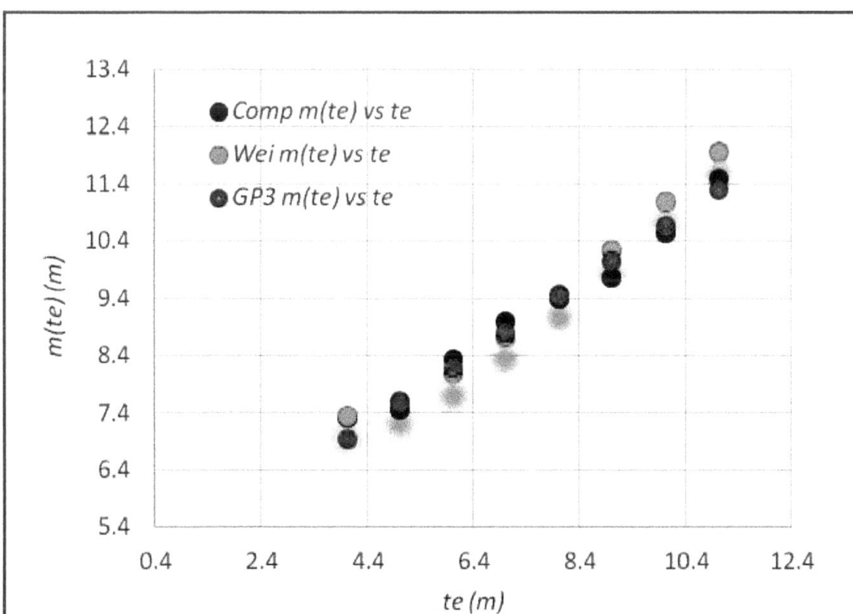

Fig. 5. Computed average conditional exceedances (black dots) versus *CMF*s of *Weibull* (red dots) and *GP*3 (green dots) of T_e

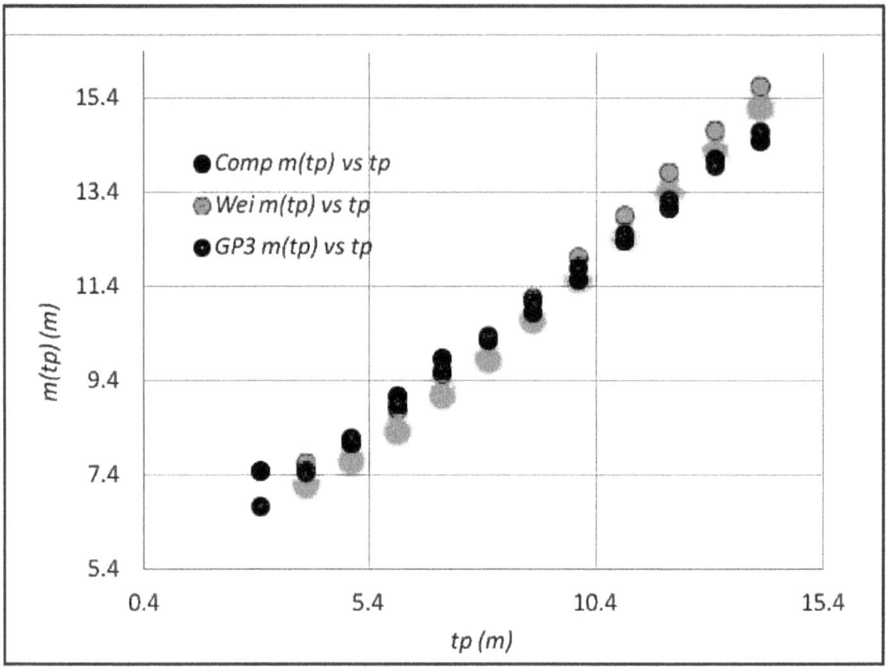

Fig. 6. Computed average conditional exceedances (black dots) versus *CMF*s of *Weibull* (red dots) and *GP*3 (green dots) of T_p

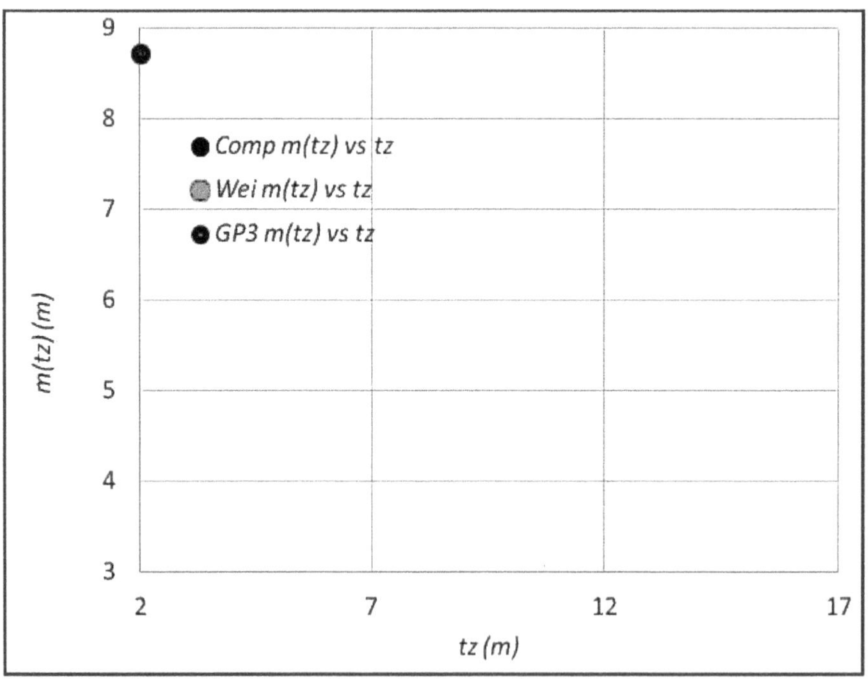

Fig. 7. Computed average conditional exceedances (black dots) versus *CMF*s of *Weibull* (red dots) and *GP*3 (green dots) of T_z

Table 8: Computed and estimated average of one third the highest H_s, T_e, T_p and T_z

Wave statistics	*Computed*	*Weibull* estimates	*GP*3 estimates
$H_{s(1/3)}$ (m)	0.85	0.88	0.88
$T_{e(1/3)}$ (s)	9.5	9.5	9.5
$T_{p(1/3)}$ (s)	10.6	10.6	10.6
$T_{z(1/3)}$ (s)	5.3	5.3	5.3

The record length of the data sample (two months) is not sufficiently large to estimate the extreme wave heights (H_{ex}) with required precision. Here an attempt is made to examine the applicability of the methodology adopted in this study to estimate extreme wave heights of designated return periods 10, 20, 50 and 100 years and the error bounds of the peak periods associated with it (Table 9).

The extreme energy periods, peak periods and zero up-crossing periods that can occur in 10, 20, 50 and 100 year span of time are also estimated (Table 10). It is to be noted that extreme wave heights are not associated with extreme peak periods. T_p associated with $H_{1/3}$ are quite short periods. The T_e and T_p estimates by *Weibull* distribution seem to be overestimated when compared with the recorded maximum of the above periods from the data, whereas *GP*3 estimates more cautiously and deliver reliable values for the above periods. The estimates of zero up-crossing periods by the models are inconsistent.

Conclusion

Significant wave heights (H_s) and the associated mean energy periods (T_e), peak periods (T_p) and zero up-crossing periods (T_z) off Kozhicode, southwest coast of India are independent during pre-monsoon (April-May) season. Conditional mean functions (*CMF*) and *Z*- goodness-of-fit test statistic are efficient tools to identify the appropriate probability model to yield sufficiently accurate estimates of required wave statistics including the extreme quantiles. Both *Weibull* and generalised *Pareto* distribution (*GP*3) estimates are consistent with computed values from the data for $H_{1/3}$, $T_{e\,(1/3)}$, $T_{p\,(1/3)}$, $T_{z\,(1/3)}$ and for extreme quartiles: extreme wave heights (H_{ex}), extreme T_e and T_p. Extreme T_z are cautiously and more accurately estimated by *GP*3. The error bounds of the peak periods associated with the extreme wave heights are small periods indicating that wave regimes are dominated by wind waves.

Acknowledgement

The authors are grateful to Indian National Centre for Ocean and Information Services (INCOIS) for providing the wave data used in this study.

Table 9. Upper and lower bounds of peak periods associated with H_{ex} estimates of designated return periods by *Weibull* and *GP3* distributions in April-May (pre-monsoon) 2013, off Kozhicode, southwest coast of India

	Return period (*T*)							
	10 years		20 years		50 years		100 years	
Weibull H_{ex} (m)	1.64		1.72		1.83		1.91	
	Lower bound of T_p s	Upper bound of T_p s	Lower bound of T_p s	Upper bound of T_p s	Lower bound of T_p s	Upper bound of T_p s	Lower bound of T_p s	Upper bound of T_p s
	2.4	2.7	2.5	2.8	2.6	2.9	2.6	3.0
GP3 H_{ex} (m)	1.44	1.48	1.52	1.55				
	Lower bound of T_p s	Upper bound of T_p s	Lower bound of T_p s	Upper bound of T_p s	Lower bound of T_p s	Upper bound of T_p s	Lower bound of T_p s	Upper bound of T_p s
	2.3	2.6	2.3	2.6	2.4	2.6	2.4	2.7

Table 10: 10, 20, 50 and 100 year wave period estimates by *Weibull* and *GP3* distributions in April-May (pre-monsoon) 2013 off Kozhicode, Southwest coast of India

	Return period (*T*)			
	10 years	20 years	50 years	100 years
Weibull				
T_e s	14.8	15.3	15.9	16.4
T_p s	20.0	21.0	22.2	23.1
T_z s	10.0	10.6	11.4	12.0
GP3				
T_e s	11.7	11.8	11.8	11.8
T_p s	15.9	16.0	16.2	16.2
T_z s	9.8	10.3	11.0	11.5

References

Bitner-Gregersen, E., Guedes Soares, C., Machado, U., Cavaco, P., 1998. Comparison of different approaches to joint environmental modelling. Proceedings of the 17th International Conference on Ocean, Offshore Mechanics and Arctic Engineering (OMEA′1998). ASME paper, OMAE98-1495.

Ferreira, J. A., Guedes Soares, C., 2002. Modelling bivariate distributions of significant wave height and mean wave period. Appl. Ocean Res. 24: 31-45.

Haver, S., 1985. Wave climate off northern Norway. Appl. Ocean Res. 7: 85-92.

Hosking, J.R.M., Wallis, J.R., 1997. Regional Frequency Analysis: An Approach Based on L-Moments. First Edn. Cambridge University Press, Cambridge.

Mollison, D., 1994. Assessing the wave energy resource. In: Statistics for the Environment 2: Water Related Issues (Barnett, V. and Feridun Turkman, K., Eds.), pp. 205-222, John Wiley and Sons Ltd.

Mollison, D., Buneman, O.P., Salter, S.H., 1976. Wave power availability in the NE Atlantic. Nature 263: 223-226.

Muraleedharan, G., Lucas, C., Martins, D., Guedes Soares, C., 2016a. On the peak period distributions conditioned on significant wave heights in Aguçadoura, Portugal, in: Progress in Renewable Energies Offshore (Guedes Soares, C., Eds.), pp.3-9, Taylor and Francis Group, London.

Muraleedharan, G., Lucas, C., Martins, D., Guedes Soares, C., 2016b. Modelling average conditional exceedances of significant wave heights and associated peak periods. In: Maritime Technology and Engineering (Guedes Soares, C. and Santos, Eds.), pp. 1065-1072, Taylor and Francis Group, London,.

Muraleedharan, G., Lucas, C., Martins, D., Guedes Soares, C., Kurup, P.G., 2015. On the distribution of significant wave heights and associated peak periods. Coastal Engng. 103: 42-51.

Muraleedharan, G., Mourani S., Rao, A.D., Unnikrishnan Nair, N., Kurup, P.G., 2009. Estimation of wave period statistics using numerical coastal wave model. Nat. Haz. 49, 165-186.

Muraleedharan, G., Rao, A.D., Kurup, P.G., Unnikrishnan Nair, N., Mourani, S., 2007. Modified Weibull distribution for maximum and significant wave height simulation and prediction. Coast. Engng. 54 (8): 630–638.

Ochi, M., 1998. Ocean waves-ocean technology series, Cambridge University Press, Cambridge.

Silva, D., Rusu, E., Guedes Soares, C., 2013. Evaluation of various technologies for wave energy conversion in the Portuguese nearshore. Energies. 6, 1344-1364.

Toffoli, A., Babanin, A., Onorato, M., Waseda, T., 2010. Maximum steepness of oceanic waves: Field and laboratory experiments. Geophys. Res. Lett. 37, 1-4.

Impact of Climate Change on Hydrological Cycle, Ecosystem, Fisheries and Food Security, Pages: 57–69
Edited by: B. Madhusoodana Kurup, M.R. Boopendranath, M. Harikrishnan and A.V. Shibu

CHAPTER - 5

Climattance and Carbon Tax Mechanism: A Case Study of Indian Coastal States

J.S. Pandey* and Rakesh Kumar

*CSIR-National Environmental Engineering Research Institute (NEERI)\
Nagpur-440020, India
E-mail: js_pandey@neeri.res.in

Climate Resilient Economics should ideally deal with the following key questions:

- How can different countries and states strengthen and protect their economic performance while dealing with climate impact mitigation and adaptation?
- How can scientific and technological innovations in terms of climate change research be translated into direct economic benefits and social relevance?
- What are the direct economic consequences of extreme weather events and rising sea-levels?
- What kinds of socio-political and administrative reforms are needed so as to take care of air, water and soil quality protection, food and energy security amongst all the cross-sections belonging to different segments of socio-economic gradients ?
- How do we build smart cities and smart villages which are not only digitally smart, but also smart from sustainability point of view?
- How do we develop region-specific environmental management plans which appropriately and adequately take care of the pollution and carbon emission potential of different regions by balancing them against their carbon sequestration and pollution-mitigation potential?

When it comes to proper scientific understanding of climate change (Pandey, 2015; 2018), it is pertinent to mention that the main regulators of temperature and humidity are carbon and hydrological cycles, which are the major components of bio-geo-chemical cycles. These cycles, in turn, are intertwined with many other macro and micro cycles such as nitrogen, phosphorus, sulphur, calcium and magnesium cycles. All these cycles are, in turn, controlled by ecological food-webs, food-chains and predator-prey relationships at macro and micro-biological levels. Once these relationships are disturbed, perturbed and hampered, they in turn hamper the rate and rhythm and synchronization of bio-geo-chemical cycles. Predator-prey relationships are maintained and managed by land use patterns. For example, how much land should be under industry (I), commerce (C), residence (R), forestry (F), agriculture (AG), aquaculture (AQ) etc. Like C:N:P ratio maintains and manages biological cells, environmental management is highly dependent on I:C:R ratio.

Under the circumstances, it is important to note that secret of environmental management lies in land-use-management, which not only includes terrestrial, but also aquatic ecosystems (Wafar *et al.*, 2011; Saroj *et al.*, 2016; Bassi *et al.*, 2014; Panigrahy *et al.*, 2012). These ecosystems like coastal, marine, coral reefs, wetlands, woodlands, grasslands and forests have the largest roles to play in terms of climate regulation (de Groot *et al.*, 2012; Costanza *et al.*, 2014) . So ultimately what is important from the environmental management of climate change point of view is to understand and quantify the spatial and temporal distribution of these ecosystems in a given bio-geographical zone (Fig. 1) and the limits to which one can change or perturb this distribution so that these ecosystems do not cross their resilience and resistance limits (Fig. 2).

In this context, it is very important to stress the need for generating region-specific environmental health cards (EHC) for all the industrial cities and surrounding and affected districts of our country so that based on these EHCs, we can appropriately translate state level climate action plans into city and district level plans. These environmental health cards would concentrate and focus mainly on the pollution-generation (emission-characteristics of the region) and pollution-mitigation (assimilation and sequestration potential of the region) by various ecosystems situated in the region. Like remittance, we have defined Climattance as an ecological economic measure reflecting the amount of money which accrues out of climate regulation (as ecosystem services) by relevant terrestrial and aquatic ecosystems. For this, we have followed ADAM and EVE approach (Pandey *et al.*, 2018) and collected the related and required information from various sources.

In short, we have looked into the following issues: (a) What is Climattance and how it can be quantified such that it helps in formulating district and state level climate action plans ? and (b) How ADAM and EVE Methodology would help in ecological economic quantification required for the above purpose ? Notwithstanding various uncertainties involved, these kinds of exercises are essential for preparing region-specific climate action plans (RSCA). With the additional information available, the methodology would be made more explicit and region-specific. While the present paper offers an excellent methodology for applying and regulating carbon-taxation mechanism, at the same time it opens an interesting avenue for further research in this area.

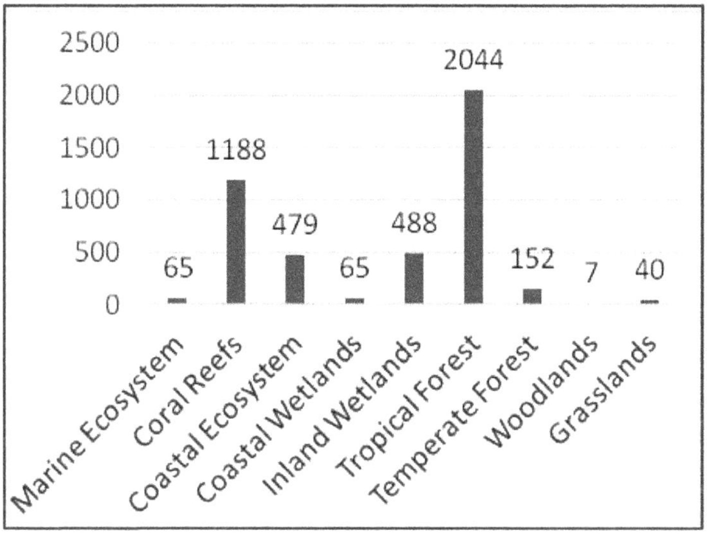

Fig. 1. Coastal regulation: monitory value (USD/ha/yr)

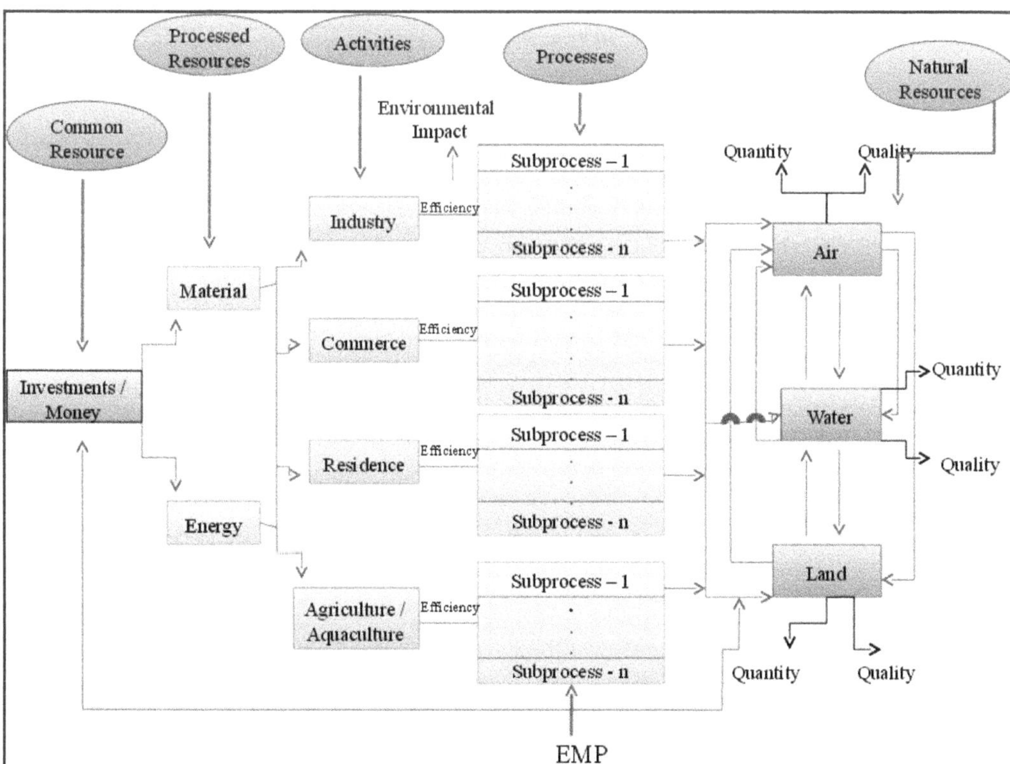

Fig. 2. Limits of change.

Evaluation and Decentralization of Climate Action Plans

There are many qualitative and quantitative ways for evaluating the performance of different climate action plans (https : // www.theclimategroup.org / sites / default / files / india_report_web_singles.pdf). The six main quantitative criteria, which are normally used are : (a) Income per capita , (b) Emissions per capita, (c) Renewable energy potential, (d) Energy efficiency of industrial output, (e) Energy efficiency in Buildings, Transport, Municipal Services, Agriculture and Power Distribution and (f) growth in forestry sector. The two qualitative criteria generally used are: (a) Climate governance and (b) Engagement with external partner.

As far as performance of our National Climate Action is concerned (https://www.prsindia.org/content/performance-national-action-plan-climate-change), different missions of our NAPCC are linked to respective Ministries responsible for their implementation, budgetary provisions, and for deciding the associated priorities and the Ministry of Environment, Forest and Climate Change (MoEF&CC) is the overall co-ordinating Ministry. However, as far as implementation of all these missions are concerned, there are various issues connected with the desired financial resources, outputs, expectations and actual budgetary provisions and allocations (GOI, 2018; 2019). For example, the National Solar Mission is expected to generate 1,00,000 MW by 2021-22, at an expected cost of Rs 6,00,000 crore. The funding has to come from internal as well as external sources.

National Mission on Sustainable Habitat is looking into the issues of energy efficiency in urban planning, building construction, solid and liquid waste management, the Atal Mission for Rejuvenation and Urban Transformation, and the Swachh Bharat Mission are expected to reduce Green House Gas emission significantly. In these missions and action plans, what is needed most is the integration of rural and urban habitats for their appropriate environmental management such a way that they automatically get integrated with National Mission for Sustainable Agriculture for taking care of farmer's welfare.

City-level Climate Action Plans

It is expected that by 2030, India will have 440 cities, which will contribute to over 70 per cent of its GDP (https://www.downtoearth.org.in/blog/energy/there-s-need-for-city-level-climate-actions-63107). While at the same time, it is important to note that more than 80 per cent of emissions come from energy consumption in industrial, commercial and residential sectors belonging to various cities. They are mainly from fuel combustion and emissions from solid waste disposal and waste water management. The effective emission reductions strategies should, therefore, include energy efficiency enhancement in buildings, effective and efficient management of solid waste and implementing continuous and enhanced use of public transport.

Thus, climate adaptation and mitigation plans and strategies are needed mostly at city level so as to achieve the international commitments in terms of "Nationally Determined Contributions" (NDCs) and "Sustainable Development Goals" (SDGs). While on one hand these plans should be easily implementable and should have focussed goals = short and long term; on the other, they should also have necessary institutional, policy and budgetary support. And for this to happen, Urban Local Bodies (ULBs) would need to be appropriately and adequately empowered with active people participation.

Need for Carbon Tax

GHG or Carbon tax (https: // science.howstuffworks.com / environmental / green-science/ carbon-tax.htm; https : // www.imf.org / external / pubs / ft / fandd / 2019 / 12 / pdf / fd1219.pdf) is a pollution tax. It taxes the use of fossil fuels, their production, distribution, etc., on the basis of their GHG-emission potential. A specified price per ton of GHG-emission is set and subsequently translated into a tax on electricity, natural gas or oil. Two direct benefits, which it has, are (a) it helps in reducing the unnecessary and avoidable consumption of fuels and (b) it makes everyone aware of the benefits of energy efficiency at every step. According to the World Bank (https://www.thebalance.com/carbon-tax-definition-how-it-works-4158043), 40 countries and 20 municipalities use the concept of carbon taxes. It takes care of 13% of global greenhouse gas emissions. In addition, there are a total of 88 countries who intend to apply carbon tax mechanism so as to achieve their Paris Agreement goals. This will take care of additional 56% of global GHG-emissions. In 2013, Britain imposed a carbon tax of $25 per metric ton of CO_2 emitted. Canada has a national carbon tax mechanism of $15-30 a ton of CO_2, which is likely to increase to $38 a ton by 2022. The revenue thus collected is proposed to be refunded to individuals through their tax bills. On a similar line if India intends to impose carbon tax and distribute the benefits to different states so that they can refund to their individual citizens, an appropriate mechanism for doing this would have to be evolved. The present paper is an effort in this direction, which demonstrates that the states can claim their share of refund or reduction in carbon-taxes based on the economic potential of the ecosystem services they provide. For this to happen, it would be worthwhile to look at the climate-economics and economic evaluation of ecosystem services.

Climate Economics and Economic Evaluation of Ecosystem Services

Different states are situated in different ecological zones and thus their ecosystem structure and functioning have different diurnal, seasonal and annual patterns (https://www.e-education.psu.edu/earth103/node/717). In general, the economic costs are divided into two categories: market and non-market values. Market values are straightaway expressed in terms of dollars, rupees etc. On the other hand, a non-market cost is normally not easily quantifiable – it is non-tangible. Damage caused to a coral-reef ecosystem is an example of a non-market cost. The sectors for which market values are easier to quantify are: (a) Agricultural impact, (b) Impact on forests, (c) Impact on fisheries, (d) Impact on Insurance sector, (e) Impact on public infrastructure, (f) Impact on energy consumption and (g) Impact on tourism and recreation. The impact on human health and ecosystem are non-market values and more difficult to quantify. Notwithstanding several uncertainties, there have been some reasonably reliable estimates in terms of ecological economic (de Groot *et al.*, 2012; Costanza *et al.*, 2014) of the ecosystem-services.

Methodology

The main question before us is if we opt for carbon-taxation, should it be done uniformly across all the Indian states or should be differentiated on the basis of the ecosystem-services, which various states provide and to differing extents. So the most appropriate way of adopting carbon-taxation system should be based on the pollution or carbon emission characteristics of different states and their pollution-mitigation and carbon sequestration potential. The monetary values (de Groot *et al.*, 2012; Costanza *et al.*, 2014) for Climate Regulation by different types of ecosystems have been depicted in Fig. 1. The best way of judging and evaluating climate-change-performance in India would be by way of creating Environmental Health Cards (EHCs) for different cities, districts and states based on their eco-geographical features; carbon and pollution emission potentials; pollution-mitigation and carbon-sequestration potential; ecosystem resistance and resilience (Fig. 2 & 3).

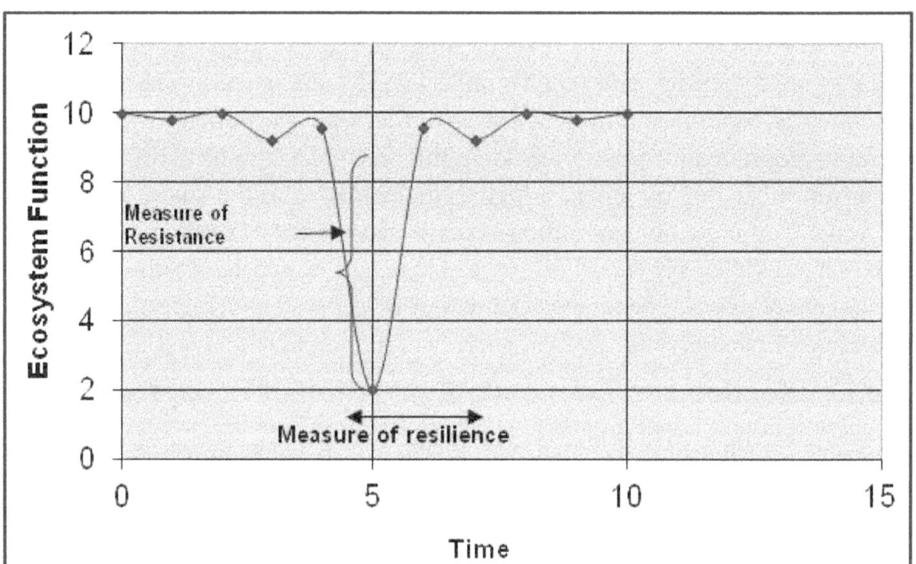

Fig. 3. Ecosystem function and resilience

Earlier for this purpose, we have applied ADAM and EVE analysis for some of the cities (Agra City), some river basins (Damodar River Basin) and some of the states like Uttarakhand. For Agra city, the issue was quantification of assimilative capacity (Fig. 4); for Damodar River Basin (Fig. 5), the issue was quantification of emission and sink (sequestration) potential (carbon cycle) of the region and for Uttarakhand region (Fig. 6), the issue was quantification of Environmental Water Demand (hydrological cycle). Essentially ADAM and EVE includes the following steps (Pandey, 2018):

- Literature review
- Appropriate selection of ecosystems, their detailed analysis and subsequent data collection
- Identifying important environmental issues and associated problems of the region
- selection of identified hot spots and their sensitivity analysis
- Identification of systems elements
- Structural interconnections (State variables, intermediate variables, parameters, exogenous forcing functions, etc.)
- Cause-effect relationships; functional relations and quantification
- Developing appropriate functions and models for their region-specific applications and site-specific analysis for quantifying ecosystem impacts and human health risks due to increased environmental pollution and climate change
- Computer programming and application
- Testing and validation

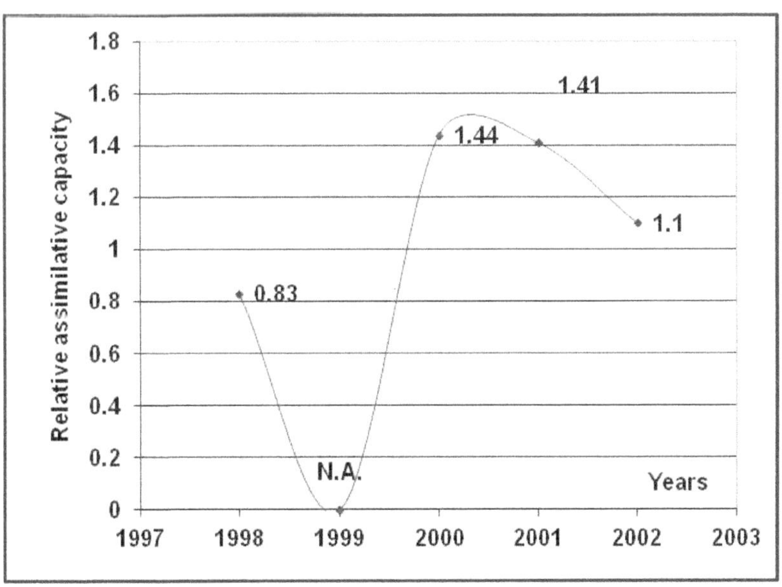

Fig. 4. Relative assimilative capacities

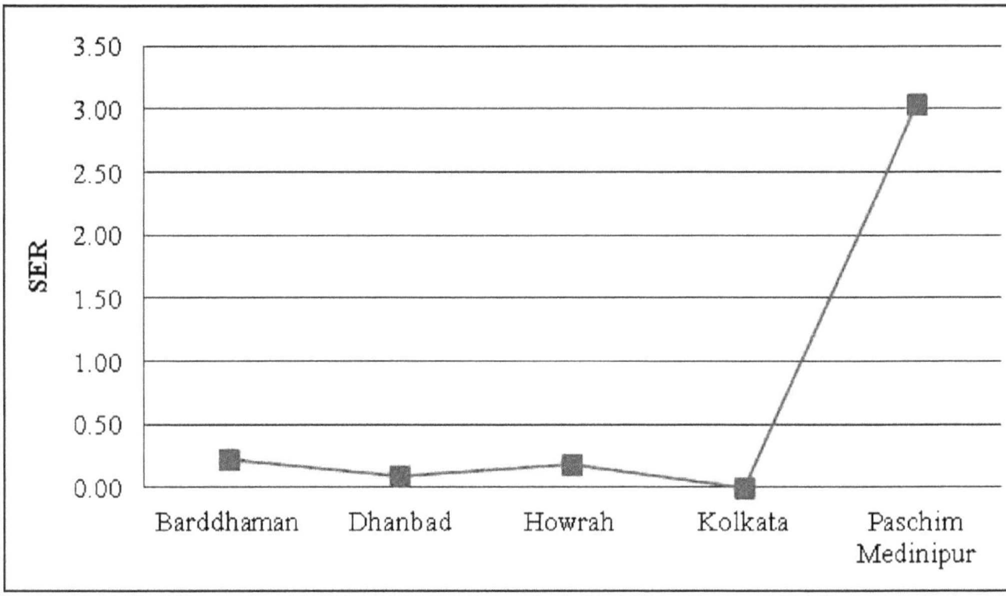

Fig. 5. Sequestration Emission Ratio (SER)

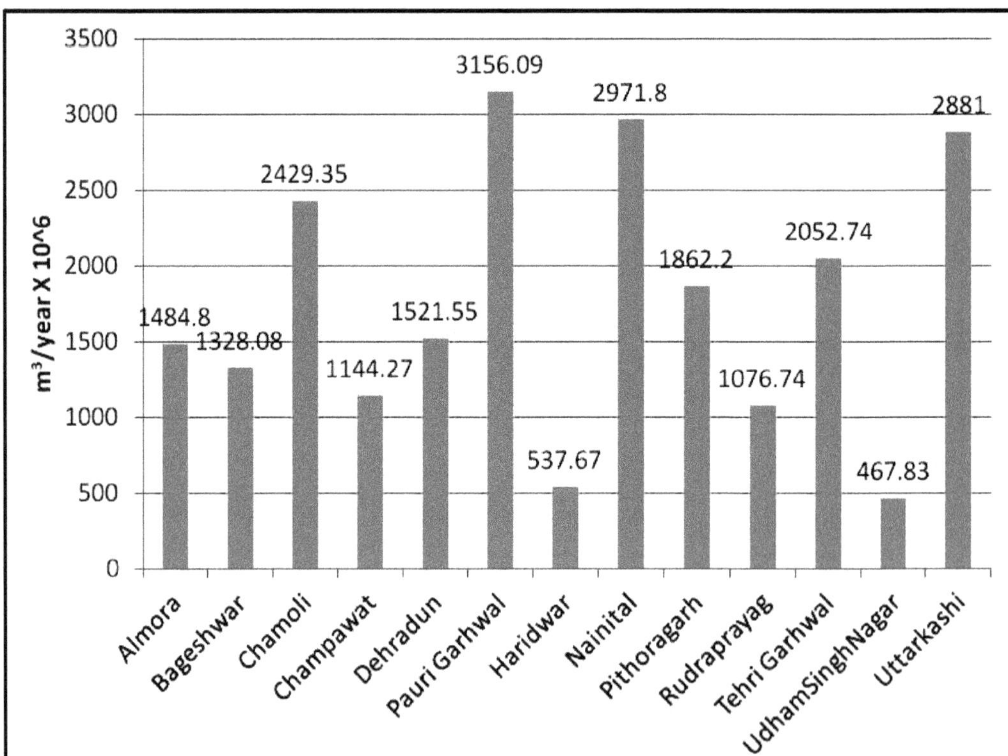

Fig. 6. Environmental water demand Uttrakhand

The monetary value of the climate regulation has mainly been quantified (Fig. 7) on the basis of coastal (Fig. 8) and mangrove ecosystems (Fig. 9) based on the values of ecosystem services provided by different types of ecosystems (https://biodiversity.europa.eu/topics/ecosystem-services; http: // www.teebweb.org / resources / ecosystem-services/; https://wle.cgiar.org/content/what-are-ecosystem-services; https://www.iucn.org/commissions/commission-ecosystem-management/our-work/cems-thematic-groups/ecosystem-services; https://community-wealth.org/sites/clone.community-wealth.org/files/downloads/article-costanza-et-al.pdf).

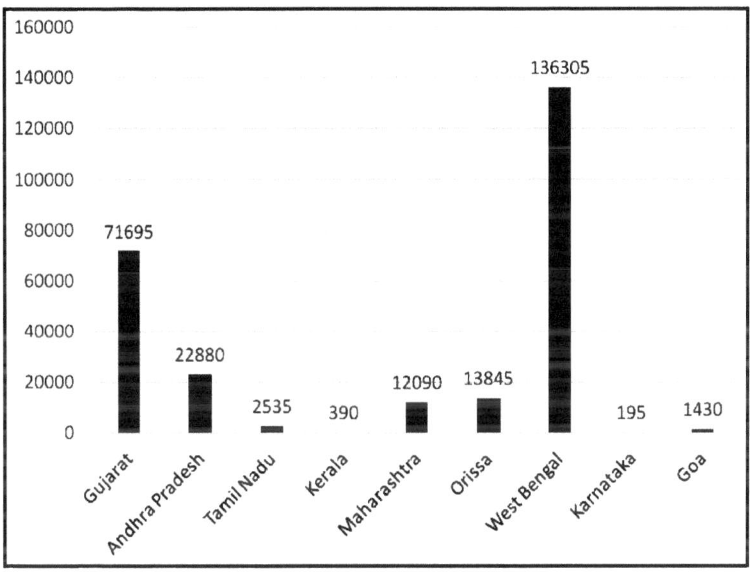

Fig. 7. Climate regulations ($/yr)

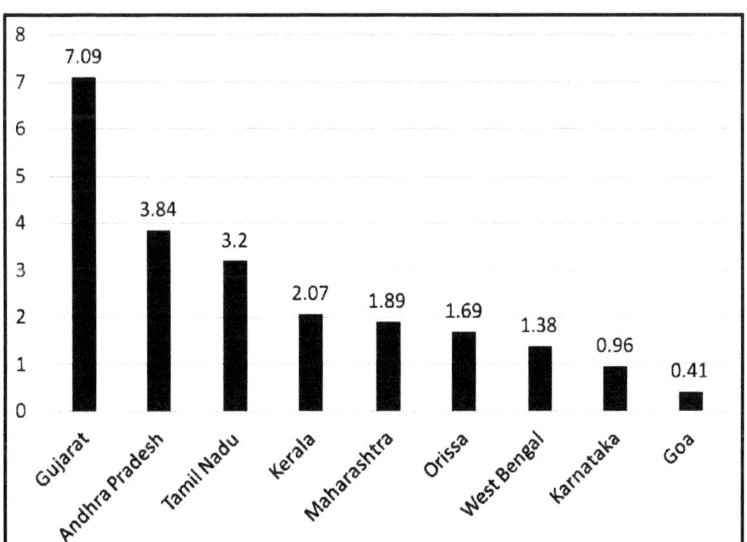

Fig. 8. Coastal area (ha) * 10 $^{(-7)}$

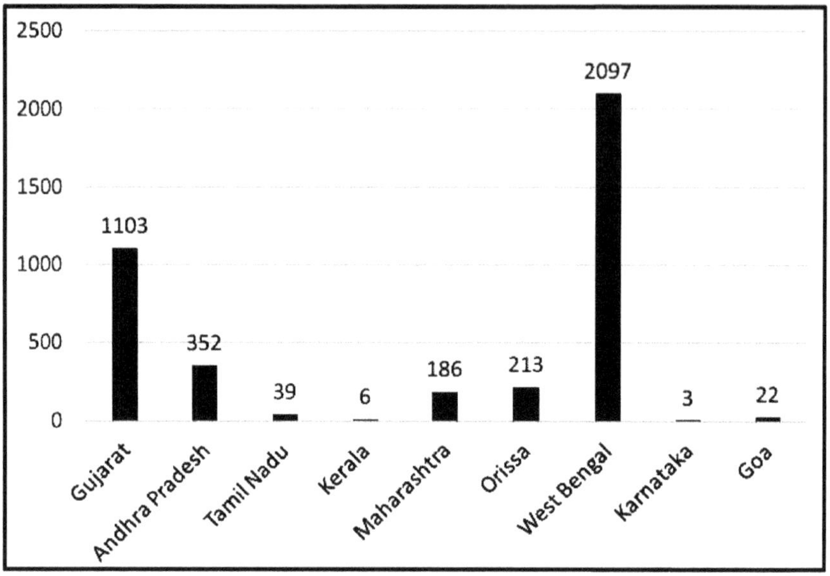

Fig. 9. Mangrove area (sq. km.)

The information on Indian coastline (its length and area) has been collected from : http:// iomenvis.nic.in/index2.aspx?slid=758&sublinkid=119&langid=1&mid=1; and https:// www.gktoday.in/gk/largest-states-by-coastline-in-india/. The information on mangroves has been collected from: https://www.mapsofindia.com/top-ten/geography/mangrove-forest.html. The information thus collected has been used for quantifying and analysing the Climattance of various Indian states. This information would be mainly useful from the point of view of taking recourse to state-specific climate action plans.

Conclusion

It is high time that Academic Institutions now engage in research-avenues, which help in finding out realistic and socio-economically relevant solutions for climate change related problems. The present paper is an attempt in this direction. These activities are part of the discipline called systems ecology. However, these exercises would need to be regularly pursued, refined, modified and applied for evolving city-specific, district-wise and state-level environmental management plans for finding appropriate solutions for region-specific climate change problems. Systems Ecology works in line with the ecological engineering principles of species-specific, community-specific, ecosystem-specific processes and models (Pandey and Khanna, 1995). Whenever needed, issues related to Ecological Footprints (Pandey *et al.*, 2001a;b), TRGs (Temporal Risk Gradients) (Pandey *et al.*, 2001c) and Ecological Economics (Pandey *et al.*, 2004) can also be gradually plugged in the present analysis. Furthermore, for generating Environmental Health Cards (EHCs) regional emission factors (Pandey *et al.*, 2007) and carbon sequestration potential can also be suitably quantified on the basis of appropriate models, which will also help in carrying out human health risk assessment (Pandey *et al.*, 1992-94; 2005) and ecosystem-impact analysis (Pandey *et al.*, 1992a) for different cities, districts and states. On the similar lines, the quantification for other terrestrial (forest land, woodland, grassland etc.) and aquatic (coastal) ecosystems like coral reefs and wetlands can be done on the basis of information available in various references.

Acknowledgement

The present article, *inter alia*, is a synthesis of earlier works done by the authors in this area. They are grateful to all those sources of information including discussions, discourses, workshops and conferences, which have helped in shaping this article by way of a systematic data analysis and appropriate synthesis and conversion of the information into useful models.

References

Bassi, N., Kumar, M.D., Sharma, A., Pardha-Saradhi, P., 2014. Status of Wetlands in India : A Review of Extent, Ecosystem Benefits, Threats and Management Strategies. Journal of Hydrology : Regional Studies 2: 1-19.

Byravan, S. and Rajan, S.C., 2012. An Evaluation of India's National Action Plan on Climate Change. Centre for Development Finance (CDF), https://www.researchgate.net/publication/ 256043498_An_Evaluation_of_India's_National_Action_Plan_on_Climate_Change.

Costanza, R., de Groot, R., Sutton, P., van der Ploeg, S., Anderson, S.J., Kubiszewski, I., Farber, S. and Turner, R.K., 2014. Changes in the global value of ecosystem services. Global Environmental Change 26: 152-158.

de Groot, R., Brander, L., van der Ploeg, S., Costanza, R., Bernard, F., Braat, L. Christie, M., Crossman, N., Gharmandi, A., Hein, L., Hussain, S., Kumar, P., McVittie, A., Portela, R., Rodriguez, L.C., Brink, P.T. and van Beukering, P., 2012. Global estimates of the value of ecosystems and their services in monetary units. Ecosystem Services 1: 50-61.

GOI, 2018. Coping with Climate Change : An Analysis of India's National Action Plan on Climate Change. 2018. Centre for Science and Environment (http://www.india environmentportal.org.in/ files/file/coping-climate-change-NAPCC.pdf)

GOI, 2019. Driving Climate Action : State Leadership in India. Report : May 2019 (theclimategroup.org/sites/default/files/india_report_web_singles.pdf)

Pandey J.S., 2015. Climate Change: Synergistic Impacts on Human and Ecosystem Health. YOJANA, December 2015: 49-52.

Pandey, J.S. and Joseph, V., 2001a. a scavenging-dependent air-basin ecological risk assessment (SABERA) - Model Applied to Acid Rain Impact around Delhi City, India. Journal of Environmental Systems 28(3) : 193-202.

Pandey, J.S. and Khanna, P., 1992a. Speed-Dependent Modeling of Ecosystem exposures from vehicles in the near-road environment. Journal of Environmental Systems 21(2): 185-192.

Pandey, J.S. and Khanna, P., 1995. Development of plant function types for studying impact of green house gases on terrestrial ecosystems. Journal of Environmental Systems 23(1): 67-82.

Pandey, J.S., 2018. Looking at Climate Change and its Socio-Economic and Ecological Implications through BGC (Bio-Geo-Chemical Cycle)-Lens: An ADAM (Accretion of Data and Modulation) and EVE (Environmentally Viable Engineering Estimates) Analysis. In: Climate Change and Agriculture in India: Impact and Adaptation, pp. 97-110, Springer, Cham

Pandey, J.S., Deb, S.C. and Khanna, P., 1997. Issues Related to Greenhouse Effect, Productivity Modelling and Nutrient Cycling : A Case Study of Indian Wetlands. Environmental Management 21(2) : 219-224.

Pandey, J.S., Joseph, V. and Kaul, S.N., 2004. A Zone-wise Ecological-Economic Analysis of Indian Wetlands. Environmental Monitoring and Assessment 98 : 261-273.

Pandey, J.S., Khan, S. and Khanna, P., 2001c. Modeling and Quantification of Temporal Risk Gradients (TRG) for Traffic Zones of Delhi City in India. Journal of Environmental Systems 28(1): 55-69.

Pandey, J.S., Khan, S., Joseph, V. and Singh, R.N., 2001b. Development of a dynamic and predictive model for ecological foot printing (EF). Journal of Environmental Systems 28(4): 279-291.

Pandey, J.S., Kumar, R. and Devotta, S., 2005. Health Risks of NO_2, SPM and SO_2 in Delhi (India). Atmospheric Environment 39: 6868-6874.

Pandey, J.S., Mude, S. and Khanna, P., 1994. Comparing Indoor Air Pollution Health Risks in India and United States. Journal of Environmental Systems 23(2): 179-194.

Pandey, J.S., Pimparkar, S. and Khanna, P., 1992b. Micro-Environmental Zones and Occupancy Factors in Jharia Coal-Field : PAH-Health Exposure Assessment. Journal of Environmental Systems 21(4): 349-356.

Pandey, J.S., Pimparkar, S. and Khanna, P., 1993. Health Exposure Assessment and Policy Analysis.International Journal of Environmental Health Research 3 : 161-170.

Pandey, J.S., Wate, S.R. and Devotta, S., 2007. Development of emission factors for GHGs and associated uncertainties. Proc. 2nd International Workshop on Uncertainty in Greenhouse Gas Inventories. International Institute for Applied Systems Analysis, A-2361 Laxenburg, Austria, 27-28 September, 2007.

Panigrahy, S., Murthy, T.V.R., Patel, J.G. and Singh, T.S., 2012. Wetlands of India : Inventory and Assessment at 1:50,000 Scale Using Geo-spatial Techniques. Current Science 102(6): 852-856.

Saroj, J., Gautam, R.K., Joshi, A. and Tehseen, P. 2016. Review of Coral Reefs of India : Distribution, Status, Research and Management. International Journal of Science, Environment and Technology 5(5): 3088-3098.

Wafar, M., Venkataraman, K., Ingole, B., Khan, S.A. and Lokabharati, P., 2011. State of Knowledge of Coastal and Marine Biodiversity of Indian Ocean Countries. PLoS ONE6 (1): 1-12.

II

Natural Hazards and Disaster Risk Management

Impact of Climate Change on Hydrological Cycle, Ecosystem, Fisheries and Food Security, Pages: 73-93
Edited by: B. Madhusoodana Kurup, M.R. Boopendranath, M. Harikrishnan and A.V. Shibu

CHAPTER - 6

A Fish Called .... Synergy

Rob Roggema

NoorderRuimte, Research Centre for the Built Environment
Hanze University of Applied Sciences Groningen
The Netherlands
E-mail: r.e.roggema@pl.hanze.nl

Many coastal zones around the Globe suffer from the 'fertility-paradox'. These areas on the one hand side belong to the most fertile areas, both on land and in the sea, of the world, which makes them extremely attractive for people to live there. At the same time, because so many people want to live in these areas, they occupy valuable fertile soil, which forms a risk for the urban and natural ecosystems. At the same time the pressure on these areas increases due to the threats of climate change, such as sea level rise, intensity of storms and hurricanes, or increased discharges of heavy rainfall from the inland.

This paradox is kept alive by prioritizing economic growth ever over the earth-features that brings us life. In the end this may mean it will destroy regenerative power the land and the sea have, it may even mean humans are no longer able to live in these areas. In this article the current policies and their problems will firstly be described. Then, a new way forward of finding the benefits for urban development through using the natural sustainability of ecological system as a primer for the urban system, will be introduced and illustrated with a couple of examples from the Netherlands and Australia. The article will conclude identifying the benefits of this way of mutual beneficial coastal development.

Current Policies and Their Problems

The way the increasing tension between living in a coastal zone vs. the degradation of the fertile land as result of this, is often tackled from a silo-ed perspective. Often urbanization is so intense the area searching for new housing, infrastructure for energy and water and creation of jobs. Other policies that are often developed completely separated from this is the water management policy, and there may be a nature or ecological vision, a plan for the social infrastructure, or a traffic and public transport management plan. In the majority of the cities these policies do not strengthen each other but are in conflict. Even more so, economic goals of growth and welfare are dominating other sectors. Investments in new infrastructure e.g. roads, industrial zones and living areas almost always dominate the interests of types of land use without a voice, such as plants, animals, water and food. In the end this is not a persistent system of urban development. The waste produced and the land occupied by these uses simply destroy so much natural values that eventually these places are not livable anymore due to illnesses or the lack of clean water, enough food and too much air pollution.

If economical interests keep on being seen as more important than the quality of the planet that gave us life in the first place, there is not much hope. When resources are discovered and profit drives the development, when choices for land-use are made in favor of industrialized agri-business, new roads and urbanization instead of giving the benefit to biodiversity, water systems and indigenous relics. These conflicts are in essence spatial. Whenever a resource is discovered the economic benefits of exploiting this source will lead to degradation of the ecological system, the water- and air-quality and human wellbeing. New housing and industrial zones occupy the landscape and diminish its natural values, reducing biodiversity, and the water-quality and -quantity. Along coasts this conflict is stronger than inland, because of the fertility-paradox, and an increasing number of people and industries moving to urbanized areas along the coast, and the fact that most industries and harbors also found their place in these zones and seem to be ever expanding. The, often, high quality of ecology, the water system, soil-quality and the beauty of the coastal landscape makes every development extra impactful, as higher values times more impact quadruple the effect.

The Great Barrier Reef in Australia is a good example of this. Due to rising temperatures, the Pacific heats up and acidifies, which has led to bleaching of large portions of the reef in 2016 and 2017 (Hughes and Kerry, 2017). This resulted in action of the Federal government, investing in research how the reef could be saved (Chen and Gartry, 2018). Parallel to this decision, the government also allowed permits to be given to open a new coalmine inland of the Great Barrier Reef (Werner and Currell, 2018). As result of new excavations, the landscape will be destroyed, water tables and watersheds disconnected, erasing valuable vegetation and eventually discharge polluted water to the exact spot where the Great Barrier Reef tries to survive. This is an exemplar of how mixed policies lead to the devaluation of natural habitats. One way to deal with this is to protest, and this is currently exactly what is happening in Queensland against the opening of the Adani-coalmine. However, the other strategy is to develop an alternative plan which bypasses the discussion whether a new coalmine must be opened (obviously not, but it still happens), or how the existing reef can be protected (which will, with the current developments, be nearly impossible). This plan (Roggema, 2017) anticipates losing the

Great Barrier Reef for a great deal (and wouldn't it be great if it doesn't?) and warming of the Pacific further south. The habitat that is being destroyed in front of the Queensland coast could be reinstalled further south, where the environmental conditions are perfect for a coral reef in the next 10-20 years. This has already been underpinned by the observations tropical fish and new coral in front of the Manly coast (Booth and Sear, 2018). Moreover, with the heating of the Pacific, increased activity of cyclones further south is also expected (Sharmila and Walsh, 2018), and this threatens the livelihoods and real estate values of the Sydney Metropolitan Area coastal zone. The response to this could be the Sydney Barrier Reef (Roggema, 2017). This is a plan (Fig. 1) that bypasses the problems with the Great Barrier Reef by anticipating the future environment off the Sydney coast by creating the conditions a new natural coral reef can emerge. This is achieved by sinking oil and gas rigs that are no longer in use to the bottom of the ocean at a maximum depth of 30 meters. On these, cleaned and environmentally safe, rigs coral parcels will be placed as to stimulate fast growing of the coral (Page *et al.*,, 2018). From this moment on the reef can grow naturally and offer the habitat for the wildlife coming from the Great Barrier Reef moving southward.

This illustration shows the process as it happens in many places around the world. The dangerous mix of economic dominance and accelerating climate change forms a danger to a serious number of people, cities and agglomerations. And while developing countries are developing mainly in an economical sense, the stage of development implies also that choices of the main direction of how an urban area wants to be in the future, still can be made. In this sense less is predetermined in developing countries and the quest is how a resilient and safe coast can be created that also offers jobs, food and a roof to its inhabitants, even when they arrive in huge numbers. The pressure of population growth accelerates the demand for energy, clean water and food, not to mention the need for housing and jobs. This all requires the infrastructure to provide it. However, especially in these fast rowing urban centers, it is a once in a lifetime opportunity to do it well, and take an ecosystem approach in which nature, fish, and water are taken as the basis to embed urban growth

Strategic Question

On the basis of the analysis of current problems in developing resilient urban livelihoods, the main issue to tackle is how to include uses without a direct economic benefit in the planning for a fast-forward urban development as are visible in many developing and also developed countries? How could an integrated future perspective be developed, in which the non-economic functions play a basic role and determine future developments? Such integration could only be achieved through a design-led approach, in combination with putting the landscape and ecology, hence forming literally the basis for other uses and intrinsically safeguard the ecological resilience of the area.

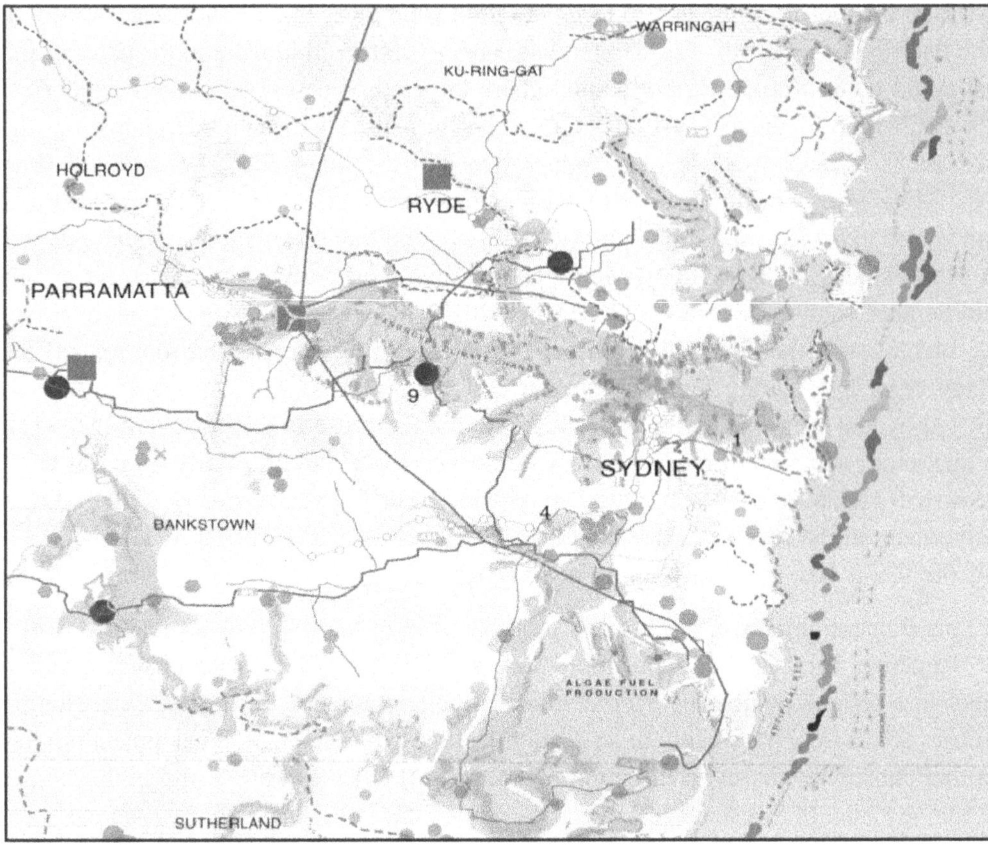

Fig.1 Map of Sydney Barrier Reef

Landscape-ecology Shapes City-building

In order to allow for the landscape to play a crucial role in urban planning and design, a methodology should be applied in which all qualities of ecology, soil, and the water system can be made available for spatial integration. The values of different aspects of the landscape can only be seen when these are all separated in distinct layers. This method is not new and has been developed in the 1960's (McHarg, 1969) when ecological exhaustion and pollution of water systems and the landscape was widespread. A design-led approach in which all ecological aspects are individually valued puts emphasis on their importance. In the Netherlands, leaning on McHarg's theory, the layer-approach was born (Van Schaick and Klaasen 2011). The landscape is subdivided in three layers: the underground/soil (abiotic), the ecology (biotic) and the occupation (urban/infrastructure). Parallel to this thinking, the value of ecology is conceptualized in the so-called Casco-concept (Sijmons, 1992). In this concept a distinction is made between high-dynamic and low-dynamic land-use. When the two types of land-uses are spatially separated ecology and water gain quality. This concept is also transformed for urban environments as the strategy of the two networks (S2N). Low-dynamic land-use, such as water retention, ecology, organic agriculture, are connected to the water network, while high-dynamic uses, such as heavy industry, energy plants, intensive agriculture, were linked to the traffic network. In between, urban living is located (Tjallingii, 1995; 2015). Meanwhile, the layer approach got enriched. First, a fourth layer was added, that of the sky, in which atmospheric functions and for instance solar energy is allocated (Tomásek, 1979; Kristinsson, 2012). To make these approaches useful in a design process, Swarm Planning (Roggema, 2012) puts the layers in a design-oriented sequence. A design process paying tribute to ecological thinking starts with the identification of networks (layer one), where these link one finds focal points (layer two), around which redundancy, or free space (unplanned) provides the flexibility to react to unprecedented developments (layer three), before the natural resources are located (layer four) and finally emergent occupation patterns are distinguished (layer five). Recently, new layers are required to understand our urban systems. Hence the cloud, IP-addresses, interfaces, users and IT-networks are added (van Timmeren and Hernandez, 2015). Still, the basis of the approach remains unchanged: it separates different typeoes of elements, starting with the given functionalities of the landscape, in which precautionary spaces must be created to allow for space-requiring surprises that cannot be predicted, on top of which humankind is building its ever more advanced systems of technological and social infrastructure.

A specific way of allowing ecological principles to form the landscape is Building with Nature (Waterman, 2010). Natural forces, such as wind, waves, currents are used to understand how the landscape would form itself, and how it can become most resilient. This knowledge is then used and applied to intervening in the system to enhance or start these processes. The purpose is to start a process of natural strengthening of the landscape without large human or technical interventions. For example, these principles are applied in the design for the Zandmotor (Sand-engine), and results show that the sea manages to increase the size of the beaches and the dunes due to dropping sand in a crucial place in the sea in front of the Dutch coast (Stive *et al.*,, 2013).

From the Eemsdelta to Western Sydney

The Sydney Barrier Reef example shows that starting a design process from a different perspective new solutions and durable future systems can be envisioned that go beyond the ordinary and well-known tracks. This type of future-thinking is needed to anticipate changes that will occur, which are uncertain. The only thing that can be predicted is that it will not be anywhere near what the past has taught us. This implies that solutions that worked well in the past will no longer suffice. And new innovative futures need to be explored. The following principles will need to be applied in order to be able to develop those solutions:

- **Landscape first:** natural values, water quality and biodiversity in the existing landscape can be taken as the starting point for design. Often, in the old paradigm, landscape was erased and new urban uses were super-positioned on top of the landscape. This degraded the quality of the ecological functions dramatically;

- **Systemic:** By taking into account the systemic features, qualities and vulnerabilities of systems operating in the landscape, such as the ecosystems, the water system, agricultural systems, energy-generating systems and infrastructural systems, and linking these with each other a sustainable basis can be formed in which these systems are enriched instead of threatened;
- **Integrated:** A sectoral approach often leads to very good solutions for one topic, but suboptimal solutions for all other aspects. When problems are tackled in an interdisciplinary way, the overall quality of an area increases, and even the individual, sectoral interests benefit from such an integrated approach;
- **Coloring outside the lines:** maybe the most important one, continuing in an old-fashioned manner will eventually prove counterproductive. In order to face new challenges innovative solutions are needed and these can only be developed with new approaches, a bit of disobedience and structurally coloring outside the lines.

Two examples, the 'Floodable Eemsdelta' in the Netherlands and the 'Western Sydney Parkland' in Australia, have applied these principles in designing for the future.

Floodable Eemsdelta

The current discourse on coastal protection for sea level rise and storm surges focuses on strengthening and heightening of protecting structures, such as levees and dikes. Accelerated sea level rise (Haasnoot *et al.*,, 2018) as predicted by Hansen and others (Hansen, 2007; Hansen *et al.*,, 2007; 2008; Hamilton and Kaiser, 2009; Lenton *et al.*,, 2008; Rahmstorf *et al.*,, 2007; Tin, 2008) raises concerns about the capability of coastal structures to withstand extreme circumstances at all times. Believing in increasingly stronger dikes to defend assets against storms and flooding imply impacts can be disastrous once these structures breach, since the value of these assets is generally high. Therefore, the plan for a Floodable Eemsdelta (Roggema, 2009; 2015) takes a different perspective.

The Eemsdelta is located in the North of the Netherlands. The topography of the area is not very dramatic, however the elevation is slightly increasing by 1.5 meters towards the coastline. The main land use is agriculture and the majority of the (small) towns are historic. The area is surrounded by the Wadden Sea and the river Eems, which makes it vulnerable sea level rise and storm surges. The coastal protection consists of a strong dike, with its weakest point just north of the town of Delfzijl. If the dike breaks the flood reaches the capital of the province, Groningen, within 36 hours. Moreover, in the area the main gas field of the Netherlands is found, which would not be functioning anymore when a flood occurs. Important economic uses are located in or near the Eems-harbor, where the generation of energy, innovative and sustainable industries, often based on the availability of energy, and the main energy grids together form a major asset.

The design for a Floodable Eemsdelta has taken sea level rise and storm surge being the main climate hazards as the point of departure. As the pace of sea level rise is uncertain and a precise moment a dike might break is unpredictable, an alternative approach bypassing these uncertainties, has been chosen.

Firstly, a collective vision has been developed emphasizing even the strongest dike will eventually break under accelerated sea level rise, and alternative coastal protection need to be investigated. The key drivers for this research by design are:

- Water will flow to the lowest points and the elevation of the landscape determines these;
- Sea level rise eventually determines the water level;
- A northwestern storm surge determines the moment sea water will eventually flow behind the coastal protection.

Secondly, a design strategy was developed (Fig. 2). Core idea is to follow expected changes rather than fighting and withstanding it. The acceptance of water being part of

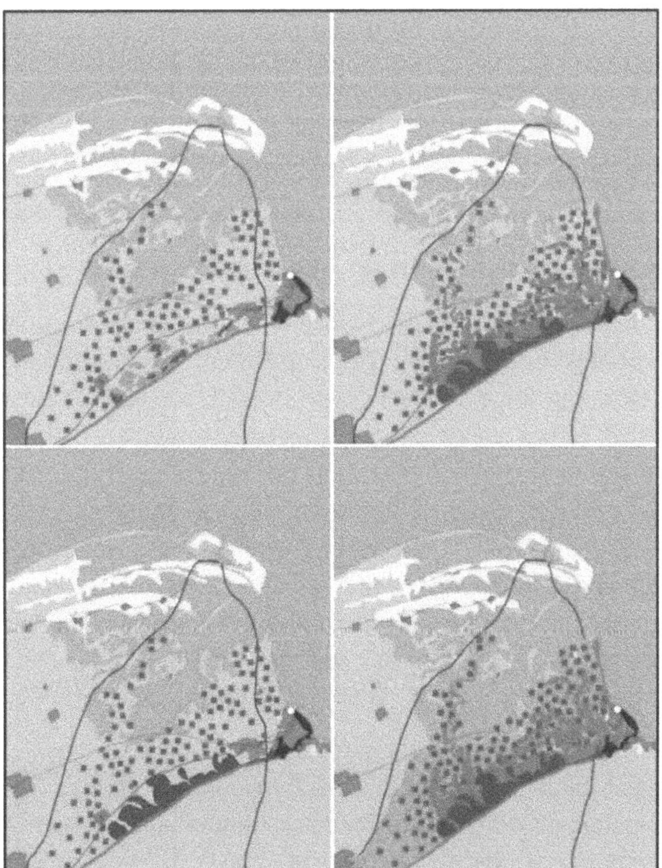

Fig. 2. Floodable Eemsdelta at 0.3, 0.6, 0.9 and 1.2 m sea level rise (Roggema, 2009; 2015)

the land and allowed behind the dike from the start of sea levels rising (as of now) makes it possible to anticipate change. The people, institutions and all land use could start to act as if this future is already a reality. When a gap in the coastal protection is created the water can slowly enter and occupy the hinterland. Whenever sea level rises to respectively 0.3, 0.6, 0.9 and 1.2 m, a larger and larger area will be impacted. Then all land use, including urban occupation will be directed by the expected change, and are adjusted to future circumstances, e.g. are floating, amphibious or waterproof.

Finally, in the third phase the adaptation process starts even when it is unknown what the future will bring exactly. The gap in the coastal protection initiates a slow transformation of the area, 'hosting' water as sea level is rising.

This design prevents damaging impacts resulting from a big disaster, because the water is treated as an ally rather than as an enemy. It is possible to accurately predict where the water will eventually flow so people, buildings and organizations can adapt at an early stage. The water enforces gradual changes but also benefits:

- The brackish conditions emerging in the landscape enriches ecological conditions;
- All new houses will face water in their direct environment, a real estate asset of great value;
- The slow pace of seawater entering the landscape gradually tames the transformation into a wet environment, so a disaster never actually happens making the area inherently safe.

Landscape-based Urbanization in Western Sydney

The urban growth happening in Western Sydney erases the landscape and historic values. In order to prevent this from happening, the different aspects of landscape must be understood. By separating these in layers every aspect can be investigated, redesigned, adjusted and revisited. Firstly, elevation, soil, vegetation, waterways, future flood-risk and ecological remnants are mapped to understand the nuanced sensitivities of life, and the reliability of water, sunlight, shade, coolness and other factors that determine the chance the existing plant- and animal life can coexist with future human occupation in the area. This takes the landscape as the basis for initial urbanization, without making future change impossible. The creeks and side streams, the contours and potential discharge of rainwater, and the natural vegetation (Fig. 3) form the basis for the Master Plan (Mushi, Shalala and Young, 2017).

Secondly, the capacity of the landscape to adjust itself to new circumstances is mapped. The Western Sydney creeks occasionally flood and after heavy rains can reach levels up to nine meters above normal. The once in one-hundred-year flood is mapped, as well as the maximum potential flood (Bewsher, 2004). These calculations, though still accepted as the prevailing ones, are reflecting climate as it was more than 15 years ago.

If we want to sketch a realistic map of future flood impacts, we would need to extend the area required to accommodate future flooding. Therefore, as an estimate we doubled the space for the maximum possible flood (Fig. 4).

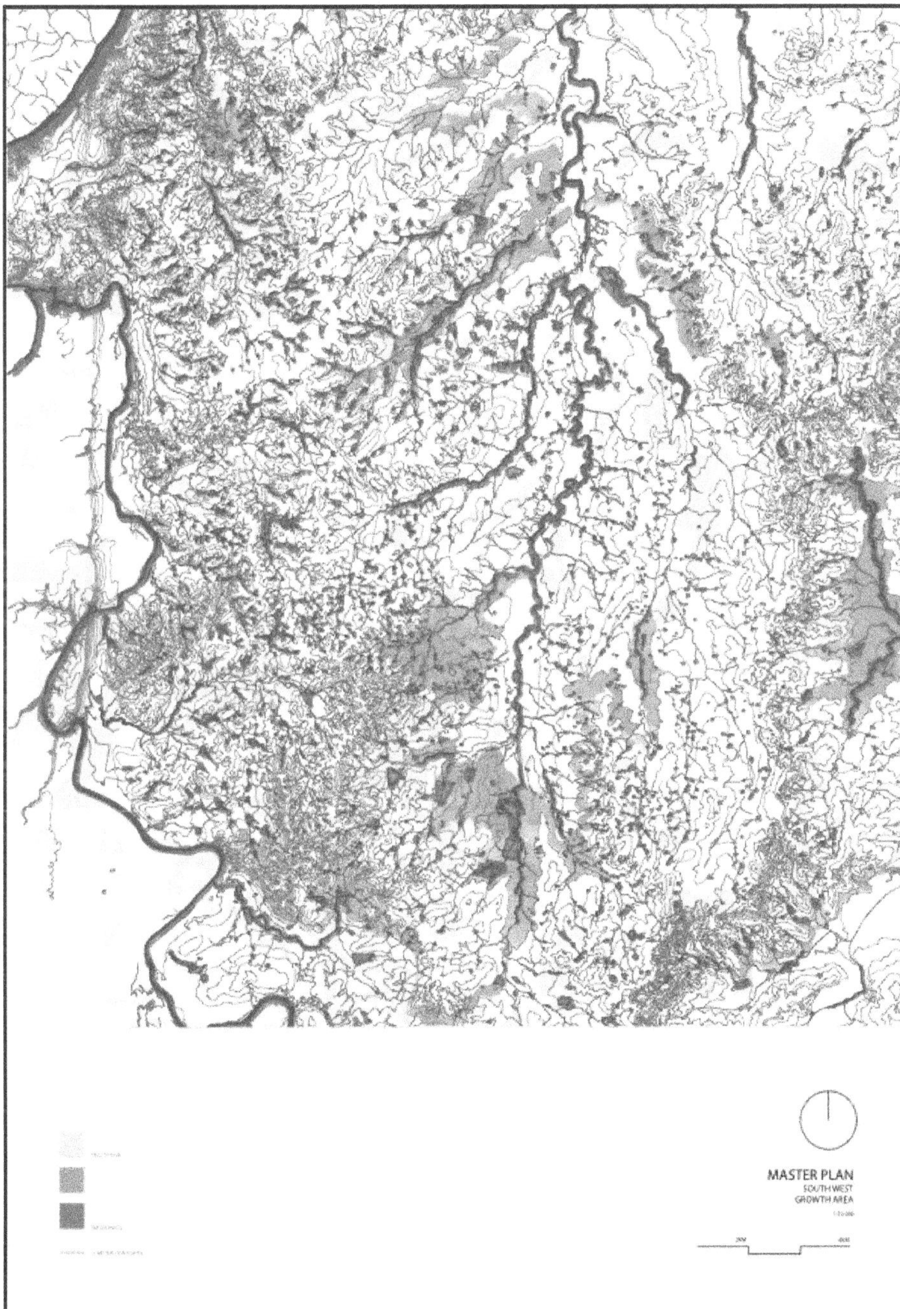

Fig. 3. Landscape with elevation, water and vegetation (Source: Mushi, Shalala and Young, 2017)

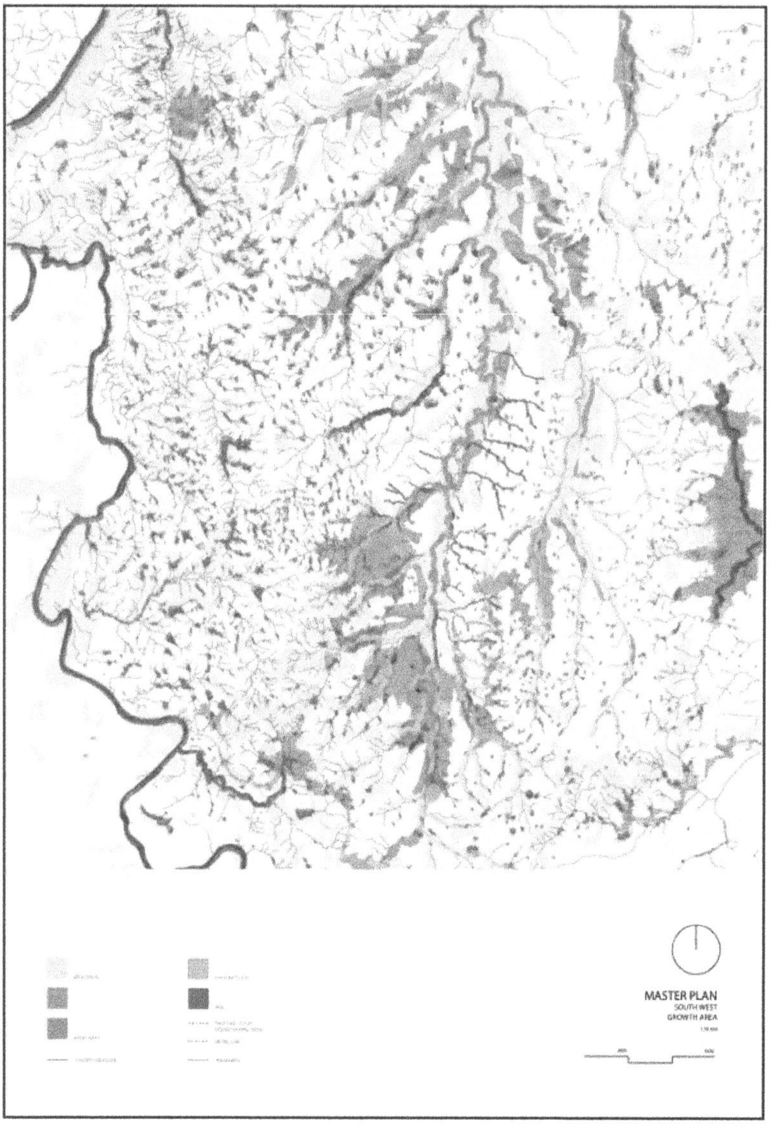

Fig. 4. Landscape with potential flood (Source: Mushi, Shalala and Young, 2017)

With temperatures already over 47 degrees Celsius and expected to rise, heat in summer is another major problem. The increase of concrete, rocks and asphalt through urbanization leads to extensive urban heat island effects. It is evident these high temperatures will impact on human health, at the same time increase the aircon use, but also complicates the growth of trees and plants. This vicious circle of less trees-increase heat-decrease trees growth-hotter impact the living conditions seriously and increases bushfire risks.

In the design for Western Sydney the creation of space for water and vegetation could cool the environment. In order to maximise cooling airflows the integration of green infrastructure deeply into the urban environment is required and only a robust green-blue system could achieve this. Such a system could also provide the space to accommodate the impacts of heavy rainfall and discharge purified household water.

Through marking the boundaries of potential flooding, the spaces required for a robust ecological system and the required space for discharging household water the landscape offers the patial framework within which the city can be developed hence creates a new lay-out for Western Sydney that intrinsically cools the environment. Vegetation, water and shade rhizomically ramify themselves in smaller branches, leaving space for smaller and separated urban fields (Fig. 5), each large enough to form a coherent community of approximately 10,000 houses.

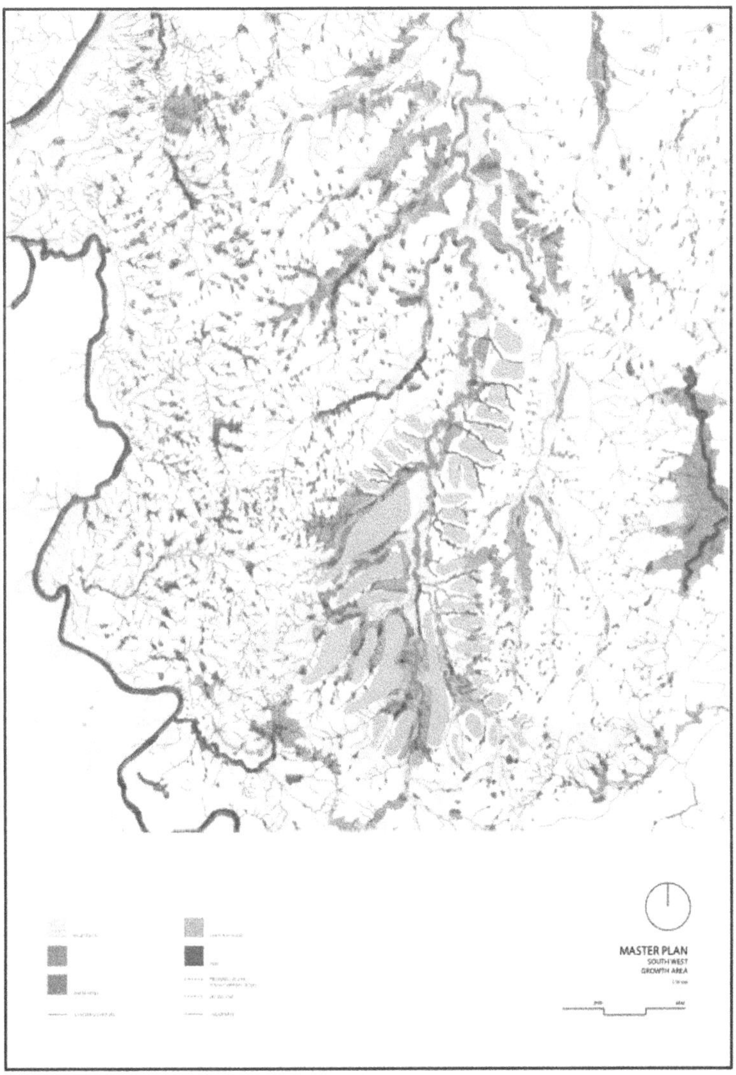

Fig. 5. Urban fields (in grey), determined by the space required to create a cooling landscape (Source: Mushi, Shalala and Young, 2017)

For every urban field, a set of rules for good urbanism should be followed. These ambitions can be reflected in six themes:

- **Establish diversity.** A lively city emerges when housing types and densities are divers, functions and land-uses are mixed and urban spaces come in different sizes and atmospheres;
- **Create an equitable city.** When the city offers equal access for everyone to housing and amenities, no matter what the background of people is, the quality of life for everyone increases. It also does justice to the different lives people have and their ethnicity, the level of education or income and their gender should not matter;
- **Strive for sustainability.** A sustainable neighborhood is climate-resilient and safe, the air and water are clean and it is not noisy. Moreover, transport and living are zero-carbon and energy efficient and the design incorporates water sensitive principles, and the place provides locally growth of food. Finally, rich and biodiverse nature and green is close to people;
- **Design a functional city.** Urban networks need to be logical and understandable. Urban precincts are accessible for people by a range of transport means not the least public transport. All amenities must be catered for, such as schools of different types, shops and restaurants;
- **Close cycles and become circular.** The flows of energy, water, nutrients and materials are minimized and when rest-flows emerge these should be recycled or reused within the same urban area. This way a city without waste can be developed, in which more clean water, local energy and food is produced than taken out of the broader environmental system;
- **Design a beautiful city.** Everyone enjoys living in an environment that is beautiful, with interesting architecture, sustainable materials and local character is respected hence used infor the benefit of the future. These factors determine the appreciation of the people for their environment.

On a larger scale the urban fields are connected to the greater Sydney Metropolitan Area. The primary connections are metro-lines linking each neighborhood and connecting Badgerys airport to the existing network (Fig. 6).

The Western Sydney Master Plan requires adjustments over time. New problems and developments will emerge, and developments need to anticipate future needs. For instance, when there is a need for ongoing urban growth densities can be increased. Only the future can tell if, and how future densification will take place, depending on changing technological, social and climatic contexts. It is very likely the final development will consist of a mix of densification typologies. The iterative process of finding optimal densification patterns (Fig. 7) starts with locating the urban fields within the boundaries of the cooling landscape, then exploring three ways to increase density (within urban fields, in valleys and on ridges) as rigid singular thought exercises, before optimal mix can be explored.

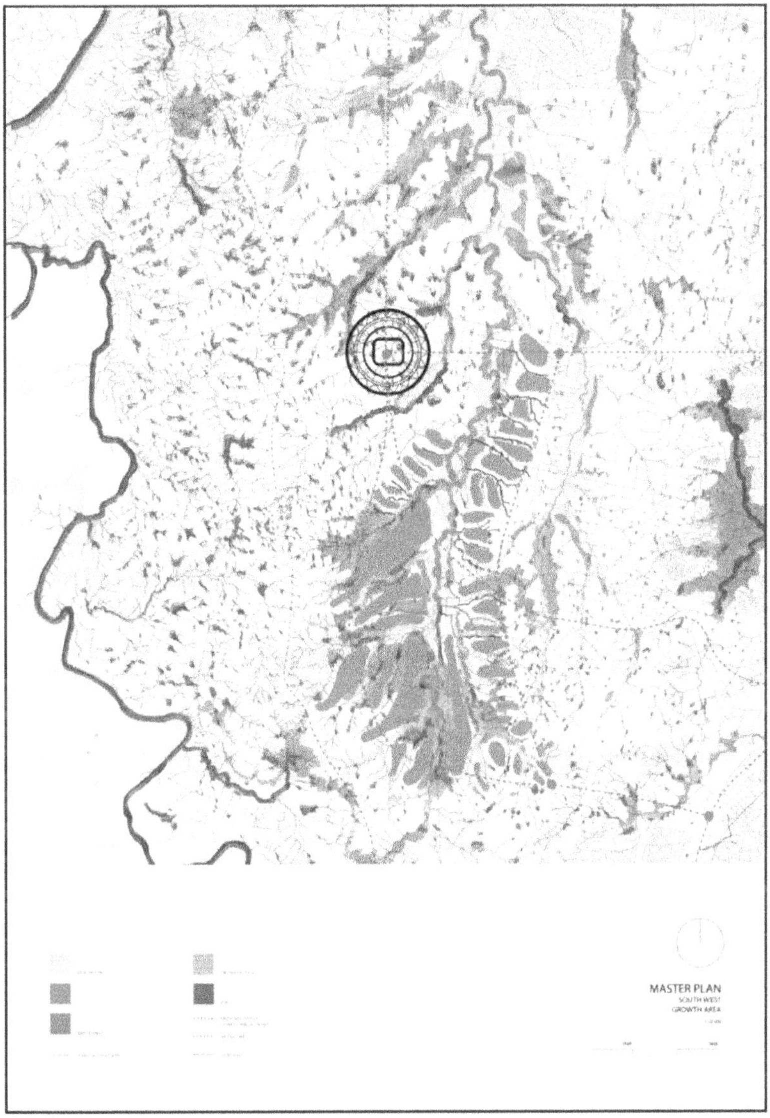

Fig. 6. Urban fields connected to the existing metro-network and Badgerys airport (Source: Mushi, Shalala and Young, 2017)

The urban design for North Bringelly (Fig. 8) locates urban fields in between gullies and valleys, leaving abundant space for the landscape to cool (Young, 2017). Within these urban field densities can be raised whenever desirable. Landscape characteristics are used to create differences. Five street typologies, depending on their relation to the gullies and the wider landscape hence the edges are designed. Every street that is angled towards the gullies is car free, so it becomes possible to walk down to the leisure areas connected to the valley and the transport and shops up the ridges, where the wider service roads and parking is located. Shared green space and private individual gardens are connected to the gully zone while amenities are located in the gully streets. The size of the gullies determines the size and shape of the neighborhood. The highest densities are found near the gullies while lower densities are realized inside the urban fields. In the future specific pockets at the edges of the ridges are selected to increase the densities further. The building lots can be subdivided built up, so the total density can potentially quadruple.

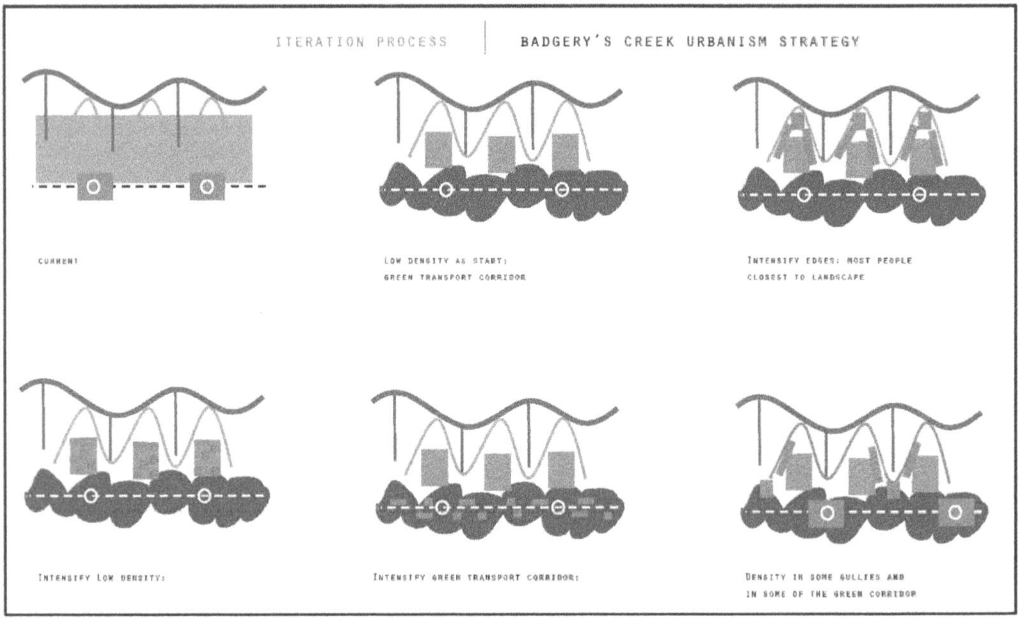

Fig. 7 Iterative process exploring the mix of densities (Source: Mushi, Shalala and Young, 2017)

'Hydraulic City' (Veringa, 2017) uses the landscape also as the driving force (Fig. 9). The discharge of rain and cleaned household water is taken as the starting point for subdividing the building blocks, which then are disconnected from each other to allow summer and winter sunlight in the houses, at the same time providing shade during hot summer days. By orienting the blocks along the contours of the gully landscape, new spaces emerge which are used to integrate shady green spaces. Both sides of the gully are connected by overbuilding the landscape and raising these building blocks above the underlying landscape, safeguarding undisturbed functioning of green and waterflows. By building over the landscape relatively high densities (Fig. 10) can be reached while the landscape is respected and keeps its natural and cultural heritage functionalities (Fig. 11).

Both detailed urban designs illustrate that the use of landscape characteristics can be combined with the ambitions to realize higher densities in a livable and walkable plan, which is close to nature, leisure and amenities.

Fig. 8. Urban Design North Bringelly (Source: Young, 2017)

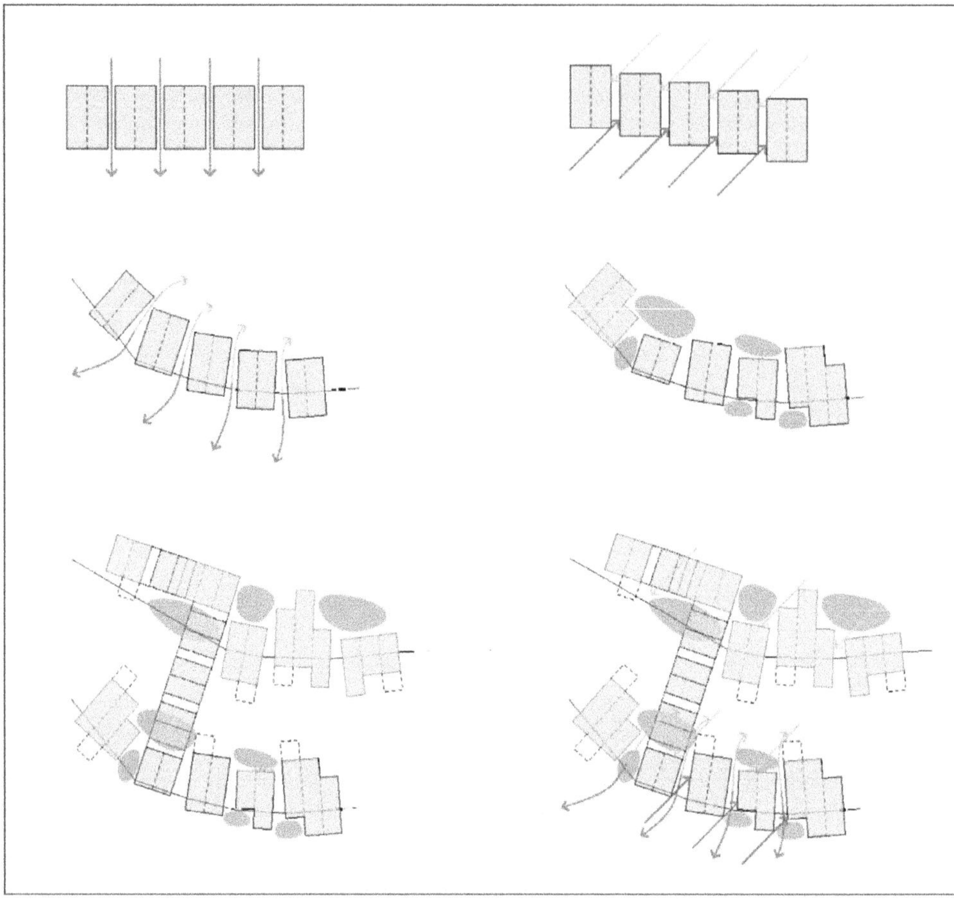

Fig. 9. Constructing Hydraulic City (Source: Veringa, 2017)

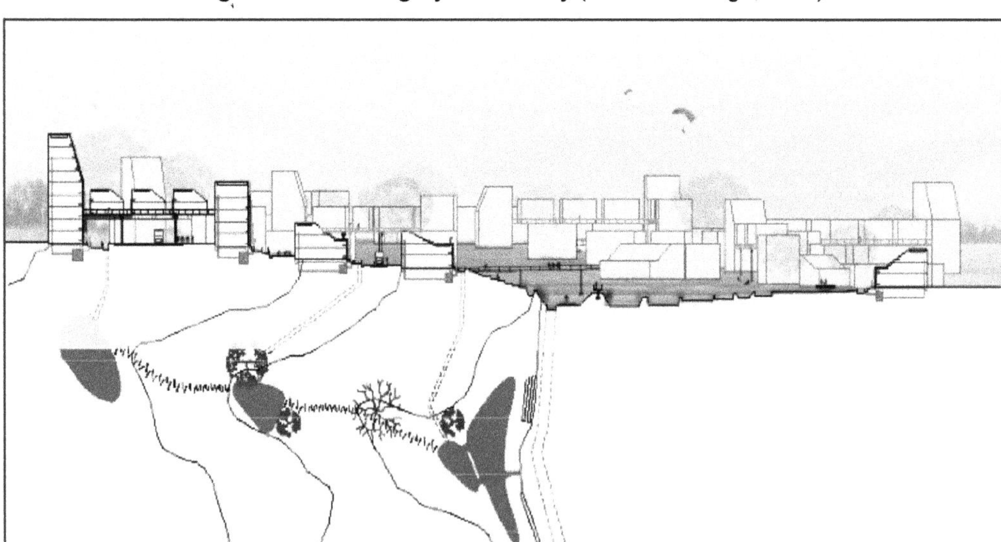

Fig. 10. Cross-section, showing high densities (Source: Veringa, 2017)

Fig. 11. Urban design Hydraulic City (Source: Veringa, 2017)

Benefits: conditions for a resilient coastal development

The benefits of taking a landscape-ecological approach to urbanism, is multiple. Several principles can be distinguished as a primer for design of future urban landscapes:

- **Integration:** By designing a multiplicity of uses integration will have to occur and stimulates beneficial combinations. In a coastal zone a fertile soil, nature-based systems such as ecological and water systems, locally grown food, green infrastructure come together, within which living environments, and productive fish and other grounds can be embedded, which on their turn are supported by circularity of flows of energy, water and materials;
- **Inclusive:** The future pressure of population growth and the need for more food, energy and clean water requires an inclusive approach to integrated development;
- **Synergetic:** Searching for the synergies between uses of intensity and dynamism, such as harbor activities, housing, urban development or infrastructure, and more vulnerable systems of ecology to sustain the overall area;
- **Open:** Design the system in an open manner so the vision for the area can be expanded. This ascertains an open mindset not excluding new solutions that were out of sight. Take into account the land as well as the sea to look for synergy and solutions;
- **Directive:** Take landscape-ecology as the prime guidance for designing the urban form, so natural systems need to be designed first, after which the urban systems are embedded;

- **Nature-based productivity:** Create a type of productivity for instance for fish breeding that is based on nature's principles. This will create a harmonious environment in which nature and productivity can be in balance, and will guarantee productivity on the long term;
- **Self-organising:** Building-with-Nature design principles will allow the powers of nature to self-organise natural elements such as sand deposition, dune-forming and enhancement of ecological values. This is not only for the coastal protection but also for the value of real estate, people's lives and nature.

These principles can be used to think about the future of coastal zones and give direction how to design the landscape. Such a landscape consists of a robust river-creek system, along which rich ecological areas can be found. The tidal influence of the sea brings the sand and sediment to raise the level of the ground and protect the coast from inundation. This self-regulating process makes it possible to live amidst the landscape, in compact urban entities that each have a high density, and optimally use their position in the landscape. The great views people in these urbanized hubs attract them and connect them to the healthy environment. At the same time the sea brings fertility and local food grows on the higher grounds using fresh water, the lower grounds using brackish water, and in the sea where aquacultures use the salt sea water to grow. The diet of the people living in these regions consists of healthy vegetables, salt-loving crops, healthy weed, fish and shellfish. People no longer depend on wheat, potatoes or rice to fill their stomachs. People and goods are transported mainly by the water on electric ships, there is mass transportation between the hubs on large ferry's which gives people the patience and tranquility to gain a moment of contemplation on the way to and from their destination. A productive coast includes also natural fisheries that are extensively found along the coast. The grounds are artificially enlarged into the sea, so an abundance of fish is living in a natural environment. The intensity of fishing is so low the population can easily restore itself every year, as over the years more and more fish can be found in front of the coast. The seaward development of fishing grounds is accompanied by a moderate and slow urbanization on new islands in front of the coast. These islands make it possible to live on a little paradise, but they also protect the coast against storm surges, and provide the circumstances where the natural fishing grounds can grow behind.

Conclusion

In conclusion, the development of our coastal systems is under a lot of pressure. Urbanisation, exploration, harbor activities and pollution take over from the once so beautiful, and ecological valuable coastal zones. This research has shown that an alternative approach to usual urbanization could give coasts a new life. Four perspectives underpin this. The first perspective is to design the landscape, its ecology and the water system first. This will bring the crucial features and qualities to the fore and prevents them from being erased by urban development. A second perspective is that it is wise to use a design methodology, allowing future explorations and visualization being made, that can be tested and evaluated. The planning and building process is not a straightforward process, but runs in circles, is iterative and brings new insights along the way. A good way to structure the process is to make use of the layer theory, in which valuable aspects of the world we live in can be separated and analysed as such, before they can be synergized. In the third perspective the interdependencies of different uses in an area are celebrated. These all belong together. There is no land without water, no city without the landscape. There are no people without their environment and there is no food without nature. The fourth perspective shows us that new problems cannot be solved using old solutions. This is doomed to fail. New alternatives should be discovered, and in order to do so creativity is needed and a healthy bit of disobedience.

References

Bewsher, 2004. South Creek Floodplain Risk Management Study and Plan for the Liverpool local government area. Bewsher Consulting Pty Ltd., Final report. Liverpool: Liverpool Council.

Booth, D.J. and Sear, J., 2018. Coral expansion in Sydney and associated coral-reef fishes. Coral Reefs, Journal of the International Society for Reef Studies **37** 995. https://doi.org/10.1007/s00338-018-1727-5.

Chen, D. and Gartry, L., 2018. Great Barrier Reef $444m budget funding awarded to small foundation without tender process. ABC News 22 May 2018. URL: https://www.abc.net.au/news/2018-05-22/great-barrier-reef-funding-labor-accuse-due-diligence/9785782

Haasnoot, M., Bouwer, L., Diermanse, F., Kwadijk, J., Van der Spek, A., Oude Essink, G., Delsman, J., Weiler, O., Mens, M., Ter Maat, J., Huismans, Y., Sloff, K. en Mosselman, E., 2018. Mogelijke gevolgen van versnelde zeespiegelstijging voor het Deltaprogramma. Een verkenning. Delft: Deltares 11202230-005-0002

Hamilton, C., and Kasser, T., 2009. Psychological adaptation to the threats and stresses of a four degree world. Oxford: Four degrees and beyond conference.

Hansen, J., Sato, M., Kharecha, P., Beerling, D., Berner, R., Masson-Delmotte, V., Pagani, M., Raymo, M., Royer, D.L., and Zachos, J.C., 2008. Target atmospheric CO_2: where should humanity aim? The open atmospheric science journal 2, 217–231 1874–2823/08.

Hansen, J., Sato, M., Kharecha, P., Russel, G., Lea, D. W., and Siddal, M. (2007). Climate change and trace gases. Philosophy Transactions society A 365, 19251954. doi:10.1098/rsta.2007.2052.

Hansen, J.E., 2007. Scientific reticence and sea level rise. Environmental research letters 2 024002 (April–June 2007) doi:10.1088/1748-9326/2/22024002. The Open Access Journal For Environmental Science. IOP Electronic Journals.

Hughes, T. and Kerry, J., 2017. Back-to-back bleaching has now hit two-thirds of the Great Barrier Reef. The Conversation published 12 April 2017. URL: https://theconversation.com/back-to-back-bleaching-has-now-hit-two-thirds-of-the-great-barrier-reef-76092

Kristinsson, J., 2012. Integrated sustainable design. Delft: Delft Digital Press.

Lenton, T., Held, H., Kriegler, E., Hall, J.W., Lucht, W., Rahmstorf, S., and Schellnhuber, H.J., 2008. Tipping elements in the earth climate system. Proceedings of the National Acedemy of Science 105(6), 1786–1793.

McHarg, I.L., 1969. Design with Nature. New York: Natural History Press.

Mushi, M., Shalala, R. and Young, J., 2017. Badgerys Urbanism. Elective Master of Architecture, UTS School of Architecture, July 2017, led by Rob Roggema and Rod Simpson

Page, C.A., Mullerb, E.M. and Vaughan, D.E., 2018. Microfragmenting for the successful restoration of slow growing massive corals. Ecological Engineering 123: 86-94.

Rahmstorf, S., Cazaneva, A., Church, J.A., Hansen, J.E., Keeling, R.F., Parker, D.E., *et al.*,, 2007). Recent climate observations compared to projections. Science, 316 (5825): 709.

Roggema, R., 2009. Adaptation to climate change, does spatial planning help? swarm planning does! In: Brebbia, C.A., Jovanovic, N. and Tiezzi, E. (Eds.), Management of Natural Resources, Sustainable Development and Ecological Hazards, pp. 161–172. Southampton: WIT Press.

Roggema, R., 2012. Swarm Planning: The development of a Methodology to Deal with Climate Adaptation. Ph.D. thesis. Delft University of Technology and Wageningen University and Research Centre, Delft, Wageningen.

Roggema, R., 2015. Smart by Nature: The use of swarm planning in creating productive and adaptive urban landscapes. In: Proceedings of the 8th Conference of the International Forum on Urbanism (IFoU), Incheon, 22-24 June 2015. DOI: 10.3390/ifou-D011

Roggema, R., 2017. The Sydney barrier reef: Engineering a natural defence against future storms. The Conversation. URL: http://theconversation.com/the-sydney-barrier-reef-engineering-a-natural-defenceagainst-future-storms-76862. Accessed: 27 July 2019

Sharmila, S. and Walsh, K.J.E., 2018. Recent poleward shift of tropical cyclone formation linked to Hadley cell expansion. Nature Climate Change **8** 730–736. http://dx.doi.org/10.1038/s41558-018-0227-5

Sijmons D., 1992. Het casco-concept, een benaderingswijze voor de landschapsplanning. Ministerie van LNV, directive NBLF, Utrecht

Stive, M.J.F., De Schipper, M.A., Luijendijk, A.P., Ranasinghe, R.W.M., Van Thiel, R.J.B., De Vries, J.S.M., Aarninkhof, S., Van Gelder-Maas C., De Vries, S., Henriquez, M. and Marx, S., 2013. The sand engine: a solution for vulnerable deltas in the 21st century? In: Proceedings Coastal Dynamics 2013: 7th International Conference on Coastal Dynamics. University of Bordeaux, Arcachon, France, 24–28 June 2013

Tin, T., 2008. Climate change: Faster, stronger, sooner, an overview of the climate science published since the UN IPCC Fourth Assessment Report. Brussels: WWF European Policy Office.

Tjallingii, S., 2015. Planning with water and traffic networks. Carrying structures of the urban landscape. Res. Urban Ser. 3(1): 57–80. doi:10.7480/rius.3.832

Tjallingii, S.P., 1995. Ecopolis: strategies for ecologically sound urban development. Backhuys Publishers, Leiden

Tomásek, W., 1979. Die Stadt als Oekosystem; Ȗberlegungen zum Vorentwurf Landschaftsplan Köln [The city as ecosystem. Considerations about the scheme of the landscape design in Cologne]. Landschaft+Stadt, 11: 51–60.

Van Schaick J. and Klaasen, I.T., 2011. The Dutch layers approach to spatial planning and design: a fruitful planning tool or a temporary phenomenon? Eur. Plan Stud. 19(10): 1775–1796

Van Timmeren, A. and Henriquez, L., 2015. Ubikquity and the Illuminated City. From smart to intelligent urban environments. Delft: TU Delft, 184 p.

Veringa, G., 2017. The Hydraulic City. Design Studio 'GO SOUTHWEST', Master of Architecture, UTS School of Architecture, Led by Rob Roggema and Rod Simpson, August-November 2017

Waterman, R.E., 2010. Integrated Coastal Policy via Building with Nature. Ph.D. thesis. Delft University of Technology, Delft

Werner, A. and Currell, M., 2019. Adani is cleared to start digging its coal mine – six key questions answered. The Conversation published 13 June 2019. URL: https://theconversation.com/adani-is-cleared-to-start-digging-its-coal-mine-six-key-questions-answered-118760

Young, J., 2017. Urban Design for North Bringelly. Master of Architecture Design Elective 'Badgerys Urbanism', led by Rob Roggema and Rod Simpson, School of Architecture, University of Technology Sydney, July 2017.

Impact of Climate Change on Hydrological Cycle, Ecosystem, Fisheries and Food Security, Pages: 95-109
Edited by: B. Madhusoodana Kurup, M.R. Boopendranath, M. Harikrishnan and A.V. Shibu

CHAPTER - 7

Multiforcing Ensemble Projection based Climate Change Information over Kerala for Adaptation Strategies

Kavirajan Rajendran[1,2*], Sajani Surendran[1,2] and Stella Jes Varghese[2]

[1]*Multi-Scale Modelling Programme, CSIR Fourth Paradigm Institute, Bangalore-560037, India*
[2]*Academy of Scientific and Innovative Research (AcSIR), Ghaziabad-201002, India*
[*]*E-mail: rajendrank@hotmail.com*

Natural climate variability affects many aspects of life, livelihoods, landscapes, and decision-making. It causes droughts, floods, heat waves, severe winds, intense storms and the battering of coastal areas. As a region that has witnessed rapid increase in population, significant alterations in land use and land cover, reduction in the supplies of water, incidence of dry spells, and other climatic changes, the state of Kerala can be considered as one of the climate-challenged regions of the country. This article summarises current understanding of climate change and impacts, and information for helping the formulation of adaptation strategies for Kerala.

Indian climate is changing due to global warming and the surface air temperature is increasing over the last few decades, with trends in temperature and summer monsoon rainfall accelerating since 1990s (Fig. 1). Consistent with the understanding that the accumulation of CO_2 and other greenhouses gases (GHGs) in the atmosphere is the fundamental cause of recent global warming, CO_2 emissions over India also show a clear increase over the last two decades (Fig. 2). The more GHGs there are in the atmosphere, the more heat is trapped and the higher the Earth's temperature becomes. Today the major cause of the increase in CO_2 emissions is anthropogenic. Anthropogenic GHG emissions also include methane (CH_4) and nitrous oxide (N_2O) and others such as hydrofluorocarbons that are released by various industrial processes. Global emissions of CO_2, similar to the emissions from India continues to rise (Fig. 3). The major sources of anthropogenic GHGs in India shows that the energy sector especially electricity and heat are responsible for the majority of GHG emissions from India (Fig. 4). Direct agriculture activities (23%) are about as important as industry (15%) and transport (6%). Contribution from land-use changes consisting mainly of the harvesting of agriculture and forestry products and the clearance of natural vegetation for agriculture and constructions is also an important source.

Kerala, the southern state of India is known for green landscape, water bodies, rolling mountains and narrow valleys. With high rainfall, chains of backwater bodies, many rivers, reservoirs, lakes, ponds, springs and wells, the state is considered by many as the land of water. When the neighbouring states are harnessing large quantum of surface and groundwater sources for irrigation, hydro power generation, domestic and industrial purposes, the scope for development of this precious resource is very limited in Kerala due to its unique characteristics. For the last three decades, Kerala is frequently facing dry spells followed by acute drinking water scarcity. The rivers hardly contain any water during the summer months; only a few reservoirs and lakes get filled up even in the monsoon and household wells and ponds are increasingly getting dry in summer. The water resource situation worsens in deficit monsoon years.

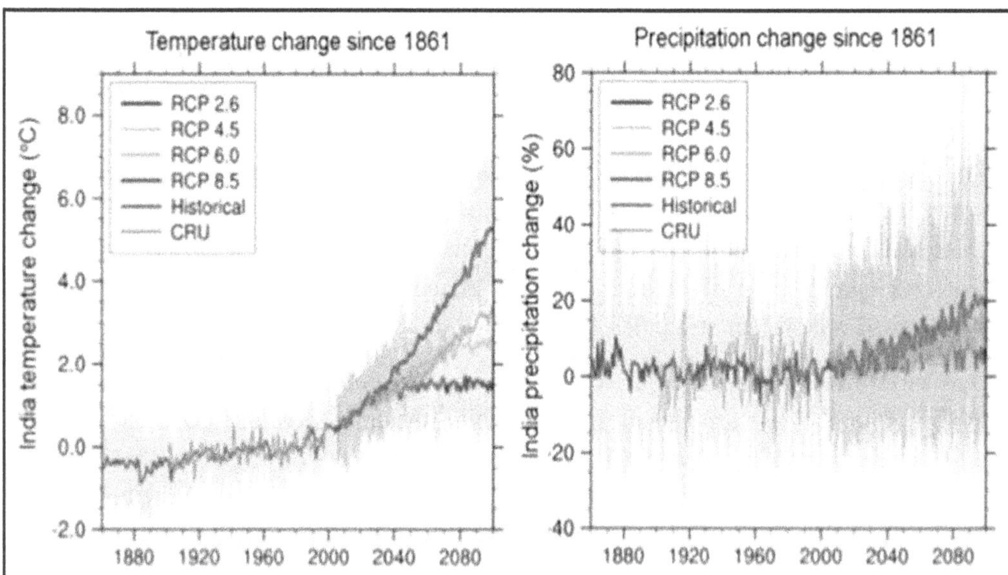

Fig. 1. 21st century climate change projections of surface air temperature (left) and summer monsoon rainfall (right) changes over India (with respect to the 1961-1990 base period) by climate models participated in IPCC 5th Assessment Report (AR5), under four global warming Representative Concentration Pathways (RCP 2.6, 4.5, 6.0 and 8.5). Observations during the historical period are based on Climatic Research Unit (CRU, Harris et al. 2014) data.

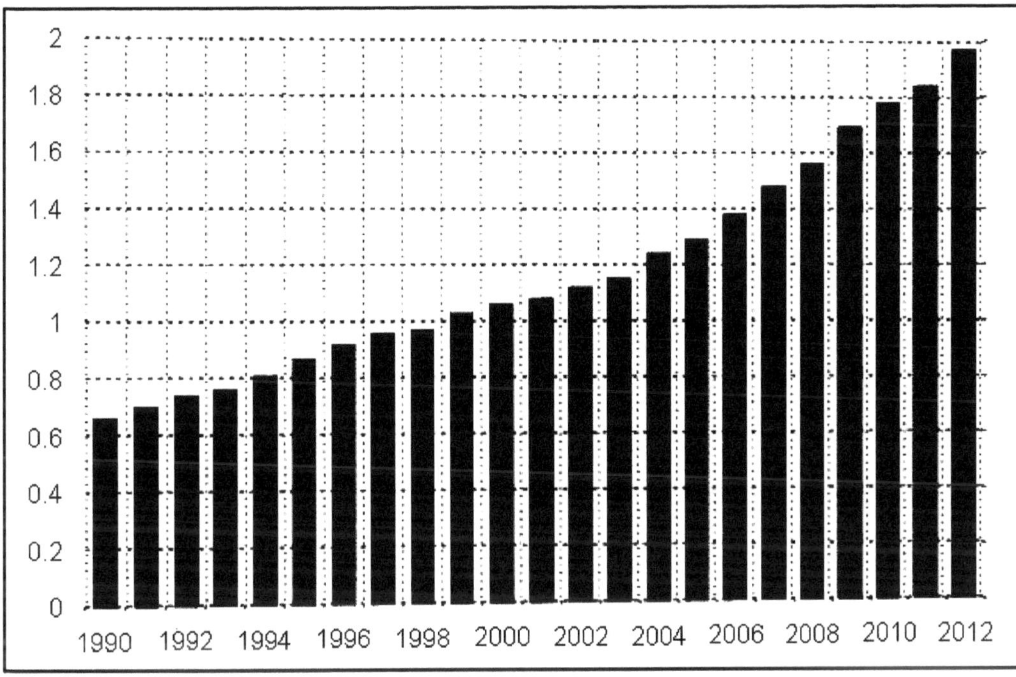

Fig. 2. CO_2 emissions (billion t) from India since 1990 based on EDGAR data.

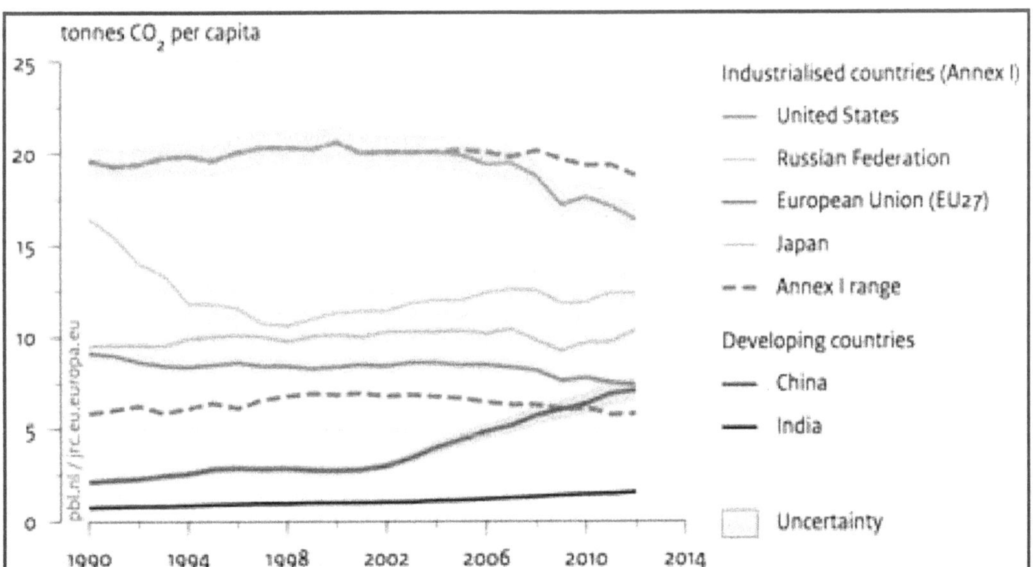

Fig. 3. CO_2 emissions (tonnes) per capita from fossil fuel use and cement production in top 5 emitters (Sources: (i) EDGAR 4.2 (1990-2011); (ii) UNPD, 2010 Annex I range (grey dashed lines): indicates the minimum and maximum per capita values spanned by major industrialized countries; Uncertainty margins: 5% for the United States, EU27 and India, and 10% for the Russian Federation and China].

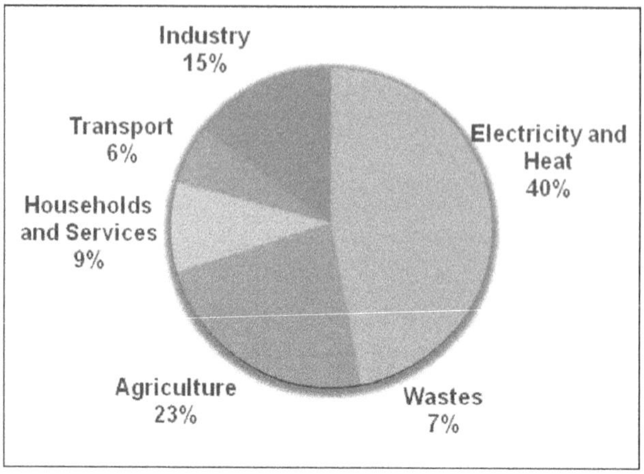

Fig. 4. GHG emissions by India (2004) by various sectors (Source: PEW Centre's International Brief 2: Climate Change Mitigation Measures in India, September, 2008).

Across the state, impacts of climate change are already in evidence. Extreme rainfall events and warm extremes have become more frequent and intense, intense dry spells during monsoon are common and patterns of rainfall are likely changing. Even if emissions of GHGs were substantially reduced now, climate would continue to change for some time and the potential consequences for humans and ecosystems will remain significant. In ecosystems, changing climate could alter the distribution patterns of plant and animal species, reduce the productivity and abundance of species, and change habitats. Sea level could be affected, threatening the natural and built environments on the coasts and in fresh water systems, especially when combined with effects of more intense coastal storms. This article presents the projections of possible climatic changes associated with global warming, impacts of climate change and some discussions on how we as a state can begin adapting to them in beneficial ways.

Future Climate Change Projections

Current state-of-the-art climate models participated in IPCC AR5 CMIP5 show greater convergence in projecting temperature change during the 21st century (Fig. 1). But, models do not agree that much as regards the trend in rainfall. Large differences exist in the projections of change in the Indian monsoon rainfall by these state of the art climate models similar to the large difference in the average monthly rainfall over the region simulated by them in present-day simulations.

To achieve accurate simulation and projection of rainfall over Kerala, parameterized physics and high-resolution processes need to be resolved with adequately high resolution in a global climate model. Hence, a global ultra-high resolution general circulation model (GCM) with a spatial grid size of about 20-km and 60 vertical levels, jointly developed by Japan Meteorological Agency (JMA) and Meteorological Research Institute (MRI), is used for climate change projections. The model corresponds to triangular truncation 959 with linear Gaussian grid (TL959) in the horizontal comprising 1920×960 grid cells (Mizuta *et al.,* 2006; Kitoh *et al.,* 2009). The model is hereafter referred to as the 20-km model.

For the present-day climate simulation, observed historical sea surface temperature (SST) by HadISST (Rayner *et al.,* 2003) from 1979 to 2003 was prescribed. A 'time-slice experiment' was performed for the future climate simulation towards the end of 21st century from 2075 to 2099 (Rajendran and Kitoh, 2008). The boundary SST dataset was prepared by superposing (i) the trend in the Multi-Model Ensemble (MME) of SST projected by the World Climate Research Program (WCRP) Coupled Model Inter-comparison Project (CMIP3) multi-model dataset, (ii) future change in MME of SST (ÄSST) and (iii) the detrended observed SST anomalies for 1979-2003 period. Future change in MME of SST was evaluated by the difference between the 20th century simulations and future simulation under IPCC RCP8.5 emission scenario. Linear trend for future climate by the Coupled GCMs (CGCMs) were also taken into account. The experimental design retains observed year-to-year variability and El Niño and Southern Oscillation (ENSO) events in future climate, but with a higher mean and increasing trend in SST.

In addition, corresponding simulations of present-day (1983–2003) and future climate at the end of the 21stcentury (2079–2099) were carried out with MRI model of triangular truncation at linear Gaussian grid of TL319 (640 × 320 and horizontal resolution of ~60 km) and analyzed. For 60-km model, multiforcing future climate change projections were made using time-slice experiments in which the model was integrated under multiple RCPs, viz. RCP2.6, RCP4.5, RCP6.0 and RCP8.5 (Stella Jes Varghese *et al.,* 2019; each scenario indicates the total radiative forcing at the end of the 21st century). Each RCP scenario has 4 realizations with 4 different SST forcing as projected by MME of 28 CMIP5 models (i.e., multiforcing projections): one with mean SST projected by MME of 28 CMIP5 models and other 3 obtained from cluster analysis based on the tropical SST change as projected by the 28 models (Mizuta *et al.*, 2014).

The validation data for the present-day rainfall simulation of the 20-km model is taken from the daily gridded rainfall data on 1o×1o grid from IMD for 1979-2003 period (Rajeevan *et al.*, 2006). This dataset is based on 1803 stations with a minimum 90% data availability during the period 1951-2003. For rainfall validation over global tropics, we have used the TRMM 3B43 (Huffman *et al.*, 2007) precipitation dataset during 1998-2008. The gridded estimates are on a calendar-month temporal resolution at a 0.25° latitude × 0.25° longitude spacing for 40°N-40°S. For rainfall validation, we have also used the Asian Precipitation Highly Resolved Observational Data Integration Towards Evaluation (APHRODITE) precipitation data set from 1951 to 2007 on 0.25° × 0.25° grid (Yatagai *et al.*, 2009).

Consideration of fine-scale processes is also critical for accurate assessment of local and regional-scale vulnerability to climate change. The 20-km model resolves features on finer spatial scales, which were till now resolved by employing high resolution regional models. But regional models have high dependency on the lateral boundary forcing and significant inability to represent regional-global scale interactions. The results at high-resolution projection aims to fill the current knowledge gap about future climate change effects by a large extent. The fidelity of a model in representing the present-day climatological features and recent climate trends is crucial for building confidence in its future projections. On regional scale, over Kerala (Fig. 5a), in spite of slight overestimation, the simulated rainfall pattern compares well with the observed JJAS rainfall in capturing the spatial variation across the state. On district level, the increase in seasonal rainfall towards the north and towards the coastal regions of Kerala is qualitatively represented in the climate model simulation (Fig. 5b).

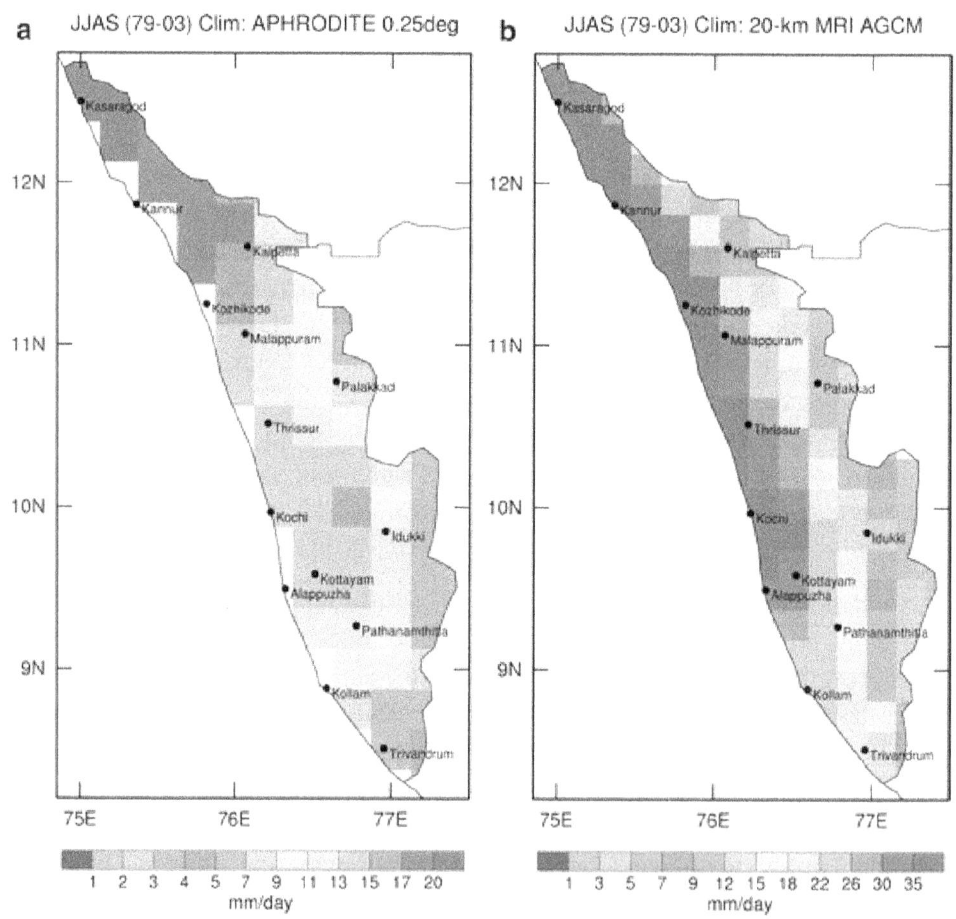

Fig. 5a. Present-day JJAS mean rainfall over Kerala from TRMM 3B43 observation (left panel) and b) from 20-km model simulation (right panel).

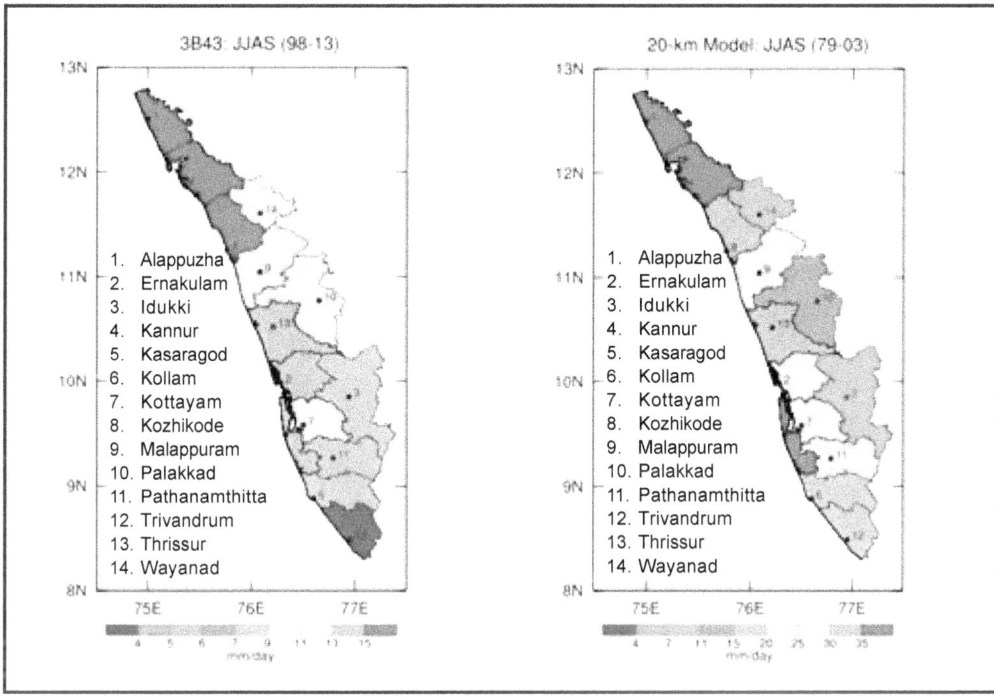

Fig. 5b. Present-day JJAS mean rainfall district-wise over Kerala from TRMM 3B43 observation (left panel) and from 20-km model simulation (right panel).

At the end of the 21st century, the model predicts a significant increase in precipitation over most of the areas of the country where rainfall is rather strong during the monsoon season (Fig. 6). However, in the southern half of the west coast of the peninsula, parts of eastern India and the Jammu-Kashmir region, future decrease in rainfall has been projected. For example, the projection for the state of Kerala (Fig. 7) shows reduction in future monsoon rainfall especially over the western coast. In addition, there is large spatial heterogeneity within the state. So far, IPCC future scenarios for the Indian summer monsoon rainfall even after using high-resolution regional climate models have projected relatively uniform climate change over the whole country. But, the ultra-high resolution global model projected changes indicate an overall intensification of the all-India monsoon rainfall with strong regional modulations in the future (Fig. 6), in response to the anticipated increase in GHG concentration and resultant widespread warming of the surface temperature. An overall increase in rainfall is seen for all the scenarios but there is regional incoherence especially over Western Ghats (WG) orographic regions and certain parts of northeast India. As warming gets intensified, the enhancement over interior India and reduction over WG gets intensified (Fig. 6).

At the end of the 21st century, 20-km model predicts wide spread but spatially varying increase in rainfall which are significant compared to the seasonal mean rainfall over many states (Fig. 6), in particular over the interior regions of Peninsular, West Central,

Central Northeast and Northeast India (~5-20% of seasonal mean). However, over the southern west coast of India (~10-15%), and some parts of Northwest India and Jammu-Kashmir (~5-10%), future reduction in rainfall is projected by the model. Over the windward side of the west coast, the projected changes indicate significant weakening of rainfall in future. In the leeward side, marked increase in future rainfall is predicted by the model. These projected changes are qualitatively similar to the recent trends observed over this region (e.g., Rajendran *et al.*, 2012). Another notable aspect is the opposite future change projected by the model in the northern parts of the Western Ghats along the west coast of Maharashtra compared to the coasts of Kerala and Karnataka.

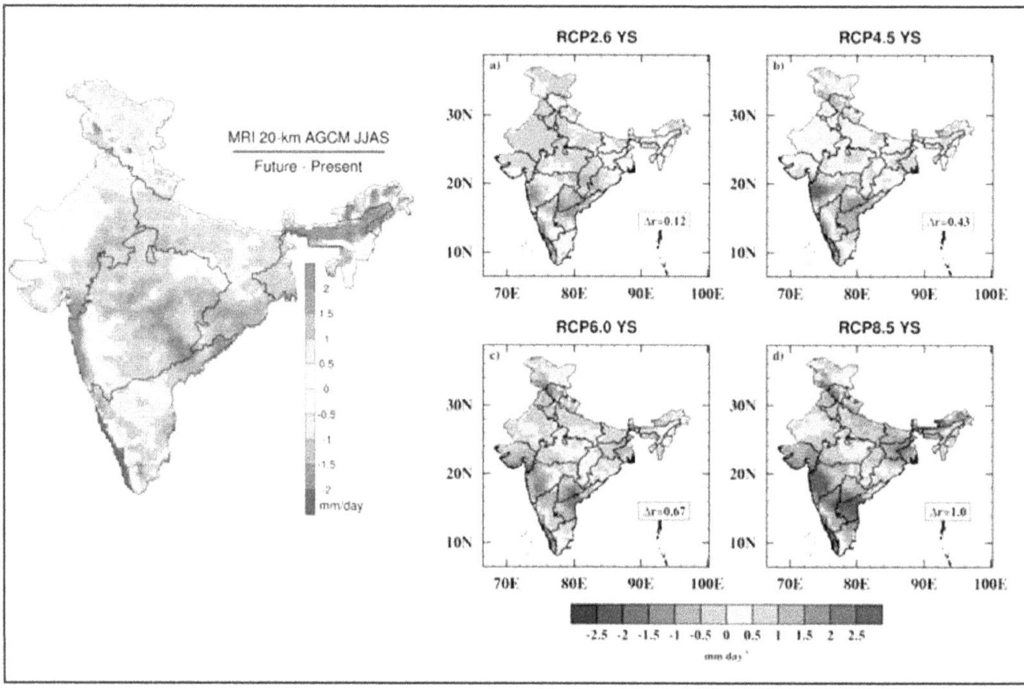

Fig. 6: JJAS mean difference in rainfall between future projection and present-day simulation of the 20-km model over India (left panel figure). Within each colour the gradient in hue corresponds to changes as percentage of mean summer monsoon rainfall (< 5%: no symbol, 5-15%: dots, 15-25%: open circle and >25%: plus). Projected future change in JJAS rainfall from 60-km multiforcing simulations under a. RCP2.6, b. RCP4.5, c. RCP6.0 and d. RCP8.5 scenarios (right four panels figure). The corresponding change averaged over the Indian region is given in the lower right of each panel

Validation with Recent Climate Trends

Recent observed rainfall trend over Kerala, shows that there is significant decrease in southwest monsoon rainfall in recent times which is consistent with the future change projected by the model (Figs. 7 & 8a). Further, during recent decades, in the present-day simulation, the model simulates a significant reduction in soil moisture over Kerala (Fig. 8b), which corresponds well with recent declining rainfall over the state (Fig. 8a). The distribution of yearly rainfall for JJAS period for three districts of Trivandrum, Kottayam and Kannur with the corresponding linear trends (Fig. 9) clearly brings out the declining trend in rainfall over Kerala. The 20-km model also shows the decreasing trend for Kerala in recent times.

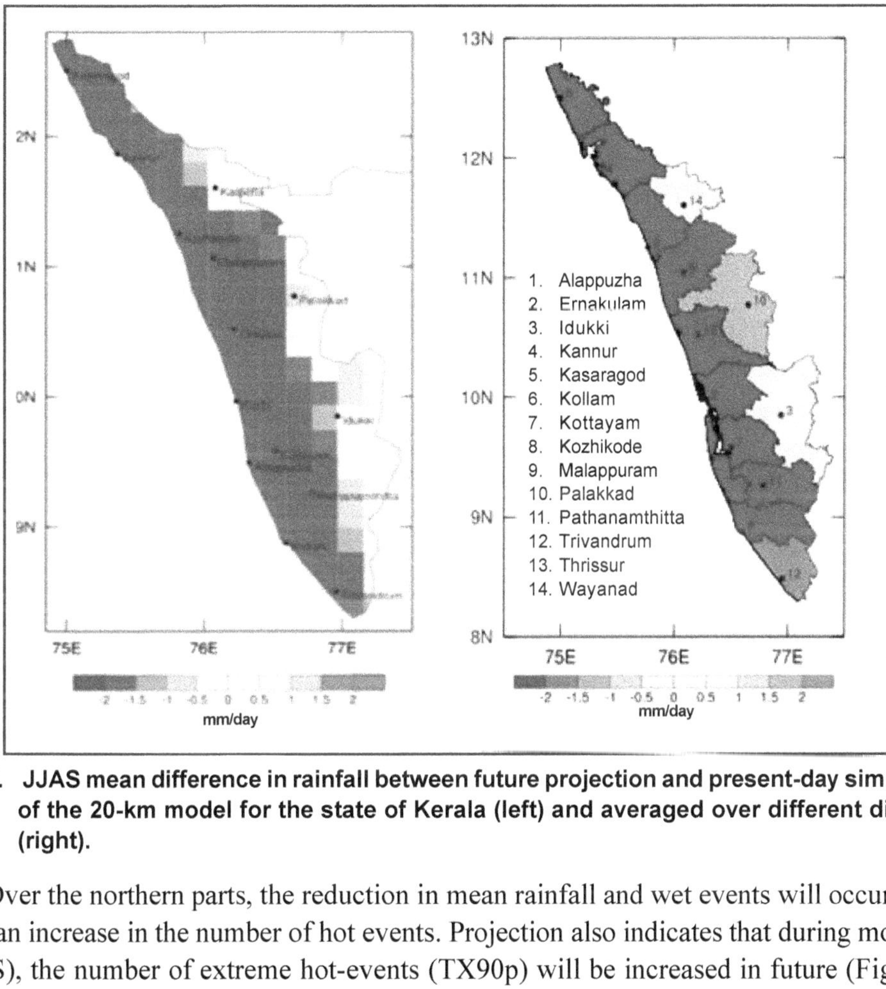

Fig. 7. JJAS mean difference in rainfall between future projection and present-day simulation of the 20-km model for the state of Kerala (left) and averaged over different districts (right).

Over the northern parts, the reduction in mean rainfall and wet events will occur along with an increase in the number of hot events. Projection also indicates that during monsoon (JJAS), the number of extreme hot-events (TX90p) will be increased in future (Fig. 10b) over the entire country. Figure 10b also shows the probability density functions (PDFs) for the daily Tmax above 90 percentile for JJAS season. The mean and standard deviation of present-day climate PDF is different from those of future simulation, indicating a clear shift towards warmer conditions at the end of the century.

It is important to study the impact of this changed climate scenario for crops in the states. The pattern of number of hot events in JJAS is caused in part by the summer surface moisture-precipitation feedback. The probability density functions (PDFs) for the daily Tmax for present-day and future monsoon season indicate a clear shift towards warmer conditions at the end of the century (Fig. 10b inset). Thus, differences between the distributions for the occurrence of warm days in the present-day climate and future warming climate, suggest that there will be warming of maximum temperatures in future. These results warrant the necessity for undertaking impact studies of such scenario for accurate assessment of local and regional-scale vulnerability to climate change.

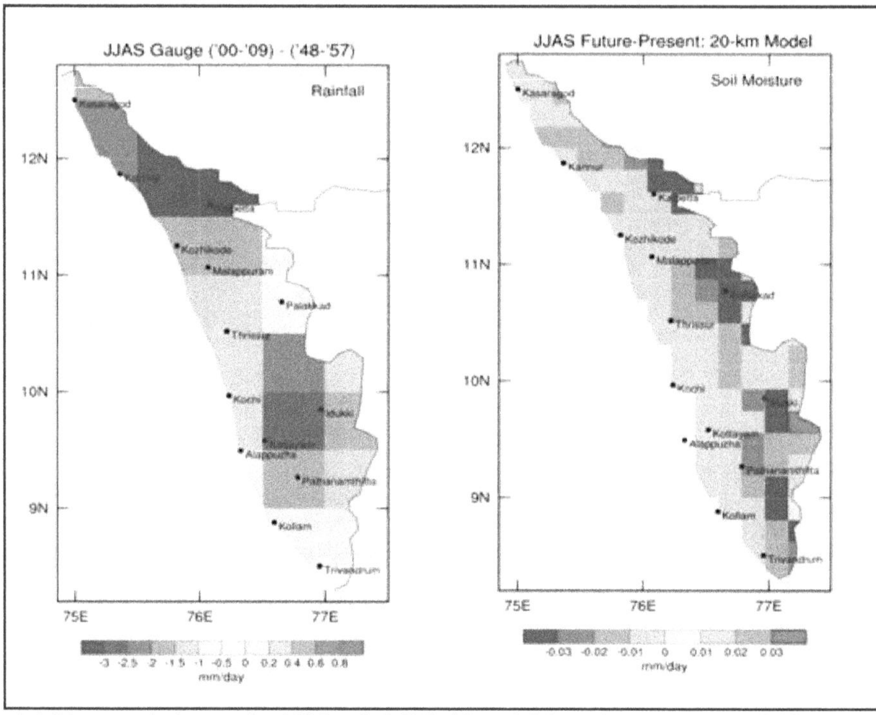

Fig. 8. (a) Observed change in JJAS rainfall (left) and (b) projected change in future summer (JJAS) soil moisture (right).

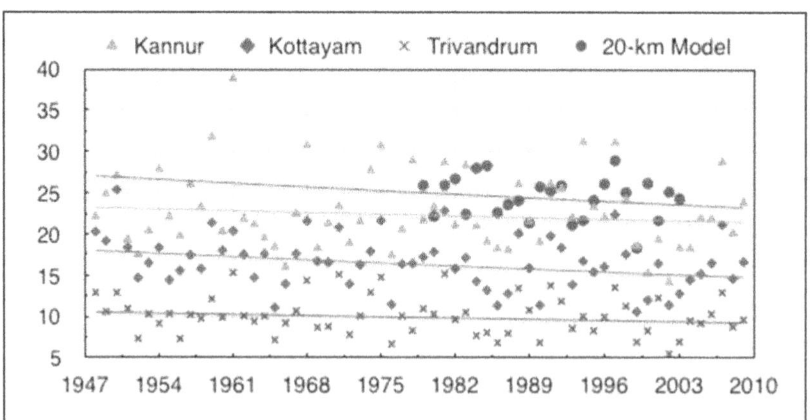

Fig. 9. Yearly JJAS mean rainfall in mm/day for Trivandrum Kottayam and Kannur districts (upper panels) with corresponding linear trends from IMD observation. Yearly JJAS rainfall simulated by the 20-km model for Kerala with trend line.

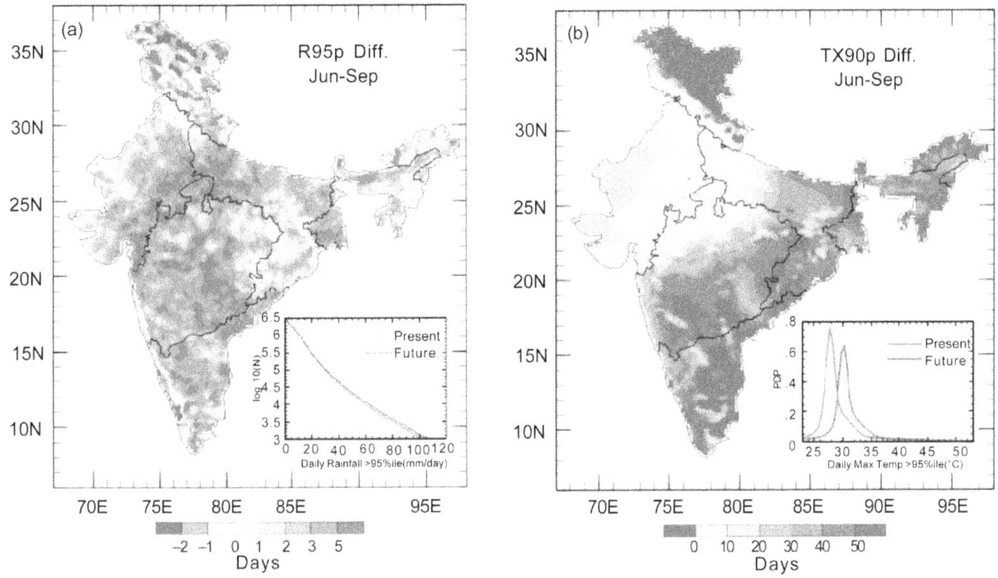

Fig. 10. **(a) Changes in number of days in JJAS with precipitation greater than 95 percentile between present and future simulations of 20-km model. Frequency distribution of daily mean precipitation above 95 percentile in JJAS (left); (b) Future changes in number of days in JJAS with daily maximum temperature (Tmax) greater than 90 percentile, between present and future simulations of 20-km model. In inset: probability density functions (PDFs) of daily T_{max} above 90 percentile in JJAS (right).**

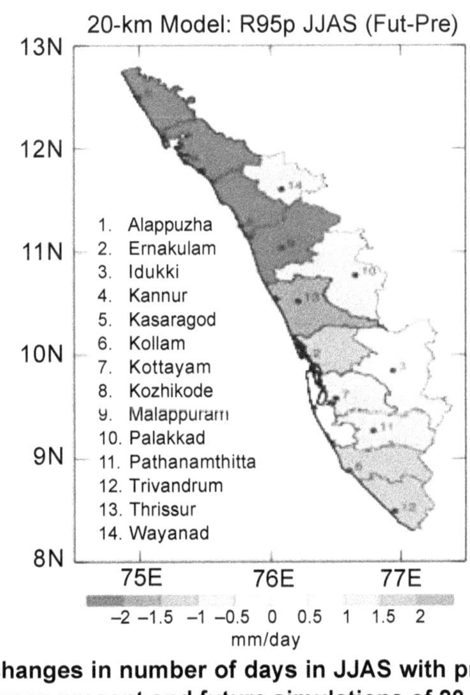

Fig. 11. **District-wise changes in number of days in JJAS with precipitation greater than 95 percentile between present and future simulations of 20-km model.**

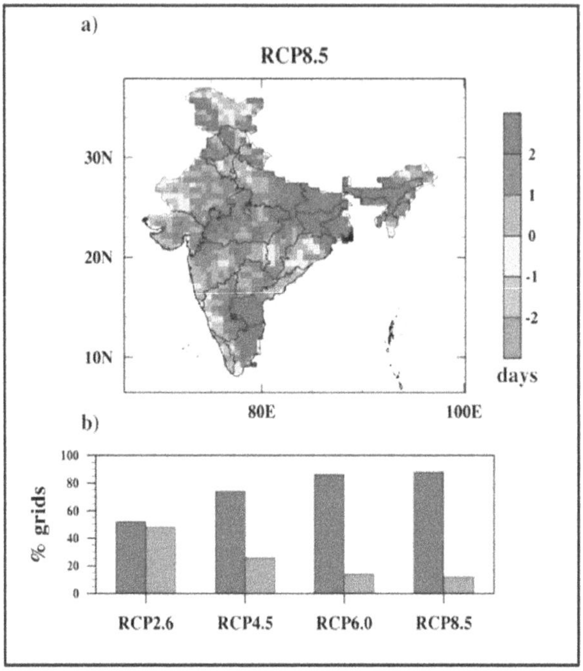

Fig. 12. (a) Projected future changes in number of days in JJAS with precipitation greater than 95th percentile (R95p). a Spatial distribution of the change for RCP8.5 scenario and (b) Percentage of grids with increased (decreased) number of days with rainfall greater (lesser) than R95p.

Probable Impacts of Climate Change

Climate scientists have confidence that the climate of the state will continue to change through the twenty-first century, in response to human induced GHG emissions, and will continue to vary in ways that can be observed in historic and paleoclimate records. However, not all aspects of the climate change or variation can be projected with equal confidence. The highest confidence is associated with projections that are consistent among climate models and with observed changes, such as those described in the above section. The magnitude and duration of future change depend mostly on the amount of GHGs emitted to the atmosphere, particularly CO_2 emitted by the burning of coal, oil, and natural gas. Much of the future change will be irreversible for centuries after substantial anthropogenic CO_2 emissions have ceased.

Probabilistic projections for the state can be summarised as:

- Warming will continue, with hotter and longer spells of hot days in summer. Temperatures will continue to increase substantially over the twenty-first century.
- Summer monsoon precipitation will decrease in the state with more dry spells.
- Extreme hot events in summer will increase and precipitation extremes during monsoon will decrease.

The juxtaposition of the state's many landscapes, mountains, valleys, plateaus, plains and coasts, affect both region's climate and its response to climate change. Whether human and natural systems are able to adapt to changes in climate will be influenced by many factors, including the complex topographic pattern of land and the associated policies and management goals. Moreover, the human population in the region will likely grow. The state's experience to date in managing and protecting its people, resources, and infrastructure is based on the historic record of climate variability during a period of relatively stable climate.

For the state, the projected climate change can possibly affect (i) terrestrial and freshwater ecosystems, (ii) coastal systems, (iii) water (as it is the limiting resource in the state and climate variability and change will continue to have substantial effects on water across the region, reduction in water supplies can lead to undesirable changes in almost all human and natural systems including agriculture, energy, industry, forestry, and recreation), (iv) human health, (v) energy (energy supplies will become less reliable as climate changes and climate change will drive increasing energy demand) and most importantly agriculture. Climate change has significant impacts on soil water, runoff, erosion, crop pests and diseases. Effects of climate change and associated variability on production of both crops and livestock could be long- lasting. Global warming will impact agriculture in many ways. For example, increase in temperature and decrease in rainfall will reduce grain yields, though the increase in CO_2 can possibly increase grain yields.

Global warming may not be the only factor that influences crop yields, for example, high air pollution can also reduce crop yields. There is a close linkage between food grain production (FGP) and monsoon rainfall anomaly where the deficit monsoon has stronger negative impact on the FGP than that of the positive impact of an excess ISMR. Thus knowledge of variation of ISMR, particularly the occurrence /nonoccurrence of the extremes continues to be extremely important.

Resilience is the capacity of a system to cope with a hazardous event responding or reorganizing in ways that maintain its essential function, identity, and structure, while also maintaining the capacity for adaptation, learning, and transformation. The history of past exposure is important to build resilience. Adaptation is the extent to which a system has adaptive capacity to absorb and rebound from any sort of impact that depends on how it copes with environmental emergencies/changes and how it improves its condition thereafter. Adaptation will alter the system or shift it to a new state. The concepts of vulnerability, resilience and adaptability are thus related in non-trivial ways. Adaptation to climate change calls for a new paradigm- one that considers a range of possible future climate conditions and associated impacts, some well outside the realm of past experience. Adaptation is thus a process that requires information on possible changes and actions from many decision makers viz. government, the private sector, non-governmental organizations, and community groups. However, current efforts are hampered by lack of reliable climate change and estimates and information about the benefits, costs, and effectiveness of various adaptation options, added by uncertainty about future climate impacts at a scale necessary for decision-making, and by a lack of coordination. Therefore, a coordinated adaptation strategy based on reliable climate change information is needed to support and coordinate decentralized efforts. As part of this strategy, the government should provide technical and scientific resources, incentives for local and state authorities to begin adaptation planning, guidance across jurisdictions, shared lessons learned, and support of scientific research to expand knowledge of impacts and adaptation.

Acknowledgements

The authors acknowledge the modeling groups, the PCMDI and WCRP for their roles in making available the WCRP CMIP5 multi-model dataset. Research grants from MoES National Monsoon Mission Phase-II project (GAP-1013) and Department of Environment and Climate Change, Government of Kerala (R-1-166) and collaborative research project with MRI-Japan and DCCC (IISc) are acknowledged. SJV thanks AcSIR for the research support.

References

Alexander, L. V., Zhang, X., Peterson, T. C., Caesar, J., Gleason, B., Klein Tank, A. M. G., Haylock, M., Collins, D., Trewin, B., Rahimzadeh, F., Tagipour, A., Ambenje, P., Rupa Kumar, K., Revadekar, J., and Griffiths, G. 2006. Global observed changes in daily climate extremes of temperature and precipitation. Journal of Geophysical Research, 111, D05109, 17 pp. doi:10.1029/2005JD006290 (2006).

Gadgil, Sulochana, and Sajani, S. 1998. Monsoon simulation in AMIP runs. Climate Dynamics, 14, 659–689

Gadgil, Sulochana and Siddhartha Gadgil, 2006. The Indian monsoon, GDP and agriculture. Economic and Political Weekly, XLI, 4887-4895

Harris, I., Jones, P.D., Osborn, T.J. and Lister, D.H. 2014. Updated high-resolution grids of monthly climatic observations the CRU TS3.10 Dataset. International Journal of Climate, 34, 623-642, dii:10.1002/joc.3711

Huffman, G.J., Adler, R.F., Bolvin, D.T., Gu, G., Nelkin, E.J., Bowman, K.P., Hong, Y., Stocker, E.F., and Wolff, D.B. 2007. The TRMM multi-satellite precipitation analysis (TMPA): Quasi-global, multiyear combined-sensor precipitation estimates at fine scales. Journal of Hydrology and Meteorology, 8: 38-55

IPCC, 2007. Climate Change 2007: The Physical Science Basis. Contribution of Working Group I to the Fourth Assessment Report of the Intergovernmental Panel on Climate Change.

Kitoh, A., Ose, T., Kurihara, K., Kusunoki, S., Sugi, M., and KAKUSHIN Team-3 Modeling Group, 2009. Projection of changes in future weather extremes using super-high-resolution global and regional atmospheric models in the KAKUSHIN Program: Results of preliminary experiments. Hydrological Research Letters, 3: 49-53

Mizuta, R., Oouchi, K., Yoshimura, H., Noda, A., Katayama, K., Yukimoto, S., Hosaka, M., Kusunoki, S., Kawai, H., and Nakagawa, M., 2006. 20-km-mesh global climate simulations using JMA-GSM model — mean climate states. Journal of Meteorological Society of Japan, 84: 165-185, doi:10.2151/jmsj.84.165

Mizuta, R., Adachi, Y., Yukimoto, S., and Kusunoki, S., 2008. Estimation of the future distribution of sea surface temperature and sea ice using the CMIP3 multimodel ensemble mean. Tech Rep. of MRI, 56, available at http://www.mrijma. go.jp/Publish/Technical/DATA/VOL 56/56.html. 28pp.

Mizuta, R., Arakawa, O., Ose, T., Kusunoki, S., Endo, H., and Kitoh, A., 2014. Classification of CMIP5 future climate responses by the tropical sea surface temperature changes. SOLA, 10: 167–71

Rajeevan, M., Bhate, J., Kale, J. D., and Lal, B., 2006. High resolution daily gridded rainfall data for Indian region: Analysis of break and active monsoon spells. Current Science, 91: 296-306

Rajendran, K., Sajani, S., Jayasankar, C. B., and Kitoh, A., 2013. How dependent is climate change projection of Indian summer monsoon rainfall and extreme events on model resolution. Current Science, 104(10): 1409–1418

Rajendran, K., Kitoh, A., Srinivasan, J., Mizuta, R., and Krishnan, R., 2012. Monsoon circulation interaction with Western Ghats orography under changing climate: Projection by a 20-km mesh AGCM. Theoretical and Applied Climatology, 110(4): 555-571, doi: 10.1007/s00704-012-0690-2

Rajendran, K., and Kitoh, A. 2008. Indian summer monsoon in future climate projection by a super high-resolution global model. Current Science, 95(11): 1360–1367

Rayner, N.A., Parker, D.E., Horton, E.B., Folland, C.K., Alexander, L.V., and Rowell, D.P., 2003. Global analyses of SST, sea ice and night marine air temperature since the late nineteenth century. Journal of Geophysical Research (Atmospheres), 108: 4407, doi: 10.1029/2002JD002670

Varghese, S.J., Sajani Surendran, and Rajendran, K., and Kitoh, A., 2019. Future projections of Indian Summer Monsoon under multiple RCPs using a high resolution global climate model multiforcing ensemble simulations, Climate Dynamics, doi:10.1007/s00382-019-05059-7 (2019).

Yatagai, A., Arakawa, O., Kamiguchi, K., Kawamoto, H., Nodzu, M.I., and Hamada, A.A., 2009. 44-year daily gridded precipitation dataset for Asia based on a dense network of rain gauges. SOLA, 5: 137–140

Impact of Climate Change on Hydrological Cycle, Ecosystem, Fisheries and Food Security, Pages: 111–123
Edited by: B. Madhusoodana Kurup, M.R. Boopendranath, M. Harikrishnan and A.V. Shibu

CHAPTER - 8

Modelling Activities for Forecasting Extreme Events

Saji Mohandas*, Raghavendra Ashrit and E.N. Rajagopal

National Centre for Medium Range Weather Forecasting,
Ministry of Earth Sciences, A-50, Sector 62, Noida-201309, India
**E-mail: saji@ncmrwf.gov.in*

Global warming has far-reaching, long-lasting and devastating effects on weather and climate, such as the increase in the severe weather events such as heat waves, droughts, and the frequency of tropical cyclones. Many previous studies have concluded that anthropogenic climate change has already increased the number and strength of some of these extreme events (Christidis *et al.*, 2011; Dai, 2006). Over the last many decades, much of the U.S. have witnessed increases in prolonged periods of excessively high temperatures, heavy downpours, and in some regions, severe floods and droughts (Doocy *et al.*, 2013; Duffy and Tebaldy, 2012; Emmanuel and Sobel, 2013). In India also the basin-scale climate has witnessed fast changes in the past couple of decades due to the human activities like deforestation, mining and other kinds of land use and land cover changes so that many of the state provinces have experienced visible changes in its rainfall pattern (Goswami *et al.*, 2006; Roxy *et al.*, 2017; Mukherjee *et al.*, 2018).

India has recently been increasingly prone to environmental catastrophes such as tsunamis and tropical cyclones - an indicator for the changing climate and weather patterns (Goswami *et al.*, 2006; Roxy *et al.*, 2017; Mukherjee *et al.*, 2018; Wang and Ding, 2006; Turner and Annamalai, 2012; Mishra *et al.*, 2012; Phillip *et al.*, 2019). The Mumbai Deluge of 26th July 2005, Leh Cloud Burst of 6th August 2010, Uttarakhand Floods of June 2013, Urban flooding of Chennai on 2nd December 2015, Heavy rains over Kerala during August 2018 are some of the extreme events studied in detail using NWP models at NCMRWF. Tropical cyclone (TC) frequency may have reduced, but the intensities of TCs over the Bay of Bengal basin have shown an increase, viz., Severe Cyclonic Storm (ESCS) Phailin (2013) and Fani (2019) that devastated coastal Odisha. In May

2015, India was struck by severe heat wave conditions that led to the deaths of at least 2,500 people across the country. Kerala, located in southern peninsular India and lying close to the Arabian Sea is one of the most ecologically sensitive provinces of the country and has been, in the recent decades, experiencing anomalous changes in the rainfall pattern with this naturally rain-surplus state seeing many years of deficient precipitation and more frequent high-intensity rainfall events. Kerala coast had not faced any tropical cyclones for decades but experienced the furry of TC Ockhi in 2017. Mishra *et al.* (2018b) showed that the mean monsoon precipitation over Kerala has significantly declined while the air temperature showed an increase from 1951-2017. The Western Ghats while extends all along the west coast of peninsular India, is more prominent over Kerala with a couple of peaks exceeding elevations of 2km above sea level.The impact of extreme weather over Kerala is immense, owing to unique terrain, high population density and the complexity of dealing with the socio-ecological-economic aspects of climate change. This paper describes a few recent extreme weather events based on NCMRWF high resolution global and regional models. The next section gives a brief description of the modeling systems and Section 3 discusses a few recent case studies of extreme weather events, followed by the concluding section.

Description of the NCMRWF Seamless Prediction System

NCMRWF mandate is to (1) continuously develop advanced numerical weather prediction systems, with increased reliability and accuracy over India and neighbouring regions, (2) development of Data Assimilation (DA) systems for both Global Forecast System (GFS) and Unified Model (UM) and (3) development of a Seamless prediction system. UK Met Office Unified Modelling (UM) based system is adopted at NCMRWF which is a seamless prediction system with the same dynamic framework across all scales from cloud-resolving to climate scale (Walters *et al.*, 2017; Kumar *et al.*, 2018). The science part of the NCMRWF Unified Modelling (NCUM) system has traceable difference between different scales, configurations or applications. This seamless prediction system is used for all types of weather-related applications and warnings, including tropical cyclones, monsoon depressions, thunderstorms and squall lines, mesoscale convective systems, gusty winds, lightning, fog and air quality. NCMRWF provides predictions of extreme weather events and guidance for informed preparedness to the various users including IMD. NCMRWF also provides support for various sectors, (viz., hydrology, wind & solar energy, etc.) with customized forecast products. Apart from the support for operational activities, the main activity is the research on the frontiers of seamless weather and climate modelling and data assimilation for improved prediction system across all timescales.

Fig. 1 depicts the seamless prediction system of NCMRWF starting from the Coupled model system at 60km resolution to Delhi Fog Model (DM) at 330m horizontal resolution. Table 1 shows the details of the configurations of the deterministic operational NCUM models at NCMRWF. The seamless prediction system covers the coupled model at 60km resolution, global deterministic (NCUM) and ensemble prediction (NEPS) systems at 12km resolution, regional deterministic and ensemble prediction systems at 4 km resolution and convective scale models at 1.5km and 330m resolutions. The operational global model resolution was upgraded to 12 km from June 2018 and before that, it was at 17 km, with 70 vertical levels describing the atmospheric layers upto 80km above the earth's surface. With the increased horizontal resolution and increased quantity of satellite data being assimilated, the performance of the global modelling system (NCUM-G) had a significant improvement as shown by the verification scores. There was no change in the resolution of regional version (NCUM-R) which continued to be at 4 km from 2015 onwards, however from June 2018, the model now covers a larger area of south-east Asian region including most of the BIMSTEC countries. The NCUM-R has the 'Regional Atmospheric and Land version 1' with a tropical science configuration with modifications of land use land cover based on NRSA data and orography based on SRTM data (Jayakumar *et al.*, 2017; Bush *et al.*, 2019). The convective scale model at 1.5km and 330 metre resolutions are only run for special occasions and seasons over relatively smaller domains over of India. The Delhi Fog Model (DM) is set up for the Delhi and the surrounding region with the inner domain of 330m resolution embedded in the outer domain of 1.5km, which in turn is nested to the global model. All configurations of the regional model have 80 vertical levels reaching upto 38.5km above the surface.

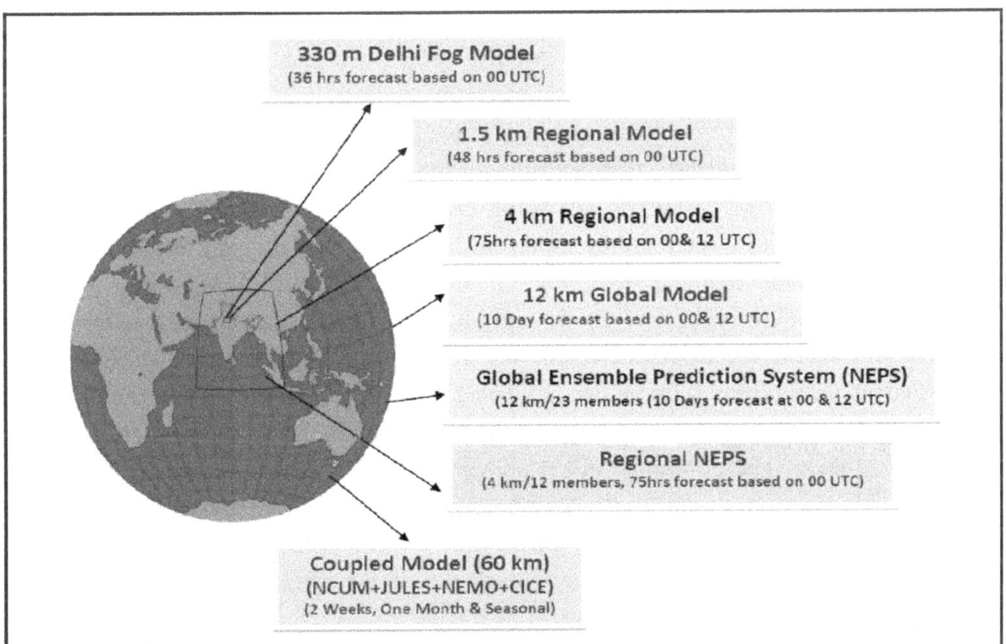

Fig. 1. The NCUM Seamless Prediction System at NCMRWF

Table 1: The Description of NCUM deterministic models at NCMRWF

Model	Domain	Grid & output	Time step
NCUM-G, (N1024)	Global, 12 km 70 levels	2048 × 1536 × 70 1.1 TB	5 min (10 days)
NCUM-R, (4KM)	BIMSTEC, 4 km 80 levels	1200 × 1200 × 80 440 GB	2 min (3 days)
NCUM-R, (1.5KM)	Local, 1.5 k 80 levels	500 × 500 × 80 39 GB	60 sec (3 days)
NCUM-R, (330m)	Delhi, 330 km 80 levels	300 × 300 × 80 440 GB	12 sec (36 days)

Case Studies with High Resolution Models

Kerala Floods (2018)

Two back-to-back prominent low-pressure areas (LPA) occurred over the head Bay of Bengal (BoB) during the episode, which show intensification on the 9th August as well as another on 15th August as seen in the NCUM analysis (Fig. 2). When viewed from a larger perspective, a couple of other tropical cyclonic circulations can be seen simultaneously over the South China Sea and the West Pacific areas, during the period. Fig. 2 shows the wind (ms^{-1}) and geopotential (m) at 850hPa level from NCUM analysis and NCUM day-7 forecast valid for 15 August 2018, which shows a very good match with the general circulations prevailing during the peak period. Westerly low-level jet (LLJ) is found to be extremely strong over the Central Arabian Sea throughout this episode and the intense westerly flow meandered the southern tip of the Indian peninsula going all along over the coastal belts of south-east Asia up to the South China Sea and beyond. This intense flow seems to be modulated by a chain of aligned tropical cyclones over the South China Sea and West Pacific apart from the LPA over the BoB and sustaining this conveyor belt like flow throughout the episode. This process is termed as "Remotely aligned intense tropical circulations (RAITC)" by Mohandas *et al.* (2020) and is described in the current section.This LLJ is feeding the general circulation into the continuous chain of vortices thereby keeping the low-level jet active and rainfall sustained for a couple of days. The combined effect appears to create a strong hemispherical flow of westerly jets right from the west Arabian Sea and moving eastwards to the South China Sea. This is the main cause of the long duration of the intense and relentless activity for a couple of days together, creating an emergency-like situation leading to the deluge.

The RAITC effect helps to bring 'an additional supply of moisture from the western North Pacific (WNP) region to the Kerala' during the episode, as the cyclones over WNP provided a strong moisture supply to the mid-tropospheric layer. A component of this moisture in the middle layers was injected into the wind flow in the south-westward direction. The relatively weakened WNP cyclonic circulations ensured the release of more moisture from its cyclonic loop and provided additional moisture supply to the LPA over BoB caused by the relative modulations of the aligned typhoons enabling tangential moisture transport to the south-westward direction. This would result in the northern ‘arc’ of LPA with enhanced moisture (due to enhanced WNP moisture supply). As the LPA intensifies and grows large in the radial extent during the period of peak activity, and ultimately gets displaced more westward, the additional supply of moisture from its northern ‘arc’ reaches Kerala through its western wing of its outer circular bands. This westward displacement of very strong LPA has another effect of deflecting the strong LLJ towards the Kerala coast, striking some of the highest peaks of the Western Ghats in Kerala.The convergence of these two sources of moisture and the orographic effect of the peaks in Kerala explains the continuous torrential rainfall activity for 2 days or more. Further in-depth analysis and detailed mechanisms are explained in Mohandas *et al.* (2020).

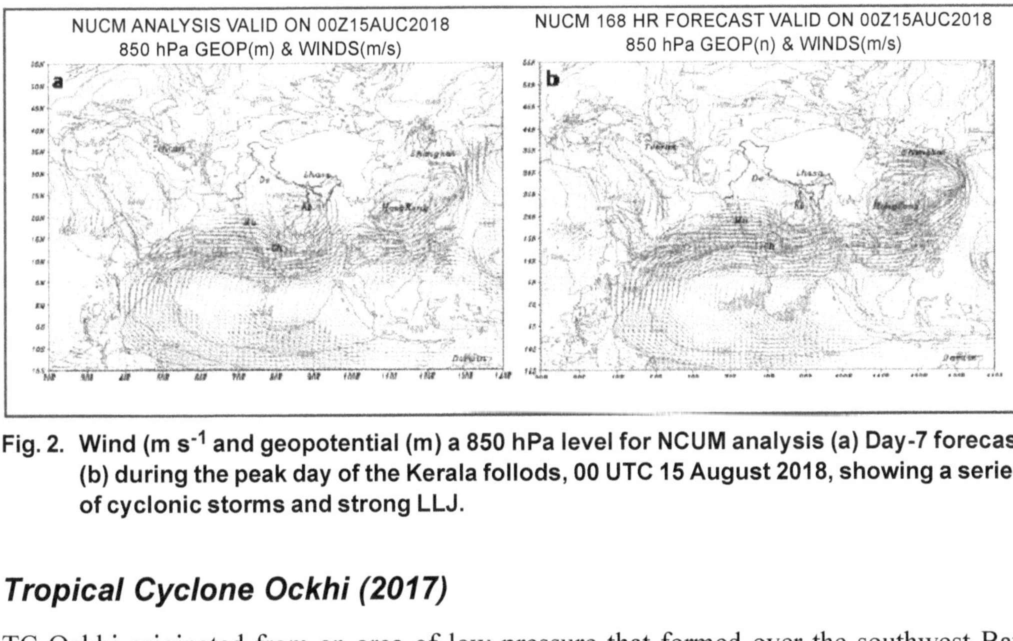

Fig. 2. Wind (m s^{-1} and geopotential (m) a 850 hPa level for NCUM analysis (a) Day-7 forecast (b) during the peak day of the Kerala follods, 00 UTC 15 August 2018, showing a series of cyclonic storms and strong LLJ.

Tropical Cyclone Ockhi (2017)

TC Ockhi originated from an area of low pressure that formed over the southwest Bay of Bengal on 28th November 2017. Initially, as a depression off the southeast coast of Sri Lanka on 29th November, it intensified fast into a deep depression and moved towards the Arabian Sea with a track very close to southwest Kerala coast causing heavy casualties mainly among the fisherman community out in the sea. Ockhi intensified into a cyclonic storm on 30th November and its further movement along the Arabian Sea (AS) and recurvature was very well predicted by NCUM global and regional models (Fig. 3). This is the only case in the recent period where none of the models could predict the cyclogenesis more than 2 days in advance before its development on 30th November. However, the day-5 forecast of NCUM from the initial condition of 00 UTC 25th November

2017 showed the low-pressure system intensifying and crossing the Kerala landmass over to AS, but no such movement was predicted in the subsequent days (Fig. 3c). Fig. 3 shows more or less fairly accurate track predictions by NCUM global and regional models. The time series of Direct Position Error (DPE) show the lowest errors for NCUM-R compared to NCUM-G and mean NEPS, though, after 42 hours of lead time, NCUM-G has the least error.

An attempt to simulate the cyclogenesis of TC Ockhi was carried out using a 17km resolution NCUM-G by taking the initial condition of 00 UTC 27November 2017. Experiments were conducted (1) by using the analyzed SST of 00 UTC 25 November 2017, (2) by reducing the SST in a box region over the warm pool region over southeast BoB, (3) by reducing the entrainment factor in convective parameterization by 25% and (4) by increasing the entrainment factor by 25%. However, none of the experiments could produce a strong cyclonic system development with a northward movement crossing over

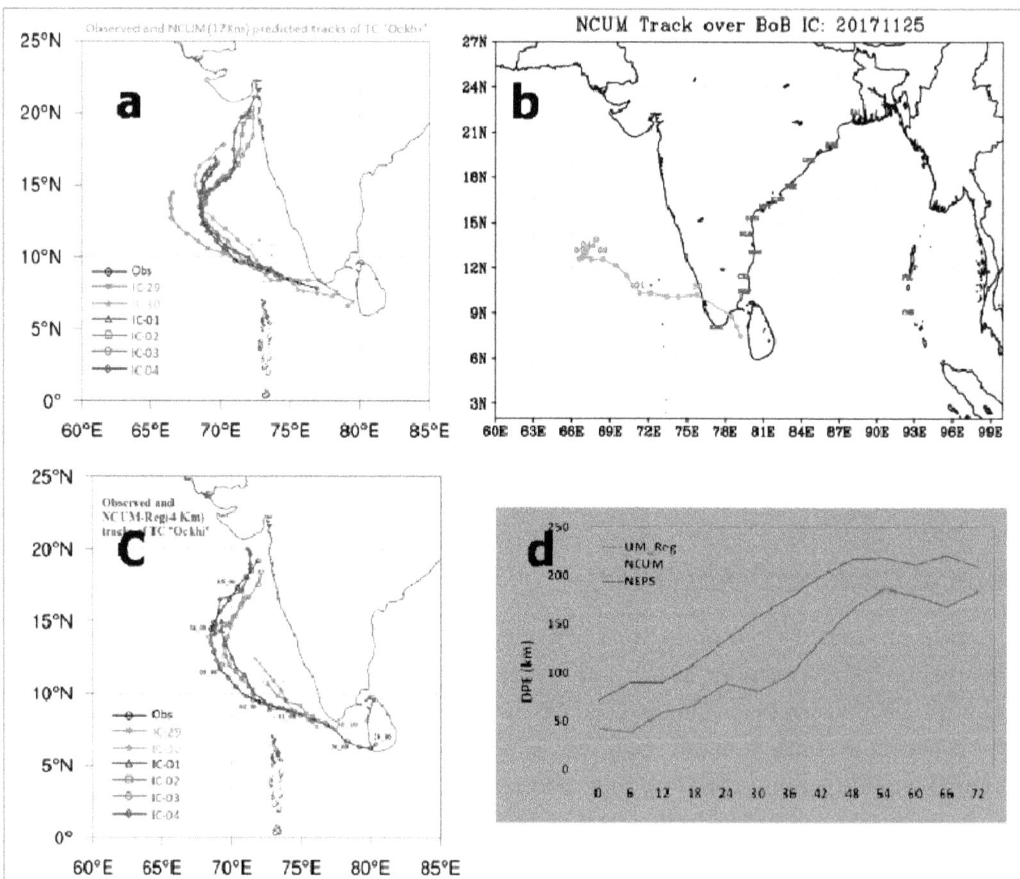

Fig. 3. The tracks of Ockhi cyclone by NCUM global (a) and NCUM regional (b), along with the prediction from 25 November 2017 (b). The DPEs of global, egional and NEPS track predictions with respect to forecast lead time (d).

to the Arabian Sea. From these experiments, it seems that a spike in the SST over south-east BoB is having an impact on the model predictions of another anomalous cyclonic activity over BoB. Its faster northward movement or intensification in quick succession of the Ockhi event might have suppressed the possibility of even any delayed developments over AS. The removal of warm SST pool over BoB had a minor positive impact on the BoB cyclonic activity by delaying its development. Increasing the entrainment rate is also found to have a minor positive impact on reducing the intensity of the BoB system.

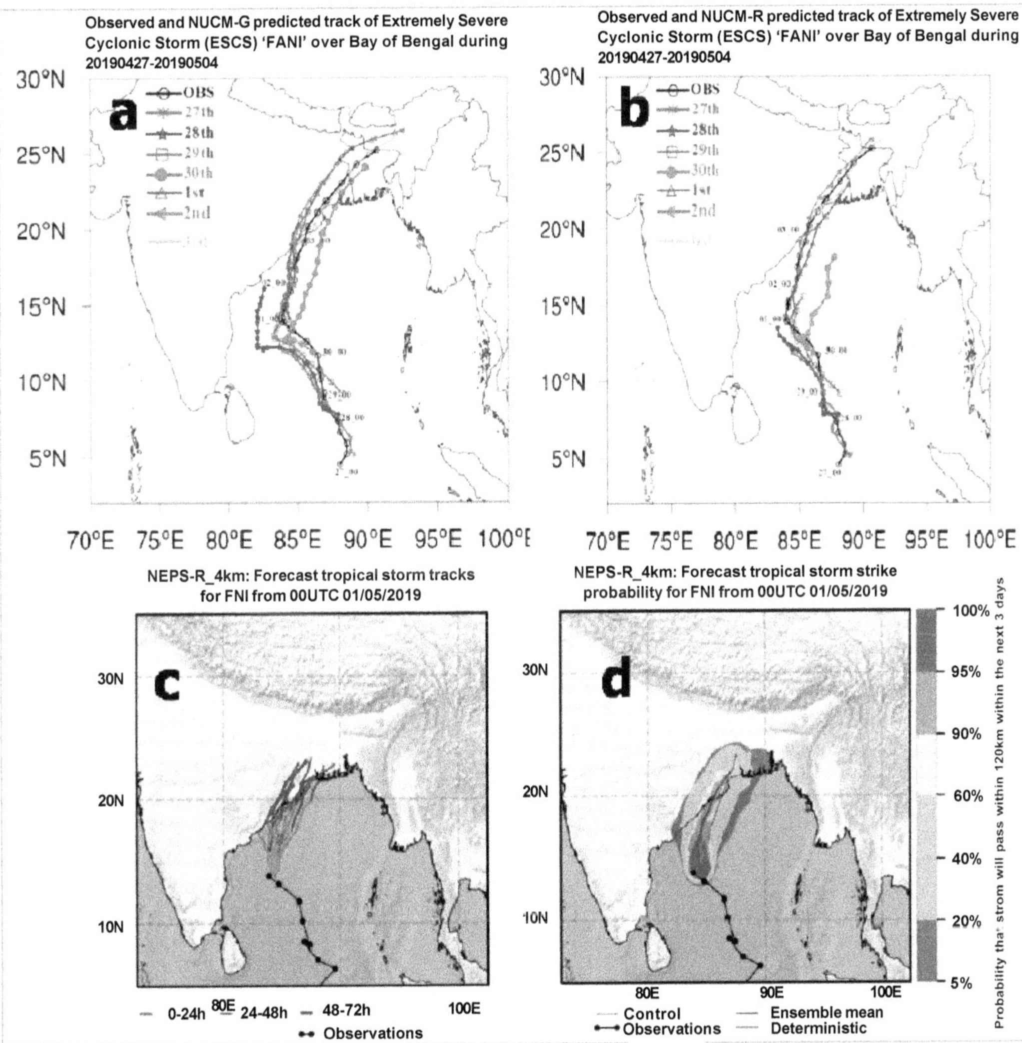

Fig. 4. **The tracks of Fani cyclone by NCUM global (a), NCUM regional (b), NEPS ensemble member tracks (c) and NEPS strike probability (d).**

Tropical Cyclone Fani (2019)

Tropical cyclone Fani was well predicted by most of the models in terms of both cyclogenesis and the recurvature sufficiently well in advance. Fig. 4 shows the predicted tracks of NCUM global and regional models for TC Fani along with the NEPS tracks and the strike probability. Fig. 5 shows the circulations at 850 hPa level for forecasts starting with 00 UTC 30 April 2019 upto 5 May 2019 along with the best centre estimates by IMD indicated with stars, which shows that the recurvature is fairly well predicted, with a slower movement of the system at the time of landfall than observed. Fig. 6 shows the time evolution of RMSE of maximum sustained wind speed along with the percentage skill for NCUM-R over NCUM-G, alongwith the time evolution of rapid intensification. Overall the RMSEs are less for NCUM-R compared to NCUM-G and though the average correlation coefficient is relatively more for NCUM-G. After 30 hours of prediction, the skill of NCUM-R is found to be higher compared with NCUM-G. Initially, the NCUM-R is found to have more rapid intensification compared to the observed estimates, whereas NCUM-G is having a better match until 30 hours, after which the intensity in NCUM-G does not peak while NCUM-R shows the peak intensification matching with the observations.

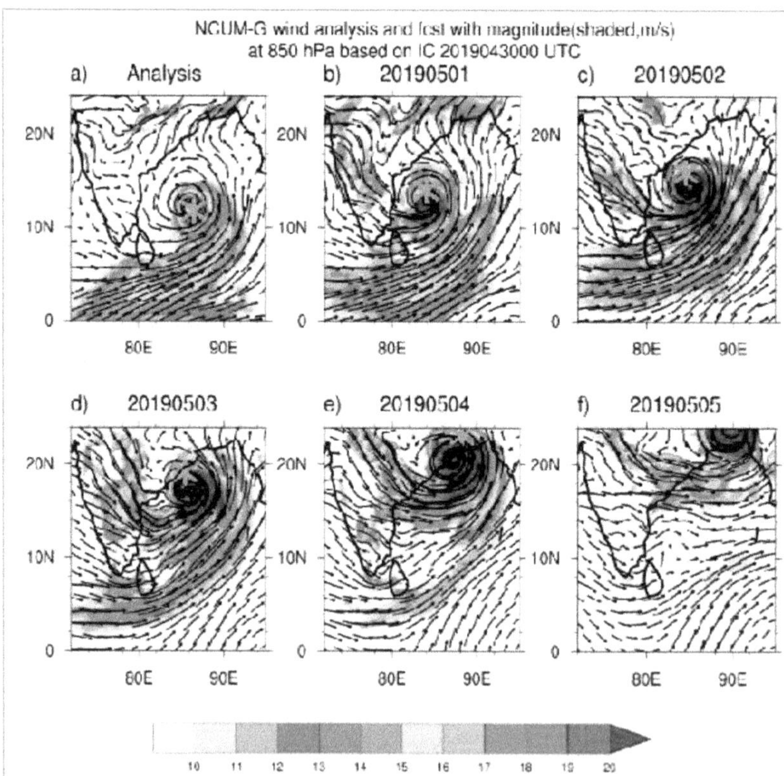

Fig. 5. The analysis positions of TC Fani with 850 hPa level circulations and starting 30 April 2019 (a) and up to 5 May 2019, along with the best centre estimatoo by IMD indicated with stars.

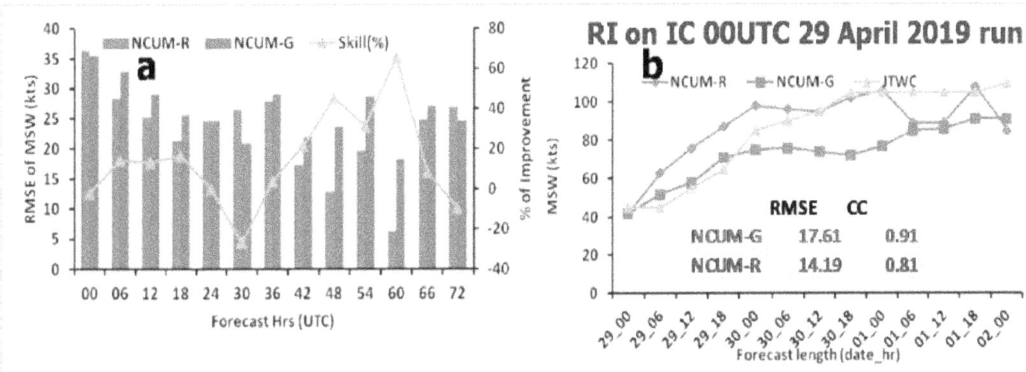

Fig. 6. the bar diagrams showin the RMSEs and percentage skill (a), and the rapid intensification (b) with forecast lead time for NCUM-R agains NCUM-G with respect to maximum sustained wind (knots) for TC Fani.

Mesoscale Convective System (2 May 2018)

This severe Mesoscale Convective System (MCS) was studied using NCUM-R at 4 km resolution to understand the impact of resolution on the extreme weather evolving out of a large scale synoptic system (WD) with induced mesoscale signatures embedded in it. The MCS that developed over northwest India on 2nd May 2018, moved southeastwards over the north of Delhi. This system affected the western parts of Himalayas and the moist downdraft and the mountain gravity waves generated caused a second lobe to develop and move southeastwards over the sub-Himalayan Gangetic belts south of Delhi, parallel to the first lobe. It reached Agra and caused heavy casualties due to the associated thunderstorm and wind squalls. The movement of the MCS was marked in the maximum wind gust plots over 24-hour period as a pattern looking like splayed fingers (Fig. 7a). The MCS was found to be led by moving gust fronts oriented perpendicular to the direction of movement, generating strong dust raising winds which appeared like dust storms. These are very well simulated by the NCUM-R compared to the NCUM-G, which in turn shows a weaker system with most of the rainfall just over the Himalayan slopes. The simulation by NCUM-R also helped understand the mechanism and remote teleconnection of MCS with the dust storm over Rajasthan which occurred during the same time. Wet deposition due to the moisture incursion in the leading edge of the cold pool outflow in the north by the MCS movement and the high relative humidity brought in from the Arabian Sea in the south creating a narrow channel of accumulated dust over Rajasthan causing severe dust storm. This study shows the importance of convective scale models or very high resolution models in realistically simulating the localized extreme events caused by the large scale organized systems.

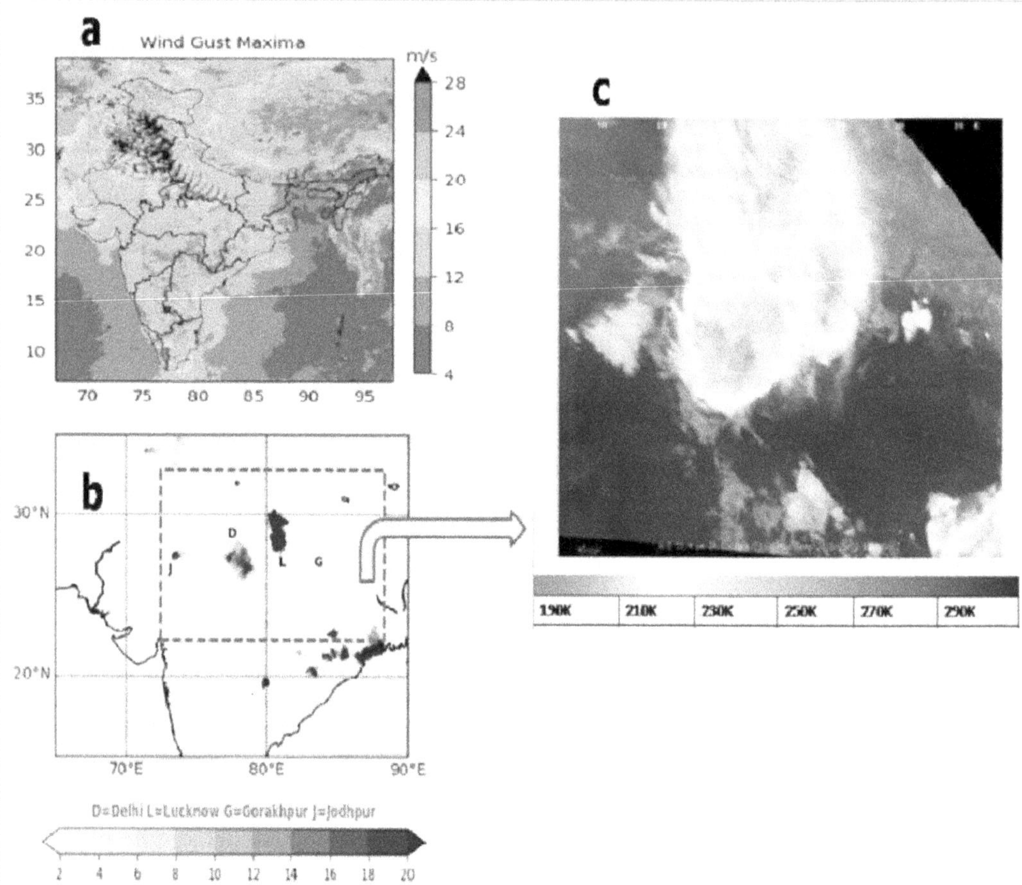

Fig. 7. **Precipitation and satellite images showing the two lobes of the MCS event propagating south-eastwards over North India using maximum wind gust in a day (a), the 24 hourly accumulated precipitation (b) and the Satellite infrared images (c), on 2 May 2018.**

Summary and Ongoing Activities

The seamless approach adopted by the NCUM modeling system with various resolutions of global, regional and convective scale models at NCMRWF has rendered an effective system for extreme weather prediction over the Indian region. With the upgradation of the NCUM modelling system at a very high resolution of 12km, the performance of models has shown marked improvement. The combination of global, regional and convective scale models along with the ensemble prediction systems can be employed for the prediction of all types of extreme weather events. The current study discussed the examples of a few recent extreme weather events and the performance of the global (12 km) and regional (4 km) operational at NCMRWF.

The case studies discussed not only illustrate some of the applications of NCUM seamless system in forecasting extreme weather events but also attempted to analyze and construct the sequence of events leading to the events. The diagnostic study of the Kerala floods case, fairly well simulated by NCUM using the model analysis, has produced new insights into the mechanism and remote teleconnection of the extreme rainfall over a localized region with the very active south-east Asian Oceanic and Pacific warm pool region. The NCUM was able to predict the genesis and track of most of the recent cases of tropical cyclones with a reasonable level of accuracy, though TC Ockhi was a rare exception for genesis prediction in terms of signals for sufficient warning. The study of the north Indian MCS event using NCUM-R has also brought a better understanding of the relationship with far-flung dust storms over Rajasthan, which cannot be very well resolved in the global model. Special products are being generated for severe weather and aviation warning from NCUM-R (4 km) model operationally. Very high resolutions of cloud-resolving models (1.5 km and 330 m) are being run at NCMRWF for some seasons and for special applications, which include Sri Amarnath Ji Yatra domain over the western Himalayas and Delhi Fog Model (DM).

NCMRWF is also participating in the collaborative research and development activities for Regional Model Evaluation and Development (RMED) with other UM partners and the next-generation exa-scale modelling system development. Research is going on at the frontiers of convective scale models to improve the skill in the simulation of many processes resolved at sub-kilometer resolutions. The convective scale configurations of NCUM are also successful in demonstrating the importance of resolution on the more realistic simulation of meso-scale convective systems and orography-induced rainfall forced by the large scale synoptic systems.

References

Bush, M., Allen, T., Bain, C., Boutle, I., Edwards, J., Finnenkoetter, A., Franklin, C., Hanley, K., Lean, H., Lock, A., Manners, J., Mittermaier, M., Morcrette, C., North, R., Petch, J., Short, C., Vosper, S., Walters, D., Webster, S., Weeks, M., Wilkinson, J., Wood, N., Zerroukat, M., 2019. The first Met Office Unified Model/JULES Regional Atmosphere and Land configuration, RAL1. Geosci. Model Dev. Discuss. doi.org/10.5194/gmd-2019-130 (under review).

Christidis, N., Stott, P.A. and Brown, S.J., 2011. The role of human activity in the recent warming of extremely warm daytime temperatures. Journal of Climate 24: 1922-1930, doi:10.1175/ 2011JCLI4150.1.

Dai, A., 2006. Recent climatology, variability, and trends in global surface humidity. Journal of Climate 19: 3589-3606. doi:10.1175/JCLI3816.1.

Doocy, S., A. Daniels, S. Murray, and T. D. Kirsch, 2013: The human impact of floods: A historical review of events 1980-2009 and systematic literature review. PLOS Currents Disasters, doi:10.1371/currents.dis.f4deb457904936b07c09daa98ee8171a.

Duffy, P.B., and Tebaldi, C., 2012. Increasing prevalence of extreme summer temperatures in the US. Climatic Change 111: 487-495. doi:10.1007/s10584-012-0396-6.

Emanuel, K., and Sobel, A., 2013. Response of tropical sea surface temperature, precipitation, and tropical cyclone-related variables to changes in global and local forcing. Journal of Advances in Modeling Earth Systems 5: 447-458. doi:10.1002/jame.20032.

Goswami, B.N., Venugopal, V., Sengupta, D., Madhusoodanan, M.S. and Xavier, P.K., 2006. Increasing trend of extreme rain events over India in a warming environment. Science 314(5804): 1442-1445. doi:10.1126/science.1132027.

Jayakumar, A., Sethunadh, J., Rakhi, R., Arulalan, T., Mohandas, S., Iyengar, G.R., and Rajagopal, E.N., 2017. Behavior of predicted convective clouds and precipitation in the high-resolution Unified Model over the Indian summer monsoon region. Earth and Space Science 4: 303–313. doi.org/10.1002/2016EA000242.

Kumar, S., Jayakumar, A., Bushair, M.T., Buddhi Prakash, J., George, G., Lodh, A., Indira Rani, S., Mohandas, S., George, J.P. and Rajagopal, E.N., 2018. Implementation of high resolution NCUM Analysis-Forecast system in Mihir HPCS. NCMRWF Technical Report. NMRF/TR/01/2018, Aug. 2018, 17 p. https://www.ncmrwf.gov.in/NCUM-Report-Aug2018_final.pdf.

Mishra, V., Shah, R., Azhar, S., Shah, H., Modi, P., Kumar, R., 2018b. Reconstruction of droughts in India using multiple land surface models (1951-2015). Hydrol. Earth Syst. Sci., 22: 2269-2284. doi: 10.5194/hess-22-2269-2018.

Mishra, V., Smoliak, B.V., Lottenmaier, D.P., Wallace, J.M., 2012.A prominent pattern to year-to-year variability in Indian summer monsoon rainfall.Proc. Natl. Acad.Sci. USA, 109, 7213-7217, doi: 10.1073/pnas.1119150109.

Mohandas, S., Francis, T., Singh, V., Jayakumar, A., George, J.P., Sandeep, A., Xavier P. and Rajagopal, E.N., 2020. NWP perspective of the extreme precipitation and flood event in Kerala (India) during August 2018, Dynamics of Atmospheres and Oceans (under review).

Mukherjee, S., Aadhar, S., Stone, D. and Mishra, V., 2018. Increase in extreme precipitation events under anthropogenic warming in India. Weather and Climate Extremes 20: 45-53. doi:10.1016/j.wace.2018.03.005.

Philip, S., Sparrow, S., Kew, S.F., van der Wiel, K., Wanders, N., Singh, R., Hassan, A., Mohammed, K., Javid, H., Haustein, K., Otto, F.E.L., Hirpa, F., Rimi, R.H., Islam, A.K.M.S., Wallom, D.C.H. and van Oldenborgh, G.J., 2019. Attributing the 2017 Bangladesh floods from meteorological and hydrological perspectives. Hydrol. Earth Syst. Sci., 23: 1409-1429. doi.org/10.5194/hess-23-1409-2019.

Roxy, M.K., Ghosh, S., Pathak, A., Athulya, R., Mujumdar, M., Murtugudde, R. and Rajeevan, M., 2017. A threefold rise in widespread extreme rain events over central India. Nature Communications 8: Article number: 708. doi:10.1038/s41467-017-00744-9.

Turner, A.G., and Annamalai, H., 2012. Climate change and the South Asian monsoon. Nat. Clim. Change, 2: 587-595. doi:10.1038/nclimate1495.

Walters, D., Boutle, I., Brooks, M., Melvin, T., Stratton, R., Vosper, S., Wells, H., Williams, K., Wood, N., Allen, T., Bushell, A., Copsey, D., Earnshaw, P., Edwards, J., Gross, M., Hardiman, S., Harris, C., Heming, J., Klingaman, N., Levine, R., Manners, J., Martin, G., Milton, S., Mittermaier, M., Morcrette, C., Riddick, T., Roberts, M., Sanchez, C., Selwood, P., Stirling, A., Smith, C., Suri, D., Tennant, W., Vidale, P.L.,Wilkinson, J., Willett, M., Woolnough, S. and Xavier, P., 2017. The Met Office Unified Model Global Atmosphere 6.0/6.1 and JULES Global Land 6.0/6.1 configurations. Geosci. Model Dev. 10: 1487- 1520. doi.org/10.5194/gmd-10-1487-2017.

Wang, B. and Ding, Q., 2006. Changes in global monsoon precipitation over the past 56 years. Geophys. Res. Lett. 33: L06711. doi: 10.1029/2005GL025347.

III

Aquaculture

Impact of Climate Change on Hydrological Cycle, Ecosystem, Fisheries and Food Security, Pages: 127–136
Edited by: B. Madhusoodana Kurup, M.R. Boopendranath, M. Harikrishnan and A.V. Shibu

CHAPTER - 9

Reducing the Carbon Footprint from Pond Aquaculture in a Changing World

M.C.J. Verdegem[1*] and K.A. Kabir[1,2]

[1]*Wageningen Institute of Animal Sciences, Aquaculture and Fisheries Group, Wageningen University and Research, Wageningen, The Netherlands*
[2]*Sustainable Aquaculture Program, WorldFish Bangladesh*
[*]*E-mail marc.verdegem@wur.nl*

Fish play an important role in meeting the current and future food demands within the global food system. Fish, from aquaculture and fisheries combined, provide at least 15% of the animal protein demand of 4.3 billion people (Lynch, 2017) and contributes 3.11% of global GDP (FAO, 2018). Fish demand increases with population growth. Supply of fish from capture fishery has been static for the last decades, meaning that current increases in fish supplies are coming from aquaculture. Consumption of fish from aquaculture as part of global supply grew from 6% to 53% in the last 50 years. Aquaculture is the fastest growing food sector and in 2016 had a total annual production of 80 million tonnes, valuing ~180 billion USD (FAO, 2018).

This rapid growth of aquaculture significantly improved food and nutrition security, economic growth and employment creation. However, this growth is also responsible for several environmental impacts. Most of the environmental impacts are local or regional and are very much context specific. Environmental impacts include chemical pollution, depletion or enrichment of nutrients, habitat loss, disease, introduction of exotic species, inefficient energy and biotic resource use, etc. (Enriksson & Elletier, 2013). In contrast to most environmental impacts from aquaculture, GHG emission is a global phenomenon, common across all aquaculture production systems. Carbon dioxide, methane and nitrous oxide are the three major gas referred as GHG. The carbon footprint is one of the popular indicators used to evaluate the total GHG released during the complete life cycle of certain products and services, from the gathering of raw materials to manufacturing, assembly, and transportation, as well as use, disposal and waste treatment (Williams, 2009).

GHG emission continues to rise and is a major threat for sustainability. According to IPCC (2014), emission from agriculture is 4th largest contributor to GHG emissions behind electricity and heat production, industry and transportation, responsible of 14% of total emission. The contribution of aquaculture to the total emission from agriculture is still little but is a major concern as aquaculture is one of the fastest growing agricultural based activities. Besides carbon dioxide, methane emission from water bodies is naturally occurring, therefore expansion of aquaculture means expansion of methane emission. Ongoing intensification concurring with increased reliance on formulated feed input, will further increase emissions of carbon dioxide and nitrous oxide as well. According to Paris agreement (COP 21) all signed parties agreed to reduce GHG emissions by 2030. That means the production needs to be more efficient so that we can achieve the expected doubling of yield without increasing the carbon footprint which is very challenging.

The aim of this review is to present a general overview on how GHG emission is measured, what are the main types of GHG emission from aquaculture, and where to reduce GHG emission sufficiently to create space to double aquaculture production by 2050. The second part is to theoretically analyse the potential of the nutritious pond concept in addressing the emission of the three major GHG gas.

How is GHG emission measured?

The most popular way of measuring GHG emission is by life cycle assessment (LCA) of a particular commodity where emissions throughout the value chain are considered. An example of LCA on a salmon farm (Pelletier *et al.*, 2009) indicated that 93% of GHG emission is from feed and feed related activities such as feed ingredients from crop, fisheries (fishmeal), livestock (meat and bone meal), feed manufacturing and transport. Here, more than 85% of the emission is coming from feed ingredients, with fishmeal alone responsible for 52%. However, in this analysis, post-harvest losses were not included. Another LCA on an ecological shrimp farm in China (Chang *et al.*, 2017) that included post-harvest and waste treatment had electricity (29%), feed (24%), indirect raw materials which are mainly used for pond preparation and water quality maintenance (22%), waste treatment (11%) and transport and refrigerant (11%) as main contributors to GHG emission. If we compare the above two assessment outcomes, we can conclude that framing the boundaries of the LCA, and species and farming system choice result in large differences in LCA output.

There is also variation within the same species and farming system. For example, yield from semi-intensive black tiger shrimp farm can range from 3000 to 7000kg/ha/year (Ahmed, 2013; Faruque *et al.*, 2017). With changing the yield the intensity of input use changes causing differences in GHG emission related to amount and types of input. On the other hand some emissions from waterbodies are natural and do not change much with the culture intensity. However, this context-specific changes are difficult to capture in LCA. Again for feed use, there are feeds with 25-40% protein content, with large variation in ingredients used that provide protein. If fish meal is the major part of the protein, the emission will be higher compared to plant protein (Henriksson *et al.*, 2014). If the ingredients are locally sourced or imported, also affects the total carbon footprint of the final product. Shrimp, which is mainly exported, has a large footprint from the processing (freezing) and shipment while carps which are often locally consumed do have a much smaller portion of emission during post-harvest handling.

One of the major alarming GHG component from aquaculture is N_2O (Nitrous oxide) which can vary a lot based on the amount of protein and their utilization. An efficient nitrogen cycle will lead to less emission and an increase in temperature will reduce N_2O emission from aquaculture (Paudel *et al.*, 2015). So, the climate also influences emission. CH_4 (methane) emission is naturally occurring from waterbodies. Conversion of wetlands to aquaculture ponds often requires deepening of the pond. CH_4 emission is higher in shallow water bodies compared to deeper aquaculture ponds (Peacock, Audet, Jordan, Smeds, & Wallin, 2019). The trade-off in emission between a natural waterbody and an aquaculture operation should also be considered in LCA. And the values of emission from different inputs and management practices calculated in LCA are also based on some assumption which can largely vary (John *et al.*, 2018). Due to these limitations, some scientists prefer estimation of emission directly from the farming system (Vasanth *et al.*, 2016).

Estimation of direct GHG emission cannot capture the total emission from the complete product development phase but presents a more accurate value coming directly from the farming system. Such a direct measurement on GHG emission from whiteleg shrimp (*Litopenaeus vannamei*) ponds in India reports 4.56 kg CO_2e/kg shrimp yield (Vasanth *et al.*, 2016) which is 35% less than the reported (6.94 kg CO_2e/kg) emission for the same species using LCA (Chang *et al.*, 2017). There are few studies measuring directly GHG emission as well as using LCA to do a meta-analysis on comparison of the methods. We can conclude that direct emission from aquaculture ponds is a fraction of the emission from the product value chain. Moreover, direct emission measured from an open water aquaculture system, such as cage culture, is not a real representation as well.

How can we reduce GHG emission?

In general, five major segments of GHG emission are electricity, feed, indirect material cost, waste treatment and post-harvest processing, including transport. Ecological intensification by growing fish/shrimp within the carrying capacity of the pond, or using renewable energy such as solar power for farm operation, and using more efficient equipment, can minimize reduce energy use. Also for many species, aeration or water pumping are not required. Hence, reducing GHG emissions through by adapting farming systems design can be explored.

Emission from fish transportation and refrigeration can also be minimized by diversified siting of processing and distribution plants nearer to demand areas and improving efficiency. Of course relocation of production for all species is not feasible. Eleven percent of globally traded fish are exported to Africa (Source: FAO online) while this continent has plenty of water resources to grow fish. Instead of inter-continental trade, regional trade can minimize GHG emission.

However, the most challenging part is addressing the emission from feed, indirect raw materials and waste treatment. With the current growth rate and the projected doubling requirement, feed use will inevitably increase near to double, leading to increase in the waste which is mainly driven by the feed and also with increased feed and waste, to keep the water quality balanced more medicine, chemicals or equipment will be required (Fenchel *et al.*, 2012). Emission from feed depends mainly on two factors – the types of ingredient used and utilization efficiency of the feed. For example, fishmeal has more GHG emission compared to rice bran. But, fishmeal provides protein in the diet while rice bran is mainly for energy and comes as a co-product. If we add protein from plant based sources such as soybean meal, the emission is comparable. However, this change in ingredient composition can influence the utilization efficiency of the feed and might raise the overall emission. Increased inclusion of plant based ingredients in fish feed is a growing trend . Often inclusion of plant based ingredients lead to increased waste production (Tran-Tu *et al.*, 2018), meaning that more emission will come from the waste management section. The core moto for reducing emission is "Reduce, Reuse, and Recycle". So grow more with less by increasing input use efficiency. In addition, reuse and recycle resources within the production system or in the linked production system (Ahmed and Glaser, 2016). One example can be growing crops with waste water or using the bottom sludge as fertilizer for crops.

What is nutritious pond concept (NPC)?

The nutritious pond concept is providing nutrient inputs to ponds, aiming for metabolic waste/faeces composition resulting from feeding that is easily degradable by microbiota and hence stimulates the natural food web of the pond. The enhancement effect on the food web includes algae, heterotrophic and chemoautotrophic microbiota, zooplankton and benthos in the water column and sediment. A concept of dual purpose feed was developed, which feeds the fish directly and indirectly, as feed and natural food, respectively (Fig. 1). It follows the core moto of GHG emission reduction - "Reduce, Reuse, and Recycle". Because of enhanced contribution of natural food we can grow more with less. The uneaten feed and feed driven waste is largely reused/recycled within the pond for enhancing the natural food. It aims to reach the maximum production potential within the carrying capacity of the pond and might even be able to increase the carrying capacity by efficiently functioning the biogeochemical cycle in the pond.

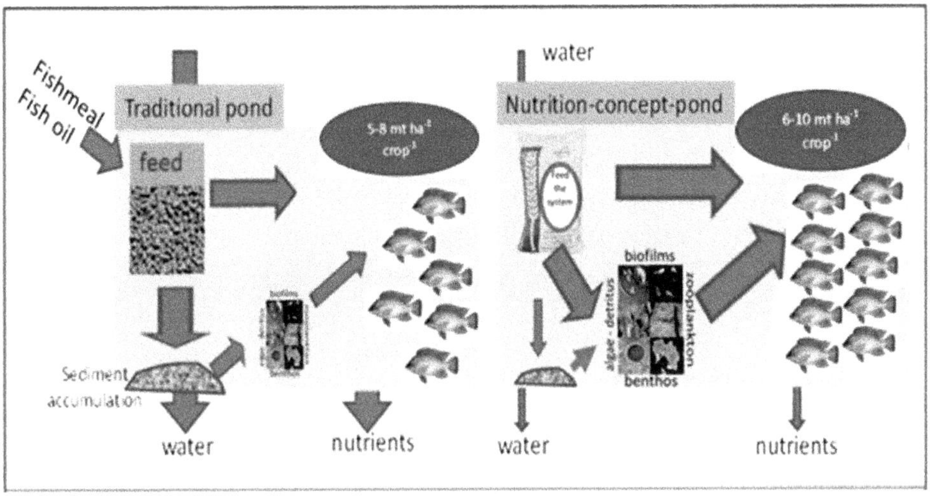

Fig. 1. Conceptual drawing visualizing the difference between a "traditional pond" to a "nutritious pond".

How nutritious pond concept helps reducing carbon footprint?

N_2O emission

N_2O emission takes place from incomplete nitrification and denitrification. Total emission of N_2O depends on nitrogen (N) input mainly via feed and fertilizer and output in the fish biomass. The more N is included the higher N_2O emission(Hu, Lee, Chandran, Kim, & Khanal, 2012). Therefore, reduction in the feed N input will reduce N_2O emission. A NPC-feed contains less protein compared to conventional diets (Kabir *et al.*, 2019a) and thus less N_2O emission assuming that the production systems are the same, except the feed.

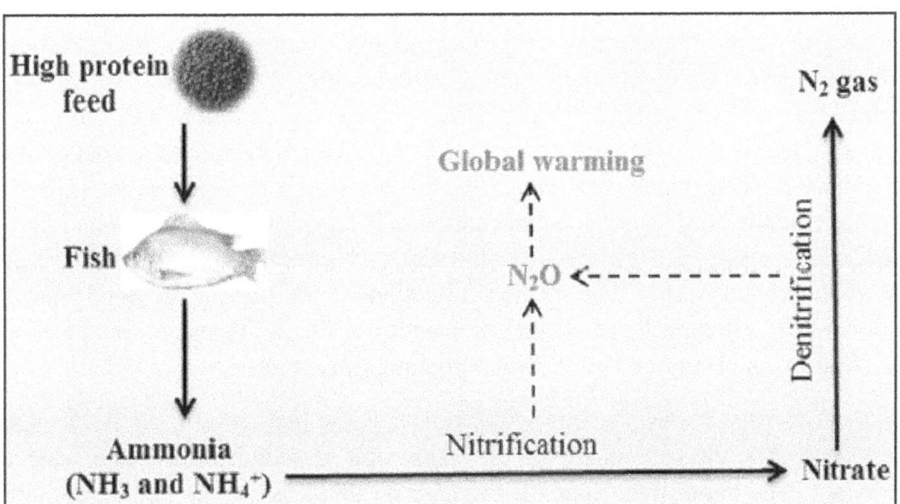

Fig. 2. Conceptual diagram on N_2O emission from feed N (Source: Hu *et al.*, 2012)

In addition to total N input, the efficiency of N use in the system also partly regulates the N_2O emission from aquaculture. In the conventional diets, 25% of feed N is converted to fish biomass and subsequently rest passes through the aquatic nitrogen cycle and 20% becomes N_2O and N_2 (Hu *et al.*, 2012) where the part of N_2O is very little. In semi-intensive tilapia ponds using NPC-feed the N retention in fish biomass was above 70% (Kabir *et al.*, 2019b). Therefore, application of NPC-feed in pond aquaculture can reduce 2.85 folds of direct N_2O emission from the farming system.

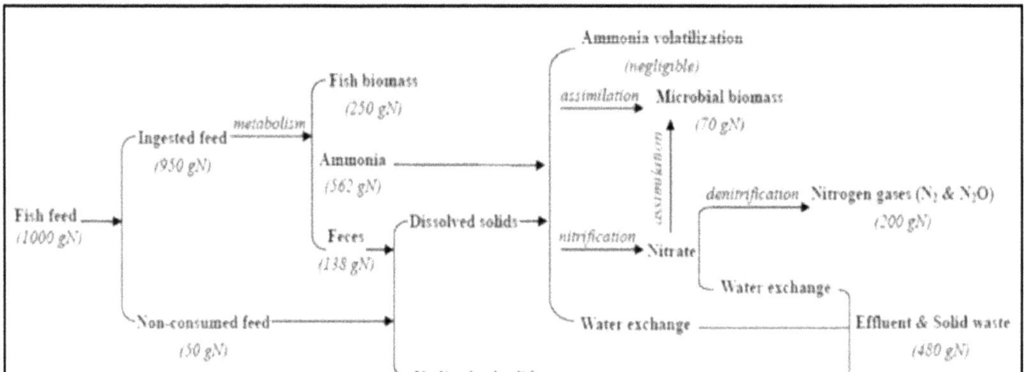

Fig. 3. Principal transformation process of nitrogen in aquaculture system (Assuming a nitrogen input of 1000gN, and the nitrogen transformation in each pathway is based on the statistical analysis of literature published; Source: Hu *et al.*, 2012)

CH_4 emission

Aquaculture ponds are potentially large sources of atmospheric methane (CH_4) that can exacerbate climate change. A thorough understanding of various CH_4 biogeochemical processes occurring in the ponds is essential for the prediction and management of CH_4 emissions arising from aquaculture. Studies on shrimp ponds in Min River Estuary and Jiulong River Estuary in China indicated that CH_4 concentration is higher in bottom water compared to middle and surface water and these concentration gradually increases with the progress of the culture cycle (Yang *et al.*, 2019). CH_4 fluxes across the sediment water interface and CH_4 oxidation rate in the overlying water significantly varied between the ponds from two different delta indicating that the soil properties and overlying water influences CH_4 dynamics in aquaculture ponds. Similarly the enhancement of the food web in pond aquaculture with NPC-feed also is partly regulated by the soil and water properties. However, the methanotrophic bacteria, enhanced by the application of NPC-feed in the pond aquaculture breaks down CH_4 via type I and type II methanotrophs (Fig. 4) and produces microbial biomass (Hanson and Hanson, 1996), while reducing CH_4 emission.

However, to what extent the NPC-feed enhances the methanotrophic bacteria and as a consequence at what percentage of CH_4 emission is reduced from the aquaculture ponds has not been quantified. Specific research on this agenda can generate interesting information on carbon footprint management from aquaculture as by quantity CH_4 is the 2nd largest GHG emission from aquaculture.

CO_2 emission

A cumulative respiratory CO_2 emission from all living biota from aquaculture is responsible for direct emission of CO_2 from the farming system. As many of the biota ends up into fish biomass and productivity increases per unit of feed input it is assumed that CO_2 emission per unit of fish/shrimp produced are less compared to the conventional approach. However, there is no clear pathway to justify this claim. Direct GHG emission measurement from the farms under NPC and conventional approach can provide insight on it.

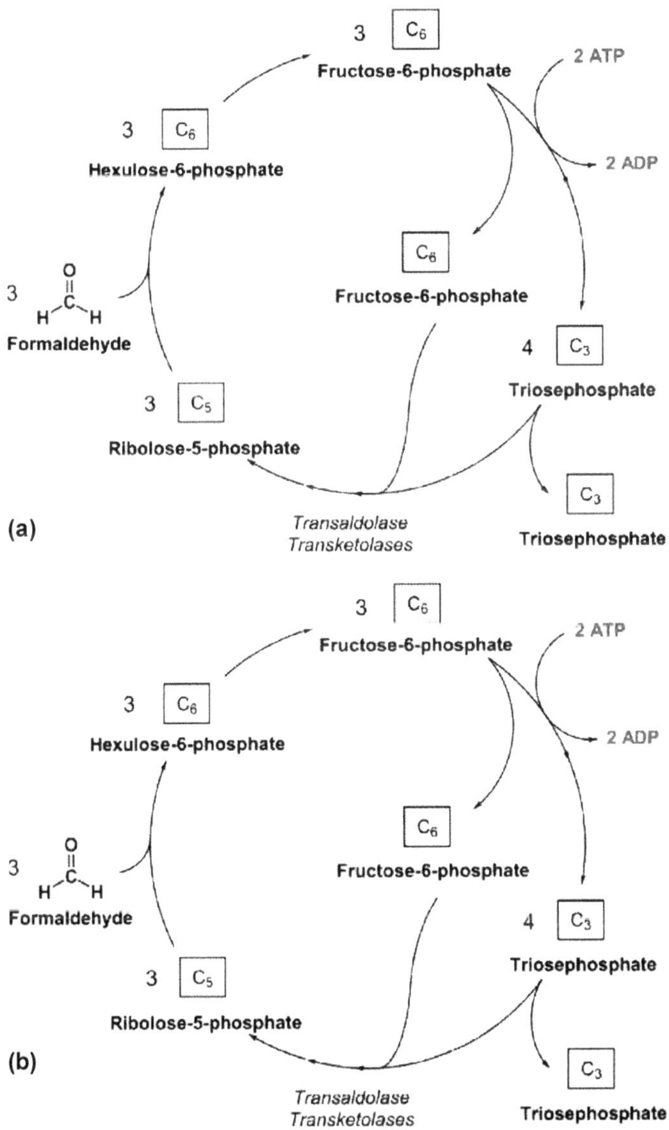

Fig. 4. Methane Oxidation by Methanotrophs: (a) is Ribulose monophosphate (RuMP) pathway in type I methanotrophs and (b) is Serine pathway in type II methanotrophs
(Source: Hanson and Hanson, 1996)

Waste management

Application of NPC allows reuse and recycle of nutrients in a productive way within the system. Kabir et al (2019b) found that the amount of nutrients present at the end of the production cycle is equal or less than the amount present at the start. As major portion of waste is recycled and reused within the pond during the culture period, the likelihood of GHG emission from waste treatment is less under the nutritious pond concept compared to a conventional approach. However, this has not been quantified yet.

Use of aqua-chemicals

Application of NPC keeps good balance between nutrient application, uptake and waste generation. As a result the pond environment remains healthy. Comparisons on aqua-chemicals used in conventional tilapia or shrimp farms are much less compared to the NPC demonstration for tilapia by Kabir *et al.* (2019b). However, the difference in GHG emission due to lesser application of aqua-chemicals has not yet been quantified.

Conclusion

Application of the nutritious pond concept has great potential to reduce direct GHG emission from (farming systems of) pond aquaculture. In addition, in feed except for the crude protein ingredients, others are often co-products from agriculture and cause less GHG emission. In the NPC-feed the inclusion of protein ingredients are significantly less and co-products are more. Assessment of product lifecycle of NPC is therefore also likely to reduce GHG emission. However, all the arguments about the potential of NPC in reducing carbon footprint is presented based on theoretical framework and very little evidence is yet generated to support the claims. Therefore, research on quantification of GHG emission from NPC by both direct emission estimation and LCA is required.

References

Ahmed, N., 2013. Linking prawn and shrimp farming towards a green economy in Bangladesh: Confronting climate change. Ocean and Coastal Management, 75: 33–42. doi:10.1016/j.ocecoaman.2013.01.002

Ahmed, N. and Glaser, M., 2016. Can "Integrated Multi-Trophic Aquaculture (IMTA)" adapt to climate change in coastal Bangladesh? Ocean & Coastal Management, 132: 120–131. doi:10.1016/j.ocecoaman.2016.08.017

Chang, C., Chang, K., Lin, W. and Wu, M., 2017. Carbon footprint analysis in the aquaculture industry/ : Assessment of an ecological shrimp farm. Journal of Cleaner Production, 168: 1101–1107. doi:10.1016/j.jclepro.2017.09.109

COP 21. 2015. The key points of the Paris Agreement

Enriksson, P.A.J.G.H. and Elletier, N.A.L.P., 2013. Life Cycle Assessments and their Applications to Aquaculture Production Systems. doi:10.1007/978-1-4614-5797-8

FAO, 2018. The State of World Fisheries and Aquaculture 2018 - Meeting the sustainable development goals. FAO, Rome.

FAO Online. GLOBEFISH - Information and Analysis on World Fish Trade. http://www.fao.org/in-action/globefish/fishery-information/resource-detail/en/c/338418/

Faruque, G., Sarwer, R. H., Karim, M., Phillips, M., Collis, W. J., Belton, B. and Kassam, L., 2017. The evolution of aquatic agricultural systems in Southwest Bangladesh in response to salinity and other drivers of change. International Journal of Agricultural Sustainability, 15(2): 185–207. doi:10.1080/14735903.2016.1193424

Fenchel, T., Blackburn, H., King, G.M. and Blackburn, T.H., 2012. Bacterial biogeochemistry: the ecophysiology of mineral cycling. Academic Press.

Hanson, R.S. and Hanson, T.E., 1996. Methanotrophic bacteria. Microbiological Reviews, 60(2): 439–471. doi:10.1002/0471263397.env316

Henriksson, P.J.G., Zhang, W., Nahid, S.A.A., Newton, R., Phan, L.T., Dao, H.M., Zhang, Z., Jaithiang, J., Andong, R., Chaimanuskul, K., Vo, N.S., Hua, H.V., Haque, M.M., Das, R., Kruijssen, F., Satapornvanit, K., Nguyen, P.T., Liu, Q., Liu, L., Wahab, M.A., Murray, F.J., Little, D.C. and Guinée J.B., 2014. Final LCA case study report. Results of LCA studies of Asian Aquaculture Systems for Tilapia, Catfish, Shrimp, and Freshwater Prawn. SEAT Deliverable Ref: D 3.5, 165 p.

Hu, Z., Lee, J.W., Chandran, K., Kim, S. and Khanal, S.K., 2012. Nitrous oxide (N_2O) emission from aquaculture: A review. Environmental Science and Technology, 46(12): 6470–6480. doi:10.1021/es300110x

IPCC, 2014. Climate Change 2014: Synthesis Report to the Fifth Assessment Report of the Intergovernmental Panel on Climate Change. Retrieved from http://ar5- syr.ipcc.ch.

John, P., Henriksson, G., Belton, B., Jahan, K.M. and Rico, A., 2018. Measuring the potential for sustainable intensification of aquaculture in Bangladesh using life cycle assessment, 115(12): 2958–2963. doi:10.1073/pnas.1716530115

Kabir, K.A., Schrama, J.W., Verreth, J.A.J., Phillips, M.J. and Verdegem, M.C.J., 2019a. Effect of dietary protein to energy ratio on performance of Nile tilapia and food web enhancement in semi-intensive pond aquaculture. Aquaculture, 499: 235-242. doi: https://doi.org/10.1016/j.aquaculture.2018.09.038

Kabir, K.A., Verdegem, M.C.J., Verreth, J.A.J., Phillips, M.J. and Schrama, J.W., 2019b. Effect of dietary protein to energy ratio, stocking density and feeding level on performance of Nile tilapia in pond aquaculture. Aquaculture, 511. doi:10.1016/j.aquaculture.2019.06.014

Lynch, A.J., Cowx, I.G., Fluet-Chouinard, E., Glaser, S.M., Phang, S.C., Beard, T.D. Jr., Bower, S.D., Brooks, J.L., Bunnell, D.B., Claussen, J.E., Cooke, S.J., Kao, Y.C., Lorenzen, K., Myers, B.J.E., Reid, A.J., Taylor, J.J. and Youn, S., 2017. Inland fisheries – invisible but integral to the UN Sustainable Development Agenda for ending poverty by 2030. Global Environmental Change, 47: 167–173. www.sciencedirect.com/ science/ article/pii/S0959378017303965

Paudel, S.R., Choi, O., Khanal, S.K., Chandran, K., Kim, S. And Lee, J.W., 2015. Effects of temperature on Nitrous oxide (N_2O) emission from intensive aquaculture system. Science of the Total Environment, 518–519, 16–23. doi:10.1016/j.scitotenv.2015.02.076

Peacock, M., Audet, J., Jordan, S., Smeds, J. And Wallin, M.B., 2019). Greenhouse gas emissions from urban ponds are driven by nutrient status and hydrology. Ecosphere, 10(3), e02643. doi:10.1002/ecs2.2643

Pelletier, N., Tyedmers, P., Sonesson, U., Scholz, A., Zeigler, F., Flysjo, A., Kruse, S., Cancino, B., Silverman, H., 2009. Not all salmon are created equal: life cycle assessment (LCA) of global salmon farming systems. Environ. Sci. Technol. 43(23): 8730–8736

Tran-Tu, L.C., Hien, T.T.T., Bosma, R.H., Heinsbroek, L.T.N., Verreth, J.A.J. and Schrama, J.W., 2018. Effect of ingredient particle sizes and dietary viscosity on digestion and faecal waste of striped catfish (*Pangasianodon hypophthalmus*). Aquaculture Nutrition, 24(3): 961–969. doi:10.1111/anu.12632

Vasanth, M., Muralidhar, M., Saraswathy, R., Nagavel, A., Dayal, J.S., Jayanthi, M., Lalitha N., Kumararaja, P. and Vijayan, K.K., 2016. Methodological approach for the collection and simultaneous estimation of greenhouse gases emission from aquaculture ponds. Environmental Monitoring and Assessment, 188(12). doi:10.1007/s10661-016-5646-z

Yang, P., Lai, D. Y. F., Yang, H., Tong, C., Lebel, L., Huang, J. And Xu, J., 2019. Methane Dynamics of Aquaculture Shrimp Ponds in Two Subtropical Estuaries, Southeast China: Dissolved Concentration, Net Sediment Release, and Water Oxidation. Journal of Geophysical Research: Biogeosciences, 124(6): 1430–1445. doi:10.1029/2018JG004794

Williams, A.S., 2009. Life Cycle Analysis: a Step by Step Approach. Illinois Sustain- able Technology Center, Champaign, IL.

Impact of Climate Change on Hydrological Cycle, Ecosystem, Fisheries and Food Security, Pages: 137–144
Edited by: B. Madhusoodana Kurup, M.R. Boopendranath, M. Harikrishnan and A.V. Shibu

CHAPTER - 10

Aquaculture-Environment Interaction: Strategies and Technologies for Impact Mitigation

V. Zonno

DiSTeBA – Università del Salento and TECNOSEA srl,
Via Provinciale Lecce-Monteroni, 73100 Lecce, Italy
E-mail: vincenzo.zonno@unisalento.it

The reduction of wild fish catches and the rising demand for sea food protein, by world human population, during the past few decades, have been compensated by a rapid increase of aquaculture production (Valderrama & Anderson, 2010; Dauda *et al.*, 2013). Nowadays, the rearing of aquatic organisms provides almost the 50 % of all world sea food for human consumption and the prevision is that aquaculture will continue to grow more rapidly than all other animal food-producing sectors. According to the Food and Agriculture Organization of the United Nations (FAO), by 2030 aquaculture will reach a global production (excluding aquatic plants) of 110 millions tons (FAO, 2018). However, the main challenge of the present and future aquaculture is sustaining a constant increase of production while minimizing the impact on the environment (Navarrete-Mier *et al.*, 2010; Longo *et al.*, 2019). All aquaculture stakeholders are, therefore, committed to make the best to safeguard the ecosystems.

Major environmental impacts of aquaculture have been associated with high-input high-output intensive systems. Fed aquaculture, such as net cage fish farming, releases nutrients in both particulate and dissolved forms, mainly due to the feed residuals and the waste produced by the fish, to the surrounding waters and to the underlying sediments, regardless of whether the aquaculture is practiced in freshwater lakes, rivers, estuaries, coastal or off-shore marine waters (Roque d'Orbcastel *et al.*, 2009; Cripps & Bergheim, 2000; Corner *et al.*, 2006; Amirkolaie, 2008). Similarly, land-based pond or tank aquaculture releases large loads of nutrients, in flow-through or partially-closed recirculating production systems (Lefebvre *et al.*, 2001; Van Rijn, 2013). This, of course, poses some concerns on the potential environmental risks of the activity and on its social acceptability, sustaining the urgent need of innovative processes and technology for fish farming.

After a brief description of the type of waste produced by an aquaculture system, this presentation reviews some available technologies and protocols to improve the waste management of a fish farm and describes sustainable options for waste reuse and valorisation.

Waste from Aquaculture Systems

Aquafarms, like any other production enterprise, discharge waste, which consist of settable or suspended solids and dissolved nutrients. When determining the amount of waste a system will generate, the amount of feed offered is an essential factor. In a properly managed intensive farm, rearing carnivorous finfish species (salmon, seabream, seabass, trout), approximately 30 % (dry matter) of the feed used, will become particulate solid waste (Lupatsch & Kissil, 1998; D'Orbcastel *et al.*, 2008). This solid matter, in the form of faeces and waste feed, constitutes the major concern in the management of waste water in aquaculture.

Faecal matter constitutes, more or less, compact particulate material that may settle in the tank, where the fish are held, or in a sedimentation tank, or accumulate on the seabed, in cage systems (Ballester-Moltó *et al.*, 2017). The chemical composition (C, N, P and trace elements) and physical characteristics such as size, density, water content and impact resistance of faeces, depend on factors such as feed composition and fish species, as well as the phase of development (Aquaetreat, 2007; D'Orbcastel *et al.*, 2008).

Of the suspended solids, the small-size organic particles may settle near or far from the farm, depending upon the local hydrodynamic regime (Sarà, 2007) and a small proportion of this organic material is composed of fine organic particles (for example, dust from fish feeds) that largely remain in suspension (Tsapakis *et al.*, 2006).

Soluble compounds (dissolved or soluble waste) are excreted in many forms through the gills, kidneys, and with faeces, depending on the quality of the feed, the fish species and their phase of development (Miller & Semmens, 2002; D'Orbcastel *et al.*, 2009). The soluble wastes consist of organic or inorganic compounds that may be taken up by organisms, undergo chemical interactions in the water column or in association with surfaces, or simply persist in the water column if they are non-reactive (Black, 1998).

Waste Treatment

Looking at the need for fish farmers to minimize nutrient release and optimize the water use, including recovery, disposal and re-use of solid waste, in recent years, innovative effluent treatment systems, applicable to all types of land-based fish farms, open and closed systems, fresh water and marine operations, regardless of species, have been developed, tested and experimentally validated at different sites in Europe.

The designed systems have proven to be cost-effective and allow efficient solid removal and sludge thickening. The first step is a mechanical filtration (normally obtained by a drum or disk filters, Fig. 1), in which all large settleable or suspended particles, both organic and inorganic, are removed; this step produces a high flow of filtered water, where the reduction of solid concentration varies between 30 and 70 %, (Table 1) and a very low flow, of concentrated waste water, containing 1 to 3 g of solids per liter. The latter flow, also called the pre-concentrated effluent, needs further treatment with additional concentration systems (settling and coagulation-flocculation) (Fig. 2 and Table 2) to obtain, on one hand, a concentrated sludge (10-30 % of dry matter content) and, on the other hand, an effluent containing high concentrations of mineral and organic soluble substances (in particular nitrogen and phosphorus). The final sludge dewatering and thickening process occurs through the employment of different devices, as settling tanks, geo-tubes and belt filters (Fig. 3), while further soluble substances removal from the nutrient enriched waste water, occurs through packed or moving bed biofilters, wetlands, both natural or constructed, and algal ponds. The same thickening system has been successfully applied to treat the sediment released by a sea cage farm. In this case, a specifically designed dispositive (a customized funnel attached to the sinker tube), allows the collection of a great part of the solids that accumulate below the cage (Fig. 4 (a) and (b). The overall efficiency of the systems has been checked by characterizing all flows. In particular, effluent quality has been screened by measuring key physical and chemical parameters, any potential disadvantageous elements, trace elements, organic carbon, polychlorinated biphenyls (PCB), polyaromatic hydrocarbons (PAH), microbiological parameters (such as total microbial count, fecal coliforms, streptococci, salmonella, clostridia) have also been measured as well as BOD (Biochemical Oxygen Demand). The quality of the recycled water has been tested by rearing sea bream, sea bass and meagre juveniles in it, with excellent results in terms of growth performance and well-being.

Fig. 1. Micro-screen filter

Table 1. Efficiency of mechanical microfiltration applied to marine land based aquaculture effluent.

Parameters	Effluent	Filtered water	Efficiency (%)
SS (mg/l)	20-25	7-15	70
Ammonium (μM)	4.572	1.698	38
Nitrite (μM)	2.382	1.912	–
Nitrate (μM)	39.783	32.084	–
N-Inorganic (μM)	46.738	38.471	17.68
N-Organic (μM)	113.143	2.036	98.20
N-Total (μM)	159.880	40.507	74.66
P-Inorganic (μM)	0.20	0.180	10.00
P-Organic (μM)	17.72	4.118	76.76
P-Total (μM)	17.92	4.298	76.01

Fig. 2. Concentrated waste water and coagulation-flocculation process

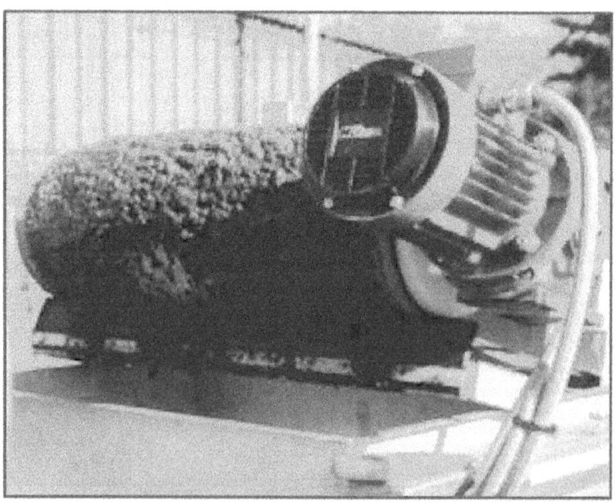

Fig. 3. Thickened sludge on belt filter

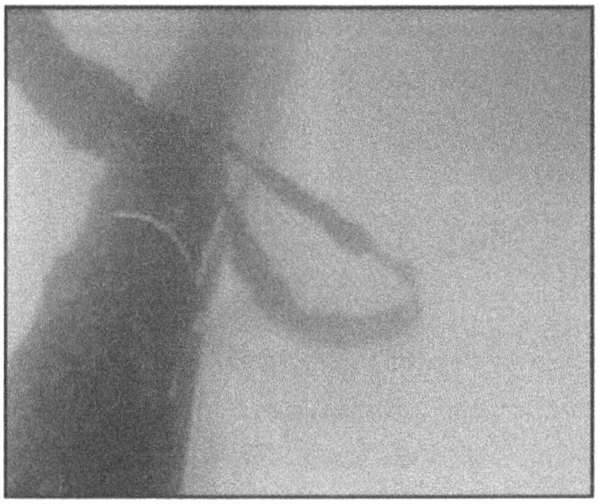

(a)

(b)

Fig. 4. (a) Dispositive for solid collection attached to the sinker tube of a fish cage; (b) Sediments accumulated under the cage.

Recycling aquaculture waste

Protocols and methods have been set up for the re-use and/or disposal of waste and by-products. Fish sludge may be described as the 'solids' part of the waste stream in a fish farm. The water content of sludge depends on the system used to separate the solid and liquid fractions.

The analysis of several sludge samples, obtained using different thickening systems and coagulation-flocculation reagents (Table 2), confirmed the nature and origin of the final sediment, prevalently constituted by carbon, nitrogen, potassium and calcium (Table 3). Significant has been also the presence of important minerals such as iron, zinc, copper and magnesium, while the presence of potential pollutants, such as heavy metals, PCBs and PAHs, has been excluded. Very promising, in terms of environmental and social acceptability, is the use of natural flocculants.

Table 2. List of coagulants and flocculants for aquaculture waste treatment

Coagulants	Flocculants	
	Synthetic Polymers	Natural Polimers
Iron chloride	Poly-acryl-amide (PAM) family	Waxy com starch
Zinc sulphate	Poly-diallyl-dimethyl-ammonium chloride (poly-DADMAC)	Cationic starch
Zinc chloride	Starch derivatives (Flocgel)	Oxidized starch (anionic)
Iron sulphate		Cellulose
Copper sulphate		Alginic acid
Magnesium sulphate		Guar gum
Aluminum sulphate		Potato starch Mouringa oil Cactus tree extract Maltodextrin Chitosan

A set of field agriculture tests has demonstrated that the use of marine fish stabilized sludge, as soil amender to grow tomatoes and other salt tolerant plants, could promote a significantly increased production when compared with other commercial and organic fertilizers.

Conclusion and Perspectives

Aqua farmers well know that the protection of the environment means saving their business. In this view, the application of good management practices is the first and the most efficient strategy to produce safe and high-quality seafood. Secondly, the application of an "easy-to-handle" technology is able to produce cleaned water that can be safely discharged in the environment. In addition, aiming at a "Circular Economy" based aquaculture, the treated effluent can be recycled inside the process units, with evident economic advantages, while the thickened and stabilized sludge can be recovered and re-used as soil amender or for other purposes. This, in turn, will give enough income to cover the investment and running costs.

In the short and medium term, the application of an integrated system (dispositive for solid collection, both on land and in off-shore cages, effluent water treatment, sludge conditioning, thickening and reuse), at full production scale, will definitely demonstrate the effectiveness of this technology to afford one of the main challenges of future aquaculture. In this direction goes the realization of innovative cage fish farms, along the Mediterranean coast of Morocco, promoted by the group of researchers and private Companies that have been actively involved in the development of the systems, described here.

Acknowledgement

Research activities were supported by EU (Aquaetreat and Aquagris projects), MIUR (Italy) under PON_02 program (ECO_P4 project) and Regione Puglia under Regional Cluster program (SEA project).

References

Amirkolaie K., 2008. Environmental Impact of Nutrient Discharged by Aquaculture Waste Water on the Haraz River. Journal of Fisheries and Aquatic Science, 3: 275-279. Aquacultural Engineering, 40 (2): 79-86.

Aquaetreat, 2007. Manual on effluent treatment in aquaculture: Science and Practice. SMEs (Collective Research) scheme (Contract N. COLL-CT-2003-5003 05).

Ballester-Moltó, M., Sanchez-Jerez, P., Cerezo-Valverde, J. and Aguado-Giménez, F., 2017. Particulate waste outflow from fish-farming cages. How much is uneaten feed? Marine Pollution Bulletin 119(1): 23-30.

Black, K.D., 1998. The environmental interactions associated with fish culture. In: Biology of Farmed Fish, Black, K.D. and Pickering, A.D. (Eds.). pp. 284-326, Sheffield Academic Press, Sheffield.

Corner, R.A., Brooker, A.J., Telfer, T.C. and Ross, L.G., 2006. A fully integrated GIS-based model of particulate waste distribution from marine fish-cage sites. Aquaculture 258: 299-311.

Cripps, S.J. and Bergheim, A., 2000. Solids management and removal for intensive land-based aquaculture production systems. Aquacultural Engineering, 22: 33-56.

D'Orbcastel, E. R., Blancheton, J.P., Boujard, T., Aubin, J., Moutounet, Y., Przybyla, C. and Belaud, A., 2008. Comparison of two methods for evaluating waste of a flow through trout farm. Aquaculture 274: 72-79.

D'Orbcastel, E. R., Person-Le-Ruyet, J., Le Bayon, N. and Blancheton, J.P., 2009. Comparative growth and welfare in rainbow trout reared in re-circulating and flow through rearing systems. Aquacultural Engineering 40(2): 79-86.

Dauda A.B., Folorunso, L.A. and Dasuki, A., 2013. Use of probiotics for sustainable aquaculture production in Nigeria. JASR 13 (2): 35-45.

FAO, 2018. The State of World Fisheries and Aquaculture 2018 - Meeting the Sustainable Development Goals. FAO, Rome. License: CC BY-NC-SA 3.0 IGO.

Lefebvre, S., Bacher, C., Meuret, A. and Hussenot, J., 2001. Modeling approach of nitrogen and phosphorus exchanges at the sediment–water interface of an intensive fishpond system, Aquaculture 195(3–4): 279-297.

Longo, S.B., Clark, B., York, R. and Jorgenson, A.K., 2019. Aquaculture and the displacement of fisheries captures. Conservation Biology, 33: 832-841.

Lupatsch, I. and Kissil, G.W., 1998. Predicting aquaculture waste from gilthead seabream (*Sparus aurata*) culture using a nutritional approach, Aquatic Living Resources 11(4): 265-268.

Miller, D. and Semmens, K., 2002. Waste management in aquaculture. Aquaculture Information Series, pp. 1-10 #AQ02-1(January).

Navarrete-Mier, F., Sanz-Lázaro, C. and Marín, A., 2010. Does bivalve mollusc polyculture reduce marine finfish farming environmental impact? Aquaculture 306(1–4): 101-107.

Sarà, G., 2007. Ecological effects of aquaculture on living and non-living suspended fractions of the water column: A meta-analysis. Water Research 41: 3187–3200.

Tsapakis, M., Pitta, P. and Karakassis, I., 2006. Nutrients and fine particulate matter released during sea bass (*D. labrax*) farming. Aquatic Living Resources 19: 69-75.

Valderrama, D. and Anderson, J.L., 2010. Market interactions between aquaculture and common property fisheries. Journal of Environmental Economics and Management 59: 115–128.

Van Rijn, J., 2013. Waste treatment in recirculating aquaculture systems. Aquacultural Engineering 53: 49-56.

Impact of Climate Change on Hydrological Cycle, Ecosystem, Fisheries and Food Security, Pages: 145–150
Edited by: B. Madhusoodana Kurup, M.R. Boopendranath, M. Harikrishnan and A.V. Shibu

CHAPTER - 11

Carbon Sequestration in Aquaculture System

A. Vijay Kumar[1], Vidya Shree Bharti*, Raj Mukhopadhyay[2], and V. Hari Krishna[1],

Chittaranjan Raul[1] and Satya Prakash[1]

[1]*Aquatic Environment and Health Management Division, ICAR-Central Institute of Fisheries Education, Versova, Andheri (W), Mumbai-400 061, India*

[2]*ICAR- Central Soil Salinity Research Institute, Karnal, Haryana-132001*

**E-mail: vidya.bharti@icar.org.in*

Carbon dioxide (CO_2) is one of the most potent GHGs, accounting for nearly 60% of the overall radiative forcing of the atmosphere. Global atmospheric CO_2 concentration has increased to 405 ppm in 2017, exceeding the pre-industrial levels by about 40% (World Meteorological Organization, 2018) with a steady increase of approximately 0.50% per year over the past century (IPCC, 2013). World agriculture accounted for an estimated emission of 5.1–6.1 Pg CO_2 equivalents year contributing 10–12% to the total global anthropogenic GHG emissions in 2005. Methane (CH_4) is second only to CO_2 in volume but has 72 times more global warming potential than CO_2 on a mass basis. CH_4 and CO_2 together, contribute to approximately 80% of the overall radiative forcing of the atmosphere due to anthropogenic influence (Myhre *et al.*, 2013). The increased level of GHGs already increased the average global temperature of the earth by 1°C above the pre-industrial baseline and 2015-18 are the four warmest years of the record (World Meteorological Organization, 2018). The International Panel on Climate Change (IPCC) predicts that, by end of the 21st century, the atmosphere may contain up to 570 ppm CO_2, causing a rise in the mean global temperature of around 1.9°C. This drastic increment in global climate change causes extreme weather events, prolonged periods of drought and floods, shifting of climatic zones, uneven rainfall, sea level is expected to rise between 0.19 and 0.58 meters.

Soil Carbon Sequestration

To tackle this drastic phenomenon, there was much attention towards reducing the emissions of GHGs, and also to sequester CO_2 in longer pools. The Paris Climate Agreement recommended a voluntary plan of "4 Per Thousand" (4PT) to seques-ter C in world soils at the rate of 0.4% annually to 0.4 m (1.3 ft) depth (UNFCC, 2015). The basic process of C sequestration in the terrestrial biosphere involves the transfer of atmospheric CO_2 into plant biomass through photosynthesis and conversion of biomass into stable SOC through for-mation of organo-mineral complexes. Carbon sequestration is an important strategy in combating rising carbon dioxide concentration in the atmosphere. There is much attention is given towards increasing the soil carbon sequestration potential and adopting proper management practices Soil is the largest terrestrial carbon sink which contains 1100–1600 Pg. sequestered C which is more than twice the amount of C in vegetation (560 Pg.) and thrice the amount of carbon in the troposphere (750 Pg.) (Brar *et al.*, 2013). The majority of the soil C held in the form of Soil organic carbon (SOC) which is an important component for the functioning of various ecosystems and also for maintenance of soil health which intern helps in improve and sustain food production. SOC has a major influence on soil structure, water holding capacity, cation exchange capacity, nutrient retention, control of soil erosion and nutrient leaching. The increase of soil organic carbon can be done by the supply of soil organic matter in the form of organic wastes, manures, and crop residues. This is assumed that SOM comprises 58% C which upon microbial mineralization adds to SOC and improves the soil physical (soil structural stabilization), chemical (buffering and changes in soil pH), and biological properties (substrate and supply of nutrients for microbes).

Aquaculture and Carbon Sequestration

Aquaculture is classified as an extensive, semi-intensive and intensive culture based on various management practices adopted. Extensive culture usually involves stocking of fish fry or fingerlings in aquatic systems whereas semi-intensive and intensive culture practices involve various human interventions like drying, plowing, liming, high-density stocking, and feeding, etc. intensive and semi-intensive culture practices mainly involve feeding of aquatic organisms with nutrient-rich feed to increase the production. However, only a small portion of the feed input is converted into fish biomass, with the feed utilization efficiency being as low as ~4.0%–27.4% (Chen *et al.*, 2015). Most of the added feeds end up accumulating within the aquaculture systems (Chen *et al.*, 2015; Yang *et al.*, 2017a) or are being discharged into adjacent bodies of water (Chen *et al.*, 2015). These ponds retain a large quantity of organic carbon from residual feed and feces and over the period either it is mineralized to CO_2 or CH_4 or it may add to soil organic carbon pool and get stabilizes as longer passive carbon pools. Manuring and fertilization are another important aquaculture management practice which also adds to ponds soil organic carbon pools and stabilizes labile fractions of carbon to longer passive pools. Feeding and manuring are not only adding its labile fractions of carbon to sediment organic carbon but upon mineralization, they release inorganic nutrients like NO_3 and PO_4^- which boosts up CO_2 fixation by aquatic algae by process of photosynthesis. These two practices can lead to controlled and effective eutrophication which exhibits the significant potential to mitigate CO_2 emission as eutrophication drastically enhances primary productivity leading to enhanced utilization of CO_2 through autotrophic activities (Moss *et al.*, 2011; Pacheco *et al.*, 2014). On the other hand, feeding and manuring, and algal blooms also have ill effects by promoting CH_4 emission and leads to qualitative degradation of aquatic bodies (Pacheco *et al.*, 2014). Aquaculture ponds functioning is mainly depending upon daily feeding and photosynthetic activities of phytoplankton (Chen *et al.*, 2015). Taking into account the total carbon assimilated in dead planktons, unused fertilizers, unconsumed fish feed, and excreta, the freshwater aquaculture ponds have a potential of sequestering ~13.1Mt C yr^{-1} (globally) (Boyd *et al.*, 2010). And in India aquaculture ponds cover an area of 0.79 million hectares (Mha), and have a large potential to bury carbon (C) in sediments (Adhikari *et al.*, 2012). Adhikari *et al.* (2012) estimated C sequestration rate from sediment of polyculture system (Indian major carps plus scampi) in aquaculture ponds in Orissa and it varied from 1463 ± 70 to 1530 ± 80Kg ha^{-1} yr^{-1} and in shrimp culture ponds it varied from 863 ± 40 to 864 ± 56kg $ha^{-1}yr^{-1}$ and scampi culture ponds 1099 ± 75 kg $ha^{-1}yr^{-1}$.

Biochar Amendment and Carbon Sequestration

Biochar is the carbonaceous product obtained by the thermal degradation of organic materials in the absence of oxygen. Biochar is the "charred organic matter, produced with the intent to deliberately apply to soils to sequester C and improve soil properties (Lehmann and Joseph, 2009). Biochar is a highly porous, carbon (C)-rich solid residue of pyrolysis of biomass and has recently been suggested for applications in the soil as a means of mitigating global climate change. Biochar application has been shown as one promising means of reducing the atmospheric CO_2 concentration slowing down the rate at which photosynthetically fixed carbon (C) is returned to the atmosphere and also reduces the emissions of other greenhouse gases (Fig. 1 and 2). Biochar apart from acting as a stable form of carbon it also known for its improvement of soil nutrient retention and water holding capacity and also acts as buffering agent and helps in the improvement of soil productivity and helps in enhanced photosynthesis activity and carbon fixation.

Biochar's climate-mitigation potential primarily stems from its highly recalcitrant nature towards chemical and microbial oxidation and will persist in soil for hundreds to thousands of years. Biochar is a highly heterogeneous material with a chemical composition that varies widely depending on feedstock and pyrolysis conditions. The decomposition rate of biochar varies significantly with experimental durations, feedstocks, pyrolysis temperatures, soil pH and clay contents. Biochar produced at lower temperatures were reported to mineralize faster than those produced at higher temperature with increased aromaticity and degree of aromatic condensation.

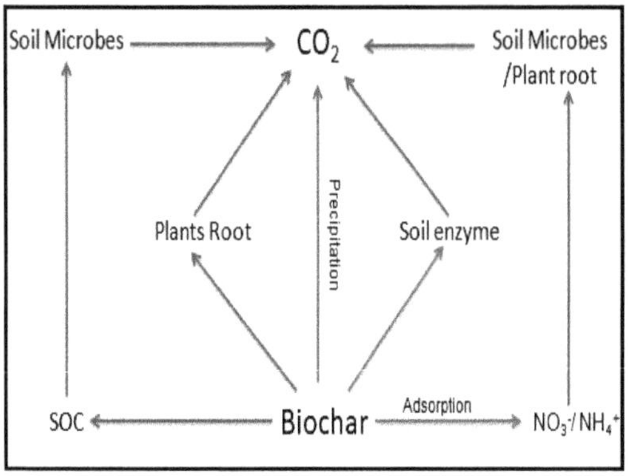

Fig. 1. Biochar inhibits CO_2 emission from soil.

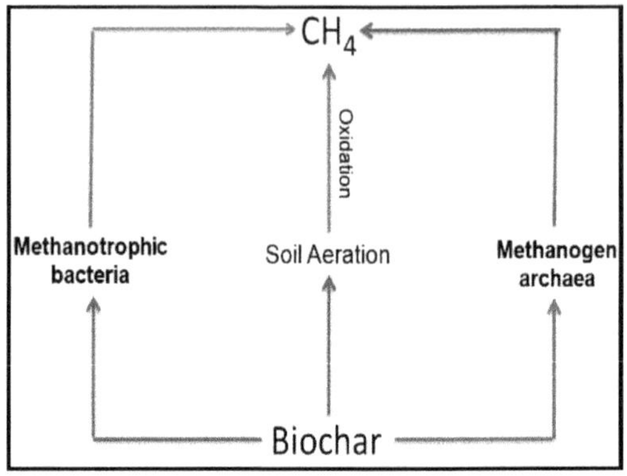

Fig. 2. Biochar inhibits CH_4 emission from soil.

Mohan *et al.* (2018) synthesized biochar from Corn Stover and rice husk biomass at 550 and 650°C respectively, and he observed their effects on CO_2 efflux from the soil system. The application of both biochar's (Fig. 3) significantly reduced the emissions of CO_2 compared to control and only biomass application and also higher reduction was observed at higher concentrations of biochar. Similar work conducted by Bhaduri *et al.* (2016) used peanut shell biochar at the rate of 0, 2.5%. 5% and 10% w/w of saline soil and reported that biochar could increase soil organic carbon on the application of highest rate of biochar addition (10%), and potentially restore the saline soils by less C mineralization and more sequestration of soil carbon.

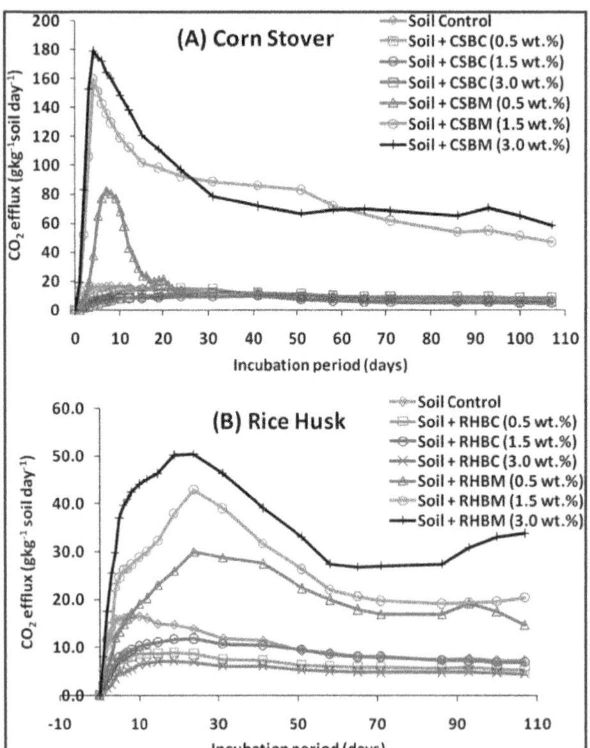

Fig. 3. Effectiveness of biochar synthesised from corn stover and rice husk biomass in reducing the emissions of CO_2 compared to control (Source: Mohan *et al.*, 2018)

Carbon Sequestration by Microalgae

Microalgae, generally considered auto phototrophs, are the primary oxygen-evolving photosynthetic microorganisms on the earth, whose carbon consumption accounts for almost 50% of the global CO_2 fixation (Zhou *et al.*, 2017). The ability of microalgae to grow heterotrophically or mixotrophically is significant and important because this allows microalgae to sequester organic carbons present in wastewaters, which can eventually emit to the atmosphere if broken down by bacteria. Algae, a major biomass in marine and fresh water environments, are most important CO_2 fixers in aquatic ecosystems. Due to their rapid growth rates, tolerance to varying environmental conditions algae can fix greater amounts of CO_2 per land area than higher plants. Chlorophyll in photosynthetic algae captures light energy, which is used to convert simple molecules (CO_2 and H_2O) into carbohydrates (sugars and starches) with the release of O_2. The fixed carbon would be assimilated to biomass, the biomass composition might relate to CO_2 absorption. From the CO_2 uptake rates, algae had higher CO2 uptake than plants. Many researchers reported positive correlation between lignin content of plants and CO2 absorption but earlier reports that algae not contain lignin until Martone *et al.* (2009) discovered the existence of lignin in algae. Yaich *et al.* (2011) found 1.56% lignin in the red alga *Ulva lactucato,* which is a contradiction; algae had a lower lignin content, but had the highest CO_2 uptake rate (Tsai *et al.*, 2017). Consequently, forests had the best CO_2 uptake rates among all plants, but these rates were much lower than that of algae (Tsai *et al.*, 2017).

References

Adhikari, S., Lal, R. and Sahu, B.C., 2012. Carbon sequestration in the bottom sediments of aquaculture ponds of Orissa, India. Ecological Engineering 47: 198-202.

Bhaduri, D., Saha, A., Desai, D. and Meena, H.N., 2016. Restoration of carbon and microbial activity in salt-induced soil by application of peanut shell biochar during short-term incubation study. Chemosphere 148: 86-98.

Boyd, C.E., 2010. Aquaculture ponds hold carbon. Global aquaculture advocate, monthly digital magazine, March/April, pp. 23-24.

Brar, B.S., Singh, K. and Dheri, G.S., 2013. Carbon sequestration and soil carbon pools in a rice–wheat cropping system: effect of long-term use of inorganic fertilizers and organic manure. Soil and Tillage Research, 128, pp.30-36.

Chen, Y., Dong, S., Wang, Z., Wang, F., Gao, Q., Tian, X. and Xiong, Y., 2015. Variations in CO_2 fluxes from grass carp *Ctenopharyngodon idella* aquaculture polyculture ponds. Aquaculture Environment Interactions, 8, pp.31-40.

Lehmann, J. and Joseph, S., 2009. Biochar for environmental management: an introduction. In: Lehmann, J., Joseph, S. (Eds.), Biochar for Environmental Management: Science and Technology. Earthscan, London, pp. 1-12.

Martone, P.T., Estevez, J.M., Lu, F., Ruel, K., Denny, M.W., Somerville, C., Ralph, J.,2009. Discovery of lignin in seaweed reveals convergent evolution of cell-wall architecture. Curr. Biol. 19: 169–175.

Mohan, D., Abhishek, K., Sarswat, A., Patel, M., Singh, P. and Pittman, C.U., 2018. Biochar production and applications in soil fertility and carbon sequestration–a sustainable solution to crop-residue burning in India. RSC advances, 8(1): 508-520.

Tsai, D.D.W., Chen, P.H. and Ramaraj, R., 2017. The potential of carbon dioxide capture and sequestration with algae. Ecological Engineering 98: 17-23.

UNFCC, 2015. COP21 Climate Agreement. Paris, France: United Nations Framework Convention on Climate Change.

Yaich, H., Garna, H., Besbes, S., Paquot, M., Blecker, C., Attia, H., 2011. Chemical composition and functional properties of Ulva lactuca seaweed collected in Tunisia. Food Chem. 128, pp.895–901.

Yang, P., Zhang, Y., Lai, D.Y., Tan, L., Jin, B. and Tong, C., 2018. Fluxes of carbon dioxide and methane across the water–atmosphere interface of aquaculture shrimp ponds in two subtropical estuaries: The effect of temperature, substrate, salinity and nitrate. Science of the Total Environment, 635: 1025-1035.

Zhou, W., Wang, J., Chen, P., Ji, C., Kang, Q., Lu, B., Li, K., Liu, J. and Ruan, R., 2017. Bio-mitigation of carbon dioxide using microalgal systems: advances and perspectives. Renewable and Sustainable Energy Reviews, 76, pp.1163-1175.

Impact of Climate Change on Hydrological Cycle, Ecosystem, Fisheries and Food Security, Pages: 151–161
Edited by: B. Madhusoodana Kurup, M.R. Boopendranath, M. Harikrishnan and A.V. Shibu

CHAPTER - 12

Climate Change Impact on Inland Fisheries and Aquaculture - Sri Lanka

K.B.C. Pushpalatha

National Aquaculture Development Authority of Sri Lanka 1/41, New Parliament Road, Pelawatte, Battaramulla, Sri Lanka
E-mail: pushpakbc@gmail.com

The Intergovernmental Panel on Climate Change (IPCC, 2007) defines climate change as "a change in the state of the climate, that can be identified by changes in the mean and or the variability of its properties and that persists for an extended period, typically decades or longer" Living and coping with uncertain impacts of climate change is no longer a choice; it is an imperative. Sri Lanka is a negligible contributor to global warming. However, as a nation, Sri Lanka also is vulnerable to the impacts of climate changes. Fisheries and aquaculture play an important role in Sri Lanka's economy, nutrition, food security, providing employment/livelihood opportunities and strengthening rural economy. This paper will review the observed trends in climate change in Sri Lanka, identify possible climate change impacts on inland fisheries and aquaculture and potential adaptive measures.

Sri Lanka is an island in the Indian Ocean just to the North of the Equator, with a land area of 65610 km^2. Its topography consists of a central highland (altitude of 300 m amsl) surrounded by an extensive low land area. The long-term mean annual temperature in the lowlands is 27^0 C, while it is less in the highlands depending on the elevation. Precipitation is the major parameter that gives rise to variability of climate conditions during annual cycle. There are three major sources of rainfall (RF) in Sri Lanka, namely, monsoonal, convectional and depressional. Country has mean annual RF around 1861mm with significant variation in regional distribution (Senaratne *et al.*, 2009). The spatial pattern of precipitation is strongly influenced by topography and two seasonal wind regimes (Chandrapala,1996), South-West monsoon (May to September) and North-East monsoon from December to February. There are two inter-monsoonal periods from March to April and from October to November.

Cyclones and depressions affect the local climate significantly. They occur frequently in Bay of Bengal area and contribute to RF in surrounding local areas. They can develop into catastrophic events occasionally, claiming for losses of human life and physical property. Sri Lanka come under the influence of these extreme events mainly during the October – November period and they mainly affect Eastern and Northern areas and occasionally inflict damage all over the island. Based on the amount of rain Sri Lanka has been divided into three main climatic zones (Fig. 1); the wet zone (>2500mm) dry zone <1750mm) and intermediate zone (1750 – 2500mm).

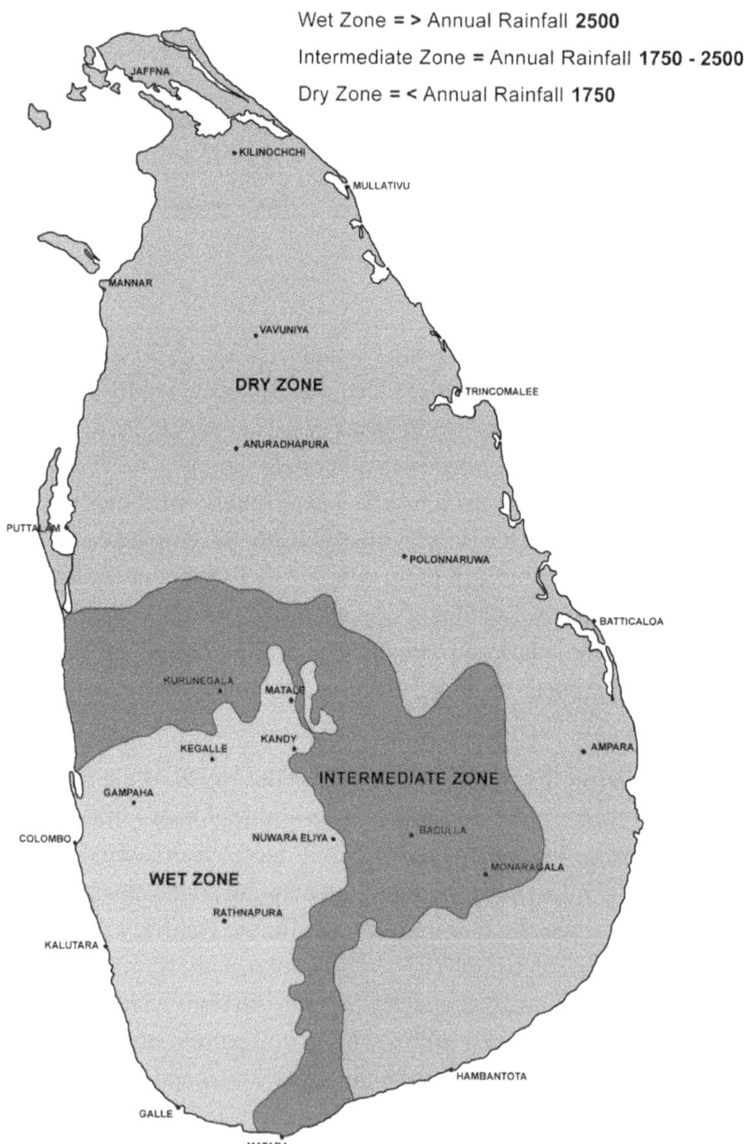

Fig. 1. Climatic Zones of Sri Lanka

Observed Changes in Climate

Temperature

Annual mean air temperature anomalies have shown a significant warming trend throughout the country during the latter half of this century (Basnayake *et al.*, 2002). Sri Lanka's 100 year warming trend from 1896 to 1996 is 0.003^0 C per year, while it is 0.025^0 C per year for the 10 year period from 1987 - 1996 (Fernando, 1997), indicating faster warming in more recent years. Projections for Sri Lanka based on studies in Sri Lanka and in general for South Asia under different scenarios suggest warming ranging from 0.9^0 C to 4^0C by 2100. (Cruz *et al.*, 2007; Kumar *et al.*, 2006). According to model projections, mean annual temperature of the country will increase steadily, throughout the century and by 2100, mean annual temperature will rise by 2.4^0C than the baseline (1961-1990) (Senaratne *et al.,* 2009). According to World Bank (2018) Sri Lanka's annual temperature could rise by 1.0^0C to 1.5^0C by 2050, even if carbon emission reduction measures are taken as recommended by Paris Agreement of 2015. If no measures are taken average temperature could increase up to 2.0^0C. Projected climate changes include increase in mean annual temperature of between 0.8^0C and 2.0^0C by 2060 (USAID, 2018).

Rainfall

There is no significant trend in Sri Lanka's mean annual rainfall (MAR) during the century although higher variability is evident (Jayatilake *et al.*, 2005). However, more recent data show a decreasing trend. Anyhow Rainfall projections for Sri Lanka within this century appear to be confusing and contradictory, while most of the studies project higher MAR under the range of IPCC scenarios (De Silva, 2006; Basnayake *et al.*, 2004; Cruz *et al.*, 2007; Kumar *et al.*, 2006) some project lower MAR. Some studies have predicted higher South West monsoonal rainfall and lower North East monsoonal rainfall (De Silva, 2006; Basnayake and Vithanage, 2004). Further, De Silva (2006) projected reduced rainfall in dry zone areas such as Anuradhapura, Batticaloa and Trincomalee. De Silva (2006) projected a 2% increase in rainfall in the Intermediate zone. Model projections suggest that annual precipitation in the country will increase in the future (Senaratne *et al.,* 2009). Projection of change in precipitation vary, with some predicting decreases and some increases, but generally indicate increase in variability.

Sea Level rise

The current rate of sea level rise in coastal areas of Asia is reported to be 1-3 mm/year (Cruz *et al.*, 2007). Further accordingly to them there are evidence to suggest an accelerated rate of sea level rise over the past decade 3.1 mm/yr compared to that over the 20th century (1.7 to 2.4 mm/yr). Sea level rise is one of the most widely discussed effects of climate change. Being an island with a densely populated low -lying coastal belt running around the country, Sri Lanka's vulnerability to sea level rise is quite apparent (Senaratne *et al.*, 2009; USAID, 2018). Despite, but only a few scientific effort have so far been made to assess the level of threat imposed by this issue. One study has estimated the land losses and area of inundation (Hettearachchi and Samarawickrama, 2009). It has been estimated that sea level rise of 1m will result in inundation of 91.25 km^2. USAID (2018) predicted a total sea level rise of between 0.2 and 0.6 meters by mid-century, compared to 1971-2010 levels.

Extreme Climate events

The intensity and frequency of extreme climate events, such as floods, droughts and landslides have increased in recent past (Imbulana *et al.,* 2006; Herath and Ratnayake, 2004). USAID (2018) predicted increased cyclone frequency and intensity, increased frequency and severity of floods, drought incidence and landslides (USAID, 2018).

Importance of fisheries

Fisheries is important to Sri Lanka in respect of food security, nutrition of the people, elimination of poverty, providing livelihood/ employment, earning of foreign exchange and strengthening national and rural economies. Sri Lanka is a fish-eating nation and 55% of animal protein consumption are met through consumption of fish. Per capita fish consumption is around 16.0 kg in recent years. Contribution of fisheries and aquaculture to GDP varies between 0.8 to 1.8% during the period 2002- 2018. Sri Lanka earned around us$ 275 million annually in the recent past through export of fishery products. There are around 275000 people actively involved in fishing and aquaculture. Around 2.6 million people are involved in fishing and aquaculture and other related livelihoods.

Fish production

Sri Lanka's annual total fish production in 2018 is 527,380 tons. Contribution of marine fisheries towards the total fish production is 83.3%, and the rest (16.7%) comes from inland fisheries and aquaculture.

Inland Fisheries, Culture-based Fisheries and Aquaculture

Resources

Sri Lanka has a coastline of 1817 km. There are 103 perennial rivers most of which drain from central highlands into the Western, Southern and Western coasts. In addition, country is rich with multitude of reservoirs with a cumulative extent of about 300,000 ha, which can support productive Inland Fisheries and Culture Based Fisheries. On average, they range from few ha to 4000 ha in size and depending on their hydrological regimes; they are either perennial or seasonal. Perennial reservoirs are divided into three broad categories. Large, medium and minor. Built mainly for the supply of water for paddy cultivation, these reservoirs are secondarily used for inland fisheries and culture-based fisheries (CBF). Area under villus or flood plains is 40,000 ha. There are about 160,000 ha of brackish water lagoons and estuaries.

Importance

Inland fisheries and aquaculture subsectors produced around 88010 m tons of fish in 2018, of which more than 85% of the production obtained from inland fisheries and CBF from reservoirs. Production of farmed shrimps is around 8000 tons. Inland fisheries, culture-based fisheries and aquaculture play an important role in providing quality fish at affordable prices, thus providing valuable protein complement to the starchy diet and contributing towards elevation of nutritional standards among rural populations. It has been reported that fish consumption in inland districts such as Anuradhapura and Polonnaruwa are high and 85 - 90% of fish consumed are inland fish. Further, this subsector including ornamental fish farming provide direct employment and livelihood for more than 50,000 people, especially in rural areas. Involvement of women is increasing, particularly, in the ornamental fish farming.

Inland fisheries and culture-based fisheries in perennial reservoirs

At present fish production from perennial reservoirs contribute significantly to inland fisheries and aquaculture fish production in Sri Lanka. Government has introduced CBF into most of the perennial reservoirs in recent years and this has resulted in significant increases in fish production in these reservoirs. Main features of the introduction of CBF in perennial reservoirs are stocking with suitable fish species, such as Nile tilapia, catla, Rohu, Mrigal and freshwater prawn; co-management of CBF with the involvement of all stakeholders; management of CBF including introduction of measures to prevent unauthorized and illegal fishing and ensure sustainability of CBF by establishing a fund by charging a commission from each kg of fish landed to continue with stocking. Chandrasoma *et al.* (2015) reported average fish production of 353 kg/ha/yr and 310 kg/ha/yr in minor and medium perennial reservoirs in Sri Lanka.

Culture-based fisheries in seasonal reservoirs

There are about 10,000 seasonal reservoirs scattered throughout the dry zone of Sri Lanka and most of these reservoirs in 10 - 50 ha size range. These reservoirs fill up during North East monsoon (Dec. - Feb.) and retain water for 6 - 10 months. These reservoirs dry up during (Aug. - Oct.) period. Fish culture in these reservoirs are undertaken by communities living around reservoirs. Average production of around 750 kg/ha/yr are obtained from these reservoirs. CBF in seasonal reservoirs is an environmentally friendly practice as the only external input is seed and it is an excellent example of a low-input secondary use of water resources for fish production (De Silva *et al.*, 2006; Amarasinghe and Nguyen, 2009).

Shrimp Farming

Shrimp farming is mainly practiced in coastal areas of the North western province and gradually expanding to areas in the North and the East of the country. At present main species cultured is *P.monodon*. *P.vannamei* has been introduced to Sri Lanka in 2019 and it is anticipated that more farmers will undertake farming of *P.vannamei* in years to come. Sri Lanka produced around 8000 tons of cultured shrimps in 2018.

Other coastal aquaculture activities

Other coastal aquaculture activities, which are getting popular and expanded are farming of sea cucumber in pens in shallow seas, farming of sea bass in cages, installed in lagoons and the sea, sea weed (*Kappaphycus* sp.) farming in shallow seas in the North and mud crab farming in ponds.

Freshwater fish farming in ponds

Mono sex Nile tilapia culture are being carried out in earthen ponds and tilapia farming is expanding in particularly in the western province of the country.

Ornamental fish Industry

Ornamental fish industry is a growing industry with huge potential for expansion in Sri Lanka. Activities include collection from the wild (in particularly marine species) and breeding and farming of fresh water species. Ornamental fish collection, fish breeding and farming are small scale operations providing livelihoods. Sri Lanka exported ornamental fish earning around US$ 15.0 million annually during last few years and although not quantified, there is a high demand for ornamental fish in the local market.

Climate Change Related Impacts on Inland Fisheries/CBF and Aquaculture

Climate change related possible impacts and potential adaptation measures are summarized in Table 1.

Table 1: Climate change related possible impacts and potential adaptation measures for inland fisheries/ CBF and aquaculture sub-sector in Sri Lanka.

Climate change parameters	Impacts	Potential adaptive measures
• Increase in temperature • Prolonged drought • Lesser rainfall in certain dry zone districts • Sea level rise • Increased frequency and intensity of rainfall and floods • Variability and unpredictability	• Farmed food fish and ornamental fish species exposed to temperatures over and above the optimum temperature range • Negative effect on hatching of culturable species • Negative effect on maturity of culturable species • Culturable fish get stressed due to above optimal water temperatures results in higher incidence of diseases • Water retention period changed due to drought • Less amount of water retained in irrigation reservoirs results in low fish production from inland fisheries and CBF • Increase salinity in lagoons/estuarine waters • Salt water intrusion to inland areas • Frequent spilling of reservoirs results in mass scale escape of fish • Lowering of salinities in shrimp ponds • Flooding of shrimp farming area • Uncertainty in culture periods in CBF in seasonal reservoirs; difficulties in planning the process including supply of fish seed. • Effect on maturity of brood-stock of culturable carp species which are related to rainy seasons; variability in peak breeding seasons	• Deepening ponds • Identify and use aquatic species with higher temperature tolerance; • Selective breeding and genetical improvements in culturable species for higher temperature tolerance; • Adjust crop calendars to avoid periods with temperature peaks; • Modification to hatchery systems to maintain temperature within optimum range; • Shift hatcheries of culturable freshwater fish species to higher elevations of the country, where temperature are low; • Deepening brood-stock ponds; • Brood-stock ponds to be re-located in low temperature areas • Better management to reduce stress; bio security measures; change to more resistant species/ strains • Use faster growing fish species/strains lower stocking densities. • Increase efficiency of water users; promote integrated water resources management; de-silting/improve maintenance of existing reservoirs; construction of new reservoirs and trans basin diversion to meet demand. • Use aquatic species suitable for higher salinities; identify suitable species. • Promote aquaculture using high salinity tolerant species; Increase availability of fish seed of suitable species such as milk fish, saline tilapia etc. • Install fish escape prevention devices • Use of aquatic species which can tolerate wide range of salinities • Use ponds with high dikes; • Improve drainage of flood waters to minimize effect on shrimp ponds • Use faster growing species/strains; use lesser stocking densities; Increase efficiency of water use by primary users; promote water efficient farming methods/ crops; promote integrated water resources management; mobilize stakeholders in support of climate adaptation. • Induced maturation methods to be determined and introduced.

Although the impact or effects of climate change on the fisheries sector has not been assessed or quantified widely, it is expected to have a wide range of impacts on the production, availability and breeding patterns of aquatic life (Jayathilake, 2008). Sector vulnerability Profile - Aquaculture and Fisheries prepared by the Ministry of Environment (2010), Sri Lanka indicate possible impacts of sea level rise, frequent storm surges and coastal flooding which can effect coastal aquaculture and possible impacts of changes in rainfall regimes and prolonged drought on inland fishery and culture-based fisheries. According to Shelton (2014), climate changes will affect fisheries and aquaculture via acidification, changes in sea temperatures and circulation patterns, the frequency and severity of extreme events, and sea level rise and associated ecological changes. Both direct and indirect impacts include impact on targeted population range and productivity, habitats and food webs as well as impacts on fishery and aquaculture costs and productivity and fishing community livelihoods and safety (Daw *et al.*, 2009; Badjeck *et al.*, 2010). Also some new opportunities and environments may be created with sea level rise and flooded coastal agricultural lands may provide new areas for aquaculture opportunities (Shelton, 2014). NACA (2012) conducted a study to identify potential impacts of climate change on CBF in seasonal reservoirs of Sri Lanka and according to the study high drawdown and greater fluctuations of water levels in seasonal reservoirs affect CBF activities negatively.

Conclusion

Although much have been talked about for years on impacts or effects of climate change and adaptations on fisheries and aquaculture, not much have been done or assessed or quantified them. It has become essential to expedite action to study and cope with climate change impacts. A climate change secretariat has been established under the Ministry of Environment and National climate change adaptation strategy (NCCAS) have been formulated. NCCAS have been structured into five thrust areas and fisheries related aspects are dealt under the thrust area "minimize climate change impacts on food security". Some of the strategies proposed under this thrust area related to fisheries sector are promote research on climate impacts/adaptive measures and pilot test and scale up community level fisheries adaptation methods. One of the key factors that should be considered in formulating adaptation strategies is the cost involved and all action should be taken to keep the costs low, to ensure the fish prices are kept at levels, competitive and affordable. Although such strategies have been proposed not much have been done on these lines in respect of fisheries sector. It is essential to create awareness among communities on potential climate change impacts and adaptation strategies and involve them in relevant research and testing. Major constraints for implementing programs related to climate change aspects in fisheries and aquaculture sector are lack of commitment, non-availability of funds for the purpose, weaknesses in institutional arrangements and lack of capacity in relevant institutions. As most of the climate change impacts on inland fisheries are common for Asian countries, in particularly south Asian countries, it is essential that greater cooperation among countries in Asia/South Asia in research and testing activities related to climate change preventing duplication of efforts. Hence regional cooperation is essential to cope with the challenges of the climate changes on inland fisheries and aquaculture sector.

References

Amarasinghe, U.S. and Nguyen, T.T.T., 2009. Enhancing rural farmer income through fish production: Secondary use of water resources in Sri Lanka and elsewhere. In: S.S. De Silva and F.B. Davy (eds) Success Stories in Asian Aquaculture. pp. 103-130. Springer Science, Netherlands. ISBN 978-90-481-3085-6 ISBN 978-90-481-3087-0. http://library.enaca.org/emerging_issues/ success_stories/success-storiesasian aquaculture.pdf

Badjeck, M.C., Allison, E. Halls, A. and Dulvy, N., 2010. Impacts of climate variability and change on fishery based livelihoods. Marine Policy, 34: 375–383.

Basnayake, B.R.S.B. and Vithanage, J.C., 2004. Future climate scenarios of rainfall and temperature for Sri Lanka. Proceedings of the 60thAnnual Session of Sri Lanka Association for the Advancement of Science (SLASS). Section E1, 212 p.

Basnayake, B.R.S.B., Fernando, T.K., Vithanage, J.C., 2002. Variation of air temperature and rainfall during Yala and Maha agricultural seasons. Proceedings of the 58th Annual Session of Sri Lanka Association for the Advancement of Science (SLASS). Section E1, 212 p.

Chandrapala, L. 1996. Long term trends of rainfall and temperature in Sri Lanka. In: Climate Variability and Agriculture , Abrol, Y. P.; Gadgil, S.; Pant, G. B. (eds.). New Delhi, India: Narosa Publishing House. pp 153-162.

Chandrasoma. J., Pushpalatha, K.B.C. and Fernando, W.A.J.R., 2015. Impact of introduction of culture-based fisheries on fish production in perennial reservoirs of Sri Lanka. In: Sena S. De Silva, Ingram, B.A. and Wilkinson, S. (Eds.). Perspectives on culture-based Fisheries Development in Asia, pp 83-90. Network of Aquaculture Centres in Asia - Pacific, Bangkok, Thailand.

Cruz, R.V., Harasawa, H., Lal, M., Wu, S., Anokhin, Y., Punsalmaa, B., Honda, Y., Jafari, M., Li, C., Huu Ninh, N. 2007. Asia. Climate Change 2007: Impacts, Adaptation and Vulnerability. Contribution of Working Group II to the Fourth Assessment Report of the Intergovernmental Panel on Climate Change. In: Parry, M.L., Canziani, O.F.; Palutikof, J.P., van der Linden, P.J., and Hanson, C.E. (Eds.), pp 469-506, Cambridge University Press, Cambridge, UK.

Daw, T., Adger, W.N., Brown, K. and Badjeck, M.C., 2009. Climate change and capture fisheries: potential impacts, adaptation and mitigation. In: Cochrane, K., De Young, C., Soto, D. and Bahri, T. (Eds.) Climate change implications for fisheries and aquaculture: overview of current scientific knowledge, pp. 107–150, FAO Fisheries and Aquaculture Technical Paper No. 530. FAO, Rome.

De Silva S.S., Amarasinghe U S., and Nguyen T.T.T. (Eds), 2006. Better-Practice Approaches for Culture-based Fisheries Development in Asia. ACIAR Monograph No. 120, 96 p.

De Silva, C.S., 2006. Impacts of climate change on water resources in Sri Lanka. Paper presented at the 32nd WEDC International Conference, November 13-17, 2006, Colombo, Sri Lanka.

Fernando, T. K. 1997. Recent variations of rainfall and air temperature in Sri Lanka. Proceedings of the 53rd Annual Session of Sri Lanka Association for the Advancement of Science (SLASS). Section E1. Pp. 318.

Herath, A., 2007. Impact of ADB Projects on Agricultural Policies: A Joint Impact Assessment for the Asian Development Bank, Centre for Poverty Analysis. Working Paper Series No. 15, 104 p.

Herath, S., Ratnayake, U.R., 2004. Monitoring rainfall trends to predict adverse impacts - a case study from Sri Lanka. Global Environmental Change 14: 71-79

Hettiarachchi, S.S.L. and Samarawickrema, S.P., 2009. A strategic approach towards planning and management of impacts of sea level rise in Sri Lanka, Paper presented at the National Workshop on Mainstreaming Climate Change for Sustainable Development in Sri Lanka in Dambulla from 19 – 21 August, 2009, Institute of Policy Studies in collaboration with the Ministry of Environment and Natural Resources.

Imbulana, K.A.U.S., Wijesekera, S., Neupane, B. (Eds.), 2006. Sri Lanka National Water Development Report. UN.WATER/WWAP/2006/11, Sri Lanka. Ministry of Agriculture, Irrigation and Mahaweli Development, 221 p., ISBN:955-8395-01-3

IPCC, 2007. Climate Change 2007: Synthesis Report. Contribution of Working Groups I, II and III to the Fourth Assessment Report of the Intergovernmental Panel on Climate Change, Core Writing Team, Pachauri, R.K. and Reisinger, A. (Eds.). Geneva, Switzerland, 104 p.

Jayatilake, A., 2008. Climate Change due to Global Warming: A Global Challenge in Sri Lanka Perspective, Economic Review, June/July 2008.

Jayatillake, H.M., Chandrapala, L., Basnayake, B.R.S.B., Dharmaratne, G.H.P., 2005. Water resources and climate change. Proceedings of Workshop on Sri Lanka National Water Development Report. Wijesekera, N.T.S., Imbulana, K.A.U.S., Neupane, B. (Eds.) Paris, France: World Water Assessment Programme (WWAP).

Kumar, K.R., Sahai, A.K., Krishna Kumar, K., Patwardhan, S.K., Mishra, P.K., Revadekar, J.V., Kamala, K. and Pant, G.B., 2006. High-resolution climate change scenarios for India for the 21st century. Current Science 90(3): 334-345.

Ministry of Environment, 2010. Sector variability profile- agriculture and fisheries, climate change vulnerability in Sri Lanka. Supplementary document to the National Climate Change Adaptation Strategy for Sri Lanka, 2011-2016. Climate Change Secretariat, Ministry of Environment. Sri Lanka.

NACA, 2012. Impact of climate change on culture-based fisheries in seasonal reservoirs of Sri Lanka and the resilience capacities of rural communities - case study report, NACA, Bangkok. www.enaca.org.

Senaratne A., Perera, N. and Wickramasinghe K. 2009. Mainstreaming climate change for sustainable development in Sri Lanka: Towards a national agenda for action. Working paper series No.14, Institute of Policy Studies, Colombo, Sri Lanka.

Shelton, C., 2014. Climate change adaptation in fisheries and aquaculture – compilation of initial examples. FAO Fisheries and Aquaculture Circular No. 1088, FAO, Rome, 34 p.

USAID, 2018. Climate Change and Adaptation Strategy 2012-2016. www.climatelinks.org.

World Bank, 2018. Building Sri Lanka's resilience to climate change. www.worldbank.org.

Impact of Climate Change on Hydrological Cycle, Ecosystem, Fisheries and Food Security, Pages: 163–170
Edited by: B. Madhusoodana Kurup, M.R. Boopendranath, M. Harikrishnan and A.V. Shibu

CHAPTER - 13

Impact of Climate Change and Global Warming on Aquaculture and Fisheries: An Insight

H. Shivananda Murthy

(Former Dean, College of Fisheries, KVAFS University, Mangalore)
"Lavanya", Kaikunje East, B.C. Road-574219, Dakshina Kannada Dist., Karnataka, India.
E-mail: hsmurthy05@gmail.com

The fisheries and aquaculture sector supports the livelihoods of between 10 and 12 percent of the world's population. In the last five decades its production has significantly outpaced population growth (FAO, 2016), and increasing its contribution to food security and nutrition. It is often mentioned that the fisheries and aquaculture sector is extremely dynamic and used to deal with change, as historical patterns of changes in marine resources demonstrate (Baumgartner et al, 1992). The magnitude and uni-directionality of future climate-driven changes demand greater preparedness to combat and mitigate the expected changes. The warming of the climate system is unequivocal. The atmosphere and ocean have warmed, the amounts of snow and ice have diminished and sea level has risen. The uptake of additional energy in the climate system is caused by the increase in the atmospheric concentration of carbon dioxide (CO_2) and other greenhouse gases (GHGs). Aquatic systems that sustain fisheries and aquaculture are undergoing significant changes as a result of global warming and projections indicate that these changes will be accelerated in future. In many regions, climate change is affecting precipitation and melting of snow and ice, altering hydrological systems and affecting water resources in terms of quantity and quality. The absorption of increasing amounts of CO_2 produced due to anthropogenic activities by the oceans results in acidification of waters, with potentially detrimental impacts on shell-forming aquatic life.

Climate change affects communities and livelihoods in fisheries and aquaculture, and efforts to adapt to and mitigate climate change must therefore be human-centred. Most food producing inland fisheries are found in developing countries and are largely located in the tropics. Some of the poorest, most food insecure countries in the world are disproportionately dependent upon inland fisheries for nutritional and food security. Climate change will lead to changes in freshwater habitats and the fish assemblages that they support. There is a wide range of physiological and ecological impacts on both fish and the freshwater ecosystems supporting inland fisheries related to water temperature, water availability and flow, and other ecological perturbations. Direct and indirect climate change factors may result in favourable, unfavourable or neutral changes in aquaculture, in the short or long-term and at different spatial scales.

Some Issues with Respect to Global Warming Related to Fisheries

The Global warming and climate change has affected the environment, in particular fisheries and aquaculture, to substantial extent. Although, this phenomenon is not any sudden change, but over time and it has deleterious effects. The icebergs are melting at a faster rate than earlier. CO_2 concentrations have increased by 40 percent since pre-industrial times, primarily from fossil fuel emissions and secondarily from net land use change emissions. It is thus extremely likely that human influence has been the dominant cause of the observed warming since the mid-twentieth century. The ocean has absorbed 93 percent of this additional heat and sequestered 30 percent of the emitted anthropogenic CO_2. Over the period 1901 to 2010, global mean sea level also rose by 0.19 m (Church *et al.*, 2013). Aquatic systems that sustain fisheries and aquaculture are undergoing significant changes as a result of global warming and projections indicate that these changes will be accentuated in the future.

Series of changes in atmospheric concentrations of GHGs are used to modeland project future climates; most of these scenarios indicate that a large fractionof anthropogenic climate change is irreversible for centuries to come even after complete cessation of anthropogenic CO_2 emissions. In many regions, climate change is affecting precipitation and melting of snow and ice, altering hydrological systems and affecting water resources in terms of quantity and quality. Projections show that rainfall can be expected to increase in equatorial areas and decrease elsewhere. Temperature of water bodies is increasing across the globe, which results in more pronounced stratification of the water column, with more dramatic consequences for freshwater systems than for oceans because of their shallowness and lower buffering capacity.

Levels of dissolved oxygen decrease with increased temperature, and oxygen minimum zones in the oceans have expanded over the last decades, both in coastal and offshore areas. This trend is expected to continue. Global and local ocean circulation is changing, with a weakening of patterns such as the Gulf Stream and the California Current upwelling, and an increase in upwelling in other areas, such as in the Canary, the Humboldt and the Benguela Current systems. Responses are still heterogeneous and predictions have low confidence. The absorption of increasing amounts of anthropogenic CO_2 by the oceans results in acidification of waters, with potentially detrimental impacts on shell-forming aquatic life; water acidity has increased by 26 percent since the industrial revolution and this trend will continue, especially in warmer low- and mid-latitudes (FAO, 2019). Tropical web and primary production in the oceans has been projected to decrease by 3% to 9% by 2100; in freshwater systems observations vary depending on the area, but overall, forecasts are highly uncertain for both marine and freshwater systems because primary production is an integrator of changes in light, temperature and nutrients.

In inland water systems, an increase of water temperature is expected to occur in most areas, as a result of an increase of air temperature. This is linked to the relatively shallow nature of surface freshwaters and their susceptibility to atmospheric temperature change. It is reported that the river basins on all continents found that an increase of up to 1.8 °C in water temperature is expected, with geographical heterogeneities, including areas where the increase is expected to be minor, such as in the Lower Mekong river basin. There is a high confidence that rising water temperatures will lead to shifts in freshwater species distributions and exacerbate existing problems in water quality, especially in those systems experiencing high anthropogenic loading of nutrients (IPCC, 2014).

In the recent years, sea level has increased by an average of 3.1 mm/year as a result of climatic and non-climatic factors (Dangendorf *et al.*, 2017). The rate of increase shows a high variability across regions, with values up to three times the global average in the Western Pacific or null or negative values in the Eastern Pacific. Sea level has already risen by a global mean of 0.19 m over the period 1901 to 2010. It is estimated that between 2000 and 2100, the projected global mean sea level rise (SLR) will very likely (90 percent probability) reach between 0.5 m and 1.2 m under RCP8.5, 0.4 m to 0.9 m under RCP4.5, and 0.3 m to 0.8 m under RCP2.6 (Kopp *et al.*, 2014). There is a high certainty that the sea level will rise in 95 percent of the ocean area; however, there will be a significant regional heterogeneity in the SLR and thus in its consequences (IPCC, 2014).

Important Changes and Factors that Affect Fisheries and Aquaculture

The impact of climate change and global warming on communities and livelihoods in fisheries and aquaculture, and efforts to adapt to and mitigate climate change must therefore be human-centred. Climate adaptation strategies must emphasize the need for poverty eradication and food security, in accordance with the Paris Agreement, the United Nations 2030 Agenda for Sustainable Development and other international instruments, such as the 'Voluntary guidelines for securing sustainable small-scale fisheries in the context of food security and poverty eradication' (FAO, 2015). Measures to eradicate poverty and provide food security for people in fishing and aquaculture communities are also instrumental for climate change adaptation, and should be integrated in the formulation and implementation of national adaptation plans.

Climate change adaptation for building resilience must be multi-dimensional and multi-sectoral to help people out of poverty and to prevent them from descending further into it. Capacity at national, regional and local levels of governance should be mobilized to facilitate adaptation to climate change for the poor and vulnerable The economies of many countries and small developing states around Indian Ocean, are strongly dependent on large scale commercial industrial tuna fisheries and small-scale coastal fisheries. Climate change has already had observable impacts on fisheries throughout the Indian Ocean. Mounting evidence indicates that climate change is causing tuna species to shift spatial distributions and range, and is expected to have knock-on effects on coral reef fisheries in the Indian Ocean.

Warmer habitats with declining productivity, increased human pressure on less resilient coastal ecosystems, and major changes in distribution of large pelagic species will induce strong social and economic impacts in the region. To mitigate the changing spatial distribution, the mobile industrial fleets targeting tuna and billfish would relocate outside this region. For locally operated domestic fleets in the Indian Ocean, innovation in fishing techniques and gears will be a requirement to adapt and mitigate the impact of evolving habitat. Whenever catch or effort allocation systems are in place for migrating species, they should be adjusted periodically to account for the spatial patterns of target species. Monitoring, information dissemination and awareness of vulnerabilities to climate change in the Indian Ocean are currently weak and should be strengthened through community-level vulnerability assessments, integrated monitoring programmes that are effective and feasible with existing capacities, complemented by implementation of an ecosystem approach to fisheries.

Existing adaptive measures that could benefit the sector in the region include: (i) Developing and implementing fishery-specific adaptive management plans;(ii) Implementation, or strengthening, of effective systems of marine protectedareas (MPAs); (iii) Implementation of an ecosystem approach to fisheries, the FAO Code of Conduct for Responsible Fisheries and Voluntary guidelines on small-scale fisheries; and (iv) A sustainable, inclusive blue economy approach, partnerships at the local, regional and international levels will be key to achieving capacity development for effective management of marine and coastal fisheries, and building adaptive capacity and resilience to climate change in the region.

India, Pakistan, Bangladesh, Myanmar and Maldives are among the countries most affected by climate change and extreme weather events. There is strong socio-economic pressure from population growth, presence of dams and irrigation needs, heavy metal and waste pollution, habitat modification and destruction, illegal fishing, and insufficient infrastructure, skills credit and welfare. Inland fisheries captures are high and interlinked with the marine environment in countries where fisheries are important for employment, exports and income generation.

Artisanal and subsistence fisheries are important in the region; however, the number of artisanal boats is decreasing probably as a result of over-exploitation and overcapacity. Maximum potential of oceans and seas productivity is likely to decrease in those regions because of climate change, but that potential is not being achieved at present because of overfishing. This suggests that climate change impacts on specific fisheries can be partly mitigated by improving fisheries management.

In many countries, government institutions have long-term plans that recognize the likely impacts of climate change. With limited resources, they are trying to establish and enforce fisheries management measures which can increase the resilience of communities to climate change. Lack of resources and scientific knowledge is often addressed through internationally funded cooperative research and support programmes that encourage ecosystem- and community-based approaches and science-driven management advice. International fisheries safety and sustainability certifications are a successful driver of change and have the potential to improve infrastructure, people training, commodity values, diversify markets and support the ecological sustainability of the activities of associated industries.

Increased economic stability of fishermen and fish farmers and associated industries through easy and low-cost access to insurance, welfare, credits and minimum wages is being promoted through government and non-government agencies.

Certain Key Issues with Respect to Inland Fisheries and Aquaculture

Inland fisheries provide important contributions to global food demands. Most food producing inland fisheries states are found in developing countries and are largely located in the tropics. Some of the poorest, most food insecure countries in the world are disproportionately dependent upon inland fisheries for nutritional and food security. Globally, freshwater ecosystems that support the majority of inland fisheries are subject to a variety of anthropogenic pressures reflecting global change including over-extraction of water, over-exploitation of fish, introduction of non-native species, pollution, habitat degradation (including fragmentation) and increases in human population. The impacts of climate change will interact with many of these factors. Climate change will lead to changes in freshwater habitats and the fish assemblages that they support: only a few of these effects are expected to be beneficial to inland fisheries especially those based on native fish populations.

Freshwater ecosystems have relatively low buffering capacity and are therefore relatively sensitive to climate-related shocks and variability. There is a wide range of physiological and ecological impacts on fish and the freshwater ecosystems, aquaculture tanks and ponds related to water temperature, water availability and flow, and other ecological perturbations. Given the scale of direct and indirect impacts of global change, the adaptive capacity of all temperate, tropical and subarctic freshwater ecosystems and existing inland fisheries is relatively low. Direct or indirect climate change impacts may see considerable shifts in species compositions, but overall productivity might be sustained because of the high diversity and resilience typically shown by tropical systems and many invasive fish species.

Production of aquatic animals to meet demand will come mainly from aquaculture in the foreseeable future. It is, therefore, essential that we understand the effects of climate change on the sector from the broad to the finest spatial scales. Direct and indirect climate change drivers may result in favourable, unfavourable or neutral changes in aquaculture, in the short- or long-term and at different spatial scales. Unfavourable changes are likely to predominate in the developing countries, adversely affecting investment and sector growth. Several adaptation measures are already available to mitigate impacts of negative changes or increase resilience. They must be considered in accordance with multi-sectoral national strategies. Knowledge gaps in science, institutional and socio-economic change and policies hamper more effective adaptation of aquaculture, especially in the developing countries.

Possible negative assessments of aquaculture to climate change show that a number of countries in both high and low latitudes are highly vulnerable. In general, negative impacts are directly associated with governance, from national governments to farm level. Global assessments of vulnerability must be complemented by investigations at more localized levels, where specific aquaculture practices, environmental conditions and interactions with stakeholders and communities are taken into account.

Long-term climate-driven trends, e.g. increases in temperature and salinity, are more readily addressed than increasing climate variability and extreme events. With regard to the former, there is time to plan and implement adaptation measures (e.g. development and adoption of strains better adapted to increasing salinity conditions) while it is more difficult to plan for surprises and short-term events, such as storm surges. Adaptation strategies must, however, encompass these short-term, which also facilitates understanding, and use inclusive, bottom-up approaches involving stakeholders. Vulnerability reduction depends on broader adaptation measures beyond the aquaculture sector and there is a strong need to integrate aquaculture management and adaptation into watershed and coastal zone management. Ultimately, it is at the farm level where vulnerability reduction efforts converge; and negative impacts and vulnerability assessments should be as fine-grained as resources allow in order to be relevant to farmers. Capacity building in addressing vulnerability and improving adaptation to climate change, especially among target stakeholders, is an investment that more than pays for itself. Specific measures to reduce aquaculture vulnerability in accordance with the ecosystem approach to aquaculture include: (i) Improved management of farms and choice of farmed species, (ii) improved spatial planning of farms that takes climate-related risks into account, (iii) improved environmental monitoring involving users, and (iv) improved local, national and international coordination of prevention and mitigation actions.

Conclusion

Impacts of climate change and global warming on fisheries and aquaculture are of short-term and long-term. In view of expected increase in temperature and adverse environmental effects, it is likely that sensitive fish and shellfish may delay onset of breeding and spawning process, which is very much affected by physiological, neurological and endocrinal hormonal systems in certain species. Some species may migrate to altogether different environmental areas to overcome the unfavourable conditions on a short term basis. It is known that breeding, spawning and multiplication of fish and shellfish is temperature dependent through endocrinal and neuro-hormonal systems. There is an urgent need on national and global levels to take measures to mitigate the possible harmful effects of global warming and climate change in fisheries and aquaculture.

References

Baumgartner, T.R., Soutar, A. and Ferreira-Bartrina, V., 1992. Reconstruction of the history of Pacific sardine and northern anchovy populations over the past two millennia from sediments of the Santa Barbara basin, California. California Cooperative Oceanic Fisheries Investigations (CalCOFI) Report, 33: 24–41.

Church, J.A., P.U. Clark, A. Cazenave, J.M. Gregory, S. Jevrejeva, A. Levermann, M.A. Merrifield, G.A. Milne, R.S., Nerem, P.D. Nunn, A.J. Payne, W.T. Pfeffer, D. Stammer and A.S. Unnikrishnan, 2013: Sea Level Change. In: Climate Change 2013: The Physical Science Basis. Contribution of Working Group I to the Fifth Assessment Report of the Intergovernmental Panel on Climate Change [Stocker, T.F., D. Qin, G.-K. Plattner, M. Tignor, S.K. Allen, J. Boschung, A. Nauels, Y. Xia, V. Bex and P.M. Midgley (Eds.)]. Cambridge University Press, Cambridge, United Kingdom and New York, NY, USA.

Dangendorf, S., Marcos, M., Wöppelmann, G., Conrad, C.P., Frederikse, T. and Riva, R., 2017. Reassessment of 20th century global mean sea level rise. PNAS 114 (23): 5946-5951. https://doi.org/10.1073/pnas.1616007114

FAO, 2015. Voluntary Guidelines for Securing Sustainable Small-Scale Fisheries in the Context of Food Security and Poverty Eradication, FAO, Rome, 18 p. ISBN 978-92-5-108704-6

FAO, 2016. The State of World Fisheries and Aquaculture 2016. Contributing to food security and nutrition for all. FAO, Rome, 200 p.

FAO, 2019. Impacts of climate change on fisheries and aquaculture: Synthesis of current knowledge, adaptation and mitigation options. FAO Fisheries and Aquaculture Technical Paper 627. FAO, Rome., 654 p. ISBN: 9789251306079.

IPCC, 2014. Climate Change 2014: Synthesis Report, Intergovernmental Panel on Climate Change, 116 p.

Kopp, R.E., Hay, C.C., Little, C.M. and Mitrovica, J.X., 2015. Geographic variability of sea-level change. Current Climate Change Reports 1: 192–204.

Impact of Climate Change on Hydrological Cycle, Ecosystem, Fisheries and Food Security, Pages: 171–184
Edited by: B. Madhusoodana Kurup, M.R. Boopendranath, M. Harikrishnan and A.V. Shibu

CHAPTER - 14

Farming System Adaptation and Natural Herbal Oil Formulations against Various Emerging Diseases of Shrimp

Rajeev Kumar Jha[2], Haig Yousef Babikian[1], Kristina[1], and Sarayut Srisombat[2]

[1]*Asclepius Pharmaceutical Sciences, Singapore*
[2]*PT Central Proteina Prima, Indonesia*
E-mail: jha.fish@gmail.com

The Shrimp farming has the potential to be a long term sustainable industry for many tropical countries. Indeed, this industry has had a long and successful history in Asia and has always led world production (Flegel, 1997). The productivity from this industry reaches 1 million metric tons in 2004 of cultivated shrimp with a market value of billions US Dollars per year (Flegel, 2006). However, in 1982 shrimp production has declined slowly. A global shrimp survey by the Global Aquaculture Alliance (GAA) in 2001 revealed a rough overall loss to disease of approximately 22% in a single year, this translated into an estimate of about US$1 billion loss. The rapid growth of this industry has increased the number of epizootics associated with infectious diseases. With the respect disease agents, the GAA survey revealed that 60% of losses were attributed to viruses and about 20% to bacteria, and other 20% caused by fungi, parasites and unknown agents (Flegel *et al.*, 2008).

One of the infectious diseases began to affect *Penaeus chinensis* in China in 1993 and subsequently spread to Japan, Taiwan, and other Asian countries. This epizootics was identified as bacilliform virus and referred to family *Nimaviridae,* known as White Spot Syndrome Virus (WSSV). Main characteristic of this viral infection is the emergence of white spots on the exoskeleton especially on the carapace and last abdominal segment in size from a few mm to 1 cm or more in diameter (Pradeep *et al.*, 2012). WSSV-infected shrimp also show lethargic behavior, reduction in food consumption, loose cuticle and reddish to pink body discoloration (Pradeep *et al.*, 2012).

A new disease emerged in 2009, called 'Early mortality syndrome' which later on named as "Acute Hepatopancreatic Necrosis Disease" (AHPND), has been reported to have begun in China and cause significant losses and expanded rapidly to other countries. By 2010, the range of affected farms in China has expanded, and by 2011, EMS was confirmed in Vietnam and Malaysia and reached Thailand in 2012 (Tran *et al.*, 2013). This disease was responsible for almost 33% of shrimp production loss. The shrimp production during the first quarter of 2013 was 63,500 tons while it was 94,400 tons for the same period in 2012, indicating a production decline of approximately 30,900 tons due to these outbreaks (Joshi *et al.*, 2014). EMS refers to an unusual, acute mortality in shrimp within the first 50 days after stocking cultivation ponds and causing serious production losses in affected areas and is also impacting employment, social welfare, and international markets (Tran *et al.*, 2013). In early 2013, the Aquaculture Pathology Laboratory at the University of Arizona was able to isolate the causative agent of AHPNS in pure culture which has been identified as a member of the *Vibrio* clade (Soto-rodriguez *et al.*, 2015; Tran *et al.*, 2013). The gross signs of AHPND are evident by pond-side examination of affected shrimp accompanied by dissection and examination of the hepatopancreas (HP). The hepatopancreas of infected shrimp will show severe lesions atrophy that exhibits unique histopathology at the acute stage of the disease, consisting of massive sloughing of HP epithelial cells. Furthermore, other species from *Vibrio* can caused degeneration of HP tissues and produce luminous shrimp with accompanying behavioral changes and lack of appetite and led to mortality.

Another disease occur in shrimp farming industry is IMNV (Infectious Myonecrosis Virus) which has reported firstly in Brazil in 2004 and by 2006, it was spread East Java Island as well as in other south east Asian countries. The disease progression is slower and not threatening as other viral disease, but there may be high mortality due to adverse environmental conditions. The virus infects all the life stages of shrimp including post larvae, juveniles, and adult, but the mortality was observed in the juveniles. This was responsible for economic losses that valued at approximately US$20 million (OIE, World Animal Health Organization, 2009) in 2003. Severely infected shrimps become moribund and show nonspecific behavioral signs such as reducing of feed intake and lethargy during or soon after stressful events such as netting, feeding, sudden changes in water temperature, sudden reduction in water salinity (Prasad *et al.*, 2016).

In an attempt to reduce the infection of pathogen on cultivated shrimp, several strategies have been employed by shrimp farmers. Cleaning of pond bottoms by installing central drains in the middle of the pond that are equipped with suction pumps for periodic waste removal has been practiced, but it was found this strategies is costly (Soowannayan *et al.*, 2019). The uses of antibiotics to overcome the problems have been implemented for many years. Nevertheless, there are problems in the aftermath of the excessive uses of antibiotics. The compounds can depress the growth of some crustacean, disturb consumer's health and lead to bacterial resistances. The shrimp farmers and technicians have modified and improved the existing techniques in hatchery, water disinfection treatment, sanitation and in culture practices. More emphasis has been given on the pathogen free and bio-secured live feed to shrimp brood stock and post larvae. The culture ponds are supplied with pathogen free larvae and cleaned water for secured shrimp production. The latest strategy is to prevent the disease infection in animal and spread of disease in environment using natural sources. In this study, aim was to test the effect of Natural Blend Herbal Formulation which is commercially named as Pondguard (KKP RI no. D 16060285–HBC), is a water supplement with anti-pathogenic properties. A successful field trial was conducted to demonstrate its performance to have better production and quality shrimp by improving the overall health of the animal.

Materials and Methods

The pond trial was conducted in the 4 commercial ponds i.e. two each for control and Pondguard ponds at the same farm in East Java, Indonesia. The average size of each pond was 0.2 ha with the average depth of 1.2 meter. The ponds were HDPE lined with central drainage and reservoir. The water was sourced using borewell with average salinity of 25 ppt. The trial was conducted from June,2019 to October, 2019.

The pond preparation and water preparation were done by following the standard protocols of PT. Central Proteina Prima, Indonesia. In short, the water was collected in the reservoir, disinfected by chlorination and fertilizer was applied for plankton growth. The post larvae were procured from a biosecured hatchery of PT. CP Prima stocked at the rate of 100 pieces per square meter. The critical water quality parameters monitored were, dissolved oxygen, salinity, water temperature, total organic matter (TOM), and total ammonia nitrogen (TAN), alkalinity, plankton population and turbidity etc. Water quality parameters were checked once a week after stocking. Shrimp growth parameters sampling were performed by cast net on weekly basis after one month of stocking. The culture period was of 110 days with two times partial harvesting prior to final harvesting of shrimp.

The Pondguard (KKP RI No. D 16060285–HBC) consists of blend natural oils, like, Lavender oil, Pine oil and Eucalyptus oil. It is a product of PT. Asclepius Indonesia registered under Ministry of Fisheries, Indonesia. The Pondguard was applied at a dose 0.4 ppm on 7 days and 3 days prior to pond stocking. The Pondguard application was 0.2 ppm once a week during culture period. The measured quantity was mixed with 10 liter of pond water and sprayed over the pond during noon.

The shrimp were provided IRAWAN brand of feed from PT. Central Proteina Prima. The standard feeding table was followed with four times daily feeding to the shrimp.

The known disease like, pathogenic Vibrio, White sport syndrome virus, Infectious myonecrosis virus and EHP were monitored by PCR in every two weeks during culture period. The post larvae were screened for the same pathogens prior to stocking.

Results and Discussion

Pondguard effect on water quality

The water quality parameters of both control and Pondguard applied ponds were in the similar range. It shows that there was no negative impact of Pondguard on the water quality parameters. The TAN (total ammonia nitrogen) was in the range throughout the culture period. It is the total amount of nitrogen in the forms of NH3 and $NH4^+$ (Hariati *et al.*, 2017) that can accumulate rapidly as a result of the high metabolic excretion rate of shrimp and feed loading. The concentration of TAN must be less than 0.5 mg/L for optimum shrimp growth (Shan and Obbard, 2001). Excessive TAN can adversely affect the productivity and result in adverse impact on pond waters. Seen from figure below, the TAN concentration in treatment group was 0.573 ppm, while in control groups was 0.622 ppm and it was concluded that the use of Ponguard could reduce up to 7.87%. This result shows a promising treatment to reduce TAN concentration on pond water in a simple and effective way (Table 1).

Total Plankton were observed to be an excellent indicator of environmental conditions and aquatic health within ponds. Plankton was sensitive to changes in water quality, they respond to low dissolved oxygen levels, high nutrient levels, toxic contaminants, poor food quality or abundance, and predation. A ponds with good environmental conditions can be derived by observing at the plankton indicators, such as their biomass, abundance, and species diversity (Casé *et al.*, 2008). As it was desirable that ponds water should support plankton growth, the Pondguard treatment group show a higher plankton density compared to the control group (Fig. 1). Total plankton in control group was 1,643,704 ind/ml while in treatment group was 2,138,366 ind/ml. It can be concluded that pond treated with Pondguard has a better water quality than in control group so that is possible for plankton to grow better as well (Fig. 2).

Another weekly water quality parameters measured were TOM (Total Organic Matter) and Total Alkalinity showed in Fig. 3 below. TOM (Total Organic Matter) in control group was 223 ppm, while in Pondguard treatment group was 234 ppm, or increases 4.7% than the control group. Pantjara *et al* (2015) stated that the highest dissolved organic matter content of in the pond was caused by some dissolved organic material from organic waste (Pantjara *et al.*, 2015). Total Alkalinity present in control group was 140 ppm, while in Pondguard treatment group has reduced to 141 ppm. Total alkalinity is the total amount of sodium bicarbonate (HCO_3^-) and carbonate (CO_3^{2-}) in water, converted into mg/L (ppm) of calcium carbonate (CaCO3). Total alkalinity measurement result from both groups were suitable with the standard range for shrimp culture. Supriatna (2017) recommended total of alkalinity in shrimp pond waters was 80-200 ppm (Supriatna *et al.*, 2017). Alkalinity is a pH buffer that provides CO_2 for photosynthesis of algae and plants in water.

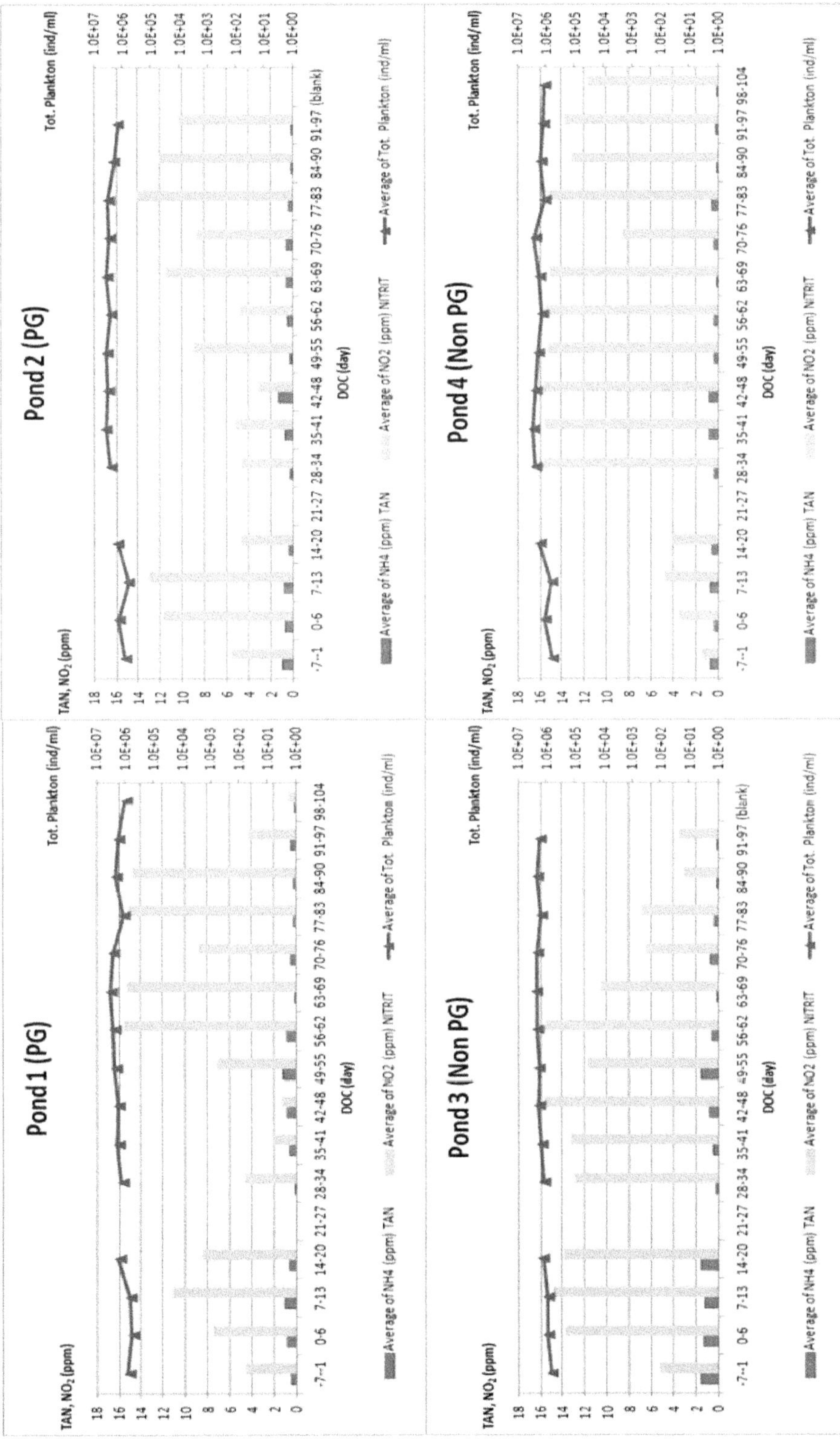

Fig. 1. Total Plankton, TAN and Nitrite in pond water during culture

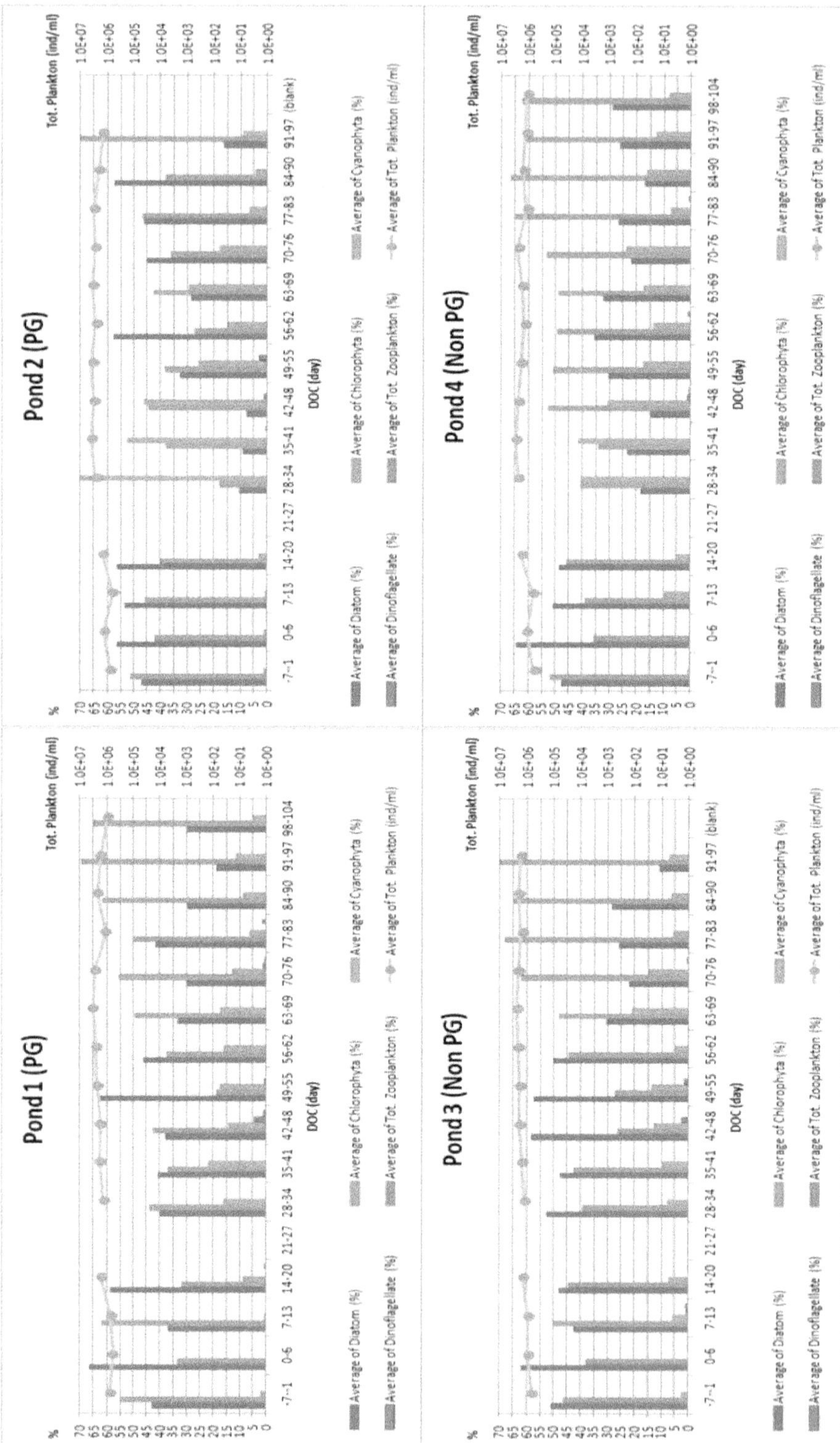

Fig. 2. Plankton community in water during culture

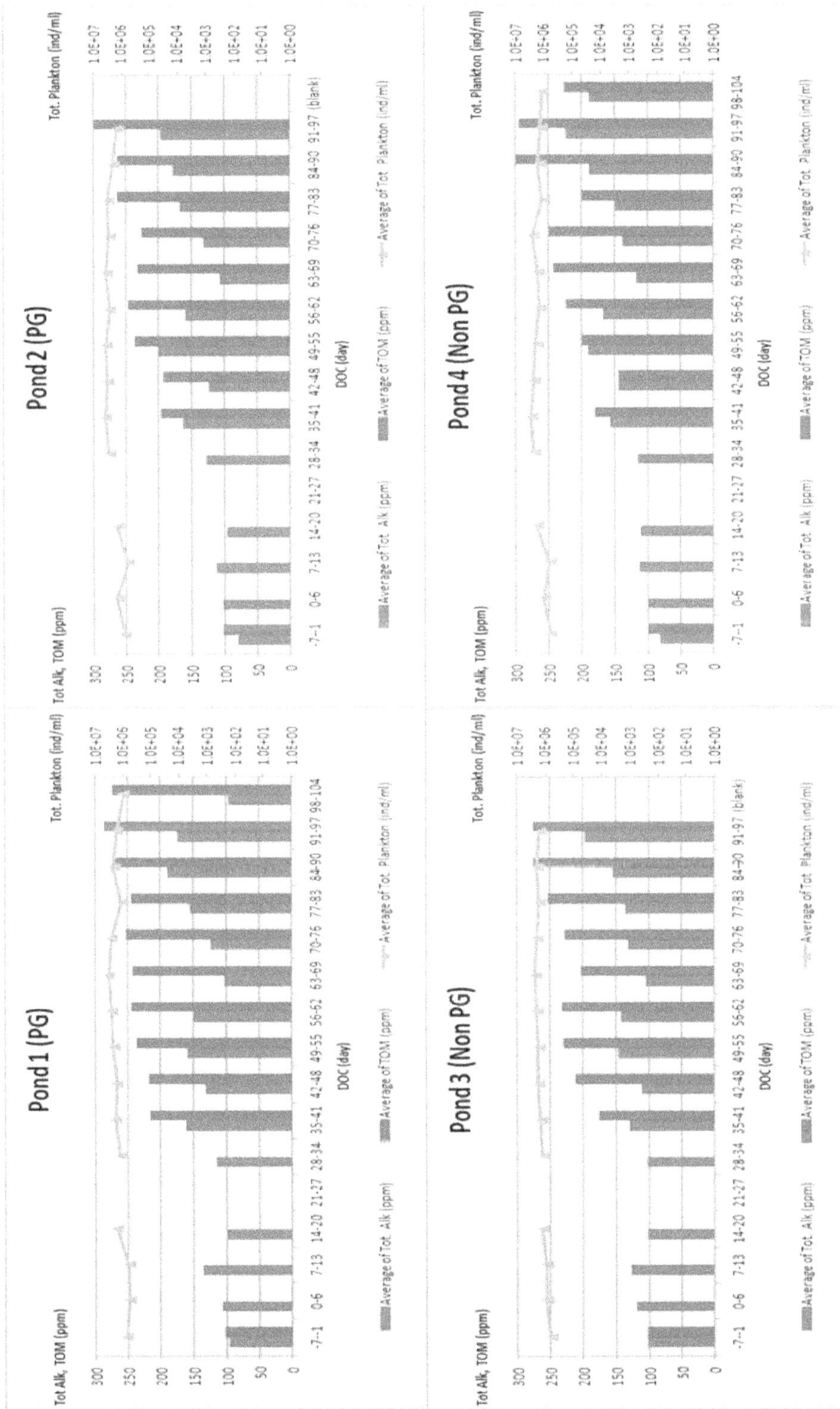

Fig. 3. Total Plankton, Total Alkalinity and TOM in water during culture

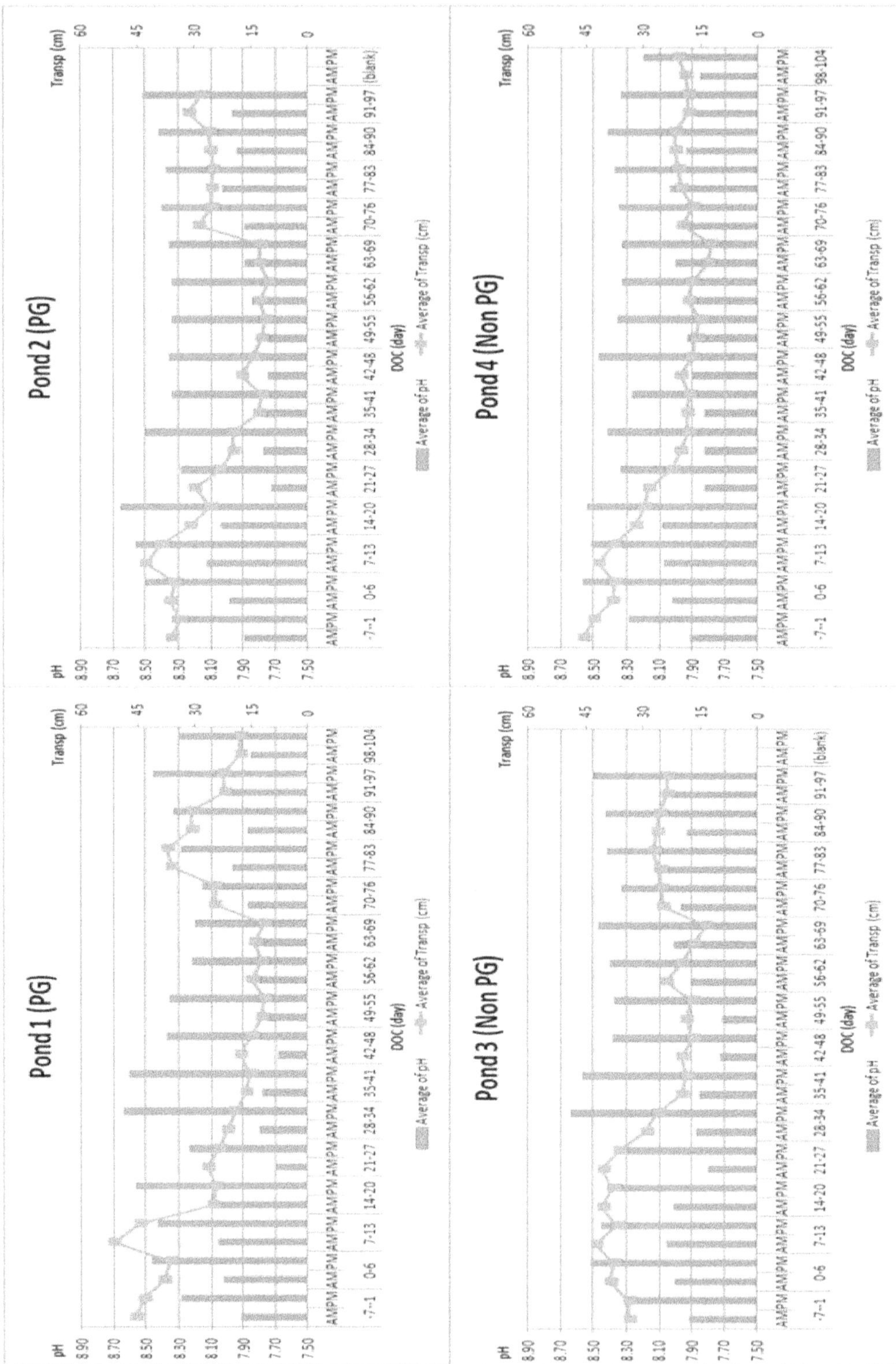

Fig. 4. Water transparency and pH (AM & PM) in water during culture

Fig. 4 shows water quality parameters measured were pH, and water transparency (Trp). Mean pH in control group was 8,17. Mean pH in Pondguard treatment group was 8,14. Mean water transparency was 12 – 45 cm in control groups and 10 – 51 cm in Pondguard treatment group. Water environments with pH of 7.5-8.8 are appropriate for shrimp aquaculture production (Hernández *et al.*, 2011; Supriatna *et al.*, 2017). Supriatna et al states that the optimal water transparency for shrimp culture ranges from 35 cm to 45 cm depth. Water transparency in the culture pond depends on the availability of zooplankton or phytoplankton and suspended solid particles (Supriatna *et al.*, 2017)

Pondguard and Disease presence

Pondguard has shown anti-bacterial (Khoa *et al.*, 2020; Babikian *et al.*, 2020; Babikian *et al.*, 2016; Jha *et al.*, 2017; Jiaravanon *et al.*, 2016) and anti-viral (Jha *et al.*, 2016; Babikian *et al.*, 2017) properties in lab scale trials in Indonesia and Vietnam. Both the control and PG treated ponds were monitored for different pathogens. Both the groups were free of virus (Table 1).

Table 1: Disease Monitoring Result by PCR Screening

Disease	Pondguard treatment						Control					
Monitoring	Pond 1			Pond 2			Pond 3			Pond 4		
Parameters	DOC 1	DOC 33	DOC 58	DOC 1	DOC 33	DOC 58	DOC 1	DOC 33	DOC 58	DOC 1	DOC 33	DOC 58
Pond Mud												
WSSV	(-)	(-)	(-)	(-)	(-)	(-)	(-)	(-)	(-)	(-)	(-)	(-)
IMNV	(-)	(-)	(-)	(-)	(-)	(-)	(-)	(-)	(-)	(-)	(-)	(-)
EHP	(-)	(-)	(-)	(-)	(-)	(-)	(-)	(-)	(-)	(-)	(-)	(-)
EMS	(-)	(-)	(-)	(-)	(-)	(-)	( )	( )	(-)	(-)	(-)	(-)
Shrimp												
WSSV	(-)	(-)	(-)	(-)	(-)	(-)	(-)	(-)	(-)	(-)	(-)	(-)
IMNV	(-)	(-)	(-)	(-)	(-)	(-)	(-)	(-)	(-)	(-)	(-)	(-)
EHP	(-)	(-)	(-)	(-)	(-)	(-)	(-)	(-)	(-)	(-)	(-)	(-)
EMS	(-)	(-)	(-)	(-)	(-)	(-)	(-)	(-)	(-)	(-)	(-)	(-)
Pond water												
EMS	(-)	-	-	(-)	-	-	(-)	-	-	(-)	-	-

Pondguard effect on shrimp growth performance

Shrimp growth performance were measured twice during the study at the harvesting time with the parameters measured were mean body weight (MBW), survival rate (SR), feed conversion ratio (FCR), total productivity, and health performance of shrimps. The results are given in Table 2 below. Mean body weight of control group was 21 g and MBW of treatment group was 20 g, concluded that MBW of PG treatment group was higher 1.5%

Table 2: Shrimp production performance

Protocol	POND GUARD				CONTROL			
Unit 2 Pond Address	Pond 1		Pond 2		Pond 3		Pond 4	
Hatchery	REMBANG		REMBANG		REMBANG		REMBANG	
PL Code	C13.1.5		C13.1.5		C13.1.5		C13.1.5	
Pond Type	Full PE		Full PE		Full PE		Full PE	
Area (m2)	1000		1000		1000		700	
PWA (HP)	5		4		5		4	
Stocking Date	9-May-19		9-May-19		9-May-19		9-May-19	
SD (pcs/m2)	106		106		106		116	
Init Biomass (kg)	0.53		0.53		0.53		0.41	
Mortality Incident_DOC (pcs)			62 (34)				94 (4487)	
Partial Harvesting Date	16-Jul-19	31-Jul-19	16-Jul-19	31-Jul-19	16-Jul-19		16-Jul-19	31-Jul-19
DOC (days)	69	84	69	84	69		69	84
Shrimp size	71	53	76	55	56		80	61
Biomass (kg)	161.0	176.6	151.6	174.3	105.9		114.3	150.9
Population (pcs)	11448	9381	11519	9598	5922		9200	9265
Tot. Biomass (kg)	337.55		325.94		105.9		265.19	
Tot. Pop (pcs)	20829		21117		5922		18465	
MBW (g)	16.21		15.43		17.88		14.36	
Final Harvesting Date	16-Aug-19		13-Aug-19		13-Aug-19		16-Aug-19	
DOC (days)	100		97		97		100	
Shrimp size	42.6	53	43.6	59	38.7	55	49	64
Biomass (kg)	992.33	8.81	970.3	33.5	796.7	31.1	763.79	16.65
Population (pcs)	42273	467	42305	1977	30832	1711	37426	1066
Tot. Biomass (kg)	1001.14		1003.8		827.8		780.44	
Tot. Pop (pcs)	42740		44282		32543		38491	
MBW (g)	23.42		22.67		25.44		20.28	
Shrimp Production								
Tot. Biomass (kg)	1338.69		1329.74		933.7		1045.63	
Tot. Pop (pcs)	63569		65399		38465		56956	
SR (%)	60		62		36		70	
MBW (g)	21		20		24		18	
FCR	1.38		1.33		1.27		1.52	
Acc.feed (kg)	1847.6		1765.6		1186.10		1591.85	
Productivity (kg/hp)	267.74		332.44		186.74		261.41	
Salmofan Score (min-max)	21	23	20	23	20	21	20	22

than the control group. It was correlated with the feed intake (FCR) where the PG treatment group's FCR are lower 3.13% than the control group. FCR of PG treatment group was 1.35 and the control group was 1.40. The survival rates of shrimp in PG treatment group are higher than the control group shown in Fig. 5 below. Survival rate of PG treatment group was 60.78% and in control group was 55,4 %. Total shrimp production of PG treatment group is also higher (10%) than the control group. The productivity of PG ponds was 13.3 ton/Ha and the Control ponds 12.1 ton/Ha.

Increasing in total productivity of shrimp in PG treatment group indicated as a result of the high survival rate. It was also confirmed by the health performances test of shrimp from all ponds. All of culture ponds and reservoir were negative to WSSV. The ponds were also detected negative to EMS, IMNV, and EHP during the experiment period. The high value of shrimp's body weight are affected the total productivity. In addition, color measurements of cooked shrimp based on Salmon Fan Scale showed that shrimp from PG applied Pond performed one score higher than the control ponds and increase the shrimp cost. The overall result of shrimp production performance measurements, concluded that the uses of Pondguard has a positive impact on shrimp growth performance.

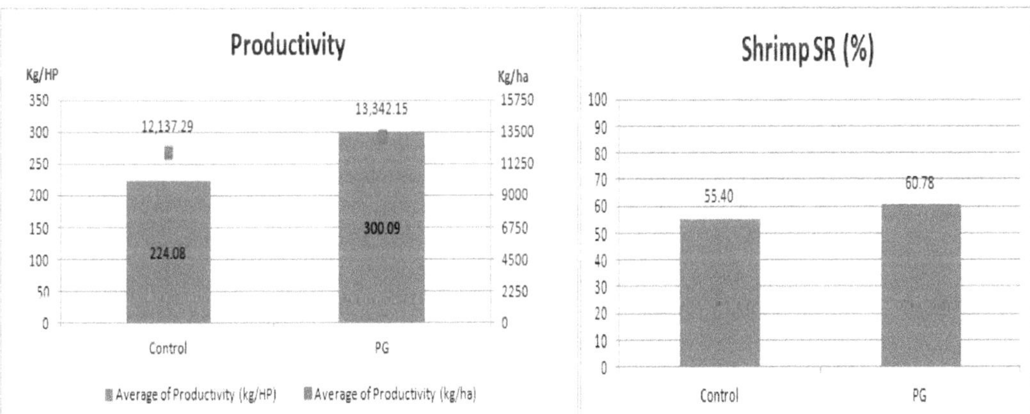

Fig. 5. Survival rate and productivity of shrimp

Conclusion

Application of Pondguard on shrimp pond, resulted in a positive impact on water quality parameters. The result of water quality parameters measurements was suitable for shrimp culture and higher than control group. The dominant water qualities which affected the shrimp production were total plankton, total ammonia nitrogen (TAN), total organic matter (TOM), total alkalinity, water pH and transparency. The shrimp body weight of Pondguard applied ponds (21 g) are better than control ponds (20 g). The FCR of Pondguard ponds (1.35) are lower than the control ponds (1.40). The shrimp's survival rate (SR) of PG ponds was 60.78% and in control group was 55.4 %. Total shrimp production of Pondguard treatment group was also higher (13.3 ton/ha) than the control group (12.1 ton/ha). The application of Pondguard also increased the shrimp colour based on Salmon Fan Scale one score higher than control pond.

References

Babikian, H.Y., Jha, R.K., Oanh, D.T.H. and Phu, T.Q., 2019. Study on the efficacy of Pondguard in improving clinical performance of white leg shrimp (*Penaeus vannamei*) in an AHPND bacterial challenge model. American Journal of Biomedical Science and Research 5(3): 80-81.

Babikian, Y.H., Babikian, H., Jha, R.K., Wisoyo, S.D., Asih, Y., Srisombat, S. and Jiaravanon, B., 2016. Efficacy of Natural Herbal Formulation against Acute Hepatopancreatic Necrosis Disease (AHPND) causing *Vibrio parahemolyticus* in *Penaeus vannamei.* Veterinary Medicine-Open Journal, November 2016. http://dx.doi.org/10.17140/VMOJ-1-109.

Babikian, Y.H., Babikian, H.Y., Jha, R.K., Wisoyo, S.D. Asih, Y., Srisombat, S. and Jiaravanon, B., 2017. Effectiveness of natural herbal oil formulation against infectious Myonecrosis Virus in *Penaeus vannamei.* Multidisciplinary Advances in Veterinary Science 1(2): 50-56.

Case, M., Leça, E.E., Leitao, S.N., Sant Anna, E.E., Schwamborn, R., de Moraes Junior, A.T., 2008. Plankton community as an indicator of water quality in tropical shrimp culture ponds. Mar. Pollut. Bull., 56: 1343–1352. https://doi.org/10.1016/j.marpolbul.2008.02.008

Flegel, T.W., 1997. Special topic review/ : Major viral diseases of the black tiger prawn (*Penaeus monodon*) in Thailand. World Journal of Microbiology and Biotechnology 13: 433–442.

Flegel, T.W., 2006. Detection of major penaeid shrimp viruses in Asia, a historical perspective with emphasis on Thailand. Aquaculture 258: 1–33. https://doi.org/10.1016/j.aquaculture.2006.05.013.

Flegel, T.W., Lightner, D.V, Lo, C.H.U.F., Owens, L., 2008. Shrimp Disease Control/ : Past , Present and Future. Dis. Asian Aquac. VI. Fish Heal. Sect. Asian Fish. Soc. 355–378.

Hariati, A.M., Aulaniam, Herawati, E.Y., Wiadnya, D.G.R., 2017. Effect of immobilized bacteria on TAN suppression, survival rate, and biomass harvest of pacific white shrimp, *Litopenaeus vannamei*, grown in biofloc culture system. Nat. Environ. Pollut. Technol. 16: 995–1002.

Hernández, J.J.C., Fernández, L.P.S. and Pogrebnyak, O., 2011. Assessment and prediction of water quality in shrimp culture using signal processing techniques. Aquac. Int. 19: 1083–1104. https://doi.org/10.1007/s10499-011-9426-z

Jha, R.K., 2017. Efficacy of natural herbal formulation against AHPND/EMS and WSSV in *Penaeus vannamei.* Paper presented in Aqua Aquaria, May 14-16, 2017 at Mangalore, India, organized by MPEDA, Government of India.

Jha, R.K., Babikian, Y.H., Babikian, H.Y., Wisoyo, S.D., Asih, Y., Srisombat, S. and Jiaravanon. B., 2016. Effectiveness of natural herbal oil formulation against white spot syndrome virus in *Penaeus vannamei.* Journal of Pharmacognosy and Natural Products 2: 123. doi: 10.4172/2472- 0992.1000123.

Jha, R.K., Jiaravanon, B., Babikian, Y.H., Babikian, H.Y., Srisombat, S., Wisoyo, S.D. and Asih, Y.,. Development of natural herbal oil formulation as an anti-White Spot Syndrome Virus agent in *Penaeus vannamei.* BIT's 5th Annual World Congress of Aquaculture and Fisheries-2016, Nov 4-6, 2016, Qingdao, China.

Jiaravanon, B., Babikian, Y.H., Babikian, H.Y., Khoa, L.V., Iswadi, and Jha, R.K. Efficacy of Anti-AHPND formulated Feed in Vannamei Shrimp in Vietnam. Third International Conference on Aquaculture & Fisheries. Sep 29-Oct 1, 2016, London, UK.

Joshi, J., Srisala, J., Truong, V.H., Chen, I.-T., Nuangsaeng, B., Suthienkul, O., Lo, C.F., Flegel, T.W., Sritunyalucksana, K. and Thitamadee, S., 2014. Variation in *Vibrio parahaemolyticus* isolates from a single Thai shrimp farm experiencing an outbreak of acute Hepatopancreatic necrosis disease (AHPND). Aquaculture 428–429: 297–302.

Khoa, L.V., Thaanh, T.L., Babikian, H., Jha, R.K., 2020. Short communication. Hiiau quaa cuaa Pondguard trong viiac phoang tro bianh hoaai tua gan tuay ceep (AHPND) trin tom nuoi. KHOA HOAC - KYA THUEAT TSVN 6 (325) THUAì HAI 16.3.2020 so6(325):so8(231).qxd 3/13/2020.

Pantjara, B., Syafaat, M.N., Kristanto, A.H., 2015. Effect of dynamical water quality on shrimp culture in the integrated multitrophic aquaculture (IMTA). Indones. Aquac. J. 10: 81. https://doi.org/10.15578/iaj.10.1.2015.81-90

Pradeep, B., Rai, P., Mohan, S.A., Shekhar, M.S. and Karunasagar, I., 2012. Biology, host range, pathogenesis and diagnosis of white spot syndrome virus. Indian J. Virol. 23, 161–174. https://doi.org/10.1007/s13337-012-0079-y

Prasad, K.P., Shyam, K.U., Banu, H., Jeena, K. and Krishnan, R., 2016. Infectious Myonecrosis virus (IMNV) – An alarming viral pathogen to Penaeid shrimps. Aquaculture 477: 99–105. https://doi.org/10.1016/j.aquaculture.2016.12.021

Shan, H. and Obbard, J.P., 2001. Ammonia removal from prawn aquaculture water using immobilized nitrifying bacteria. Appl. Microbiol. Biotechnol., 57: 791–798. https://doi.org/10.1007/s00253-001-0835-1.

Soowannayan, C., Chandra Teja, D.N., Yatip, P., Mazumder, F.Y., Krataitong, K., Unagul, P., Suetrong, S., Preedanon, S., Klaysuban, A., Sakayaroj, J. and Sangtiean, T., 2019. Vibrio biofilm inhibitors screened from marine fungi protect shrimp against acute hepatopancreatic necrosis disease (AHPND). Aquaculture 499: 1–8. https://doi.org/ 10.1016/ J.AQUACULTURE.2018.09.004

Soto-rodriguez, S.A., Gomez-gil, B., Lozano-olvera, R., Betancourt-lozano, M. and Morales-covarrubias, M.S., 2015. Field and experimental evidence of *Vibrio parahaemolyticus* as the causative agent of acute Hepatopancreatic necrosis disease of cultured shrimp (*Litopenaeus vannamei*) in Northwestern Mexico. Appl Environ Microbiol. 81(5): 81, 1689–1699.

Supriatna, Marsoedi, Hariati, A.M. and Mahmudi, M., 2017. Dissolved oxygen models in intensive culture of whiteleg shrimp, *Litopenaeus vannamei*, in East Java, Indonesia. AACL Bioflux 10: 768–778.

Tran, L., Nunan, L., Redman, R.M., Mohney, L.L., Pantoja, C.R., Fitzsimmons, K. and Lightner, D.V, 2013. Determination of the infectious nature of the agent of acute hepatopancreatic necrosis syndrome affecting penaeid shrimp. Dis. Aquat. Organ., 105: 45–55.

Impact of Climate Change on Hydrological Cycle, Ecosystem, Fisheries and Food Security, Pages: 185–194
Edited by: B. Madhusoodana Kurup, M.R. Boopendranath, M. Harikrishnan and A.V. Shibu

CHAPTER - 15

Molecular Identification and Characterization of Sugar and Peptide Carriers in the Giant Tiger Prawn, *Penaeus monodon*

V. Zonno[1*], A. Danieli[1], A. Romano[2], R. Acierno[3], R.A. Sreepada[4], R. Khandeparkar[4], R. Schiavone[1] and M. Maffia[1]**

[1] *DiSTeBA – Università del Salento, Prov. Lecce – Monteroni-73100 Lecce, Italy*
[2] *Division of Neuroscience–IRCCS S. Raffaele Scientific Institute-Milano, Italy*
[3] *TECNOSEA srl–Ecotekne Centre-Prov. Lecce – Monteroni-73100 Lecce, Italy*
[4]*Aquaculture Lab., CSIR-NIO, Dona Paula, Goa-403004, India*
[*]*E-mail: vincenzo.zonno@unisalento.it;* [**]*michele.maffia@unisalento.it*

For medical reasons, gastro-intestinal studies have been mainly focused on mammalians. However, most of Metazoa, more than 95%, are invertebrates that present very complex digestive organs. Shrimps are important invertebrates for aquaculture economy and, as for every aquatic animal, need a high dietary protein. Other nutrients, such as carbohydrates, have also an important physiological role. Some studies, in different shrimp species, have demonstrated that the dietary inclusion of lipid and carbohydrates might compensate the reduction of food protein content, without prejudicing the growth (Andrews *et al.*, 1972; Bautista, 1976; Colvin, 1976; Sedgwick, 1979). Functional investigations in crustaceans, *Penaeus japonicus* and *Homarus americanus,* have shown sodium-dependent mechanisms of sugars transport in hepatopancreas (Ahearn *et al.*, 1985; Verri *et al.*, 2001; Vilella *et al.*, 2003). Moreover, sodium-dependent glucose uptake was demonstrated in the gut of prawn *Macrobrachium rosenbergii* (Ahearn & Maginnis, 1977) and of the blue crab *Callinectes sapidus* (Chu, 1986). The presence of a dipeptide transporter has been also demonstrated in the lobster hepatopancreas (Thamotharan & Ahearn, 1996).

In spite of numerous functional studies, there is limited knowledge, at the molecular level, of genes and proteins involved in the mechanisms of nutrients uptake, even though the whole genome sequencing data of Litopenaeus *vannamei* is now available (https://www.ncbi.nlm.nih.gov/). In the present study, the expression and the tissue distribution of two transport proteins involved, respectively, in sugar and peptide absorption, have been investigated in the intestine, hepatopancreas and gills of *P. monodon*.

Material and Methods

Individuals of the giant tiger prawn *P. monodon* were collected at a local shrimp farm in Goa (India). Intestines, hepatopancreas and gills were rapidly removed, washed in an isotonic solution, transferred into an appropriate volume of RNA later solution (Life Technologies) and then stored at -20°C. Then the samples were shipped, in dry ice, to the Laboratory of General Physiology-Di.S.Te.B.A.-University of Salento, Italy. Here they were stored at -20°C until RNA extraction. Total RNA was extracted with TRIzol Reagent (Life Technologies) following the manufacturer's protocol and homogenized by a Polytron homogenizer (Kinematica, Switzerland). First-strand synthesis was performed from 1 μg total RNA using High TM Capacity RNA-to-cDNA Master Mix (Applied Biosystems).

Amplification of partial shrimp SGLT-like and PepT-like sequences was obtained by PCR with PCR Master Mix (Promega) and degenerate primers designed on conserved regions of known sequences of SGLT1 and PepT1, respectively.

The PCR reactions, for SGLT-like sequences, were conducted for 30 cycles (1 cycle: 45 sec. at 94°C, 60 sec. at 54°C, 3 min at 72°C) preceded by an initial denaturation step at 94°C for 2 min. and a final step of elongation for 7 min at 72°C.

The PCR reactions, for PepT-like sequences, were conducted for 24 cycles (1 cycle: 40 sec. at 94°C, 45 sec. at 63-53°C, 40 sec. min at 72°C) preceded by an initial denaturation step at 94°C for 2 min and a final step of elongation for 3 min at 72°C. The annealing step was conducted starting at 63°C up to 53°C decreasing of 1°C for each cycle.

The PCR products, diluted 1:30, were amplified by nested PCR. The SGLT-like products were amplified by new primers under the same PCR parameters while for PepT-like products, the nested PCR reactions were conducted for 35 cycles (1 cycle: 40 sec. at 94°C, 45 sec. at 52°C, 40 sec. min. at 72°C) preceded by an initial denaturation step at 94°C for 2 min. and a final step of elongation for 3 min at 72°C.

The full length of SGLT-*like* was obtained by 5-RACE and 3-RACE kits (Invitrogen) overlapping the initial middle fragment with the 5RACE and 3RACE sequences.

The obtained PCR products were cloned in pGEM-T Easy vector (Promega). The DNA sequence of the PCR products was identified by ABI PRISM 310 genetic analyzer following the manufacturer's protocol (Applied Biosystems, Foster City-USA).

In silico analysis. Shrimp SGLT-like and PepT-like amino acid sequences were deduced using the open reading frame (ORF) finder program (https://www.ncbi.nlm.nih.gov/orffinder/) and molecular weight calculation of protein by Expasy tool (https://web.expasy.org/compute_pi/). Sequence similarity search was performed by BLAST (htpps://blast.ncbi.nlm.nih.gov/Blast.cgi) while multiple sequence alignments were obtained by Clustal Omega (https://www.ebi.ac.uk/Tools/msa/clustalo/). Topology analysis was conducted by Protter tool (http://wlab.ethz.ch/protter).

Results and Discussion

SGLT-like sequence

Starting from *P. monodon* intestine mRNA we obtained the full sequence of shrimp SGLT-like transporter consisting of 2544 nucleotides. The deduced amino acid sequence (Fig. 1) corresponds to a hypothetical protein of 736 residues possessing an MW of 80,7 kDa. Similarity with other members of the SGLT family, obtained by BLAST analysis, is shown in Table 1, while multi-alignments performed by Clustal Ù, is reported in Fig. 2. Conserved residues, involved in the function of the transporter, are evidenced. The hypothetical topology of SGLT-like protein is reported in Fig. 3. Starting from *P. monodon* hepatopancreas mRNA, a partial region of 387 nucleotides was obtained and reported in Fig. 4, together with the corresponding amino acid sequence. No PCR amplification products, coding for an SGLT-like protein, were obtained from *P. monodon* gills mRNA.

```
MDSIPRGLVPVVVDATQDSDAFLGRATSASTDAPEGNEYDRRLPNWVDFLVLALYFAFVLAIGLYSSWRTKRGSVSSYFL
ASRNMHWIPVGASLFASNIGSGHFIGLAGSGAASGIGIGAFELNAIYVLMLLGWIFVPVYMSSGVYTMPEYLRERFGGQR
IRIYLSVLALILYVFTKISADLYAGALFIQTALDKGDSTEWLYLSILILLAIAALFTITGGLTAVIWTDFVQTILIIIGA
FILMGMSFHAVGGYNALVTDYFHAIPTNTVYSADNLTNCGEPPAYAMDSFRSIAPGKSDLPWTGMLFGVTISGIWYWCTD
QVIVQRTLSSKNMVHAKAGCVLAAALKFLPMFLLVFPGMAARILFTDRVACVVPKECEVVCGSTAGCTNIAYIELVLRLL
PTGLCGLMLAVMMAALMSSLTSIFNSASTIFTIDIWTRIRRGATDIELLIVGRAFVLVLVVISVIWIPVIQGSANSQLFV
YIQSISSFLSPPICAVYLLAIFWPRTTEPGAFWGLMIGLIIGLIRFIMEYAFVVPPCGSTEQDPRPEWIKIIVGNVHYLH
FGCILFVIVLAVTVVVSYLTEPIPEECLHRLTFWTRGSTEVRVSIREWLNRKKNAAAYSNEVVAENGGGEVVPETEEGGT
EVVSEKKQKYAERSSSVSSSGEDLPLWRRALNCVCGVETQKVSQAQEEDPEEHLTPEQKATRAAEFLEEKKPWTNVVDGA
SILVMCIAAFMWGYYA
```

Fig. 1. Putative amino acid sequence of *P. monodon* intestinal SGLT-like transporter

PepT-like sequence

Starting from mRNA samples of shrimp intestine and hepatopancreas, we amplified a PepT-like transcript showing, in both tissues, an identical partial sequence of 430 nucleotides, corresponding to a protein region of 143 amino acids (Fig. 5).

On the other hand, starting from shrimp gills, we obtained a putative PepT-like transcript of 426 bp and a deduced amino acids sequence of 141a.a (Fig. 6).

BLAST analysis has given the per cent identity with other peptide transporters members, reported in Table 2 (a) and (b).

Table 1: Percent identity of SGLT-like sequence from intestine of *P. monodon*

Organism	% Identity	N. Accession
Penaeus vannamei (sodium/glucose cotransporter 5-like isoform)	88,6	XP_027237377.1
Folsomia candida (sodium/glucose cotransporter 4-like isoform)	59,7	XP_021966846.1
Penaeus vannamei (sodium/glucose cotransporter 4-like isoform)	61,1	XP_027237504.1
Hylella azteca (sodium/glucose cotransporter 4-like isoform)	59,7	XP_018025308.1
Folsomia candida (sodium/glucose cotransporter 1)	55,9	OXA38115.1
Tridacna squamosa (sodium/glucose cotransporter 1-like protein)	56	ARQ84971.1
Limulus polyphemus (sodium/glucose cotransporter 5-like isoform)	56,3	XP_013789754.2
Myzuhopecten yessoensis (sodium/glucose cotransporter 4-like)	58	XP_021345702.1
Ixodes scapularis (sodium/glucose cotransporter 1 isoform)	55,9	XP_029842000.1
Ixodes scapularis (sodium/glucose cotransporter 5 isoform)	55,9	XP_029841998.1
Pongo abelii (sodium/glucose cotransporter 1 isoform)	49,2	XP_024095648.1

There are several experimental data on the presence of sodium-dependent sugar transports in crustacean hepatopancreas (Blaya *et al.*, 1998; Verri *et al.*, 2001; Vilella *et al.*, 2003; Sterling *et al.*, 2009; Sterling & Ahearn, 2011). On the converse, it's very controversial the presence and the physiological role of intestinal isoforms of sugar carrier transporters. At this regard, experimental pieces of evidence for the functional expression of sugar transporters are reported in the prawn *Macrobrachium rosenbergii* (Ahearn & Maginnis, 1977), in the blue crab *Callinectes sapidus* (Chu, 1986) and the lobster *Homarus americanus* (Obi *et al.*, 2011) intestines. Our experiments demonstrate, for the first time, the presence in *Penaeus monodon* intestine of an SGLT-like transcript similar to isoform 5 and, with less similarity, to isoform 4 and isoform 1 (around 60%, 50%, respectively, table 1). The presence of a similar transcript also in the *P. monodon* hepatopancreas suggests the tissue expression of an SGLT-like protein. No PCR amplification product of SGLT-like transcripts was found at the level of *P. monodon* gills, excluding the functional expression of the sugar transporter at respiratory epithelium site. Besides, bioinformatic analysis has let to identify putative functional sites of intestinal SGLT-like protein; these regions are conserved in phylogenetically distant organisms. The residues of glycine (G) 115,220,221 and arginine (R) 155 are conserved in *P. monodon*, confirming their essential role in members of the SGLT family (Turk & Wright, 1997). The Lys (K) at position 347 involved in the interaction with sodium ions is conserved (Panayotova-Heiermann *et al.*, 1998) and the region among 182 and 193 positions, that represents a part of pore for sodium (Lo & Silvermann, 1998), seems rather conserved during evolution

```
SGLT-like P. monodon        MDSIPRGLVPVVVDATQDSDAFLGRATSASTDAPEGNEYDRRLPNWVDFLVLALYFAFVL    60
SGLT2_H.sapiens             ME------------E--HTEA---------GSAPEMGAQKALIDNPADILVIAAYFLLVI    37
SGLT1_H.sapiens             MD------------S--STWSP------KTTAVTRPVETHELIRNAADISIIVIYFVVVM    40
SGLT1_D.rerio               MG------------A--DYFGF------SLLRNDVKRNTTIYVNNPADISVIVIYFLVVL    40
SGLT4_H.sapiens             MSKELAAMGP---GA--SGDGV------RTETAPHIALDSRVGLHAYDISVVVIYFVFVI    49
SGLT4_D.rerio               --------MP----A--SPEPV-------TATPEPEEVPAKFTLEAADIAVVVVYFVFVL    39
SGLT5_H.sapiens             ----------------------------MAANSTSDLHTPGTQLSVADIIVITVYFALNV    32
SGLT5_D.rerio               ----------------------------MAFNSTTKFFALSQSFSVSDIIVIGAYFLLNV    32
                                                                           *: ::  ** . :

SGLT-like P. monodon        AIGLYSSWRTKRGSVSSYFLASRNMHWIPVGASLFASNIGSGHFIGLAGSGAASGIGIGA    120
SGLT2_H.sapiens             GVGLWSMCRTNRGTVGGYFLAGRSMVWWPVGASLFASNIGSGHFVGLAGTGAASGLAVAG    97
SGLT1_H.sapiens             AVGLWAMFSTNRGTVGGFFLAGRSMVWWPIGASLFASNIGSGHFVGLAGTGAASGIAIGG    100
SGLT1_D.rerio               AVGVWAMVRTNRATVGGFFLAGRSMVWWPIGASLFASNIGSGHFVGIAGTGAAAGIAIGG    100
SGLT4_H.sapiens             AVGIWSSIRASRGTIGGYFLAGRSMSWWPIGASLMSSNVGSGLFIGLAGTGAAGGLAVGG    109
SGLT4_D.rerio               AVGIWSSIRANRGTVGGYFLAGRSMTWWPIGASLMSSNVGSGLFIGLAGTGAAGGLAVGG    99
SGLT5_H.sapiens             AVGIWSSCRASRNTVNGYFLAGRDMTWWPIGASLFASSEGSGLFIGLAGSGAAGGLAVAG    92
SGLT5_D.rerio               AVGIWSSCRVSRNTLSGYFLAGRDMAWWPIGASLFASSEGSGLFIGLAGTGAAGGIAVAG    92
                            .:*:::   ..* ::..:***.*.* * *:****::*. *** *:*:**:***.*:.:..

SGLT-like P. monod          FELN-------------------------AIYVLMLLGWIFVPVYMSSGVYTMPEYLRER    155
SGLT2_H.sapiens             FEWN-------------------------ALFVVLLLGWLFAPVYLTAGVITMPQYLRKR    132
SGLT1_H.sapiens             FEWN-------------------------ALVLVVVLGWLFVPIYIKAGVVTMPEYLRKR    135
SGLT1_D.rerio               FEWN-------------------------ALVVVIILGWLFVPIYIKAGVVTMPEYLKKR    135
SGLT4_H.sapiens             FEWNMRKSRSGGDRGIHPRSHGRTGVRSQATWLLLALGWVFVPVYIAAGVVTMPQYLKKR    169
SGLT4_D.rerio               FEWN-------------------------AAWVLIALGWIFVPVYISAGVVTMPEYLRKR    134
SGLT5_H.sapiens             FEWN-------------------------ATYVLLALAWVFVPIYISSEIVTLPEYIQKR    127
SGLT5_D.rerio               FEWN-------------------------ATYALLALAWVFVPVYVSSGIVTMPEYLGRR    127
                            ** *                         *   :: *.*:*.*:*: : : *:*:*: .*

SGLT-like P. monodon        FGGQRIRIYLSVLALILYVFTKISADLYAGALFIQTALDKGDSTEWLYLSILILLAIAAL    215
SGLT2_H.sapiens             FGGRRIRLYLSVLSLFLYIFTKISVDMFSGAVFIQQALGW-----NIYASVIALLGITMI    187
SGLT1_H.sapiens             FGGQRIQVYLSLLSLLLYIFTKISADIFSGAIFINLALGL-----NLYLAIFLLLAITAL    190
SGLT1_D.rerio               FGGQRIRIYLSVLSLCLYVFTKISADMFAGAIFINQALGL-----NIYLAVVILLLITAL    190
SGLT4_H.sapiens             FGGQRIQVYMSVLSLILYIFTKISTDIFSGALFIQMALGW-----NLYLSTGILLVVTAV    224
SGLT4_D.rerio               FGGQRIRIYMSVLSLILYILTKISTDIFSGALFIQVSLGW-----DLYLSTVILLAVTAL    189
SGLT5_H.sapiens             YGGQRIRMYLSVLSLLLSVFTKISLDLYAGALFVHICLGW-----NFYLSTILTLGITAL    182
SGLT5_D.rerio               FGGERIRMYLSALSLLLSVFTKISTDLYSGALFVQVCLGW-----NLYLSTVLMLVVTAL    182
                            :**.**::*:* *:* * ::**** *:::**:*:: .*.       :* :    * :: :

SGLT-like P. monodon        FTITGGLTAVIWTDFVQTILIIIGAFILMGMSFHAVGGYNALVTDYFHAIPTNTVYSAD-    274
SGLT2_H.sapiens             YTVTGGLAALMYTDTVQTFVILGGACILMGYAFHEVGGYSGLFDKYLGAATSLTVSEDPA    247
SGLT1_H.sapiens             YTITGGLAAVIYTDTLQTVIMLVGSLILTGFAFHEVGGYDAFMEKYMKAIPTIVSD---G    247
SGLT1_D.rerio               YTVTGGLAAVIYTDTLQTIIMVVGSFILMGFAFNKVGGYENFRDSYMTAIPSVIG-----    245
SGLT4_H.sapiens             YTIAGGLMAVIYTDALQTVIMVGGALVLMFLGFQDVGWYPGLEQRYRQAIPNVTVP----    280
SGLT4_D.rerio               YTIAGGLTAVIYTDALQTVIMVIGAFVLMFIAFDKVGWYEGLLVQYEKAAPALTVP----    245
SGLT5_H.sapiens             YTIAGGLAAVIYTDALQTLIMVVGAVILTIKAFDQIGGYGQLEAAYAQAIPSRTIA----    238
SGLT5_D.rerio               YTIAGGLAAVIYTDTLQTFIMIIGAVILTITAFNKIGGYSNLESVYLQAVPSKIIP----    238
                            :*::*** *:::** :**.::: *: :*   .*. :* *  :   *  *

SGLT-like P. monodon        ---NLTNCGEPPAYAMDSFRSIAPGKSDLPWTGMLFGVTISGIWYWCTDQVIVQRTLSSK    331
SGLT2_H.sapiens             VGNISSFCYRPRPDSYHLLRH--PVTGDLPWPALLLGLTIVSGWYWCSDQVIVQRCLAGK    305
SGLT1_H.sapiens             NTTFQEKCYTPRADSFHIFRD--PLTGDLPWPGFIFGMSILTLWYWCTDQVIVQRCLSAK    305
SGLT1_D.rerio               -TNISESCYTPRADSFHIFRD--PINGDLPWPGLLFGLTIQAGWYWCTDQVIVQRCLSAK    302
SGLT4_H.sapiens             ----NTTCHLPRPDAFHILRD--PVSGDIPWPGLIFGLTVLATWCWCTDQVIVQRSLSAK    334
SGLT4_D.rerio               ----NTTCHLPRSDAFHIFRD--PVTGDIPWPGLIFGLTVLATWVWCTDQVIVQRSLSAK    299
SGLT5_H.sapiens             ----NTTCHLPRTDAMHMFRD--PHTGDLPWTGMTFGLTIMATWYWCTDQVIVQRSLSAR    292
SGLT5_D.rerio               ----NTTCHLPRHDAMHLFRD--PVHGDLPWPGMTLGLTIIATWYWCTDQVIVQRSLSAK    292
                                   *  *   : . :*   *  .*:** .: :*:::   * **:******* *:.:

SGLT-like P. monodon        NMVHAKAGCVLAAALKFLPMFLLVFPGMAARILFTD----------------RVACVVPK    375
SGLT2_H.sapiens             SLTHIKAGCILCGYLKLTPMFLMVMPGMISRILYPD----------------EVACVVPE    349
SGLT1_H.sapiens             NMSHVKGGCILCGYLKLMPMFIMVMPGMISRILYTE----------------KIACVVPS    349
SGLT1_D.rerio               NLSHVKAGCILCGYLKLLPMFLMVFPGMISRVLYPD----------------EIACVDPD    346
SGLT4_H.sapiens             SLSHAKGGSVLGGYLKILPMFFIVMPGMISRALFPD----------------EVGCVDPD    378
SGLT4_D.rerio               NLSHAKAGSVLGGYLKVFPMFFVVMPGMISRALYPD----------------EVACVDPD    343
SGLT5_H.sapiens             DLNHAKAGSILASYLKMLPMGLIIMPGMISRALFPGAHVYEERHQVSVSRTDDVGCVVPS    352
SGLT5_D.rerio               SLSHAKGASIFASYLKLLPMMFIILPGMISRALYPD----------------TVACVDPE    336
                            .: * *...:: . **. ** ::::*** :* *:                   :.** *.

SGLT-like P. monodon        ECEVVCGSTAGCTNIAYIELVLRLLPTGLCGLMLAVMMAALMSSLTSIFNSASTIFTIDI    435
SGLT2_H.sapiens             VCRRVCGTEVGCSNIAYPRLVVKLMPNGLRGLMLAVMLAALMSSLASIFNSSSTLFTMDI    409
SGLT1_H.sapiens             ECEKYCGTKVGCTNIAYPTLVVELMPNGLRGLMLSVMLASLMSSLTSIFNSASTLFTMDI    409
SGLT1_D.rerio               ECKAVCGAEVGCTNIAYPKLVVDLMPNGLRGLMLSVMLASLMSSLTSIFNSASTLFTMDI    406
SGLT4_H.sapiens             VCQRICGARVGCSNIAYPKLVMALMPVGLRGLMIAVIMAALMSSLTSIFNSSSTLFTIDV    438
SGLT4_D.rerio               ECQKICGAKVGCSNIAYPKLVVELMPVGMRGLMIAVMMAALMSSLTSIFNSSSTLFTMDI    403
SGLT5_H.sapiens             ECLRACGAEVGCSNIAYPKLVMELMPIGLRGLMIAVMLAALMSSLTSIFNSSSTLFTMDI    412
SGLT5_D.rerio               ECVKVCGAEVGCSNIAFPKLVIELMPSGLRGLMIAVMMAALMSSLTSIFNSSSTLFTMDI    396
                             *   **: .**:***:  **: *:* *: ***::*::*:*****:*****:**:**:*:

SGLT-like P. monodon        WTRIRRGATDIELLIVGRAFVLVLVVISVIWIPVIQGSANSQLFVYIQSISSFLSPPICA    495
SGLT2_H.sapiens             YTRLRPRAGDRELLLVGRLWVVFIVVVSVAWLPVVQAAQGGQLFDYIQAVSSYLAPPVSA    469
SGLT1_H.sapiens             YAKVRKRASEKELMIAGRLFILVLIGISIAWVPIVQSAQSGQLFDYIQSITSYLGPPIAA    469
SGLT1_D.rerio               YTKIRPKAKEKELMIAGRLFMLLLIGISIAWIPIVQTAQSGQLFDYIQSITSYLAPPIAA    466
SGLT4_H.sapiens             WQRFRRKSTEQELMVVGRVFVVFLVVISILWIPIIQSSNSGQLFDYIQAVTSYLAPPITA    498
SGLT4_D.rerio               WQRIRPRASEKELMVVGRVFILLLVALSIVWIPVIQTANSGQLFDYIQAITSFLSPPITT    463
SGLT5_H.sapiens             WRRLRPRSGERELLLVGRLVIVALIGVSVAWIPVLQDSNSGQLFIYMQSVTSSLAPPVTA    472
SGLT5_D.rerio               WKKYRRGASEKELLLVGRIVTVILVVISVVWIPILQSANSGQLYVYIQSVTSYLAPPVTA    456
                            : : *  : : **::.**   : :: :*: *:*::* : ..**: *:*:::* *.**: :

SGLT-like P. monodon        VYLLAIFWPRTTEPGAFWGLMIGLIIGLIRFIMEYAFVVPPCGSTEQDPRPEWIKIIVGN    555
SGLT2_H.sapiens             VFVLALFVPRVNEQGAFWGLIGGLLMGLARLIPEFSFGSGSCVQPSACP------AFLCG    523
SGLT1_H.sapiens             VFLLAIFWKRVNEPGAFWGLILGLLIGISRMITEFAYGTGSCMEPSNCP------TIICG    523
SGLT1_D.rerio               VFALAIFCKRVNEPGAFYGLCVGLLVGLARMITEFAYGTGSCVNPSNCP------TIICG    520
SGLT4_H.sapiens             LFLLAIFCKRVTEPGAFWGLVFGLGVGLLRMILEFSYPAPACGEVDRRP------AVLKD    552
SGLT4_D.rerio               VFIMAIFWGRVNEQGAFWGLMVGLVVGMVRMIMEFVYGTPSCGETDLRP------SLLKD    517
SGLT5_H.sapiens             VFVLGVFWRRANEQGAFWGLIAGLVVGATRLVLEFLNPAPPCGEPDTRP------AVLGS    526
SGLT5_D.rerio               IFVMAVFWKRTNEAGAFWGLMVGLVVGLTRMVLEFAFPPPRCGVFDPAP------SVLRS    510
                            :: :.:*  *..* ***:**  ** :*  *:: *:      *   .  *       .: .

SGLT-like P. monodon        VHYLHFGCILFVIVLAVTVVVSYLTEPIPEECLHRLTFWTRGSTEVRVSIREWLNRKKNA    615
SGLT2_H.sapiens             VHYLYFAIVLFFCSGLLTLTVSLCTAPIPRKHLHRLVFSLRHSKEEREDLDADEQQGSSL    583
SGLT1_H.sapiens             VHYLYFAIILFAISFITIVVISLLTKPIPDVHLYRLCWSLRNSKEERIDLDAEEENIQEG    583
```

Fig. 2. Multi-alignments performed by Clustal Ω.

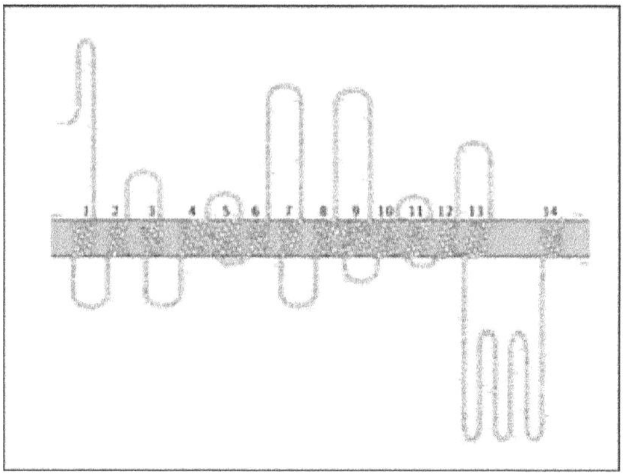

Fig. 3: Membrane topology model of intestinal SGLT-like carrier in *P. monodon*

```
5′ tccctcttcgcgagcaacatcggctctggtcacttcatcgggctggctggatcgggcgcc
   S  L  F  A  S  N  I  G  S  G  H  F  I  G  L  A  G  S  G  A
   gcgtctggcataggtattggcgccttcgagctcaatgcgatctatgtgcttatgctgctc
   A  S  G  I  G  I  G  A  F  E  L  N  A  I  Y  V  L  M  L  L
   ggttggattttcgtccctgtgtacatgagttcaggagtttacacaatgcccgagtatctt
   G  W  I  F  V  P  V  Y  M  S  S  G  V  Y  T  M  P  E  Y  L
   agggaaagatttggaggccagagaattaggatttacctctctgtcttggctctaatctta
   R  E  R  F  G  G  Q  R  I  R  I  Y  L  S  V  L  A  L  I  L
   tacgtcttcactaaaatttcggccgacctgtatgcgggggcactctttatccaaacagcc
   Y  V  F  T  K  I  S  A  D  L  Y  A  G  A  L  F  I  Q  T  A
   ctggataaaggagactctaccgaatggctgtacctcagtatccttatcctgttggccatt
   L  D  K  G  D  S  T  E  W  L  Y  L  S  I  L  I  L  L  A  I
   gcggcgctgttcactataaccggggt 3′
   A  A  L  F  T  I  T  G  G
```

Fig. 4: Partial nucleotide and aminoacid sequences of *P. monodon* hepatopancreas SGLT-like transporter

```
5′tgttcttcctcaacaggaacgacaacttgcccagttcttttccatttttacttcagtatc
   V  L  P  Q  Q  E  R  Q  L  A  Q  F  F  S  I  F  Y  F  S  I
  aatgcgggttctcttatatcaacttttgtaacaccagtcttgagagatgacgtaaagtgc
   N  A  G  S  L  I  S  T  F  V  T  P  V  L  R  D  D  V  K  C
  ttcaatgatgactgctactctttagcctttggagttcctgctgtcttaatgcttgttgcc
   F  N  D  D  C  Y  S  L  A  F  G  V  P  A  V  L  M  L  V  A
  ctaatcatcttcgtgattggatatttgactggctcatacaaaatcatccctcccgagggt
   L  I  I  F  V  I  G  Y  L  T  G  S  Y  K  I  I  P  P  E  G
  agtatgctgacgaacgtatttggatctatttggcatgcactgagaagcaaatcaaagagc
   S  M  L  T  N  V  F  G  S  I  W  H  A  L  R  S  K  S  K  S
  aaggaaaagaaggaacactggctagattatgcctcagataaatacgaaccaaagcttatc
   K  E  K  K  E  H  W  L  D  Y  A  S  D  K  Y  E  P  K  L  I
  aatgatgtcaagattctgctcaagatcctcgtgttgtacattcctcttccatttttctgg
   N  D  V  K  I  L  L  K  I  L  V  L  Y  I  P  L  P  F  F  W
  gctttgttt 3′
   A  L  F
```

Fig. 5: Partial nucleotide and amino acid sequences of *P. monodon* hepatopancreas and intestine PEPT-like transporter

```
5' tgttcttcctcaacaggaacgacaacttgcccagttcttttccatttttacttcagtatc
     V  L  P  Q  Q  E  R  Q  L  A  Q  F  F  S  I  F  Y  F  S  I
   aatgcaggttctcttatatcaacttttgtaacaccagtcttgagagatgacgtaaagtgc
     N  A  G  S  L  I  S  T  F  V  T  P  V  L  R  D  D  V  K  C
   ttcaatgatgactgctactctttagcctttggagttcctgctgtcttaatgcttgttgcc
     F  N  D  D  C  Y  S  L  A  F  G  V  P  A  V  L  M  L  V  A
   ctaattctgttcctgctcggattttcaatgtacacaatcaagcctccagagggaaacata
     L  I  L  F  L  L  G  F  S  M  Y  T  I  K  P  P  E  G  N  I
   gtagtccgtgtggggtcatgtgtcgcgcatgcactgagaagcaaatcaaagagcaaggaa
     V  V  R  V  G  S  C  V  A  H  A  L  R  S  K  S  K  S  K  E
   aagaaggaacactggctagattatgcctcagataaatacgaaccaaagcttatcaatgat
     K  K  E  H  W  L  D  Y  A  S  D  K  Y  E  P  K  L  I  N  D
   gtcaagattctgctcaagatcctcgtgttgtacattccctcttccattttctggggctt
     V  K  I  L  L  K  I  L  V  L  Y  I  P  S  S  I  F  L  G  L
   tgttt 3'
     C
```

Fig. 6: Partial nucleotide and amino acid sequences of *P. monodon* gill PEPT-like transporter

Table 2: Percent identity of SGLT-like sequence from intestine of *P. monodon*

(a) Per cent identity of PEPT-like squence from intenstine and hepatopancreas of *P. monodon*

Organism	% Identity	N. Accession
Penaeus vannamei (solute carrier family 15 2-like isoform X4)	96,5	XP_027230528.1
Zootermopsis nevadensis (peptide transport family 1 isoform X1)	64,1	XP_021917356.1
Xenopus laevis (peptide transport 2)	63,6	NP_001080398.1
Daphnia magna (peptide transport family 1)	58,6	KZS10979.1

(b) Per cent identity of PEPT-like squence from gill of *P. monodon*

Organism	% Identity	N. Accession
Penaeus vannamei (solute carrier family 15 2-like isoform X2)	99,3	XP_027230526.1
Zootermopsis nevadensis (peptide transport family 1 isoform X4)	64,3	XP_021917360.1
Amyelois transitella (peptide transport family 1 isoform X2)	63,8	XP_013189413.1
Galleria melonella (peptide transport family 1 isoform X2)	63,8	XP_026759612.1

with an identity up to 66 %; several substituted amino acids have similar chemical properties. The multiple alignments show conserved residues of cysteine (C279, C318, C377, C387 and C537) involved in disulphide bond and other interaction functionally essential for the transporter (Martin *et al.*, 1996; Xia *et al.*, 2005; Gagnon *et al.*, 2006). Moreover, residues involved in glucose interaction and recognition, are conserved: Gln (Q) 471 and 483, Thr (T) or Ser (S) 486, Arg (R) 525 (Diez-Sampedro *et al.*, 2001; Hirayama *et al.*, 2007). The computational topology analysis showed a typical structure of an SGLT family member constituted by 14 trans-membrane helical domain, an outside NH2-terminal start position and a large loop within the 13th and 14th TMH (Turk & Wright, 1997).

Previous experimental data reported a peptide transport dependent glycin-sarcosine uptake in brush border membrane vesicles isolated from lobster hepatopancreas (Thamotaran & Ahearn, 1996). Our molecular experiments confirm, for the first time in *P. monodon*, the presence, both in intestine and hepatopancreas, of a PEPT-like transcript similar to the isoform 2, with a homology up to 64 % to isoform 1. The short sequence, by the way, does not help to assess a clear identification. Surprisingly our preliminary data suggest the expression of a PEPT-like protein also in *P. monodon* gills, opening several questions about the function of this peptide absorption mechanism in the respiratory epithelium.

These results upgrade the functional and molecular knowledge on nutrient absorption mechanisms of *P. monodon*; a better understanding of intestinal physiology could be helpful to obtain a proper feed formula for shrimp farming. The improved quality of feed, based on the nutritional need of the animals, will positively affect the growth performance and well-being of the farmed shrimps thus improving their resilience to climate change pressure. This aspect, together with the reduction of waste, as the consequence of better feeding management, represents an important step toward a more sustainable and eco-friendly shrimp farming activity.

Acknowledgement

Project partially financed by EU under FP7 program (PIRSES-GA 2008-230847 PASSA project).

References

Ahearn, G. A. and Maginniss, L. A. (1977). Kinetics of glucose transport by the perfused mid-gut of the freshwater prawn *Macrobrachium rosenbergii*. *Journal of Physiology*. 271, 319-336.

Ahearn, G. A., Grover, M. L. and Dunn, R. E. (1985). Glucose transport by lobster hepatopancreatic brush border membrane vesicles. *American Journal of Physiology*. 248, R133-R141.

Andrews, J. W., Sick, L. V. and Baptist, G. J. (1972). The influence of dietary protein and energy levels on growth and survival of penaeid shrimp. *Aquaculture*. 1, 341-347.

Bautista, M. N. (1986). The response of *Penaeus monodon* juveniles to varying protein/energy ratios in test diets. *Aquaculture*. 53, 229-242.

Blaya J. A., Vázquez C. M., Muriana F. J., Ruiz-Gutiérrez V., Bolufer J. (1998). alpha-MeGlc and D-glucose transport by hepatopancreatic brush border membrane vesicles from prawn. *J. Physiol. Biochem*. 54, 1–7.

Chu K. H. (1986). Glucose transport by the *in vitro* perfused midgut of the blue crab, *Callinectes sapidus*. *J. Exp. Biol.*123, 325-344.

Colvin, P. M. (1976). Nutritional studies on penaeid prawns: protein requirements in(compounded diets for juvenile *Penaeus indicus* (Milne Edwards). *Aquaculture*. 7, 315-326.

Diez-Sampedro, A., Wright, E. M. and Hirayama, B. A. (2001). Residue 457 Controls Sugar Binding and Transport in the Na/Glucose Cotransporter, J. Biol. Chem. 276, 49188-49194.

Gagnon D. G., Pierre Bissonnette P., Lapointe J.Y. (2006). Identification of a Disulfide Bridge Linking the Fourth and the Seventh Extracellular Loops of the Na+/Glucose Cotransporter. *J. Gen. Physiol.* Volume 127 Number 2 February 2006 145–158.

Hirayama B. A., Loo D. D. F., Diez-Sampedro A., Leung D. W., Meinild A.K., Lai-Bing M., Turk E., and Wright E. M. (2007). Sodium-Dependent Reorganization of the Sugar-Binding Site of SGLT1 *Biochemistry 46,* 13391-13406.

Lo, B. and Silverman, M. (1998). Cysteine scanning mutagenesis of the segment between putative transmembrane helices IV and V of the high affinity Na/glucose cotransporter SGLT1. *The Journal of Biological Chemistry*. 273,29341-29351.

Martin, G. M., E. Turk, M. P., Lostao, C. Kerner, and E. M. Wright (1996). Defects in sodium/glucose cotransporter (SGLT1) trafficking and function cause glucose-galactose malabsorption. *Nat. Genet.,*12, 216-220.

Obi I. E., Sterling K. M., Ahearn G. A. (2011). Transepithelial D-glucose and D-fructose transport across the American lobster, *Homarus americanus*, intestine. *J. Exp. Biol.* 214, 2337–2344.

Panayotova-Heiermann M., Loo D.D.F., Lam J.T. and Wright E.M. (1998). Neutralization of conservative charged transmembrane residues in the Na^+/glucose cotransporter SGLT1. *Biochemistry*. 37, 10522-10528.

Sedgwick, R. W. (1979). Effect of ration size and feeding frequency on the growth and food conversion of juvenile *Penaeus merguiensis* de Man. *Aquaculture*. 16, 279-298.

Sterling, K. M. and Ahearn, G. A. (2011). Glucose and fructose uptake by *Limulus polyphemus* hepatopancreatic brush border and basolateral membrane vesicles: evidence for Na-dependent sugar transport activity. *Journal of Comparative Physiology B*. 181, 467-475.

Sterling, K. M., Cheeseman, C. I. and Ahearn, G. A. (2009). Identification of a novel sodium-dependent fructose transport activity in the hepatopancreas of the Atlantic lobster *Homarus americanus*. *Journal of Experimental Biology*. 212, 1912-1920.

Thamotharan, M. and Ahearn, G. A. (1996). Dipeptide transport by crustacean hepatopancreatic brush-border membrane vesicles. *The Journal of Experimental Biology*. 199,635-641.

Turk, E. and Wright, E. M. (1997). Membrane Topological Motifs in the SGLT Cotransporter Family. *Journal of Membrane Biology*. 159, 1-20.

Verri, T., Zilli, L., Bossa, D., Mandal, P. K., Ingrosso, L., Zonno, V., Vilella, S., Ahearn, G. A. and Storelli, C. (2001). D-glucose transport in decapod crustacean hepatopancreas. *Comparative Biochemistry and Physiology A*. 130, 585-606.

Vilella, S., Zilli, L., Ingrosso, L., Schiavone, R., Zonno, V., Verri, T. and Storelli, C. ((2003). Differential expression of Na /D-glucose cotransport in isolated cells of *Marsupenaeus japonicus* hepatopancreas. *Journal of Comparative Physiology B*. 173, 679-686.

Xia, X., Wang, G., Peng, Y. and Jen, J. (2005). Cys351 and Cys361 of the Na+/glucose cotransporter are important for both function and cell-surface expression. *Archives of Biochemistry and Biophysics*. 438, 63-69.

Impact of Climate Change on Hydrological Cycle, Ecosystem, Fisheries and Food Security, Pages: 195–204
Edited by: B. Madhusoodana Kurup, M.R. Boopendranath, M. Harikrishnan and A.V. Shibu

CHAPTER - 16

Rice Yield Changes in Kerala due to Climate Change

Stella Jes Varghese[1,2*], Sajani Surendran[1,2], B. Ajithkumar[3], Kavirajan Rajendran[1,2] and Akio Kitoh[4]

[1]*Multi Scale Modelling Programme, CSIR Fourth Paradigm Institute, Bangalore-560037, India*
[2]*Academy of Scientific and Innovative Research (AcSIR), Ghaziabad-201002, India*
[3]*College of Horticulture, Kerala Agricultural University, Vellanikkara-680656, India*
[4]*Japan Meteorological Business Support Center, Tsukuba, Japan*
[*]*E-mail: stella.varghese@gmail.com*

Over the last few years, a dramatic change in daily weather has been noted. This is greatly accounted by human activities, where greenhouse gases are emitted into the atmosphere at an alarming rate by burning forests, grasslands and fossil fuels. The effect of climate change is most commonly manifested as a change in agricultural output due to changes in climate variables which are responsible for growth and development of crops. The population is expected to go up to 1.6 billion by 2050 in India and therefore, there needs to be an increase in the productivity, especially of rainfed areas of India (Aggarwal *et al.*, 2008). An agrarian economy as that of India, the Gross Domestic Product (GDP) is greatly dependent on timely monsoon (Gadgil & Gadgil 2006), and other climatic parameters like temperature and solar radiation. Therefore, the impact studies of climate parameters on crop yield are inevitable.

Over Asia, rice is regarded as the major food and it is also consumed by more than half of the world (Schiller *et al.*, 2006). The major parameters responsible for rice production are temperature, rainfall, CO_2 and solar radiation (Fageria 2007). The yield and biomass under increased CO_2 concentration was studied and they found to increase (Razzaque *et al.*, 2011; Kim *et al.*, 2003). Varying results has been found under increasing temperature and increasing CO_2 concentrations. Krishnan *et al.* (2007) showed that for every 1°C of rise in temperature, the rice yield declines by 7.2% with the current trend of CO_2 concentration, but with increase in CO_2 up to 700 ppm, an average increase in yield by 31% was found over India. Many climate change scenario studies were also done

(Rajwade *et al.* (2015)) and rice yield was found to decrease in the future. Impact studies such as Satapathy *et al.*, (2014) showed a decrease in rice yield with increasing temperatures greater than 0.8°C. Swain & Thomas (2015) reported decrease in yield of rice under various climate scenarios at different locations in India. Temperature stress was found to have greater influence on anthesis in climate projection, whereas it is not so in baseline simulation. Challinor *et al.* (2007) study shows that significant correlation exists between duration of the crop and corresponding yield simulated. With the rise in temperature, an overall decline in rate of photosynthesis as well as associated biomass accumulation tends to be limited (Squire 1990). All these studies used very crude resolution climate models and that would bring in large bias in simulating regional scale features, especially over India.

For cropping simulation models, station based weather data is regarded as the major source of input data (Sultan *et al.*, 2005), but for future climate scenarios, we resort to Global Climate Model (GCM) output. GCMs in general predict a warmer and moister climate for the future, but they have drawbacks in projecting the regional and seasonal rainfall distribution due to their coarse resolution (Allen and Ingram, 2002; Chou and Neelin, 2004). Therefore, as an alternative, a very high resolution GCM is preferred. We used high-resolution climate data from Meteorological Research Institute (MRI) GCM which is found to have high reliability in simulating present day climate of India (Rajendran *et al.*, 2013; Rajendran *et al.*, 2016) particularly the observed climate trends over the Western Ghats and Kerala, the southern-most state of India (Rajendran *et al.*, 2012).

Among the various crop simulation models (CSMs), Decision Support System for Agrotechnology Transfer (DSSAT) (Jones *et al.*, 2003) has proved to be versatile in having CSM of different varieties of crops. Many climate change impact studies were done using DSSAT for rice production in India (Subash and Mohan, 2012; Mishra *et al.*, 2013) but, they all suffer from lack of sufficient high-resolution climate data.

Material and Methods

Field experiments were done in Agricultural Research Station (ARS) Mannuthy, Kerala which is located at 10°32' N and 76°20' E at an altitude of 22m above sea level. Climate data for validation was taken from Principal Agro-meteorological observatory, Vellanikkara, Department of Agricultural Meteorology, College of Horticulture, Kerala Agricultural University. Long-term climate data is given is Fig. 1 which shows the climatological averages of maximum temperature, minimum temperature, rainfall and solar radiation for the period 1994-2003. Soil of the study region is sandy clay loam type and its characteristics were taken into account while predicting the yield using crop model. The current study was done using short duration rice variety *Jyothi* (110-125 days). A split plot design with five dates of planting at 15 days interval starting from 5th June to 5th August (D1-D5) were carried out.

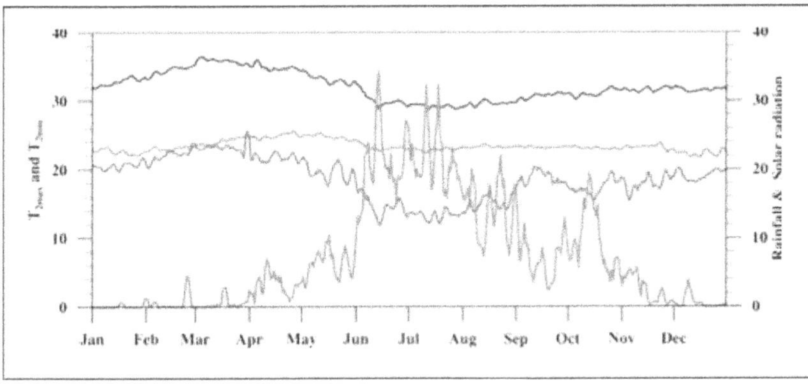

Fig. 1. Climate variables at Vellanikkara station. Climatological (1994-2003) mean annual variation of rainfall (mm/day), solar radiation (MJm^{-2}/day), maximum and minimum temperatures (°C)

Climate data is taken from MRI-GCM (Mizuta *et al.*, 2012) simulation with ~20 km horizontal resolution corresponding to triangular truncation of T_L959 (1920 x 960) with linear Gaussian grid in the horizontal. The specifications of model forcing for present day and future RCP8.5 scenario are given in detail in Rajendran *et al.*, 2013. The climate change simulation by MRI GCM produces an overall intensification of summer monsoon rainfall over India but a reduction in rainfall over the Western Ghats region, near to the area under consideration for the present study (Rajendran *et al.*, 2012, 2013, 2016; Rajendran & Kitoh 2008). This is consistent with current trend in rainfall during summer monsoon season which shows increasing trend over most parts of India at the same time with a decreasing trend over WG region with 90% significance (Guhathakurta & Rajeevan 2008; Rajendran et al., 2012). Climatology of rainfall during the summer monsoon season (June, July, August, September) as simulated by MRI GCM and observation is shown in Fig. 2. Many of the coarse resolution climate models that are being used now lacks this information. MRI 20 km thus can be considered superior to other state-of-the-art climate models. The crop yield was simulated for present day (1994-2003), and end of the century (2090-2099).

Results and Discussion

The MRI data for 1994-2003 is first validated against observation. Fig. 3 shows the climatological annual cycle of maximum temperature, minimum temperature, rainfall and solar radiation. It shows a slight underestimation of maximum temperature and overestimation of minimum temperature. The annual cycle of rainfall is closely followed by the model, but it overestimates rainfall in the summer monsoon season. Solar radiation was also found to have close match with observation.

The climate change scenario can be established by looking at the changes in climatological variables for present day (1994-2003) (referred as base period) and future

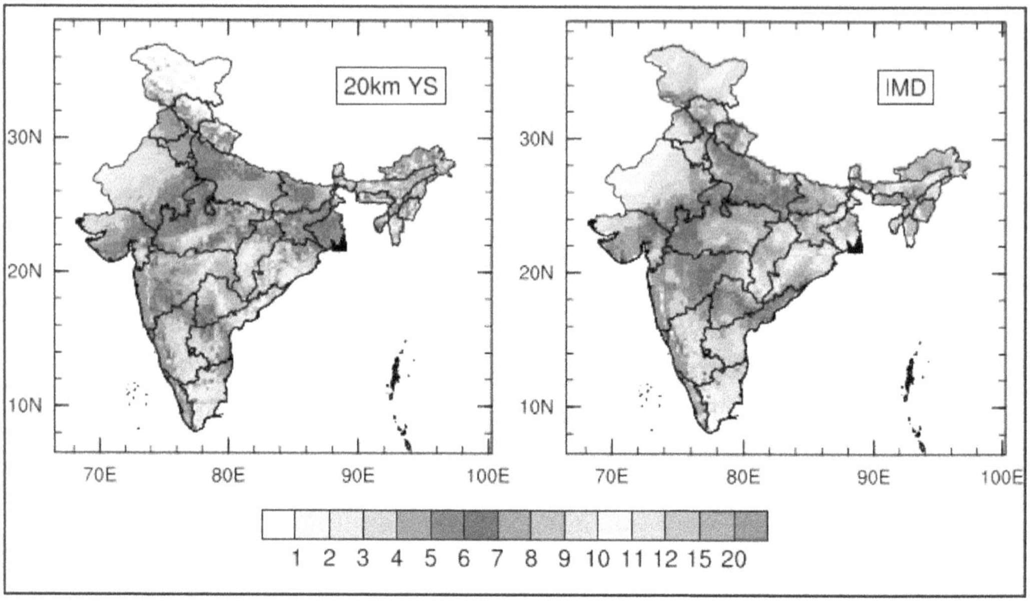

Fig. 2: June-July-August-September (JJAS) climatology of rainfall (1983-2003) from the present day simulations of 20 km model along with the observation.

RCP8.5 scenario (2090-2099) (referred as scenario). For basc period, we use observed data and for future scenario, we used bias corrected model output. Bias correction is done using the delta change approach. For variables like maximum temperature, minimum temperature and solar radiation, future climate can be derived from

$$S_{corr} = Obs + \left(\overline{S_{Fut}} - \overline{S_{pre}}\right)$$

where S is the simulated quantity with corr, fut and pre referring to bias corrected, future and present simulations respectively. *Obs* stands for actual observed value. The overbar represents climatology corresponding to the period under consideration. For rainfall, bias correction is performed as

$$S_{corr} = Obs * \frac{\overline{S_{Fut}}}{\overline{S_{pre}}}$$

The annual climatology of present day observed and bias corrected future maximum temperature, minimum temperature, rainfall and solar radiation are given in Fig. 4.

For model calibration and testing, the best cultivar coefficient is derived prior through iterative process of calibration based on crop type and location. The experimental plot yield is first compared with the simulated yield, as the final yield is regarded as most important variable to consider. The CERES Rice model within DSSAT was validated for

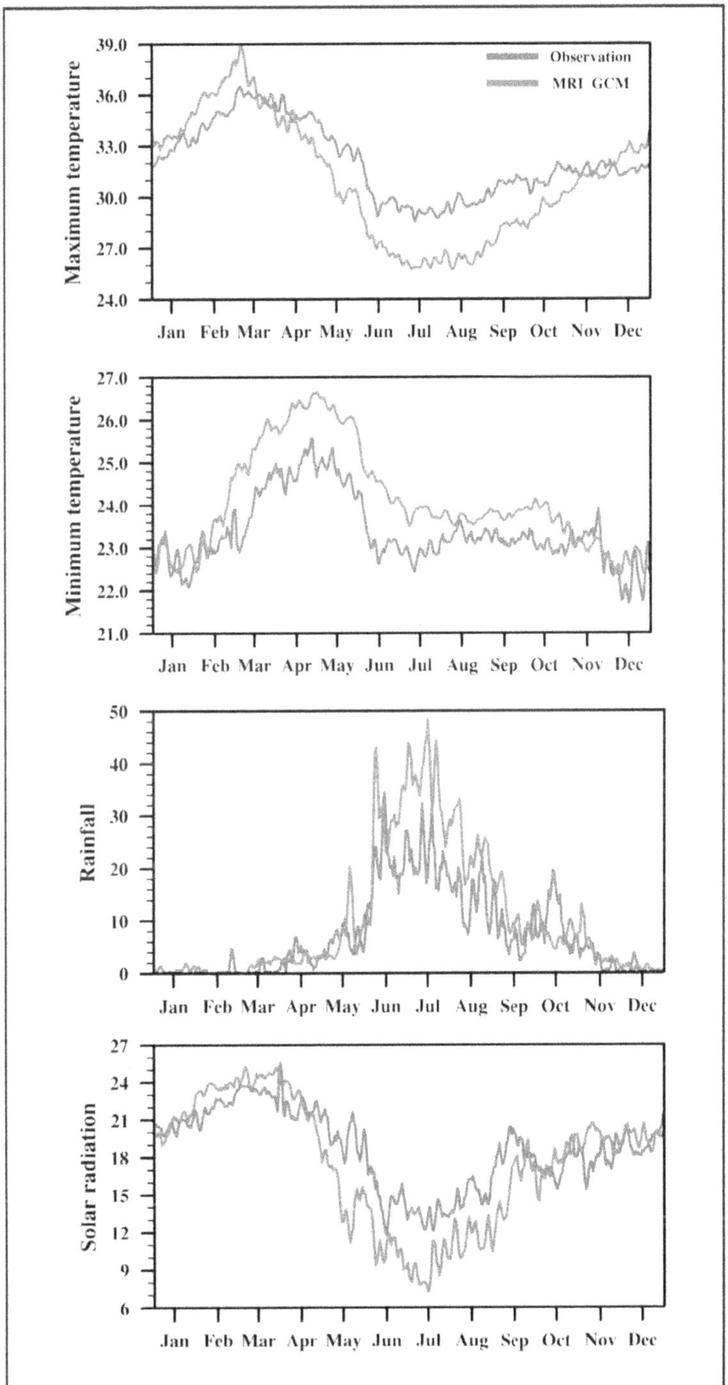

Fig. 3. Observed and simulated climate parameters for the base period over Vellanikkara. Climatological mean (1994-2003) annual variation of minimum and maximum temperatures (°C), rainfall (mm/day) and solar radiation (MJm^{-2}/day) for the base period.

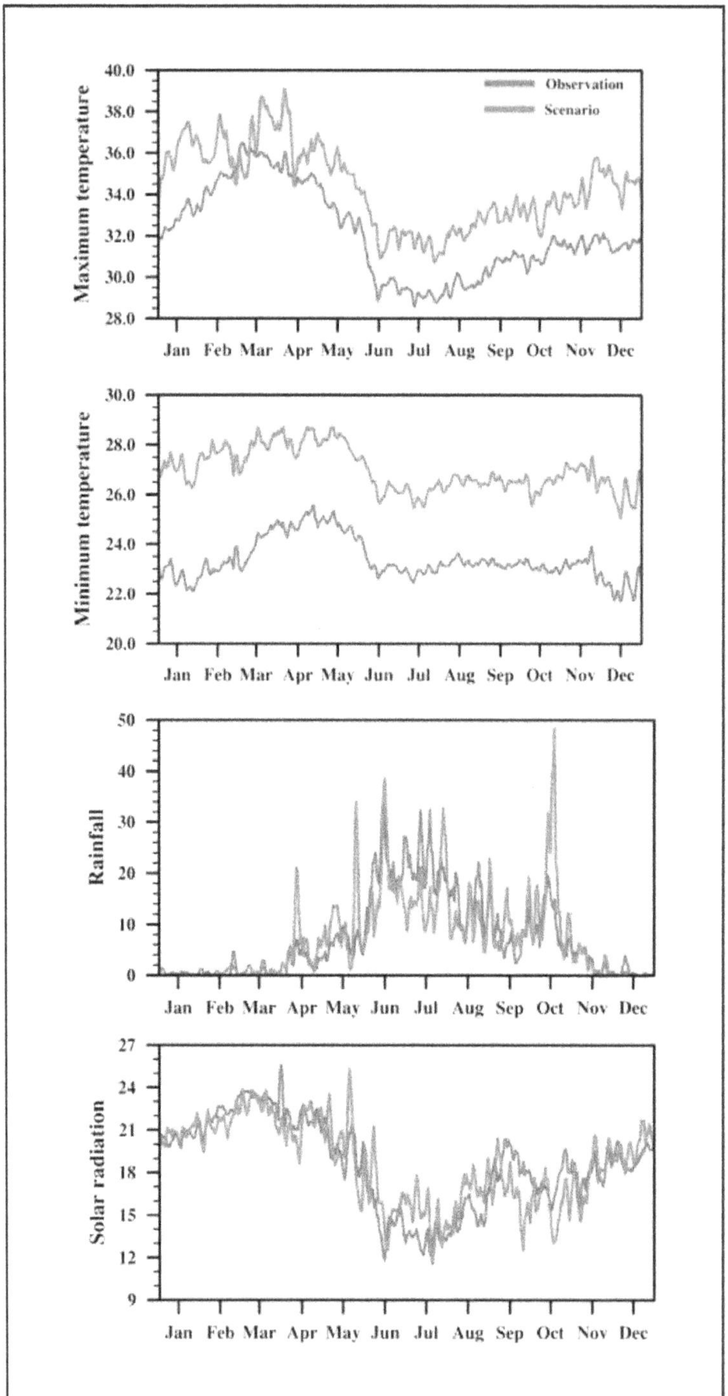

Fig. 4. Climate parameters observed for the base period (1994-2003) and projected for future RCP8.5 scenario (2090-2099) (bias corrected) over Vellanikkara. Climatological mean of annual variation of minimum and maximum temperatures (°C), rainfall (mm/day) and solar radiation (MJm^{-2}/day) for the base period and for future scenario.

the short duration variety of *Jyothi* for the wet season of 2017. For validation, change in yield between simulated and observed is analysed in terms of percentage change in yield as:

$$\text{Percentage change in yield } (\Delta Y) = \frac{Y_S - Y_O}{Y_O} \times 100$$

where Y_S is yield simulated using CERES model [in kg/ha] and Y_0 is yield observed [kg/ha] at the experimental site for the year 2017. Model was found to simulate the yield close to observed value. The Root Mean Square Error (RMSE) was found to be 660kg/ha, suggesting a moderate error in the simulated yield.

Simulations were carried out for the base period and for future scenario to calculate the respective crop growth parameters. Simulation were done to calculate the potential yield, which means, the model will itself take care of the nutrient and water requirements. This will ensure that only climate parameters are taken into the model as user input. The CO_2 concentration is kept at a constant value of 380ppm.

We found that, the panicle initiation day, anthesis day and physiological maturity date has shifted to an early period (a decrease of ~4 days, ~7 days and ~8 days respectively) in future scenario. The leaf area index was also found to decrease in the future scenario. This implies a decrease in total biomass production. Potential yield (yield at harvest maturity) also decreases in the future with the kind of climate expected for RCP8.5 scenario (a decrease of ~957 kg/ha).

Conclusions

The focus of this study was to ascertain the importance of changing climate variables on the yield of rice. Weather parameters such as maximum temperature, minimum temperature, rainfall and solar radiation were analysed for the station for base period and future scenario. A clear increase in maximum and minimum temperature was seen whereas rainfall was found to decrease. Crop simulations were then done for base period and future RCP8.5 scenario in order to find the potential yield. We found that increased temperature and decreased rainfall affected the crop duration and in turn affected the final yield. The thermal time requirement was thus reduced leading to earlier maturity of rice with less biomass and yield. Our analysis suggests for adaptive measures that needs to be taken in order to cope with reduction in rice yield. Rice varieties that are tolerant to high temperatures and that require less water may be preferred in future.

Acknowledgements

The climate model simulations were done by the Earth Simulator of JAMSTEC, under the SOUSEI and TOUGOU Programs funded by MEXT, Japan. Crop model was taken from

DSSAT software. The first author acknowledges the financial support of Council of Scientific and Industrial Research (CSIR) for the Senior Research Fellowship (SRF) and AcSIR. The authors acknowledge Agricultural Research Station, Mannuthy, Kerala, and Principal Agro Meteorological Observatory, Vellanikkara, Kerala for providing the station soil and weather data respectively. Part of this work was also supported by NCAP project of MoEFCC, Govt. of India (GAP-1009).

References

Aggarwal, P.K., Hebbar, K.B., Venugopalan, M.V., Rani, S., Bala, A., Biswal, A. and Wani, S.P., 2008. Quantifi cation of Yield Gaps in Rain-fed Rice, Wheat, Cotton and Mustard in India. Global Theme on Agroecosystems Report No. 43. International Crops Research Institute for the Semi-Arid Tropics, Patancheru 502 324, Andhra Pradesh, India., 36 p.

Allen, M.R. and Ingram, W.J., 2002. Constraints on future changes in climate and the hydrologic cycle. Nature 419(6903): 228.

Challinor, A., Wheeler, T., Garforth, C., Craufurd, P. and Kassam, A., 2007. Assessing the vulnerability of food crop systems in Africa to climate change. Climatic change, 83(3), 381-399.

Chou, C., Neelin, J.D., 2004. Mechanisms of global warming impacts on regional tropical precipitation. Journal of Climate, 17(13): 2688-2701.

Fageria, N.K., 2007. Yield physiology of rice. Journal of Plant Nutrition 30(6): 843-879.

Gadgil, S.and Gadgil, S., 2006. The Indian monsoon, GDP and agriculture. Economic and Political Weekly, 41(47), 25 Nov, 2006.

Guhathakurta, P. and Rajeevan, M., 2008. Trends in the rainfall pattern over India. International Journal of Climatology: A Journal of the Royal Meteorological Society 28(11): 1453-1469.

Jones, J.W., Hoogenboom, G., Porter, C.H., Boote, K.J., Batchelor, W.D., Hunt, L.A., Wilkens, P.W., Singh, U., Gijsman, A.J. and Ritchie, J.T., 2003. The DSSAT cropping system model. European Journal of Agronomy 18(3-4): 235-265.

Kim, H.Y., Lieffering, M., Kobayashi, K., Okada, M. and Miura, S.H.U., 2003. Seasonal changes in the effects of elevated CO2 on rice at three levels of nitrogen supply: a free air CO_2 enrichment (FACE) experiment. Global Change Biology 9(6): 826-837.

Krishnan, P., Swain, D.K., Bhaskar, B.C., Nayak, S.K., Dash, R.N., 2007. Impact of elevated CO_2 and temperature on rice yield and methods of adaptation as evaluated by crop simulation studies. Agriculture, Ecosystems and Environment 122(2): 233-242.

Mishra, A., Singh, R., Raghuwanshi, N.S., Chatterjee, C. and Froebrich, J., 2013. Spatial variability of climate change impacts on yield of rice and wheat in the Indian Ganga Basin. Science of the Total Environment 468: S132-S138.

Mizuta, R., Yoshimura, H., Murakami, H., Matsueda, M., Endo, H., Ose, T., Kamiguchi, K., Hosaka, M., Sugi, M., Yukimoto, S. and Kusunoki, S., 2012. Climate simulations using MRI-AGCM3. 2 with 20-km grid. Journal of the Meteorological Society of Japan. Ser. II, 90: 233-258.

Rajendran, K., Kitoh, A., 2008. Indian summer monsoon in future climate projection by a super high-resolution global model. Current Science 95(11):1560-1569.

Rajendran, K., Kitoh, A., Srinivasan, J., Mizuta, R. and Krishnan, R., 2012. Monsoon circulation interaction with Western Ghats orography under changing climate. Theoretical and Applied Climatology, 110(4): 555-571.

Rajendran, K., Sajani, S., Jayasankar, C.B., Kitoh, A., 2013. How dependent is climate change projection of Indian summer monsoon rainfall and extreme events on model resolution?. Current Science 1409-1418.

Rajendran, K., Sajani, S., Kitoh, A. and Varghese, S.J., 2016. Reduction of uncertainty associated with future changes in Indian summer monsoon projected by climate models and assessment of monsoon teleconnections. In: Remote Sensing and Modeling of the Atmosphere, Oceans, and Interactions VI (Vol. 9882, p. 98820D). International Society for Optics and Photonics.

Rajwade, Y.A., Swain, D.K., Tiwari, K.N., Mohanty, U.C. and Goswami, P., 2015. Evaluation of field level adaptation measures under the climate change scenarios in rice based cropping system in India. Environmental Processes 2(4): 669-687.

Razzaque, M.A., Haque, M.M., Khaliq, Q.A., Soliman, A.R.M. and Hamid, A., 2011. Effects of CO_2 and nitrogen levels on yield and yield attributes of rice cultivars. Bangladesh Journal of Agricultural Research 36(2): 213-221.

Satapathy, S.S., Swain, D.K. and Herath, S., 2014. Field experiments and simulation to evaluate rice cultivar adaptation to elevated carbon dioxide and temperature in sub-tropical India. European Journal of Agronomy 54: 21-33.

Schiller, J.M., 2006. Rice in Laos. Int. Rice Res. Inst., 457 p. ISBN: 9712202119.

Squire, G.R., 1990. The Physiology of Tropical Crop Production. CAB International.

Subash, N., Mohan, H.R., 2012. Evaluation of the impact of climatic trends and variability in rice–wheat system productivity using Cropping System Model DSSAT over the Indo-Gangetic Plains of India. Agricultural and Forest Meteorology 164: 71-81.

Sultan, B., Baron, C., Dingkuhn, M., Sarr, B. and Janicot, S., 2005. Agricultural impacts of large-scale variability of the West African monsoon. Agricultural and Forest Meteorology 128(1-2): 93-110.

Swain, D.K. and Thomas, D., 2015. Climate Change Impact Assessment and Evaluation of Agro-Adaptation Measures for Rice Producti on in Eastern India. Journal of Environmental Informatics 16(2): 94-101.

Impact of Climate Change on Hydrological Cycle, Ecosystem, Fisheries and Food Security, Pages: 205–210
Edited by: B. Madhusoodana Kurup, M.R. Boopendranath, M. Harikrishnan and A.V. Shibu

CHAPTER - 17

Breeding of *Gnathophyllum americanum*: First Success in Ornamental Aquaculture

Sheena Jose, T.T. Ajith Kumar* and Kuldeep K. Lal

ICAR - National Bureau of Fish Genetic Resources, Lucknow-226 002, U.P., India
**E-mail: ttajith87@gmail.com*

Marine ornamental shrimps are gaining popularity in aquarium industry. The ornamental shrimps belongs Stenopodidae, Hippolytidae, Rhynchocinetidae, Palaemonidae, Alpheidae, Hymenoceridae and Gnathophyllidae are playing major role in aquarium trade (Calado *et al.*, 2003). They exhibit striking coloration, delicacy and some can even control undesirable organisms in aquarium tanks. The mimetic capacity, symbiotic behavior, peculiar appearance and rarity helps them to achieve ornamental "status". Even though, there is high market value and demand, the number of tank-raised marine ornamental shrimps is still insignificant and the vast majority of the traded specimens are being collected from wild (Calado *et al.*, 2003). There is a worldwide concern by researchers, traders, collectors and hobbyists on the sustainability of the marine ornamental industry, hence a long term remedy is need of the hour.

In India, ornamental aquarium trade is gaining popularity and is becoming an amusement among hobbyists. To guarantee the sustainable supply for the aquarium industry, while minimizing the negative impacts on the natural environment, developing aquaculture technology for marine ornamental shrimps is urgently needed. But, the long and variable larval durations impair the captive production on commercial scale. The key factors to improve larval quality and shorten larval duration are to identify appropriate broodstock and larval diets (Lin *et al.*, 2002). In India, *G. americanum* is reported from Andaman & Nicobar and Lakshadweep Islands (Prakash *et al.*, 2015, Kumaralingam & Raghunathan, 2016). Breeding of *G. americanum* is the first attempt in India and worldwide. The trade value of this species costs around 14-16 US$. Even though, the size of *G. americanum* is small and the larval duration is quite lengthy, an attempt on the larval rearing and captive production was attempted at Lakshadweep.

Materials and Methods

The work was conducted at the ICAR - NBFGR and DBT Germplasm resource centre for marine ornamental invertebrates, available on the premises of the Centre for Marine Living Resources and Ecology (CMLRE) Field station, Agatti Island, Lakshadweep (Fig. 1). Water from the lagoon was pumped in to the sump / tanks and kept 24 hours for sedimentation and later siphoned into storage tanks through UV filters and finally, water goes to the storage tank, which is installed with canister filter, filled with ceramic rings, activated carbon and bio-balls. This water was taken in to the brooder as well as larval rearing tanks. Conditioned water was used in the culture system. The following water quality parameters were routinely monitored; temperature, salinity, pH and DO, which were measured using water quality analysis probe (Eutech instruments, Waterproof CyberScan PCD 650). Liquid type water quality test kits (API MARINE, Saltwater Marine test Kit) designed for salt water aquariums was used to measure the nitrate, nitrite and ammonia.

Fig. 1. ICAR - NBFGR & DBT Germplasm Resource Centre for Ornamental Invertebrates, Lakashadweep

Broodstock development

G. americanum individuals were collected from wild at different Islands of Lakshadweep, adopting hand picking and snorkeling methods. Followed, they were acclimatized / quarantined for a week, before introducing to the brooder development tanks. Collection of gravid females was avoided, since they were susceptible to stress, caused mortality. Twenty numbers of healthy and bigger size male and female *G. americanum were* identified separately and used for the experiment. The presence of parasite such as bopyrid isopods can negatively affect the reproductive efficiency of marine ornamental shrimp (Calado *et al.*, 2006; 2008), so the organisms were closely and carefully examined before releasing to brooder tanks. 250 lit. water holding capacity FRP tanks were used for this purpose and it was filled with 200 litre water. At the time of introduction, average body weight and total length of the shrimps was 0.14 - 0.16 g and 1.9 - 2.2 cm, respectively. Live rocks and coral boulders were housed in the tanks to facilitate hiding / shelter to shrimps. Tanks were provided with mild aeration. Manual cleaning of tank bottom was practiced by siphoning once in the mornings, every day. Each tank received approximately 10% total water exchange daily. Presence of molts and gravid shrimps were observed and recorded. A mixed diet of clam meat and frozen *Artemia* was fed thrice daily at a rate of 5% body weight. Quantity of the feed was regulated, according to the acceptance of the feed. Tank aeration was stopped just before feeding and half an hour break is being provided. Broodstock development was noticed within three months and obtained eight individual pairs (Fig. 2.).

Fig. 2. Captive raised juveniles of *Gnathophyllum americanum*

Larval rearing

After four months rearing, spawning was noticed and the larvae were shifted to separate tanks. Once the larvae (Zoeal stage) hatched out in brooder tanks, aeration was stopped until larvae migrate to surface, so that collecting of larvae was easy. The larvae were shifted to 100L FRP tank, which was provided with mild aeration and 12 hrs photoperiod. The larvae were fed with ad-libitum of *Brachionus pilicatilis* enriched with green algae, *Chlorella marina* and followed by *Artemia* nauplii. Bottom siphoning was carried out daily to remove waste, which was practiced once in two days. Aeration was stopped 30 minutes in advance to siphoning and sponging. Normally 5-10% water was exchanged. Larva hatches out as zoea. Larval feeding commenced from first day onwards. During the initial days (1-10 days), larvae were fed with rotifer *B. pilicatilis* enriched with microalgae, *C. marina.* From 10th day to till metamorphosed juveniles, *Artemia* nauplii were given ad-libitum.

Juvenile rearing

Zoeal larvae metamorphosed to decapodid stage, before reaching juvenile stage. Juvenile stage is the one when the larvae resemble miniature adult. When the decapodid stage reached juvenile, they were shifted to 200L FRP tank. Mild aeration and natural sunlight were provided. The juveniles are fed with live *Artemia,* during initial days followed by frozen *Artemia.* Water exchange (10%) was done once in two days. Dead coral boulders / live rocks were provided as hiding substrate for juveniles in the tank. Larval duration was calculated by the number of days taken to complete the larval life and it was calculated as duration from hatching to metamorphosis, followed to juvenile. Survival rate was estimated by the average number of zoea metamorphosed to juveniles per batch.

Results and Discussion

During the experimental period, both brooder and larval rearing tanks, the water quality parameters were maintained at different level and standardized the optimum levels and followed. The temperature, 27.5-28.5°C; salinity, 35.0-36.0 ppt; pH, 8.0-8.5 and dissolved oxygen 5-5.5 ppm respectively, which is the recommended water quality parameters for the clownfish, *Amphiprion nigripes* bred in captivity at Lakshadweep (Kumar *et al.*, 2017). Nitrate concentration remained below 20 mg/L and the ammonia concentration was less than 0.1 mg/L.

Fig. 3. *Gnathophyllum americanum*

The spawning interval of *G. americanum* was 15-16 days and spawning usually occurred during the night hours. The average number of larvae per spawner was more than 100, whereas in *Lysmata amboinensis*, the average number were more than 500 (Simoes *et al.*, 1998). It was observed that, the zoeal larvae usually swim upside down and use its tail to change the direction of movement. The larvae are phototrophic and they exhibited many morphological changes throughout the larval life. After 48 hours post-hatching, molting of zoea 1 to zoea 2 stages was evident by numerous tiny exoskeletons floating on the water surface. Larvae turned to miniature adults in 110 -120 days of post-hatching, however this period was noted as 140 days in *L. amboinensis (*Fletcher *et al.*, 1995). When the zoeal stage metamorphosed to decapodid stage, it was noticed that, the larvae were clinging off the tank wall. Juveniles resembled miniature adult (Fig. 3). and they easily acclimatized with the new hiding place provided Successful raising of larvae to juveniles in captivity is still in the experimental stage as the survival rate is 35-40%.

The larval development of the family Gnathophyllidae is poorly documented. Only the zoeal stage of *G. americanum* (Gue´rin-Me´neville, 1855) and *G. elegans* (Risso,1816) were described (Bruce, 1986, Meyer *et al.*, 2014). The commercial culture technique of *G. americanum* was not addressed (Calado *et al.*, 2003) as the larval duration may high, however, the available literature also revealed that, the other highly priced marine ornamental shrimps like *L. amboinensis* and *Stenopus hispidus* also has quite longer larval duration. *L. amboinensis* took 20 weeks and *S. hispidus* acquired 17-30 weeks (Fletcher *et al.*, 1995). According to Gore (1985), the long and variable larval durations are largely due to mark-time molting, a sequence of molts without development. The reasons for the longer larval days may be due to poor nutritional quality in larval diets, management practices and sometimes the temperature conditions of water maintained in rearing tanks (Palmtag and Holt, 2001; Hettiarachchi & Edirisinghe, 2016). In the present study, even though the larvae were fed with live feeds initially, the larval duration was lengthy, however study on nutritional requirements of *G. americanum may* help in reduce the larval rearing duration. In this experiment, initial larval survival was 15-20% and gradually enhanced to 35-40%. However, no mortality was observed in juvenile stage, which is the advantage on this species. Careful handling was essential to avoid possible damages, while transferring the metamorphosed ones to grow out tanks and as well as during siphoning off the tank bottom.

Captive breeding of ornamentals and popularizing the technology for aquaculture is a leading solution that could conserve the wild livestock and also provide an alternative / additional livelihood to the entrepreneurs.

Acknowledgements

Authors are thankful to the Director, ICAR-National Bureau of Fish Genetic Resources, Lucknow for the facilities and encouragements. They are grateful to the Department of Biotechnology (DBT), New Delhi for financial support and Department of Fisheries, Union Territory of Lakshadweep and the Centre for Marine Living Resources and Ecology (MoES) for the local support.

References

Bruce, A.J., 1986. Observations of the family Gnathophyllidae Dana, 1852 (Crustacea: Decapoda). J. Crust Biol., 6: 463-470.

Calado, R., Bartilotti, C., Goy, J.W. and Dinis, M.T., 2008. Parasitic castration of the stenopodid shrimp *Stenopus hispidus* (Decapoda: Stenopodidae) induced by the bopyrid isopod *Argeiopsis inhacae* (Isopoda: Bopyridae). J. Mar. Biol. Assoc. UK. 88: 307-309.

Calado, R., Narciso, L., Araujo, R. and Lin, J., 2003. Overview of marine ornamental shrimp aquaculture. In: Marine Ornamental Species: Collection, Culture and Conservation (Cato, J.C., Brown, C.L., Eds.), pp 221-230, Iowa State Press.

Calado, R., Vitorino, A., Dinis, M.T., 2006. Bopyrid isopods do not castrate the simultaneous hermaphroditic shrimp *Lysmata amboinensis* (Decapoda: Hippolytidae). Dis. Aquat. Organ., 73: 73-76.

Fletcher, D.J., Kotter, I., Wunsch, M. and Yasir, I., 1995. Preliminary observations on the reproductive biology of ornamental cleaner prawns. Int. Zoo. Yearb., 34: 73-77.

Gore, R., 1985. Molting and growth in decapods larvae. In: Crustacean Issues - Larval Growth. Vol. 2, (Wenner. A., Ed.), pp 1-65, A.A. Balkema Publishers, Rotherdam, Netherlands.

Hettiarachchi, H.A.S.U., Edirisinghe, U., 2016. Captive breeding of fire shrimp (*Lysmata debelius*) under Sri Lankan conditions. J. Sustain. Trop. Agric. Res. 28(1): 88 - 99.

Kumar, T.T.A., Gopi, M., Dhaneesh, K.V., Vinoth, R., Ghosh, S., Balasubramanian, T., Shunmugaraj, T., 2012. Hatchery production of the clownfish, *Amphiprion nigripes* at Lakshadweep, India. J. Environ. Biol. 33: 623-628.

Kumaralingam, S. and Raghunathan, C., 2016. An account of some reef associated caridean shrimps and stomatopods of Andaman Islands. Rec. Zool. Surv. India, 116(2): 117-128.

Lin, J., Zhang, D. and Rhyne, A.L. 2002. Broodstock and larval nutrition of marine ornamental shrimp. In: Avances en Nutrición Acuícola VI. Memorias del VI Simposium Internacional de Nutrición Acuícola. (Cruz Suárez, L.E., Ricque Marie, D., Tapia Salazar, M., Gaxiola Cortés, M.G. and Simões, N., Eds.), pp 277-280, Cancún, Quintana Roo, México.

Meyer, R., Lehmann, T., Melzer, R.R. and Geiselbrecht, H., 2014. Morphology of the first zoeal stage of the Mediterranean bumblebee shrimp *Gnathophyllum elegans* studied with both light microscopy and scanning EM. J. Mar. Biol. Assoc. UK., 94(1): 151-158.

Palmtag, M.R. and Holt, G.J., 2001. Captive rearing of fire shrimp (*Lysmata debelius*). A Texas Sea Grant College Program Research Report. National Oceanic and Atmospheric Administration, U.S. Department of Commerce, Washington DC, USA, 4 p.

Prakash, S., Kumar, T.T.A., Subramoniam, T., 2015. Notes on some Indo-Pacific caridean shrimps (Crustacea: Decapoda, Caridea: Palaemonidae and Gnathophyllidae) particularly from India. Zootaxa 3914(4): 456-466.

Simoes, F., Ribeiro, F. and Jones, D.A., 1998. The effect of diet on the reproductive performance of marine cleaner shrimps *Lysmata debelius* (Bruce 1983) and *L. amboinensis* (De Man 1888), (Caridea, Hippolytidae) in captivity. Aquaculture 98 Book of Abstracts. Las Vegas, Nevada, USA, 497 p.

Impact of Climate Change on Hydrological Cycle, Ecosystem, Fisheries and Food Security, Pages: 211–218
Edited by: B. Madhusoodana Kurup, M.R. Boopendranath, M. Harikrishnan and A.V. Shibu

CHAPTER - 18

Virulence Features of *Vibrio parahaemolyticus*: A Review

Reshma Silvester[1*], Deborah Alexander[2], S. Sudha[3], M. Harikrishnan[1] and Mohamed Hatha[3]

[1]*School of Industrial Fisheries, CUSAT, Kochi-682 016, Kerala, India*
[2]*American college, Madurai-625002, Tamil Nadu, India*
[3]*Dept. of Marine Biology, Microbiology and Biochemistry, CUSAT, Kochi-682016, Kerala, India*
**E-mail: sanasilvester@gmail.com*

Vibrio species are Gram-negative rod-shape curved, non-spore forming, halophilic bacteria autochthonous to estuarine, coastal and marine environments (Letchumanan *et al.*, 2015). Many *Vibrio* species are recognized as human pathogens and have been associated with water and seafood-related outbreaks of gastrointestinal infections in humans (Austin, 2010). At present, there are more than 100 recognised species under the genus *Vibrio* (Okada *et al.*, 2010) and 12 of them are reported to be pathogenic to humans. *Vibrio parahaemolyticus* is a leading cause of seafood borne gastroenteritis in Asia as well as in other countries (Odeyemi, 2016). It is attracting increasing interest worldwide where raw or undercooked seafood is often consumed (Chao *et al.*, 2009). The pathogen colonizes the human gut *via* the gastrointestinal route through consumption of contaminated raw or undercooked seafood. The most common symptoms include gastroenteritis, diarrhoea, headache, nausea and vomiting. *V. parahaemolyticus* infections can be life threatening to the infants or immuno-compromised persons. The pandemic serovar *V. parahaemolyticus* O3:K6 that emerged in India in 1996 has since been found to cause many cases of outbreaks worldwide (Silvester *et al.*, 2015). Pathogenic bacteria have adopted a wide range of strategies to colonize and invade host cells. Potential virulent *V. parahaemolyticus* strains are generally distinguished from non-virulent strains by the presence of virulence determinants (Beshiru *et al.*, 2020). The pathogenic potential of the bacterium is not solely governed by a given virulence function (Klein *et al.*, 2014). A single factor alone does not contribute to the establishment of infection in the host cell. It is a complex process with different strains employing different strategies. In the present review we discuss the various virulence strategies adopted by *V. parahaemolyticus*.

Virulence Related Genes of *Vibrio Parahaemolyticus*

Pathogenicity of *V. parahaemolyticus* is not dependant on a given virulence function alone; rather, virulence is a complex process and different strains employ different strategies (Klein *et al.*, 2014). The virulence properties of *V. parahaemolyticus* have been known to be associated with thermostable direct haemolysin (TDH) and TDH- related haemolysin (TRH) (Honda *et al.,* 1987a;b). However, several studies reported the absence of TDH and TRH in clinical strains of *V. parahaemolyticus* (Li *et al.,* 2014; Pazhani *et al.,* 2014). These reports suggest that many other factors are also involved in pathogenicity of these strains. Recently, it was found that the type III (T3SSs) and VI secretion systems (T6SSs) also have a major role in virulence machinery in *V. parahaemolyticus* (Ham and Orth, 2012; Ben-Yaakov and Salomon, 2019).

Thermostable Direct Hemolysin (TDH) and Thermostable Direct Hemolysin-related Hemolysin (TRH) Genes

The pathogenicity of *Vibrio parahaemolyticus* is rather complex and involves multiple factors (Broberg *et al.,* 2011). Thermostable direct hemolysin (TDH) and TDH-related hemolysin (TRH) are the most frequently used indicators of pathogenicity of the organism (Raghunath, 2015). Both the TDH and TRH are tetrameric proteins (Yanagihara *et al.,* 2010; Broberg *et al.*, 2011) and they share nearly 70% homology (Kishishita *et al.,* 1992). They act as porins and helps in the efflux of divalent cations and other solutes from and influx of water molecules into intestinal cells (Takahashi *et al.*, 2000; Ohnishi *et al.,* 2011). TDH forms large pores on erythrocyte membrane, allowing both water and ions to flow through the membrane (Matsuda *et al.,* 2012). TRH also alters the ion flux by activating ion channels (Takahashi *et al.,* 2000). The change in the ion flux of the intestine results in diarrhea during infection.

Detection of TDH in *Vibrio parahaemolyticus* is conventionally studied by beta haemolysis assay on a blood agar called Wagatsuma agar (Nishibuchi *et al.,* 1985) and the presence of TRH gene in *Vibrio parahaemolyticus* by urease phenotype (Cai and Ni, 1996). Previous reports state that most of the environmental *V. parahaemolyticus* strains are not pathogenic to humans, and only 1 to 2% of the environmental strains have shown to be positive for these genes while 90% of the clinical strains carry them (Nishibuchi and Kaper, 1995; Hervio-Heath *et al.,* 2002; Robert-Pillot *et al.,* 2004). However, some investigators have reported a higher prevalence of TDH and TRH genes in *V. parahaemolyticus* from environmental samples (DePaola *et al.,* 2003; Deepanjali *et al.*, 2005; Raghunath *et al.,* 2008; West *et al.,* 2013). Reports also reveal that these genes are not present in all the clinical strains of *V. parahaemolyticus* (Meador *et al.,* 2007). Hence, they alone are not responsible for *V. parahaemolyticus* pathogenicity. A study by Jones *et al.* (2012) revealed that even in the absence of these genes, *V. parahaemolyticus* strains exhibited pathogenicity. Thus, it was clear that other virulence factors also exist which contributes to the organism's virulence.

Type III Secretion Systems in *V. parahaemolyticus*

The type III secretion systems (T3SSs) have been widely reported in many Gram-negative bacteria such as *Pseudomonas aeruginosa*, *Yersinia* sp., *Salmonella* sp., *Shigella* sp. and *Escherichia coli* (Ono *et al.*, 2006) and the bacteria use these systems to induce pathogenicity in the host cells (Kumar *et al.,* 2014 a,b). Recently, it was found that toxins secreted by T3SS have a profound role in pathogenicity of *V. parahaemolyticus* (Noriea *et al.,* 2010; Broberg *et al.*, 2011; Karunasagar *et al.,* 2012). T3SSs are needle-like apparatus that enable injection of bacterial effector proteins directly into host cells resulting in modulation of numerous host processes (Noriea *et al.,* 2010). This apparatus of T3SS consists of the basal body spanning the bacterial inner and outer membrane, a needle that connects between the bacteria and the host cells and a translocon pore that penetrates the eukaryotic cell membrane (Letchumanan *et al.*, 2014). The apparatus is made of about 20 proteins and is highly conserved among the bacteria (Ono *et al.*, 2006). Once the effector enters the host cells, it disrupts the host's immune response (Coburn *et al.,* 2007). These effectors modify the host signalling pathways and make it beneficial for the pathogen (Ham and Orth, 2012). The whole genome sequencing of *V. parahaemolyticus* RIMD 22106633 (O3:K6) revealed two non-identical and non-redundant T3SS clusters, T3SS1 and T3SS2, with one on each of its two circular chromosomes (Makino *et al.*, 2003). Thus, the organism carries two type III secretion systems namely T3SS1 and T3SS2.

Type III Secretion System 1 (T3SS1)

T3SS1 gene cluster is located on pathogenicity island of the chromosome 1 (Caburlotto *et al.*, 2009). The gene causes cytotoxicity (Ham and Orth, 2012). The effectors secreted from T3SS1 include VopQ, VopS, VopR and VPA045 (Ham and Orth, 2012; Matsuda *et al.*, 2012). All these effector proteins have a distinct role and they work in an orchestrated manner to help the pathogen in colonising the host cell and establishing an infection (Ham and Orth, 2012). The effector protein VopQ is responsible for cytotoxicity of T3SS1 (Burdette *et al.*, 2009) and VPA0450 contributes to host cell death (Broberg *et al.*, 2011). VopS plays a major role in the bacterial evasion from inflammatory responses (Higa *et al.*, 2013). VopR helps in the refolding of T3SS effector proteins after their delivery into host cells and thus has a major role in the infection of *V. parahaemolyticus* (Burdette *et al.*, 2009). ExsA is the major transcriptional regulator of T3SS1 gene expression (Zhou *et al.*, 2015). Almost all clinical and environmental strains of *V. parahaemolyticus* harbours T3SS1 (Wang *et al.*, 2015).

Type III Secretion System 2 (T3SS2)

T3SS2 is mainly associated with enterotoxicity (Ham and Orth, 2012). The gene cluster is located on the pathogenicity island in Chromosome 2 of *V. parahaemolyticus* (Ham and Orth, 2012). This location of the gene on a pathogenicity island (Vp-PAI) may facilitate the transfer of this specific DNA fragment from a pathogenic strain to non-pathogenic autochthonous microorganism in the marine environment (Caburlotto *et al.*, 2009). Till date 7 effectors proteins are characterised from T3SS2 (Ghenem *et al.*, 2020). It includes VPA1380, VopA/VopP, VopL, VopT, VopV, VopZ and VopC (Ham and Orth, 2012; Matsuda *et al.*, 2012). The regulation of T3SS2 genes are mainly regulated by two transcriptional regulators, VtrA and VtrB (Kodama *et al.*, 2010).

Type VI Secretion ⊠ystem (T6SS)

Type VI secretion systems (T6SSs) are bacterial secretion systems used by Gram-negative bacteria to outcompete their neighbours (Ben-Yaakov and Salomon, 2019). They have a bacteriophage tail-like device to inject effector proteins to the host (Coulthurst, 2013). T6SSs are tightly regulated systems and are induced by external conditions such as temperature, salinity, quorum sensing etc (Salomon *et al.*, 2013; Bachmann *et al.*, 2015). *V. parahaemolyticus* encodes two T6SSs, one on each chromosome. T6SS1 is mainly found in association with clinical isolates, while all strains of *V. parahaemolyticus* harbours T6SS2 gene (Yu *et al.*, 2012). T6SS is stated to play an active role in actin cross linking, intracellular trafficking, secretion and vesicular transport during an infection (Boyd *et al.*, 2008).

Conclusion

V. parahaemolyticus, especially the O3:K6 serotype, is an emerging foodborne pathogen causing widespread foodborne outbreaks worldwide. A detailed understanding of virulence mechanisms of *V. parahaemolyticus* is lacking and it is required to deliver a better understanding of the pathogen and its host–pathogen interactions. However, the exact mechanism by which the infection is initiated is not yet known. Therefore, further studies are needed to fill the knowledge gaps.

Acknowledgement

The author would like to thank UGC-BSR for the financial assistance and Cochin University of Science and Technology (CUSAT) for the facilities provided.

References

Austin, B., 2010. Vibrios as causal agents of zoonoses. Vet. Microbiol., 140: 310-317.

Ben-Yaakov, R. and Salomon, D., 2019. The regulatory network of *Vibrio parahaemolyticus* type VI secretion system 1. Environ. Microbiol., 21(7): 2248-2260. doi: 10.1111/1462-2920.14594.

Boyd, E.F., Cohen, A.L., Naughton, L.M., Ussery, D.W., Binnewies, T.T. and Stine, O.C., 2008. Molecular analysis of the emergence of pandemic *Vibrio parahaemolyticus*. BMC Microbiol. 8: 110.

Broberg, C.A., Calder, T.J. and Orth, K., 2011. *Vibrio parahaemolyticus* cell biology and pathogenicity determinants. Microbes and Infection 13: 992-1001.

Burdette, D.L., Seemann, J. and Orth, K., 2009. *Vibrio* VopQ induces PI3-kinase-independent autophagy and antagonizes phagocytosis. Mol. Microbiol., 73: 639-49.

Caburlotto, G., Gennari, M., Ghidini, V., MariaCarla, T. and Maria, M.L., 2009. Presence of T3SS2 and other virulence-related genes in TDH-negative *Vibrio parahaemolyticus* environmental strains isolated from marine samples in the area of the Venetian Lagoon, Italy. FEMS Microbiol. Ecol., 70: 506-514.

Cai, Y.L. and Ni, Y.X., 1996. Purification, characterization and pathogenicity of urease produced by *Vibrio parahaemolyticus*. Journal of Clinical Laboratory Analysis 10: 70-73.

Chao, G., Jiao, X., Zhou, X., Yang, Z., Huang, J., Zhou, L. and Qian, X., 2009. Distribution, prevalence, molecular typing and virulence of *Vibrio parahaemolyticus* isolated from different sources in coastal province Jiangsu, China. Food Control 20: 907-912.

Coburn, B., Sekirov, I. and Finlay, B.B., 2007. Type III secretion systems and disease. Clin. Microbiol. Rev., 20: 535-549.

Coulthurst, S.J., 2013. The Type VI secretion system – A widespread and versatile cell targeting system. Res. Microbiol.,164: 640-54.

Deepanjali, A., Kumar, H.S., Karunasagar, I., Karunasagar, I., 2005. Seasonal variation in abundance of total and pathogenic *Vibrio parahaemolyticus* bacteria in oysters along the southwest coast of India. Appl. Environ. Microbiol., 71: 3575-3580.

DePaola, A., Nordstrom, J.L., Bowers, J.C., Wells, J.G. and Cook D.W., 2003. Seasonal abundance of total and pathogenic *Vibrio parahaemolyticus* in Alabama oysters. Appl. Environ. Microbiol., 69: 1521-1526.

Ghenem, L., Elhadi, N., Alzahrani, F. And Nishibuchi, M., 2020. *Vibrio parahaemolyticus*: A review on distribution, pathogenesis, virulence determinants and epidemiology. Saudi J. Med. Sci., 5(2): 93-103.

Ham, H. and Orth, K., 2012. The role of type III secretion system 2 in *Vibrio parahaemolyticus* pathogenicity. J. Microbiol., 50: 719-725.

Hervio-Heath, D., Colwell, R. R., Derrien, J. M., Fournier, J. M. and Pommepuy, M., 2002. Occurrence of pathogenic Vibrio in coastal areas of France. J. Appl. Microbiol. 92: 1123-1135.

Higa, N., Toma, C., Koizumi, Y., Nakasone, N., Nohara, T. and Masumoto, J., 2013. *Vibrio parahaemolyticus* effector proteins suppress inflammasome activation by interfering with host autophagy signaling. PLoS Pathog. 9:e1003142.

Honda, S., Goto, I., Minematsu, I., Ikeda, N., Asano, N., Ishibashi, M., Kinoshita, Y., Nishibuchi, M., Honda, T. and Miwatani, T., 1987a. *Vibrio parahaemolyticus* infectious disease caused by Kanagawa phenomenon-negative O3: K6 originated from Maldives. Japanese Journal of Infectious Diseases 61(9): 1070-1078.

Honda, S., Goto, I., Minematsu, I., Ikeda, N., Asano, N., Ishibashi, M., Kinoshita, Y., Nishibuchi, M., Honda, T., Miwatani, T., 1987b. Gastroenteritis due to kanagawa negative *Vibrio parahaemolyticus*. The Lancet 329: 331-332.

Jones, J.L., Ludeke, C.H.M., Bowers, J.C., Garrett, N., Fischer, M., Parsons, M.B., Bopp, C.A., DePaola, A., 2012. Biochemical, serological, and virulence characterization of clinical and oyster *Vibrio parahaemolyticus* isolates. J. Clin. Microbiol., 50: 2343-2352.

Karunasagar, I., Kumar, K. B. and Nair, G.B., 2012. Epidemiology and genetics of the pandemic clone of *Vibrio parahaemolyticus*. In: Foodborne and Waterborne Bacterial Pathogens - Epidemiology, Evolution and Molecular Biology (Faruque, S.M., Ed.), pp. 185196, Caister Academic Press, UK.

Kishishita, M., Matsuoka, N., Kumagai, K., Yamasaki, S., Takeda,Y. and Nishibuchi, M., 1992. Sequence variation in the thermostable direct hemolysin-related hemolysin (*trh*) gene of *Vibrio parahaemolyticus*. Appl. Environ. Microbiol., 58: 2449-2457.

Klein, S.L., West, C.K.G., Mejia, D.M., Lovell, C.R., 2014. Genes similar to the *Vibrio parahaemolyticus* virulence-related genes TDH, TLH and vscC2 occur in other Vibrionaceae species isolated from a pristine estuary. Appl. Environ. Microbiol., 80: 595-602.

Kodama, T., Yamazaki, C., Park, K.S., Akeda, Y., Iida, T. and Honda, T., 2010. Transcription of *Vibrio parahaemolyticus* T3SS1 genes is regulated by a dual regulation system consisting of the ExsACDE regulatory cascade and H-NS. Fems Microbiology Letters 311(1): 10-17.

Kumar, B.K., Deekshit, V.K., Rai, P., Shekar, M., Karunasagar, I. and Karunasagar, I., 2014a. Presence of T3SS2â genes in TRH+ *Vibrio parahaemolyticus* isolated from seafood harvested along Mangalore coast, India. Lett. Appl. Microbiol., 58(5): 440-446.

Kumar, B.K., Deekshit, V.K., Raj, J.R.M., Rai, P., Shivanagowda, B.M., Karunasagar, I. and Karunasagar, I., 2014b. Diversity of *Vibrio parahaemolyticus* associated with disease outbreak among cultured *Litopenaeus vannamei* (Pacific white shrimp) in India. Aquaculture 433: 247-251.

Letchumanan, V., Chan, K.G., Lee, L.H., 2014. *Vibrio parahaemolyticus*: A review on the pathogenesis, prevalence, and advance molecular identification techniques. Front. Microbiol. 5: 705.

Letchumanan, V., Yin, W. F., Lee, L.H., and Chan, K.G., 2015. Prevalence and antimicrobial susceptibility of *Vibrio parahaemolyticus* isolated from retail shrimps in Malaysia. Front. Microbiol. 6: 33.

Li, Y., Xie, X., Shi, X., Lin, Y., Mou, J., Chen, Q., Lu, Y., Zhou, L., Jiang, M., Sun, H., Ma, H., Cheng, J., Hu, Q. and Qiu, Y., 2014. *Vibrio parahaemolyticus*, Southern Coastal Region of China, 2007–2012. Emerg. Infect. Dis., 20: 685-688.

Makino, K., Oshima, K., Kurokawa, K., Yokoyama, K., Uda, T., Tagomori, K., Iijima, Y., Najima, M., Nakano, M., Yamashita, A., Kubota, Y., Kimura, S., Yasunaga, T., Honda, T., Shinagawa, H., Hattori, M. and Lida, T., 2003. Genome sequence of *Vibrio parahaemolyticus*: a pathogenic mechanism distinct from that of *V. cholerae*. The Lancet 361: 743-749.

Matsuda, S., Okada, N., Kodama, T., Honda, T. and Iida, T., 2012. A cytotoxic type III secretion effector of *Vibrio parahaemolyticus* targets vacuolar H+-ATPase subunit c and ruptures host cell lysosomes. PLoS Pathog., 8(7): e1002803.

Meador, C.E., Parsons, M.M., Bopp, C.A., Gerner-Smidt, P., Painter, J.A. and Vora, G.J., 2007. Virulence gene- and pandemic group-specific marker profiling of clinical *Vibrio parahaemolyticus* isolates. J. Clin. Microbiol., 45: 1133-1139.

Nishibuchi, M., and Kaper, J.B., 1995. Thermostable direct hemolysin gene of *Vibrio parahaemolyticus*: a virulence gene acquired by a marine bacterium. Infect. Immun. 63(6): 2093-2099.

Nishibuchi, M., Ishibashi, M., Takeda, Y. and Kaper, J.B., 1985. Detection of thermostable direct hemolysin gene and related DNA sequences in *Vibrio parahaemolyticus* and other Vibrio species by the DNA colony hybridization test. Infect. Immun., 49: 481-486.

Noriea, N.F. III, Johnson, C.N., Griffitt, K.J. and Grimes, D.J., 2010. Distribution of type III secretion systems in *Vibrio parahaemolyticus* from the northern Gulf of Mexico. J. Appl. Microbiol., 109(3): 953-962.

Odeyemi, O.A. 2016. Incidence and prevalence of *Vibrio parahaemolyticus* in seafood: a systematic review and meta-analysis. SpringerPlus 5, 464 (2016). doi.org/10.1186/s40064-016-2115-7.

Okada, N., Matsuda, S., Matsuyama, J., Park, K.S., de los Reyes, C. and Kogure, K., 2010. Presence of genes for type III secretion system 2 in *Vibrio mimicus* strains. BMC Microbiol., 10: 302.

Pazhani, G. P., Bhowmik, S. K., Ghosh, S., Guin, S., Dutta, S., Rajendran, K., Saha, D. R., Nandy, R. K., Bhattacharya, M. K., Mukhopadhyay, A. K. and Ramamurthy, T., 2014. Trends in the epidemiology of pandemic and non-pandemic strains of *Vibrio parahaemolyticus* isolated from diarrheal patients in Kolkata, India. PLOS Negelct. Trop. D., 8 (5): 2815.

Raghunath, P. 2015. Roles of thermostable direct hemolysin (TDH) and TDH-related hemolysin (TRH) in *Vibrio parahaemolyticus*. Front. Microbiol., 5: 805.

Raghunath, P., Acharya, S., Bhanumathi, A., Karunasagar, I. and Karunasagar, I., 2008. Detection and molecular characterization of *Vibrio parahaemolyticus* isolated from seafood harvested along the southwest coast of India. Food Microbiol., 6: 824-30.

Robert-Pillot, A., Guenole, A., Lesne, J., Delesmont, R., Fournier, J. M. and Quilici, M.L., 2004. Occurrence of the TDH and TRH genes in *Vibrio parahaemolyticus* isolates from waters and raw shellfish collected in two French coastal areas and from seafood imported into France. Int. J. Food Microbiol., 91(3): 319-325.

Silvester, R., Alexander, D., Hatha, M., Ammanamveetil, A., 2015. Prevalence, antibiotic resistance, virulence and plasmid profiles of *Vibrio parahaemolyticus* from a tropical estuary and adjoining traditional prawn farm along the south west coast of India. Ann. Microbio., 65(4): 2141-2149. doi.org/10.1007/s13213-015-1053-x .

Takahashi, A., Kenjyo, N., Imura, K., Myonsun, Y. and Honda, T., 2000. Cl-secretion in colonic epithelial cells induced by the *Vibrio parahaemolyticus* hemolytic toxin related to thermostable direct hemolysin. Infect. Immun., 68: 5435-5438.

Wang, R., Zhong, Y., Gu, X., Yuan, J., Saeed, A.F. and Wang, S., 2015. Corrigendum: The pathogenesis, detection, and prevention of *Vibrio parahaemolyticus*. Front. Microbiol. 6: 437.

West, C.K.G., Klein, S.L. and Lovell, C.R., 2013. High frequency of virulence factor genes TDH, TRH, and TLH in *Vibrio parahaemolyticus* strains isolated from a pristine estuary. Appl. Environ. Microbiol., 79(7): 2247-2252.

Yanagihara, I., Nakahira, K., Yamane, T., Kaieda, S., Mayanagi, K., Hamada, D., Fukui, T., Ohnishi, K., Kajiyama, S., Shimizu ,T., Sato, M., Ikegami, T., Ikeguchi, M., Honda, T. and Hashimoto, H., 2010. Structure and functional characterization of *Vibrio parahaemolyticus* thermostable direct hemolysin. J. Biol. Chem., 285: 16267-16274.

Yu, Y., Yang, H., Li, J., Zhang, P., Wu, B., Zhu, B., 2012. Putative type VI secretion systems of *Vibrio parahaemolyticus* contribute to adhesion to cultured cell monolayers. Arch. Microbiol., 194: 827-35.

Zhou, X., Konkel, M. E., Call, D.R., 2010. Regulation of type III secretion system 1 gene expression in *Vibrio parahaemolyticus* is dependent on interactions between ExsA, ExsC, and ExsD. Virulence 1(4): 260-272. doi: 10.4161/viru.1.4.12318.

Impact of Climate Change on Hydrological Cycle, Ecosystem, Fisheries and Food Security, Pages: 219–225
Edited by: B. Madhusoodana Kurup, M.R. Boopendranath, M. Harikrishnan and A.V. Shibu

CHAPTER - 19

A Comparative Study on the Wetland Avifauna in the Pokkali Fields of Ernakulam District, Kerala

S. Surya Babu* and K. Raju Thomas

Department of Zoology, Mar Thoma College (M.G. University, Kottayam), Tiruvalla-689103, Kerala, India
**E-mail: suryababu2016@gmail.com*

Wetlands are peculiar ecosystems between terrestrial and aquatic environments. They are also known as 'Kidney's of landscape' (Mitsch and Gosselink, 1993). In India, about 15.26 meter hectares of area comprises of wetlands (Panigrahy *et al.*, 2012). Kerala state has the largest area under wetlands (Nayar and Nayar, 1997). The Pokkali rice is a unique, tall, saline resistant, traditional variety with Geographical Indication (GI) (Ashamol, 2014). The paddy cultivation is practised during the low saline phase (June-October/mid November) followed by shrimp cultivation during the high saline phase (October/mid November-April). It is a sustainable, organic agricultural practice carried out in the water-logged coastal regions of Ernakulam, Alappuzha and Thrissur Districts of Kerala, India (Shylaraj and Sasidharan, 1998; Suchitra and Venugopal, 2005). The Pokkali wetlands are part of the Central Asian Flyway (CAF) and Vembanad Ramsar site. This ecosystem serves as Important Bird Areas (IBA) providing habitat for many avifauna and stopover sites for migratory birds. They are unique, nutritive, biodegradable and biodiverse tidal wetland ecosystems. (Islam and Rahmani, 2004; Wetlands International, 2007).

Kerala has 16 threatened species of wetland birds (Birdlife International, 2003). Wetland birds are limiting factors of restoration success and regional biodiversity. They are important indicators of wetland quality (Kumar and Gupta, 2009).The wetland avifauna are safeguarded by the Wildlife (Protection) Act, 1972 and the Convention on International Trade of Endangered Species of Wild Flora and Fauna (CITES). They are also under the Agreement on the Conservation of African-Eurasian Migratory Waterbirds (AEWA) (JNCC, 2005; CITES, 2012). Basic data collection on these birds is essential for the implementation of the appropriate conservation strategies.

Materials and Methods

An intense study was carried out during 01 June 2016 - 31 May 2019 in the five selected Pokkali wetlands of Ernakulam District. The study stations include Valiya Kadamakkudy, Palliackkal, Thathappilly, Elamkunnapuzha and Kandakkadavu. Fortnightly visits were carried out and observations were carried out from 06.00 a.m. to 06.30 p.m. with the help of spotting scope (10 – 45 X) and 8 x 40 (Bushnell) binocular. Direct observation method was adopted (Altman, 1974). Point count method was also used to study the activity of the birds (Bibby, 1999). The wetland avifauna were identified with the help of standard reference books (Neelakantan *et al.*, 1993; Ali, 1996) and experts. Photographs were taken with help of 36 X optical zoom camera (Nikon) for the proper identification of the species. Most of the observations were carried out from a distance of 50-200 m. Care was taken not do disturb the activities of the birds.

Results and Discussion

A total of 83 wetland and wetland associated birds belonging to 9 Orders and 17 Families were recorded during the study period. The species richness was reported as highest (83) from study station 1 (Valiya Kadamakkudy) and lowest (69) from station 3 (Thathappilly) and station 4 (Elamkunnapuzha). A total of 13,20,64 individuals were recorded during the study period. On comparing, the total avifaunal population, highest number of individuals were recorded in 2016-17 and thc lowest was in 2018-19. The decrease in the number of avifauna was due to climate change and the historic flooding of 2018 (Fig. 1). As the Pokkali wetlands are part of the Central Asian Flyway (CAF) and Vembanad Ramsar site, the ecosystem serves as Important Bird Areas (IBA) providing habitat for many avifauna and stopover sites for migratory birds (Islam and Rahmani, 2004).

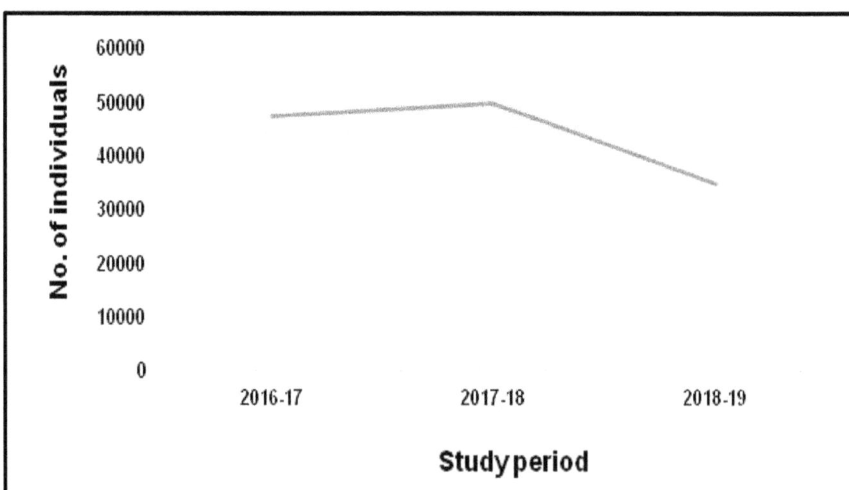

Fig. 1. Population fluctuation of avifauna during the study period

A total of 14 avifaunal species belonging to the Threatened and Near Threatened category were recorded, of which three are Threatened (T) and 11 are Near Threatened (NT). Out of the three Threatened (T) species, two are Vulnerable (V) and one is Endangered (E). The recorded avifauna represented Seven Orders and Eight Families (Table 1). Three Threatened bird species belonging to three Families and three Orders were recorded. White-necked stork *Ciconia episcopus* (Boddaert, 1783) and Greater spotted Eagle *Aquila clanga* (Pallas, 1811) were categorised as Vulnerable (V). A total of 56 White-necked Storks and three Greater spotted Eagles were recorded from the study stations. A total of three Gull-billed Terns *Gelochelidon nilotica* (Gmelin, 1789) that falls into Endangered (E) category were also recorded. About 16 threatened wetland birds were recorded by Birdlife International (2003). Out of this, three threatened birds were recorded from the study area.

A total of 11 Near Threatened (NT) birds belonging to four Orders and five Families were also recorded. The Near Threatened avifauna includes are Rufous-bellied Eagle *Hieraaetus kienerii* (G. de. sparre, 1835), Spot-billed Pelican *Pelecanus philippensis* (Gmelin, 1789), Oriental Darter *Anhinga melanogaster* (Pennant, 1769), Oriental White Ibis *Threskiornis melanocephalus* (Latham, 1790), Painted Stork *Mycteria leucocephala* (Pennanat, 1769), Ferruginous Pochard *Aythya nyroca* (Guldenstadt, 1770), Curlew Sandpiper *Calidris ferruginea* (Pontoppidan, 1763), Black-Tailed Godwit *Limosa limosa* (Linnaeus, 1758), Bar-Tailed Godwit *Limosa lapponica* (Linnaeus, 1758), Red Knot *Calidris canutus* (Linnaeus, 1758) and River Tern *Sterna aurantia* (J.E. Gray, 1831). A total of three individuals of Rufous-bellied Eagle 729 Spot-billed Pelicans, 740 Oriental Darters, 1400 Oriental White Ibis, 446 Painted Storks, 18 Ferruginous Pochard, 298 Curlew Sandpipers, 56 Black-Tailed Godwits, 10 Bar-Tailed Godwits and 461 River Terns were recorded from the study area. Only a single Red Knot was sighted from the Valiya Kadamakkudy pokkali wetland (Station 1) during 15 September 2018. This was recorded as the first report from the Ernakulam District and one of the rare reports from Southern India.

A total of 62 Threatened (T) birds and 4264 Near Threatened (NT) birds were recorded during the study period (Fig. 2). All of these species were recorded only during the migratory season except Oriental White Ibis and Spot-billed pelican. Both of these birds were recorded throughout the study period. The heronries of Spot-billed Pelicans were recorded from station 5 (Kandakkadavu). It was found that new species of migratory birds were visiting the study area every year. Greater spotted Eagle, Gull-billed Tern, Rufous-bellied Eagle, Black-Tailed Godwit, Bar-Tailed Godwit and Red Knot were not recorded in 2016. They were recorded in 2018 from the Valiya Kadamakkudy Pokkali wetland. The population of the Threatened and Near Threatened birds are decreasing due to threats like pollution, poaching, habitat loss and collection of eggs (Ali, 1996; Neelakantan *et al.*, 1993). No threats were recorded from the study area. Wetland birds are limiting factors of restoration success and regional biodiversity. They are important indicators of

Table 1: **List of Vulnerable (V), Endangered (E), and Near Threatened (NT) Avifauna recorded from the study area during the study period (01 June 2016 - 31 May 2019).**

Sl No.	Common Name	Scientific Name	Order	Family	IUCN Status
1	White-necked Stork	*Ciconia episcopus*	Ciconiiformes	Ciconiidae	V
2	Greater spotted Eagle	*Aquila clanga*	Acciptriformes	Accipitridae	V
3	Gull-billed Tern	*Gelochelidon nilotica*	Charadriiformes	Laridae	E
4	Rufous-bellied Eagle	*Hieraaetus kienerii*	Acciptriformes	Accipitridae	NT
5	Spot-billed Pelican	*Pelecanus philippensis*	Pelecaniformes	Pelecanidae	NT
6	Oriental Darter	*Anhinga melanogaster*	Suliformes	Anhingidae	NT
7	Oriental White Ibis	*Threskiornis melanocephalus*	Ciconiiformes	Threskiornithidae	NT
8	Painted Stork	*Mycteria leucocephala*	Ciconiiformes	Ciconiidae	NT
9	Ferruginous Pochard	*Aythya nyroca*	Anseriformes	Anatidae	NT
10	Curlew Sandpiper	*Calidris ferruginea*	Charadriiformes	Scolopacidae	NT
11	Black-Tailed Godwit	*Limosa limosa*	Charadriiformes	Scolopacidae	NT
12	Bar-Tailed Godwit	*Limosa lapponica*	Charadriiformes	Scolopacidae	NT
13	Red Knot	*Calidris canutus*	Scolopacidae	Scolopacidae	NT
14	River Tern	*Sterna aurantia*	Charadriiformes	Laridae	NT

wetland quality (Kumar and Gupta, 2009). Hence, the increased occurrence of these avifauna in the study area highlights the healthy nature of the ecosystem and the environment.

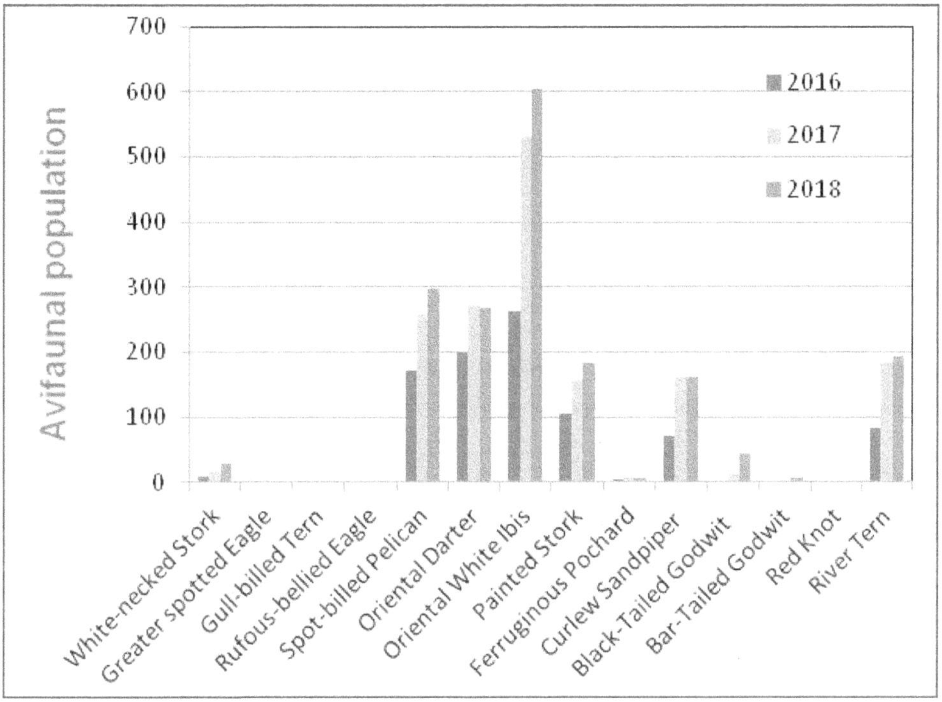

Fig. 2. Population fluctuation of Threatened (T) and Near Threatened (NT) avifauna during the study period

Conclusion

Birds are excellent indicators of ecosystem health. Their occurrence in an area highlights the climatic changes and the healthy nature of the ecosystem and the environment. The study area is free from much disturbances, chemicals and pollution. The people in these areas have great conservative feeling towards the birds. It was found that the study areas are changing as a major habitat for many wetland bird species including Threatened and Near Threatened. Hence, immediate efforts must be adopted for their protection and conservation.

Acknowledgements

Thanks to Dr. Icy K. John, Principal, Mar Thoma College, Tiruvalla for providing the facilities. We thank Dr. Sugathan, R., Research Director, Salim Ali Bird Sanctuary, Thattekkad for the encouragement and timely advice. Thanks for the financial support provided by UGC. We thank Dr. John George, M., Associate Professor (Rtd.), Mar

Thoma College, Tiruvalla, Dr. L. Krishnan, Senior Scientist (Rtd.), CMFRI, Kochi and P. Bala Narayanan, Programme Executive of All India Radio, Kochi for their valuable support. The support from Shri. Basil Peter, Shri. Mukundan Kizhakemadom and Shri. Umer Nizar for taking the photographs was helpful. Also, we are thankful to the librarians of KFRI, Peechi and CMFRI, Kochi.

References

Ali, S., 1996. The Book of Indian Birds. 12th edn. Oxford University Press, Bombay.

Altman, J., 1974. Observational study of behaviour: sampling methods. Behaviour 49: 227-267.

Ashamol, A.T.V., 2014. Effect of Pokkali cultivation and shrimp farming on the soil characteristics of two different Pokkali fields at Chellanam and Kadamakudy, Kochi, Kerala, India. International Research Journal of Environmental Sciences 3(9): 61-64.

Bibby, C.J., 1999. Making the most of birds as environmental indicators. Ostrich., 70: 81-88.

BirdLife International, 2003. Saving Asia's Threatened Birds: A guide for government and civil society. BirdLife International, Cambridge, UK. i-x, 1-246.

CITES, 2012. Annual Report 2012. CITES Management Authority in India. GOI, MoEF. 339p.

Islam, Z.U. and Rahmani, A.R., 2004. Important bird areas in India. Priority sites for conservation. First edn. Indian Bird Conservation Network, Mumbai. Bombay Natural History Society and Birdlife International (UK).

IUCN, 2012. Changes in the number of species in the Threatened categories (CR, EN, VU) from 1996 to 2012 (IUCN Red List version 2012.2) for the major taxonomic groups on the Red List. IUCN.

JNCC, 2005. Checklist of birds listed in the CITES appendices and in EC Regulation No. 338/97. 8th edn. Joint Nature Conservation Committee (JNCC), 362 p.

Kumar, P. and Gupta, S.K., 2009. Diversity and abundance of wetland birds around Kurukshetra, India. Nepal Journals online 7: 212-217.

Mitsch, W.J. and Gosselink, J.G., 1993. Wetlands. Van Nostrand Reinhold, New York.

Nayar, S. and Nayar, N.M., 1997. Wetlands. In: Thampi, K.B., Nayar, N.M., Nayar, C.S. (Eds.). The Natural Resources of Kerala. WWF State Office, Trivandrum.

Neelakantan, K.K., Sasikumar, C. and Venugopalan, R., 1993. A Book of Kerala Birds. World Wide Fund for Nature (WWF) - India, Kerala State Committee, Trivandrum, 146 p.

Panigrahy, S., Murthy, T.V.R., Patel, J.G. and Singh, T.S., 2012. Wetlands of India: inventory and assessment at 1:50,000 scale using geospatial techniques. Current Science 102(6): 852-856.

Shylaraj, K.S. and Sasidharan, N.K., 1998. VTL 5: A high yielding salinity tolerant rice variety for the coastal saline ecosystems of Kerala. Crop Science 38: 394-398.

Suchitra, M. and Venugopal, P.N., 2005. In troubled waters. Agriculture. [Online]. 36(4): 118-125. www.questfeatures.org/articles/pokkali.html[Apr.23, 2014].

Wetlands International, 2007. Waterbird Population Estimates. 5th edn. Wetlands International, Wageningen, The Netherlands.

IV

Capture Fisheries

Impact of Climate Change on Hydrological Cycle, Ecosystem, Fisheries and Food Security, Pages: 229–247
Edited by: B. Madhusoodana Kurup, M.R. Boopendranath, M. Harikrishnan and A.V. Shibu

CHAPTER - 20

Impact of Climate Change on Ecology and Fisheries of Tropical Estuaries with reference to Vembanad Estuary (India): Past Trends and Potential Projections

B. Madhusoodana Kurup[1*], M. Harikrishnan[1], A.V. Shibu[1] and Sileesh Mullasseri[2]

[1]*School of Industrial Fisheries, Cochin University of Science and Technology, Kochi-682016, Kerala, India*
School of Ocean Science and Technology, Kerala University of Fisheries and Ocean Studies, Panangad, Kochi-682506, Kerala, India
**E-mail: kurup424@gmail.com*

Wetlands cover approximately 6% of the Earth's surface and are frequently found at the interface between terrestrial and aquatic ecosystems. They are strongly dependent on water cycle and therefore, wetlands are extremely vulnerable to the effects of climate change. It is possible to predict some of the main consequences of climate change in wetlands such as modifications in hydrological regimes capable of causing intense droughts or inundations. Climate change is really affecting the ecosystem functioning and there are visible impacts such as rise in atmospheric (2-4°C) and sea surface temperatures (0.6 to 1°C). The available reports suggest that the expected sea-level rise in 2100 can be in the range 59 cm to 226 cm with a mean of 0.1-0.2 mm/year (Bamber *et al.*, 2019; Nauels *et al.*, 2017). Wetlands play an important role as natural solutions, in adapting to and mitigating the impact of climate change. Water supports life and wetlands ensure functioning of the water cycle. As 'kidneys of landscape', wetlands receive flows of water and effluents from upstream sources (Shine and Klemm, 1999). Extensive food chain and biological diversity in wetlands make them biological supermarkets, they play unique roles in the region by offering a distinctive environment for a broad category of flora and fauna and regulating water resources for utilization of indigenous community (Mitsch and Gosselink, 2007). Thus, wetlands are vital for our water requirement and food security.

The Vembanad Lake spanning between latitudes 9°28' and 10°10' N and longitudes 76°13' and 76°31' E is the biggest among the 30 backwaters in Kerala (Fig.1). The Vembanad wetland system, bordered by three districts and with a span of about 2033.02 km^2 (Gopakumar & Takara, 2009), is fed by *Achenkovil, Manimala, Meenachil, Muvattupuzha, Pamba,* and *Periyar* rivers (Kurup and Samuel,1987; Menon *et al.*, 2000). The western side of the lake is bordered by a narrow belt of land separating the lake from the Lakshadweep sea. It extends to a length of 96.5 km from Azhikode in the north to Alappuzha town in the south.

The Thanneermukkom barrage, commissioned in 1975, is a salt-water barrage constructed at narrow region of the lake at Thanneermukkom and separates the lake in to freshwater dominant southern side and salt water dominant northern side. It has 1252 m total length and 93 vents, each having a dimension of 12.15 m x 5.47 m. A twin lock at west end and a single lock at each end are provided for navigation. It was constructed to prevent saltwater intrusion during premonsoon periods and to promote double cropping of paddy in about 55,000 ha of low-lying fields in Kuttanad, the rice bowl of Kerala. A slipway was also constructed at Thottappally in 1955 for immediate and fast drainage of flood water from the lake to Arabian Sea. The Vembanad lake system is integrated to a marshy delta in south end, known as 'Kuttanad' which is formed by four river networks namely, Pampa, Manimala, Achankovil, and Meenachil. Large parts of the vast estuary lie below the sea level up to a depth of about 2.5 m, waterlogged for the most part of the year subject to flood and inundation during the monsoons and saline water intrusion during the summer months. In this paper, an attempt is made to bring out the probable impact of climate change on the ecology and fisheries of Vembanad Lake through trend analysis and model projections.

Materials and Methods

Data were compiled from secondary sources (Kurup, 1982; Kurup and Samuel, 1986; Kurup and Samuel, 1987; Kurup *et al.*,1989; 1990a; 1990b; 1992; 1993; Harikrishnan and Kurup, 2001; Dipson *et al.*, 2015; Krishna Kumar and Priyadarsanan, 2012; Jacob *et al.*, 2013; Asha *et al.*, 2014) and from Asia Pacific Data Research Centre (APDRC) database, for trend analysis. Besides, rainfall data were also collected from the Meteorological department of India. Box-Jenkins method (Box and Jenkins, 1970) was employed for analysing the trend from time-series data. This method is referred to as a systematic method of identifying, fitting, checking, and using integrated autoregressive, moving average (ARIMA) time series models. The Box-Jenkins method is appropriate for medium to long time series data. Changes in the abiotic and biotic parameters and fishery of the wetland ecosystem have been evaluated from 1970 to 2015 during pre-monsoon (February to May), monsoon (June-September) and post-monsoon period (October -January), based on published information.

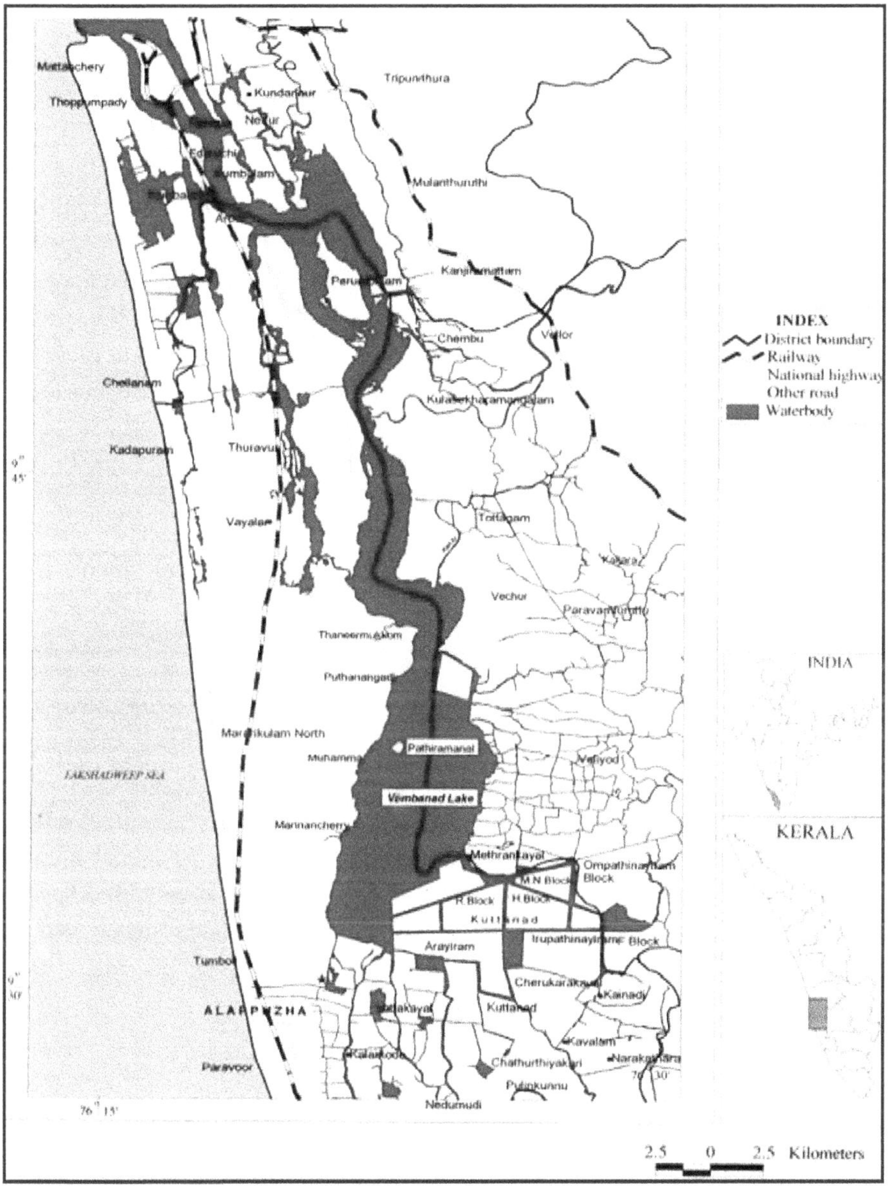

Fig. 1. Map of Vembanad Lake

Results and Discussion

Spatial changes of estuary

Wetland ecosystems, in general, are prone to both natural and anthropogenic forces that bring about perceptible changes in their physical and biological characteristics. One of most perceptible impacts on them is serious alteration in their physical outlines leading to spatial shrinking. Vembanad estuary also suffered severe shrinking in its area from 96.6

km^2 in 1944 to 81.26 km^2 in 2007 (Dipson *et al.*, 2015) (Fig. 2). According to Gopalan *et al.* (1983), the Vembanad estuary which had an area of 36,500 ha in the last century had undergone man-made shrinkage to only 23,350 ha due to reclamation for various agricultural, aquacultural, urbanisation, housing, industrialisation, retting and other activities. Nair (1990) reported that the surficial area of Vembanad backwater has shrunken to 51.0 km^2 in 15 years from 230 km^2 in 1968 to 179.25 km^2 in 1983. The pattern of surface area shrinkage of Vembanad Lake is depicted in Fig 3. There was a reduction in surface area of the lake from 177.26 km^2 in 1973 to 165.01 km^2 in 2015. The reclaimed area of 12.28 km^2 of the estuary over a period of 42 years which formed 6.53% of the estuary had been utilised for paddy cultivation, raising plantation crops and extension of Willington Island (Nair and Babu, 2016). The pattern of shrinkage of water body during the period is depicted in Fig. 4.

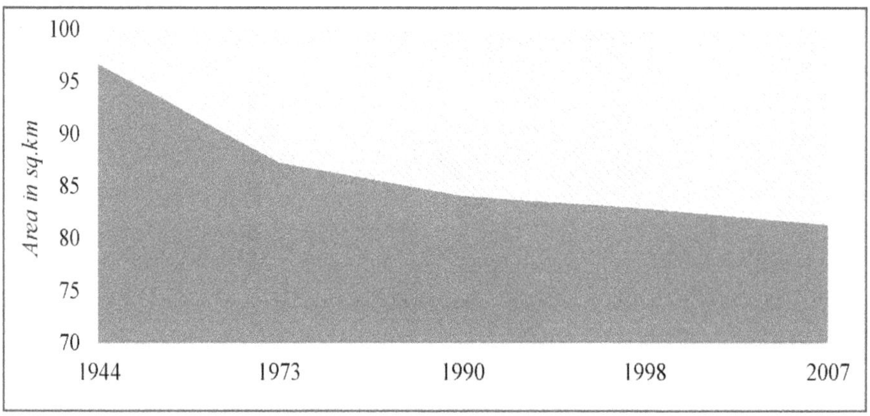

Fig. 2. Decrease in the area of the Vembanad lake system from 1944 to 2007 (Dipson *et al.*, 2015).

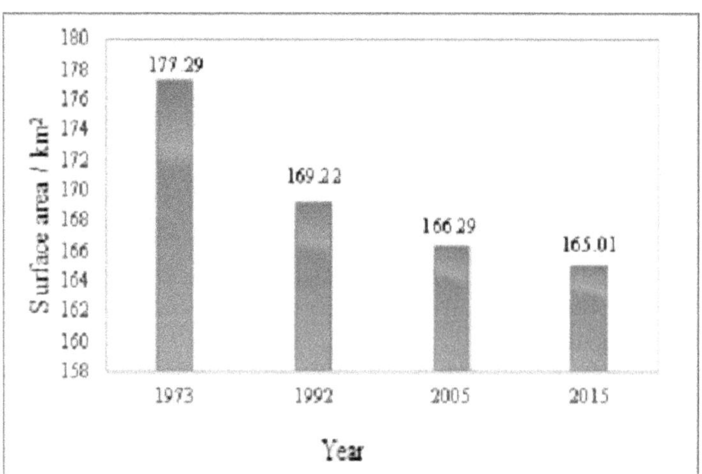

Fig. 3. Shrinkage of surface area of Vembanad Lake from 1973 to 2015 (Source Nair and Babu, 2016)

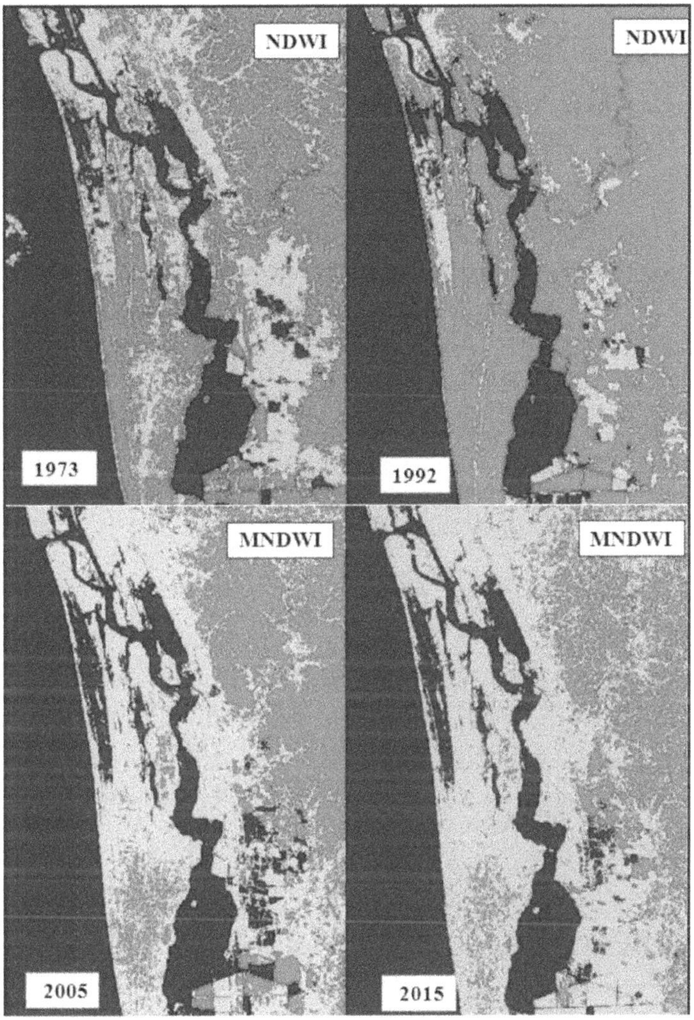

Fig. 4. The 1973 and 1992 images shows the water body (blue colour) as extracted from NDWI and 2005 and 2015 images shows estimates (blue colour) from MNDWI (Source: Nair and Babu, 2016).

Surface water temperature

Sea surface temperature in coastal area considerably influences water temperature in estuarine ecosystem and affects photosynthesis, metabolic rate and sensitivity of aquatic organisms. Variations in surface water temperature recorded from Vembanad estuary from 1993 to 2020 is depicted in Fig. 3, indicating an increasing trend over the years. Temperature recorded in the lake during pre-monsoon (hottest) period in past five decades registered a highest value of 33°C during 2011-2020 (Vincy *et al.*, 2012) while the lowest during the summer periods was recorded (32.4°C) was recorded during 1980-90 periods (Kurup *et al.*,1989). It is clearly discernible that the water temperature of the lake during the past four decades has shown an increasing trend with a rise of 0.6 °C. The ARIMA model also has shown a mean temperature rise of 0.5 °C from 1995 to 2020 (Fig. 5) which also concurs with the reported highest temperature value from the lake.

Rainfall and river flow

Freshwater discharge to the estuarine system comes from rainfall on the estuarine catchment areas. The amplified runoff can influence the temperature, salinity, and other nutrient composition of the estuary. Precipitation over Vembanad estuary (Ernakulum district), and rainfall have a significant influence in the estuarine ecosystem, the yearly precipitation for the region has shown an increasing trend and high fluctuation in the predicted precipitation (Fig. 6).

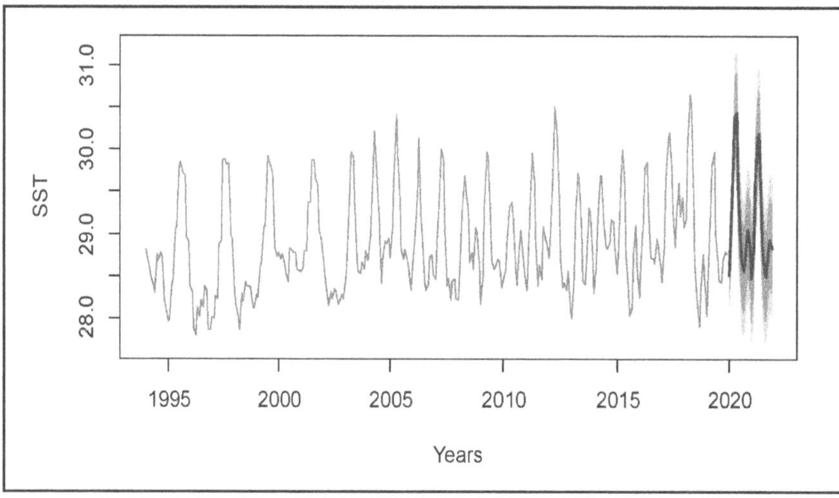

Fig. 5. Trend and forecast of coastal sea surface temperature in the Vembanad lake system.

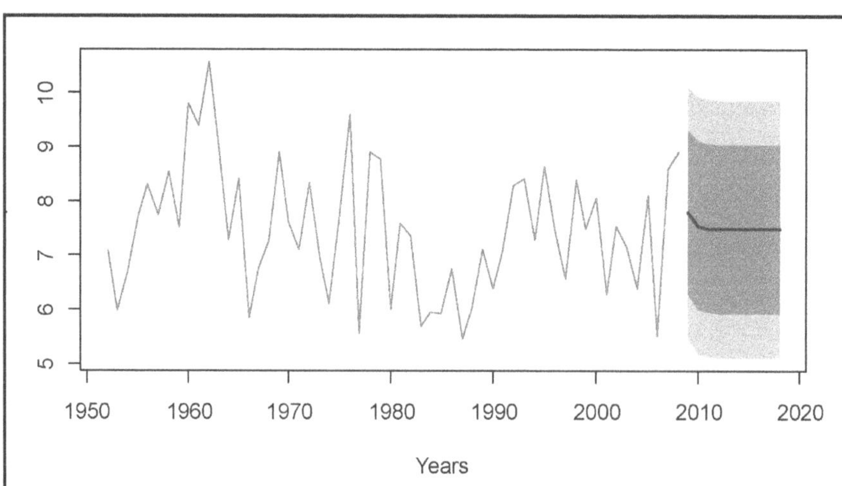

Fig. 6. Trend and forecast of precipitation (mm) over Vembanad lake system

Salinity

Biodiversity of estuarine ecosystem are seriously influenced by fluctuations in salinity which controlled the distribution and abundance of various species. The highest salinity recorded from the lake in premonsoon (hottest) periods during past five decades (1971-80, 1981-90, 1991, 2000, 2001-2010, 2011-2020) was 34.2 ppt during 2001-2010 (Jacob *et al.*, 2013) and 2011-20 (Vincy *et al.*, 2012). Salinity in post-monsoon and monsoon were also high (33.99 and 27.02 ppt, respectively) during 2001-2010 (Jacob *et al.*, 2013). The highest salinity values reported from the lake in pre-monsoon periods during the other three decades i.e. from 1970 onwards were in the order of 30.5 ppt during 1978-80 periods (Kurup and Samuel, 1987), 33.0 ppt during 1981-90 periods (ICMAM, 2002) and 33 ppt during 1998 (Nair *et al.*, 1988). While comparing the highest values recorded from the lake over the past five decades, it appears that the highest values of salinity during the pre-monsoon months was not steady and showed ups and down. While comparing the salinity values recorded during the hottest season from the lake, the values have shown a rise of 1.2 ppt from the lake during the latter decades when compared to pre 90 years. However, while analysing the trend and prediction of the salinity profile of Vembanad estuary using ARIMA model from 2012 onwards (Fig. 5), it appears that the salinity values of the Vembanad estuary have an increasing trend till 2015 with a mean value of 33 °C; however, the trend got reversed from 2016 showing a diminishing trend, thereafter. The highly fluctuating trends in salinity of the lake can well be attributed to the variation in monsoon intensity and the consequential freshwater discharge during both southwest and northeast monsoon periods. The pattern of rainfall over the periods of this study has also shown vide fluctuations with an increasing trend in recent years (Fig. 6). Further, the lake has also been prone to the after effects of massive deluges took place in 2018 and 2019. This finding also concurs with the trend observed in precipitation over the estuary, which showed an increasing trend during the same period (Fig. 6), a finding which can be attributed to the lowering of salinity during the recent years.

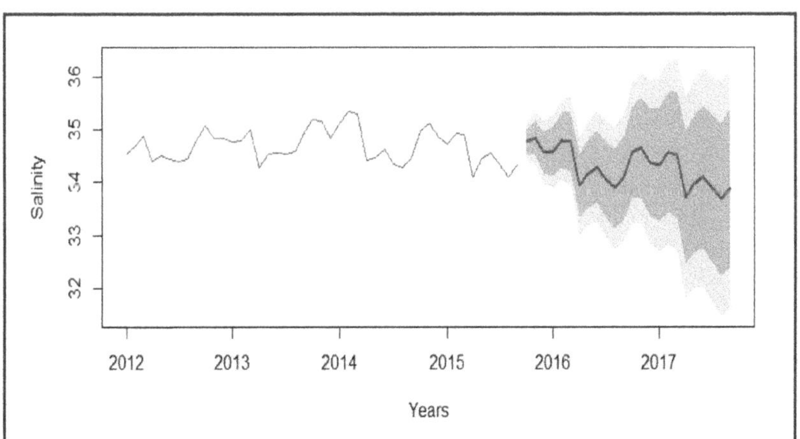

Fig. 7. Trend and forecast of salinity profile of Vembanad lake system

Other chemical parameters

During post-2000 period, the dissolved oxygen concentration was comparatively on a lower side when compared to the pre-2000 period, registering 3.69 to 5.0 mg/l during the pre-monsoon period of 2009 while it ranged from 2.16 to 4.89 mg/l during post-monsoon. Lowest pH value in pre-monsoon was recorded as 6.1 (James *et al.*, 1997) and 6.2 (Harikrishnan and Kurup, 1998) while during post-monsoon and monsoon, the lowest pH were 6.0 to 6.2 (Girishkumar, 2013). The highest pH value during the pre-monsoon was reported as 8.5 (Harikrishnan and Kurup, 2001) and the same value was also reported by Asha *et al.* (2016). In general, pH values of the lake were not subjected to extreme variations during the past four decades and ranged between 6.1 and 8.5. Sunil Kumar (2002) reported the sediment organic carbon during pre-monsoon, monsoon and post-monsoon periods as 2.01, 1.84 and 1.94%, respectively, whereas during recent years, the sediment organic matter showed an increase to 2.88% during monsoon, 3.48% during post-monsoon, and 2.77% during monsoon months (Renjith and Chandramohanakumar, 2009).

Primary productivity

Gross productivity values showed an average value of 0.99 g C/m^3/day and the net productivity of 0.51 g C/m^3/day during the period 1996-97 (Anon, 2001). Qasim *et al.* (1969) reported a gross productivity value ranging from 272 to 293 gC/m^2/year and the net productivity from 184 to 202 gC/m^2/year in the Vembanad Lake during the pre-barrage phase. According to Gopinathan *et al.* (1974), the primary productivity of the lake varied from 41 to 577 mg/m^3. However, ICMAM (2002) reported an annual mean gross and net productivity of Cochin backwaters as 0.753 C/m^3/day and 0.603 gC/m^3/day, respectively. Akram (2000) reported the productivity of the lake in the range from 5.8 mgC/m^3/day to 222.4 mg/C/m^3/day. It appears that, in general the trend in primary productivity of the lake had shown a declining trend during post-2000 periods.

Chlorophyll

A decreasing trend could be seen in the chlorophyll content of the lake (Fig. 8) which manifests a plummeting of primary productivity in the lake system.

Fish and Fisheries

Fish diversity of Vembanad Lake includes 158 species of 100 genera and 56 families (Kurup, 1982; Kurup and Samuel, 1987). Among them, 23 species are oligohaline, 79 euryhaline and 48 stenohalaine (Kurup *et al.*, 1993). Nine species of prawns, 3 species of crabs and eight species of clams are commercially important (Kurup *et al.*, 1990; Kurup *et al.*, 1993). Seven fish species, endemic to the rivers of Western Ghats, are also found in this lake system such as *Dayella malabarica, Horabagrus brachysoma, Mastacembelus guentheri, Mystus malabaricus, Mystus oculatus, Puntius filamentosus*, and *Labeo dussumieri*. Krishna Kumar and Priyadarsanan (2012) reported 67 species of finfishes and 14 species of shellfishes in the lake during 2008-11 period as evidensed from Vembanad Fish Count survey. The number of bivalves recorded from the lake was eight, among them, 6 were clams and two mussels (ICMAM, 2002) whereas 20 species of crabs were identified from the lake. Roy and Nandi (2008) recorded other faunal diversity in the Vembanad-Kol complex as 123 species of phytoplankton, 24 species of zooplankton, 8 macro-benthos, 20 crabs, 11 species of bivalves, 2 freshwater turtles and 225 bird species. Instances of sighting of spinner dolphins (*Stenella longirostris*) have also been reported from Cochin bar mouth. *Crocodilus porosus*, which were once present in the estuary, are no longer sighted.

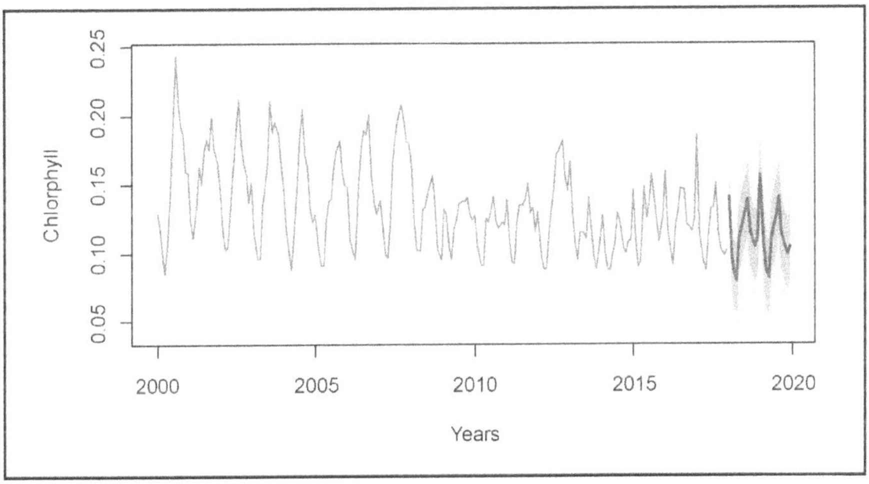

Fig. 8. Trend and forecast of chlorophyll content of Vembanad lake system

During 1975, the fishery in this water body was quantified as 14,000 tonnes (Rasalam and Sebastian, 1976), which showed a massive decline to 7200 tonnes during 1989-90, of which 93% was landed in lake north of Thaneermukom Barrage (Kurup *et al.*, 1990a). In the southern part of lake, a catch of 687 tonnes was reported in 2000-2001 (Padmakumar *et al.*, 2002). Asha *et al.* (2014) estimated an annual landing of 4387.31 t from Vembanad lake of which 480.98 t and 3906.33 t were contributed by southern and northern zone of Vembanad lake, respectively. Eighty species of finfishes, five species of penaeid shrimps, three species of palaemonid prawns and two species of crabs were identified from the lake, during the study period. The total area of live clam production in Vembanad estuary is about 1565 ha with an annual production of 50,275 tonnes (Kurup *et al.*, 1990b). The fishery of *M. rosenbergii*, which was to the extent of 305 tonnes during 1960-61 (Raman, 1967) in the lake, was reduced to about 39 tonnes by 1987-88 (Kurup *et al.*, 1989a, Harikrishnan and Kurup, 2001). Mullets (*Mugil* sp.), perch (*Lates calcarifer*), milkfish (*Chanos chanos*), catfish (*Tachysurus maculatus*), halfbeak (*Hyporhamphus* sp.),

Elopiformes (*Megalops* sp.) and pearl spot (*Etroplus suratensis*) are the commercially important estuarine finfish species in the lake. Prior to seventies, the bulk of the catch was constituted by Hilsa, oil sardines, mackerel, herrings, croakers, flatfishes and marine catfishes (Shetty, 1965). Catches of freshwater fish like *Horobagrus brachysoma, Clarius batrachus* and *Ompak bimaculatus* are also reported to have dwindled (Kurup *et al.*, 1990a). The euryhaline conditions created by early opening of Thannermukkom barrage allowing salinity influx was attributed to be the reason for high fish diversity in 2009. Total of 5 exotics viz; *Pterygoplichthys multiradiatus* (sucker catfish), *Catla catla* (catla), *Labeo rohita* (rohu), *Oreochromis mossambicus* (tilapia) and *Pangasious sutchi* (sutchi catfish) were observed in Vembanad Fish Count 2009, however, no exotic fishes were recorded in 2011 (Kumar and Priyadarsanan,2012). Kumar & Priyadarsanan (2012) reported a serious decline in the population of the dwarf puffer, *Carinotetraodon travancoricus*, which constituted 90% of experimental fishing catches in 2008, dwindled to 2% in 2009 and was very rare or absent in 2010 and 2011. It has also been reported that six brackish water fish species recorded in 2009 were not observed during 2010 and 2011 (Kumar and Priyadarsanan, 2012). Jacob *et al.* (2013) also recorded 67 species of finfishes and 14 species of shellfishes as reported earlier (Kurup and Samuel, 1987; Kurup & Samuel, 1985). However, it may be worth noticing that among the 60 species reported earlier (Kurup, 1982; Kurup and Samuel, 1985; Kurup *et al.*, 1989b) only 23 species were not found in the fish count survey while 24 new records were added (Krishna Kumar and Priyadarsanan, 2012). Furthermore, four freshwater fishesviz., *Clarias dussumieri, Ehirava fluviatilis, Macrognathus guentheri* and *Ompok bimaculatus* earlier reported by Kurup and Samuel (1985)were found missing recently in the lake. *Clarias dussumieri* has completely disappeared from the lake and might have become locally extinct. The reasons for the non-availability of these species which were true denizens of Kuttanad demand concerted investigation. Prior to eighties, marine migrant species dominated the lake (56%). On the contrary, the preponderance of limnetic migrants increased perceptibly with a remarkable decline in marine migrants (Maitra *et al.*, 2018). Accordng to Kurup *et al.* (1993), the ichthyofauna of the lake was constituted by omnivorous fishes (49%), followed by carnivores (44%) and herbivores (4%) and detritivores (3%). However, during recent decades, the carnivores declined considerably in the lake to 28% and the higher carnivorous predators constituted only 0.66% of total catch in the lake (Kumar and Priyadarshan, 2012; Maitra *et al.*, 2018). It clearly indicates that the fishery in Vembanad lake the fishery is now is in the state of "fishing down the food web".

Anthropogenic influence

The environmental and ecological changes brought about in the lake have been attributed to various anthropogenic activities such as urban development and industry. The major chemical industries in Kerala are located in Cochin, near the banks of Vembanad lake system. Around 250 industries, including major chemical industries, are located near the banks of Periyar river (Dsikowitzky *et al.*, 2014). The Vembanad lake receives a high load of partially treated industrial effluents and untreated domestic effluents (CPCB, 1996), wastes from aquaculture fields, agricultural fields, coconut husk retting yards, fish processing plants, shrimp farms and animal bone processing units (Thomson, 2002).

There is a consensus among scientists and policymakers that human activities are increasing the levels of greenhouse gases in the atmosphere and the consequential effects are sea-level rise, increase in sea surface temperature, acidification of oceans, and marine and coastal ecosystem impairments. The sea surface temperature (SST) increased along the southwest coast of India during the past 40 years by 1ºC and it has been predicted that mean SST in Indian seas would increase by 2.0 to 3.5 ºC by 2099. The spatial changes in estuarine ecosystem is evident from the recent research (Dipson *et al.*, 2015), but there is no reliable evidence of the influence of climate change on it. Sea surface temperature (SST), sea-level rise and atmospheric shifts are the main climate drivers that influence estuarine environments and ecosystems. In our analysis, the SST showed an increasing trend in the highest values recorded from the lake during the pre-monsoon periods and the same may be attributed several factors such as reduction in freshwater flow and sea level rise in the Arabian Sea, etc. An elevation in water temperature on the floodplains would cause an increase in frequency and duration of hypoxic or anoxic episodes, which might further lead to a reduction in growth rates or the reproductive success of many species. In mangroves, an increase in water temperature would influence the sea level, causing losses of these environments through coastal erosion processes. Therefore, climate change will likely cause the loss of or reduction in tropical wetlands and will challenge the adaptability of species, composition, and distribution. In the present study, the highest surface water temperature of the lake recorded from the lake have shown a rise of 0.6ºC, and this finding is in agreement with that of the general observation that in Arabian Sea the rate of SST warming is about 0.075°C/decade during the period 1940–2009 whereas during the period from 1960 to 2009, the rate of SST increase was at 0.12°C/decade (Dinesh Kumar, 2016). Climate change effects on fishes are often interpreted relative to temperature change. Temperature and salinity were found to be the primary determinants influencing the biogeography of fishes, particularly tropical species, in South African estuaries (Harrison and Whitfield, 2006). It is anticipated that estuaries and estuary-associated fishes will be affected by changes in both surface air and ocean temperatures, with anthropogenic greenhouse gases and aerosols that have contributed to an increase in global air and ocean temperatures over the last 50 years (Solomon *et al.*, 2007). In coastal areas, warming of up to 0.55°C per decade has been recorded along the Transkei and KwaZulu-Natal coastline (Rouault *et al.*, 2010). The role of wetlands in coping up with climate change is tremendous. They are valuable as sources, sinks and transformers of a multitude of biological, chemical, and genetic materials. Wetlands and especially peat-lands, store at least 550 Gt of carbon, which is almost double the amount stored in the world's forests. Although they cover only 3% of the world's land area, they contain 30% of its soil carbon (Parish *et al.*, 2008). The higher values of soil organic carbon found in sediments of Vembanad Lake fully conform to the view that wetlands serve as carbon sinks and enhance carbon sequestration.

The rainfall and deluge are major factors influencing the hydrography of this estuary. In recent trends, precipitation has increased with high fluctuations, which is also resulting in minimizing the salinity fluctuation of estuarine water. The salinity profile of the Vembanad Lake is showing an increasing trend with a mean salinity increase of 1.0 ppt over the past five decades. The annual sea level rise in the Arabian Sea is reported as 1 mm per decade and the salinity increase registered now can also be attributed to the sea level rise in the Cochin coast. While the altimeter data analysis over the 1993–2012 period reveals that the rate of sea-level rise is rather spatially homogeneous over most of the north Indian Ocean, reaching values close to global mean sea-level-rise trend (3.2 mm yr^{-1}) (Unnikrishnan *et al.*, 2015). On the contrary, the forecast of the salinity using ARIMA model has shown that the salinity is showing a decreasing trend during the forthcoming years and the same can be attributed to the monsoon intensity during the recent years and the consequential river discharge from the major rivers which are emptying into the lake. Impacts of sea level rise in coastal wetlands include loss of habitats and community structure which lead to decline in lacustrine biodiversity, invasion of exotic species, alteration in salinity profile, changes in horizontal zonation, shrinkage of estuarine and oligohaline zones. Further, such drastic situations also adversely affect traditional shrimp farming practices like pokkali culture owing to preponderance of undesirable marine organisms like jellyfish. It may be noted that life processes of estuarine fish fauna like recruitment, growth, survival, abundance and reproduction are likely to be influenced by climate change through changes in a range of environmental factors such as temperature, salinity, dissolved oxygen (Gillanders *et al.*, 2011).

Climate change is predicted to alter precipitation patterns, which will affect the quality, rate, magnitude, and timing of freshwater delivery to estuaries and will potentially exacerbate human modifications of these flows (Alber, 2002). The functioning of estuaries is strongly influenced by the magnitude and timing of freshwater runoff reaching them (Turpie *et al.*, 2002; Taljaard *et al.*, 2009). In the present investigation, the precipitation level over the estuary showed an increasing trend over the past decades, which might have changed the freshwater runoff to the estuary thus affecting the functioning of this water body. A possible reduction in rainfall can cause a decrease in the areas of mangroves and floodplains, with a consequent decline in their species diversity. Conversely, an increase in rainfall would probably cause the substitution of plant species, which would not be able to survive under new conditions for a long period.

Wetlands such as mangroves, salt marshes, and coral reefs play crucial roles in controlling flood peaks and spreading of water table. These unique ecosystems protect coastlines from extreme meteorological events and reducing storm surges and protecting coastlines. According to the Global Wetland Outlook of the Ramsar Convention, 35% of the world's wetlands have been lost since the 1970s and the global extent of wetlands is estimated to have declined between 64 and 71% in the 20th century (Davidson, 2014). Vembanad lake has also been subjected to heavy horizontal shrinkage during the past century and more than 50% of the lake has been reclaimed for various activities such as agriculture, urbanization, shrimp culture, housing, port development, etc. Wetlands are significant sources and sinks of greenhouse gases and thus play an important role in global climate change. They are particularly vulnerable to the effects of global climate change such as a rise in global temperatures, which may release large carbon stores in northern wetlands, creating a positive feedback loop. An increase in the frequency of extreme weather events, together with sea level rise, may alter the hydro-geomorphology of estuaries and result in a loss of essential estuarine habitat, such as mangroves and salt marsh, which will ultimately affect estuarine fish communities and will have implications for fisheries targeting estuary-associated species (Elliott, 2002;Clark, 2006).

There are several studies reported in many aspects of the Vembanad estuary and its changes over time. However, most of the researchers focus on anthropogenic activities and pollution. Recent trend and forecast on the chlorophyll data from the lake showed a decreasing trend, which have greater influence in the primary productivity and biota of the ecosystem. The decreasing of aquatic resources like fish, clams, etc. may have a socio-economic effect on the people, who depend on the estuary for their livelihood. There are many reports on the loss of biodiversity and fishery resources form the estuary (Menon *et al.*, 2000). Moreover, there are several reports on anthropogenic interventions on the decline of fisheries in the lake. The rapid industrialization of Cochin City, port development, tourism, water transport, pesticide, and chemical contamination are the major human interventions in the lake, which are deemed to be disastrous to the living resources. Therefore, a detailed study on the impact of climate change is required in the changing hydrological cycle, physico-chemical parameters, biota, productivity, carbon sequestration and sinking harvestable biological resources from the lake. This waterbody had designated as a Wetland of International Importance (Ramsar Site) under the Ramsar Convention by the Ministry of Environment Forests and Climate Change (MoEFCC), Government of India, in 2002. In the Coastal Regulation Notification 2011 (CRZ, 2011) of the MoEFCC, the Vembanad Lake, along with eight other ecologically sensitive water bodies of India, has been categorized under Critically Vulnerable Coastal Areas (CVCAs). This is also designated as Areas requiring Special Consideration (ASC), by duly recognizing the extent of habitat degradation, potential threat of inundation due to sea level rise, changes in the ecosystem and services, the decline in the fishery, deterioration of water quality, shrinkage of the lake over the past century, and the number of coastal communities depending for their livelihood security embarking upon various resources of the lake. This wetland sustains livelihoods of nearly 0.2 million households through backwater tourism, inland navigation, and a range of resources such as clams, shellfish, and finfish. Located at the apex of the basin, the Vembanad-Kol also regulates hydrological regimes, providing flood protection to large settlements as well as water for agriculture in the Kuttanad region – the rice bowl of Kerala.

Wetlands are particularly vulnerable to the effects of global climate change. Increases in global temperatures may release large carbon stores in wetlands, creating a positive feedback loop. Sea level rise driven by climate change is the greatest threat to coastal wetlands. Protecting wetlands is a key area to protect our planet, since 35% of the world's wetlands have been lost from the 1970s and the global extent of wetlands is estimated to have declined between 64 and 71% in the 20th century. Preserving the hydrological cycle and protecting water are essential for nature conservation since human life cannot exist without water, and a challenge ahead is to preserve and restore the functions of wetland ecosystems.

Acknowledgement

Authors wish to place on record sincere gratitude to Advanced Centre for Atmospheric Radar Research, Cochin University of Science and Technology, Cochin-682 022, India for providing vital inputs.

References

Akram, K.M.A., 2000. Contribution of Size Fraction of Planktonic Algae to Primary Organic Productivity in the Coastal Waters of Cochin, Southwest Coast of India. Ph.D. Thesis. Cochin University of Science and Technology, Cochin, India.

Alber, M., 2002. A Conceptual Model of Estuarine Freshwater Inflow Management. Estuaries, 25: 6B.

Anon, 2001. Ecology and Fisheries Investigation in Vembanad Lake. CIFRI Bulletin 07, ISSN0970616 X, 38.

Asha, C.V., Retina, I.C., Suson, P.S. and Nandan, S.B., 2016. Ecosystem analysis of the degrading Vembanad wetland ecosystem, the largest Ramsar site on southwest coast of India – Measures for its sustainable management. Regional Studies in Marine Science 8(3): 408-421. doi: 10.1016/j.rsma.2016.06.003.

Asha, C.V., Suson, P.S., Retina C.I., Bijoy Nandan, S., 2014. Decline in diversity and production of exploited fishery resources in Vembanad wetland system: strategies for better management and conservation. Open Journal of Marine Science 4: 344-357. http://dx.doi.org/10.4236/ojms. 2014.44031.

Asha, C.V., Suson, P.S., Retina, I.C. and Bijoy Nandan, S., 2014. Decline in diversity and production of exploited fishery resources in Vembanad wetland system: strategies for better management and conservation. Open Journal of Marine Science 4: 344-357.

Bamber, J.L., Oppenheimer, M., Kopp, R.E., Aspinall, W.P. and Cooke, R.M., 2019. Ice sheet contributions to future sea level rise from structured expert judgment. Proc Natl Acad Sci U S A, 116(23): 11195-11200. doi: 10.1073/pnas.1817205116.

Box, G.E.P. and Jenkins, G.M., 1970. Time Series Analysis: Forecasting and Control. Holden-Day, San Francisco.

Clark, B.M., 2006. Climate change:a looming challenge for fisheries management in southern Africa. Marine Policy 30: 84-95.

CPCB, 1996. Pollution potential of industries in coastal areas of India. Coastal Pollution Control Series, Central Pollution Control Board, New Delhi, 83 p.

CRZ, 2011. Coastal Regulation Zone Notification. Ministry of Environment and Forests, Government of India, New Delhi.

Davidson, N.C., 2014. How much wetland has the world lost? Long-term and recent trends in global wetland area. Marine and Freshwater Research, 65(10): 934-941. http://dx.doi.org/ 10.1071/MF14173.

Dinesh Kumar,, P.K., Steeven Paul, Y., Muraleedharana, K.R, Murty, V.S.N. and Preenu, P.N., 2016. Comparison of long-term variability of sea surface temperature in the Arabian Sea and Bay of Bengal Reg. Stud. Mar. Sci., 3: 67-75.

Dipson, P.T., Chithra, S.V., Amarnath, A., Smitha, S.V., Harindranathan Nair, M.V. and Shahin, A., 2015. Spatial changes of estuary in Ernakulam district, Southern India for last seven decades, using multi-temporal satellite data. J. Environ. Manage., 148: 134-142. doi: 10.1016/j.jenvman.2014.02.021.

Dsikowitzky, L., Nordhaus, I., Sujatha, C.H., Akhil, P.S., Soman, K. and Schwarzbauer, J., 2014. A combined chemical and biological assessment of industrial contamination in an estuarine system in Kerala, India. Science of the Total Environment 485-486: 348-362. doi: 10.1016/j.scitotenv.2014.03.034.

Elliott, M., 2002. An overview of the status, study and management of fishes in estuaries. In: M. Elliott and K.L. Hemmingway (Eds.), Fishes in Estuaries, pp. 53"575, Blackwell Science, Oxford.

Gillanders, B.M., Elsdon, T.S., Halliday, I.A., Jenkins, G.P., Robins, J.B. and Valesini, F.J., 2011. Potential effects of climate change on Australian estuaries and fish-utilising estuaries: a review. Marine and Freshwater Research 62: 1115–1131. doi: 10.1071/ MF11047.

Gireeshkumar, T.R., Deepulal, P.M. and Chandramohanakumar, N., 2013. Phosphorous speciation in surface sediments of the Cochin estuary. Environ. Monit. Assess., 185(3): 2535-2552. doi: 10.1007/s10661-012-2729-3.

Gopakumar, R. and Takara, K., 2009. Analysis of the bathymetry and spatial changes of Vembanad Lake and terrain characteristics of Vembanad wetlands using GIS. Paper presented at the Joint International Convention of 8th IAHS Scientific Assembly and 37th IAH Congress Hyderabad, India.

Gopalan, U.K., Vengayil, D.T., Udayavarma, P. and Krishnakutty, M., 1983. The Shrinking backwaters of Kerala. J. mar. biol. Ass. India, 25(1 & 2): 131-141.

Gopinathan, C.P., Nair, P.V.R. and Kesavan Nair, A.K., 1974. Studies on the phytoplankton of the Cochin Backwaters, a tropical estuary. Indian Journal of Fisheries, 21(2): 501-503.

Harikrishnan, M. and Kurup, B.M., 2001. Fishery of *Macrobrachiumrosenbergii* (de Man) in the Vembanad Lake and confluent rivers. Indian Journal of Fisheries 48(2): 189-198.

Harikrishnan, M. and Kurup, B.M., 2002. Ashtamudi estuary as a nursery ground for *Sardinella longiceps* (Val.). Journal of Marine Biological Association of India 44(1&2): 136-144.

Harrison, T.D. and Whitfield, A.K., 2006. Estuarine typology and the structuring of fish communities in South Africa. Environmental Biology of Fishes 75(3): 269-293. doi: 10.1007/s10641-006-0028-y.

ICMAM, 2002. Critical habitat information for Cochin backwaters - Kerala. Integrated Coastal and Marine Area Management (ICMAM), DOD, 31 p.

Jacob, B., Revichandran, C. and Naveen Kumar, K.R., 2013. Salt intrusion study in Cochin estuary-using empirical models. Indian Journal of Geo-Marine Sciences 42(3): 304-313.

James, E.J., Anitha, A.B., Divakaran, Namubdiripad, K., Joseph, E.J., Nandeswer, M.D., Nirmala, E., Padmini, V., Unni, P.N. and Venugopal, M.R., 1997. The Vembanad-Kol Wetland system in relation to dainage basin management, In: Wetlands and Integrated River Basin Management: Experiences in Asia and Pacific. ISBN 983-9663-24-0. UNEP/Wetland International - Asia Pacific, Kuala Lumpur, Malaysia.

Krishna Kumar, K. and Priyadarsanan, D.R., 2012. Fish and Fisheries in Vembanad Lake - Consolidated Report of Vembanad Fish Count 2008-2011. Alleppey, Kerala. Ashoka Trust for Research in Ecology and the Environment (ATREE).

Kurup, B.M. and Samuel, C.T., 1985. Fish and fishery resources of the Vembanad Lake In: Proceedings of Harvest and Post-harvest Technology of Fish, pp. 77-82, Society of Fisheries Technologists (India), Cochin.

Kurup, B.M. and Samuel, C.T., 1987. Ecology and fish distribution pattern of a tropical estuary. Proceedings of the National Seminar on Estuarine Management, Trivandrum: 339-349.

Kurup, B.M., 1982. Studies on the Systematics and Biology of the Fishes of Vembanad Lake. Ph.D. Thesis, University of Cochin, 683 p.

Kurup, B.M., Sebastian, M.J., Sankaran, T.M. and Rabindranath, P., 1989. Exploited fishery resources of the Vembanad Lake, Final report presented to Kuttanad Water Balance Study Project: 1-142

Kurup, B.M., Sebastian, M.J., Sankaran, T.M. and Rabindranath, P., 1990. Exploited fishery resources of Vembanad lake Part - I. Indian Journal of Fisheries 40(4): 199-206.

Kurup, B.M., Sebastian, M.J., Sankaran, T.M. and Rabindranath, P., 1990. Exploited fishery resources of Vembanad Lake - Clam fisheries. Mahasagar 23(2): 127-137.

Kurup, B.M., Sebastian, M.J., Sankaran, T.M. and Rabindranath, P., 1992. Socio economic status of fisherman population in the vicinity of Vembanad Lake. Fishery Technology 29(2):166-171.

Kurup, B.M., Sebastian, M.J., Sankaran, T.M. and Rabindranath, P., 1993. Exploited fishery resources of the Vembanad Lake: Status of residents and migrants. Fishery Technology - Special Issue, pp: 44-49, Society of Fishery Technologists, Cochin.

Maitra, S., Harikrishnan, M., Shibu, A.V., Sureshkumar, S., Ranjeet, K. and Nandan, S.B. 2018. Studies on temporal variations of exploited fishery resources and their trophic levels in a tropical estuary. Regional Studies in Marine Science 22: 61-69.

Menon, N.N., Balchand, A. and Menon, N.R., 2000. Hydrobiology of the Cochin Backwater System – A review. Hydrobiologia 430: 149-183. doi: 10.1023/A:1004033400255.

Mitsch, W.J. and Gosselink J.G., 2007.Wetlands. 4th Edn., John Wiley and Sons, Inc., Hoboken.

Nair, K.K.C., Sankaranarayanan, V.N., Gopalakrishnan, T.C., Balasubramanian, T., Lalithambikadevi, C.B., Aravindakshan, P.N. and Krishnankutty, M., 1988. Environmental conditions of some paddycum-prawn culture fields of Cochin backwaters, South west coast of India. Indian Journal of Marine Sciences 17: 24-30.

Nair, N.B., 1990. Ecobiology of Marine Fouling and Wood Boring Organisms along the Indian Coasts. Final Technical Report, DOD Project, 311 p.

Nair, P.K. and Babu, D.S., 2016. Spatial Shrinkage of Vembanad Lake, South West India during 1973-2015 using NDWI and MNDWI. International Journal of Science and Research 5(7): 1394-1401.

Nauels, A., Rogelj, J., Schleussner, C.F., Meinshausen, M. and Mengel, M., 2017. Linking sea level rise and socioeconomic indicators under the Shared Socioeconomic Pathways. Environmental Research Letters 12(11): 114002. https://doi.org/10.1088/ 1748-9326/ aa92b6.

Padmakumar, K.G., Anuradha, K., Manu, P.S., Shiny, C.K. and Radhika, R., 2002. Thaneermukkom barrage and fishery decline in Vembanad wetlands, Kerala. Paper presented at the Fifth Kerala Science Congress, 28-30 January 1993, Kottayam, Kerala, India.

Parish, F., Sirin, A., Charman, D., Joosten, H., Minayeva, T., Silvius, M. and Stringer, L. (Eds.), 2008. Assessment on Peatlands, Biodiversity and Climate Change: Main Report. Global Environment Centre, Kuala Lumpur and Wetlands International, Wageningen.

Qasim, S.Z., Wellesshaus, S., Bhattathiri, R.M.A. and Abidi, S.A.H., 1969. Organic production in a tropical estuary. Procedings of Indian Academy of Science 69B(2): 51-94.

Raman, K., 1967. Observations on the fishery of Giant Freshwater Prawn *Macrobrachium rosenbergii* (de Man). In: Proceedings of the Symposium on Crustacea, Part 2, MBAI, 12-16 January 1965, Ernakulam.

Rasalam, E.J. and Sebastian, M.J., 1976. The lime-shell fisheries of the Vembanad Lake, Kerala. Journal of Marine Biological Association of India 18: 323-335.

Renjith, K.R. and Chandramohanakumar, N., 2009. Distribution of heavy metals in the surficial sediments of a complex micro-tidal estuarine system in Southwest India. Research Journal of Chemistry and Environment 13(2): 34-44.

Rouault, M., Pohl, B. and Penven, P., 2010. Coastal oceanic climate change and variability from 1982 to 2009 around South Africa. African Journal of Marine Science 32(2): 1-10.

Roy, D. and Nandi, N.C., 2008. Checklist and distribution of brachyuran crabs of West Bengal, India. Journal of Environment and Sociobiology, 5(2):

Shetty, H.P.C., 1965. Observations on the fish and fisheries of the Vembanad backwaters, Kerala. Proceedings of National Academy of Science, India, 35(1): 115-130.

Shine, C. and Klemm, C.D., 1999. Wetlands, Water and the Law:-Using law to advance wetland conservation and wise use. IUCN Environmental Policy and Law Paper No. 38: IUCN Environmental Law Centre.

Solomon, S.D., Dawei, Q., Martin, M., ZeNa, C., Melinda, M., Avery, K.B., Tignor, M. and Miller, H.L., 2007. Climate Change 2007: The Physical Science Basis. Cambridge University Press, Cambridge.

Sunil Kumar, R., 2002. Habitat preference and environmental relations of *Hydrobia* sp., Mollusca: Gastropoda, in the intertidal subsoil of a tropical mangrove region. J. Bombay nat. Hist. Soc., 99(2): 245-249.

Taljaard, S., Van Niekerk, L. and Joubert, W., 2009. Extension of a qualitative model on nutrient cycling and transformation to include microtidal estuaries on wave-dominated coasts: southern hemisphere perspective. Estuar. Coast. Shelf Science, 85: 407-421.

Thomson, K.T., 2002. Economic and social issues of biodiversity loss in Cochin backwaters. In: Technical Report, pp. 51-82, Cochin University of Science and Technology Cochin, India.

Turpie, J.K., Adams, J.B., Joubert, A., Harrison, T.D., Colloty, B.M., Maree, R.C., Whitfield, A.K., Wooldridge, T.H., Lamberth, S.J., Taljaard, S. and Van Niekerk, L., 2002. Assessment of the conservation priority status of South African estuaries for use in management and water allocation. Water SA 28(2): 191-206.

Unnikrishnan, A.S., Nidheesh, A.G. and Lengaign, M., 2015. Sea level rise trends off the Indian coasts during the last two decades. Current Science 108(5): 966-970.

Vincy, M.V., Brilliant, R. and Pradeep Kumar, A.P., 2012. Water quality assessment of a tropical wetland ecosystem with special reference to Backwater Tourism, Kerala, South India. International Research Journal of Environment Sciences 1(5): 62-68.

Impact of Climate Change on Hydrological Cycle, Ecosystem, Fisheries and Food Security, Pages: 249–259
Edited by: B. Madhusoodana Kurup, M.R. Boopendranath, M. Harikrishnan and A.V. Shibu

CHAPTER - 21

Barriers and Opportunities for Climate Change Adaptation in Marine Fisheries in India

E. Vivekanandan

Former Principal Scientist, Central Marine Fisheries Research Institute, Kochi-682018, Kerala, India
E-mail: evivekanandan@hotmail.com

India has a coastline of 8118 km and Exclusive Economic Zone (EEZ) of 2.02 million sq.km. The continental shelf area is approximately 0.53 million sq.km. Marine fisheries is an important sector in the country, contributing significantly to the nutritional security, livelihood, employment and economy (1% to the country's Gross Domestic Product). The marine fish catch has increased from 1.55 million tonnes in 1980 to 3.69 million t in 2018, but has not increased in the last five years. The monetary value of the catch has consistently increased and reached US$ 7.5 billion at ex-dock level and US$ 11.2 billion at retail level in 2018 (CMFRI, 2019). The sector also contributes significantly to the export earnings. In 2018, marine products worth US$ 6.4 billion were exported. The sector provides employment and source of livelihood for more than one million people.

The potential of marine fish resources in the Indian EEZ has been estimated as 5.44 million t (DoF, 2018), and the economically sustainable yield is approximately 4.35 million t. Data on potential yield and catch indicate that there is scope for increase in the yield by about 0.66 million t. However, the potential yield estimates and catches mask many adverse conditions of the health of the fish stocks. Though the information is patchy, different annual reports of Central Marine Fisheries Research Institute (for example, CMFRI, 2018, 2019) indicate that (i) about one-third of the marine fish stocks is depleted; (ii) catch-per-unit effort (CPUE) of a few fisheries is showing declining trend; and (iii) mean size of a few important species has decreased over the years. There are also evidences to show the prevalence of fishing-down- marine food web, a phenomenon in which large predators are removed at a faster rate than their prey species (Vivekanandan *et al.*, 2005). The main causes for the declining health of fish stocks are overfishing, pollution, habitat degradation and climate change.

Impacts of Climate Change on Marine Fisheries in India

In its Fifth Assessment Report (AR5), the Intergovernmental Panel on Climate Change (IPCC, 2014) evaluated the changes to natural systems, and the impact on human communities and industry. While impacts varied substantially between systems, sectors and regions, many changes over the past 50 years were attributed to human-driven climate change and its impacts. Risks were observed to be increasing for natural ecosystems as climate extremes increase in frequency and intensity. In association with the changes with the natural ecosystems, the associated fauna and flora are shifting their biogeographical ranges to higher latitudes and altitudes. This will have consequences for ecosystem services and human dependence. Like other sectors, fisheries also will be impacted. The links between warming, physical changes in the oceans and ecological changes leading to effects on the economics of the fishing sector are complex. Many of the potential pathways of impact linking climate change, oceans, aquatic ecosystems and fisheries have been identified in scientific literature. A summary of this linkage is provided in Figure 1.

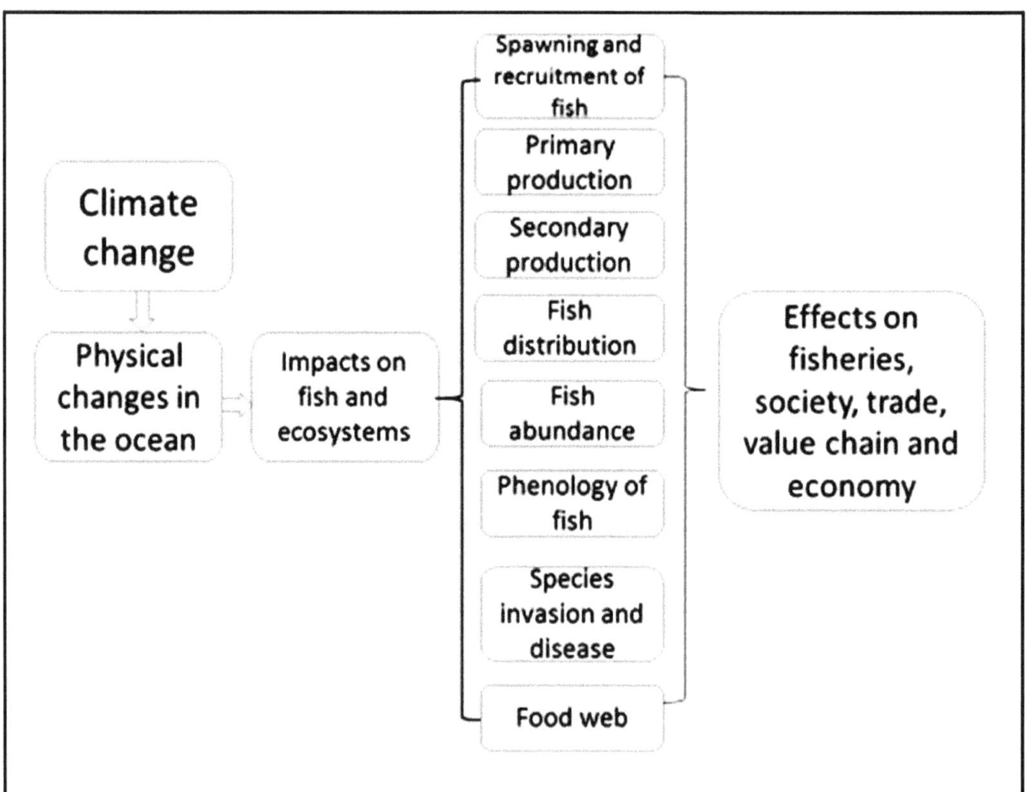

Fig. 1. Pathway from climate change to fisheries (modified from MacFadyen and Allison, 2009)

It is difficult to segregate the impacts climate change on fish stocks from that of other three stressors as these stressors act jointly on the stocks. It is also impossible to isolate the impacts of specific changes in climate from climate variability and other drivers of change. However, a few clear and conspicuous climate change impacts on marine fisheries in India are apparent (Vivekanandan, 2011). They are: (i) expansion of a few small pelagic stocks such as oil sardine and Indian mackerel towards northern latitudes along the east and west coasts of India; (ii) shifting of a few species to relatively deeper waters; (iii) phenological changes like gradual shifting of spawning season towards relatively cooler months; (iv) indications of rapid turnover of generations of phytoplankton due to warming of seawater; and (v) mass coral bleaching events followed by their demise in the next 30 to 40 years in the major reef areas.

The impacts in marine fisheries sector is mainly at two levels – within fisher communities, and at national level (Fig. 2). While changes in catch volumes, fishing locality, target species, seasonality, loss of fishing days, sea safety issues and losses to craft and gear are at fishing community level, damage to coastal fisheries infrastructure like fishing harbours, multiplier effects in the fisheries supply chain across each nodal point from producer to consumer, and payment of compensation to the losses incurred by the communities, particularly the fishing community, are at national level. The impacts on these two levels vary by region/location, and the scale of impact is dependent on the levels of adaptive capacity of individuals, communities and the governments.

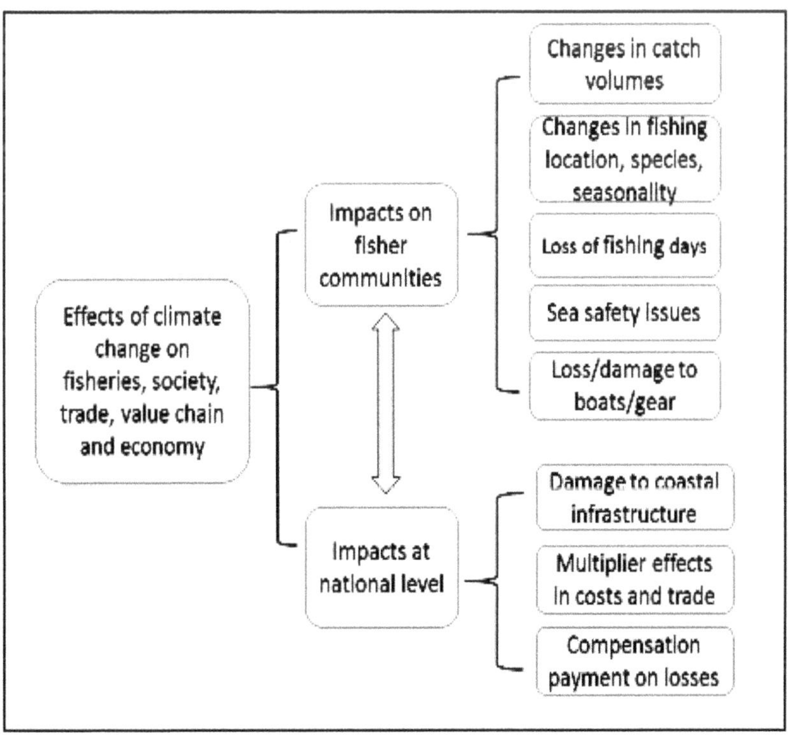

Fig. 2. Climate change impacts on marine fisheries at fisher community and national levels

Vulnerability of species

The manifestations of climate change are highly dependent on specific environmental and economic contexts in which they are located. Moreover, climate change impact will not be uniform among the species and it varies based on the habitats and biological traits of species. In this context, vulnerability assessment for species of commercial importance is necessary because it gives baseline information for improving adaptation planning, raising awareness of risks and opportunities and advancing scientific research. An extensive analysis of available information by Zacharia *et al.* (2016) on 5 exposure attributes (anomalies in sea surface temperature, rainfall, ocean current, coastal upwelling, chlorophyll concentration), 9 sensitivity attributes (fecundity, complexity in early development, growth coefficient, trophic level, longevity, length-at-first maturity, anomaly in catch-per-unit effort, exploitation rate and price) and 3 adaptive capacity attributes (horizontal and vertical distribution, duration of spawning and prey specificity) of 68 species contributing to the bulk of marine fish landings along the four coastal zones of the Indian coast has brought out important findings on the relative vulnerability of Indian marine fishes to climate change. In the present paper, a recalculation of the results presented by them was made, which has revealed the following important differences between the four zones and species:

(i) Among the four coastal zones (northwest, southwest, southeast and northeast) along the Indian coast, the exposure risks are highest along the northeast coast (Table 1). The exposure risks of species occurring along the northeast coast are 18% to 31% higher than the other three coastal zones due to higher anomalies in the physical and ecological attributes.

(ii) The higher exposure to risks is reflected in the higher number of vulnerable species along the NE coast. It has been estimated that 71.0% of the species subjected to analysis along the NE coast are vulnerable to climate change as against 30.0% of the species along the NW and SW coasts (Fig. 3).

(iii) The exception is the southeast coast where the exposure risks are low, but the highest percentage of species (74.2%) are vulnerable. Further analysis of the data presented by Zacharia *et al.* (2016) shows that adaptive capacity of the species occurring along the SE coast is substantially lower than that occurring along the NW and SW coasts, and even that occurring along the NE coast. Restricted horizontal and vertical distribution of the species as well as food web dynamics appear to lower the adaptive capacity of species occurring along the SE coast. The SE coast has wide differences in the depth contour, from very shallow Palk Bay (average depth: 17 m) to very deep Coromandel Coast where the continental shelf is within 20 km from the shoreline in locations off Cuddalore and Puducherry. These wide changes have perhaps influenced and restricted the horizontal and vertical distribution of species along the SE coast. Food web dynamics is another factor that is influenced by climate change. While the mean trophic level of exploited species is highest along the SE coast, the mean trophic level along the SE coast is reducing at a faster rate (0.04 trophic level/decade) than that along the other three coastal zones (0.03 trophic level/decade), showing that fishing down food web is occurring at a faster rate along the SE coast (Vivekanandan *et al.*, 2005). While fishing down food web occurs due to high fishing intensity, the impact of climate change has exacerbated the situation and increased the vulnerability of fish species along the SE coast.

(iv) Due to the reason mentioned above, in the case of a few species, there are large variations in the adaptive capacity and vulnerability of the same species occurring in different coastal zones.

Table 1: Score on exposure attributes along the four coastal zones of India (NW: northwest; SW: southwest; SE: southeast; NE: northeast) (recalculated from Zacharia *et al.*, 2016)

Zone	Range	Mean ± SD
NW	1.33-1.95	1.72 ± 0.22
SW	1.35-1.77	1.62 ± 0.19
SE	1.43-1.69	1.60 ± 0.12
NE	1.67-2.22	2.10 ± 0.19

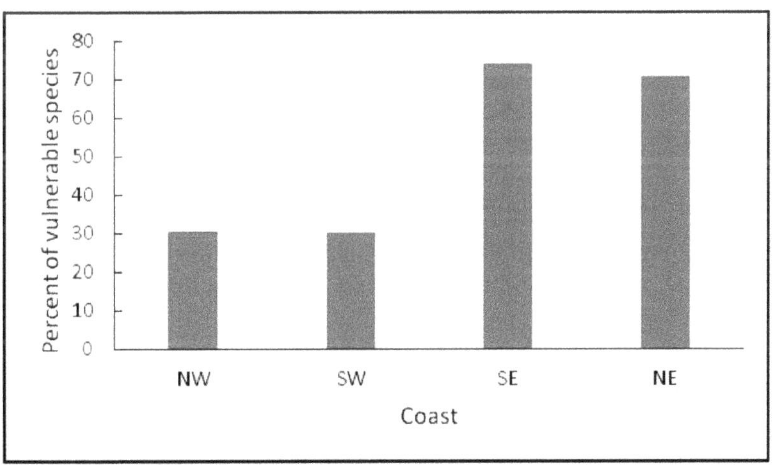

Fig. 3. Percent of vulnerable species along the four coastal zones of India (n = 68 species) (recalculated from Zacharia *et al.*, 2016)

The relative vulnerability assessment of marine fishes has shown that the level of adaptive capacity and vulnerability differs between species and geographical locations. Hence, all species will not experience uniform rate of change in abundance when the intensity of impact of climate change increases in future. The analysis further shows that fishery for the species with less adaptive capacity will diminish, leading to large changes in fishing practices.

Options for Adaptation in Marine Fisheries in India

The marine fisheries sector is globalized through trade, but production is concentrated in developing countries that have a growing share of production and trade, with a high percentage of small-scale fishers playing a major part. Hence, developing countries like India have greater responsibility in the national as well as global context to adapt to climate change. Climate change is expected to alter aquatic food prices, by affecting not only the availability of aquatic resources and the global supply, but also the cost of goods, infrastructure and services required in aquatic food production, processing and distribution activities (Vannuccini *et al.*, 2018).

While there is no distinct mechanism to directly reduce the impact of climate change on fish stocks, the key options are to reduce the impact of factors other than climate change and relieve fish stocks from other stressors. Healthy fish stocks will be able to face the challenge of climate change better than the declining stocks. There is a scope to relieve stress on fish stocks by effectively implementing the conventional fishing regulation acts such as input and output control measures, and by adopting ecosystem approach (Shelton, 2014). Ecosystem approach to fisheries management (EAFM) is being promoted globally as it has the capacity to address multiple objectives with the participation of a wide range of stakeholders. Projects like the Bay of Bengal Large Marine Ecosystem have strongly advocated implementation of EAFM in larger marine areas to address multiple objectives including climate change (Vivekanandan *et al.*, 2016; 2019).

However, the marine fisheries in India was open access until the 1990s and implementing regulatory measures met with challenges. The arguments from the fishing communities is that (i) the fishes are abundant, which is reflected as increase in catches; and (ii) the price of fish is increasing which compensates increasing cost of fishing. At present, a few input control measures such as spatial restrictions for fishing, seasonal fishing ban and mesh size regulation are in place. Recent initiatives by the government such as cap on the number of boats, licensing of boats for fishing in specific locations, etc face stiff resistance. Awareness building, engaging in dialogues, creating trust and partnership among the stakeholders are needed for addressing fishing-related as well as climate change-related issues. It is important to mainstream climate change into the existing policies and negotiations with other resource users.

Some of the on-going measures (not necessarily as climate change adaptation measures), and potential measures that are relevant for adaptation in marine fisheries of India are listed in Table 2. The list consists of 27 measures under 5 themes, namely, Policy and Management, Research, Economics and Trade, Safety Measures and Others.

Table 2: Potential adaptation options for marine fisheries in India

	Adaptation options	Potential activities
(i)	**Policy and Management**	
1	Policy and management	Fishing capacity reductions, removal of incentives for overfishing, implementing National Policy on Marine Fisheries 2017, FAO Code of Conduct for Responsible Fisheries, FAO Voluntary Guideline for Small-scale Fisheries and Marine Fishing Regulation Act can improve resilience and increase system adaptability. It will also ease both socio-economic impacts from changing fisheries and prevent overfishing.
2	Expanding and Marine Protected Areas (MPAs)	At present, about 9,600 sq.km. area is MPA in the Indian EEZ, which is only about 0.5% of the EEZ of India. The UN Sustainable Development Goal 14 (SDG-14) "Life Under Water", Indicator 14.5 ''Conserve coastal and Marine Areas'' states "By 2020, conserve at least 10% of coastal and marine areas, consistent with national and international law and based on the best available scientific information''. India has to substantially increase the MPA from the present 0.5%.
3	Ecosystem approach to fisheries management	EAFM has the capacity to address multiple objectives such as overfishing, pollution, degradation and climate change.
4	Spatial planning	Zoning of EEZ for fishing by different categories of fishing boats; balancing the subtidal areas with land-based development, mangrove plantation and shoreline protection to adapt to rising sea level
5	Combatting Illegal, Unreported and Unregulated (IUU) fishing	Effective implementation of the voluntary International Plan of Action-IUU fishing through binding commitments applicable in waters of national jurisdiction and on the high seas by enacting appropriate domestic laws.
6	Reducing capture of juveniles and low-value bycatch	Releasing fishing pressure on juveniles and improving recruitment into the fishery by following suitable measures will greatly improve the health of fish stocks.
7	Minimising subsidies	Subsidies that encourage overcapitalisation and overfishing need to be totally phased out and replaced by subsidies that support sustainable fisheries.
8	Reducing ghost fishing	As storm severity increases, it is likely that more gear will be lost. Such lost gear can cause mortality and habitat damage. Measures that can reduce their impacts may be adopted. In addition to gear retrieval programmes, gear could be designed to minimize impacts, if lost.
9	Reducing external stressors on natural systems	Reducing land-based sources of pollution (e.g. agricultural and urban runoff) and destructive fishing practices
10	Co-management	Willingness of stakeholders to adopt co-management is important in enhancing adaptive capacity. On the other hand, the threat of climate change impacts may galvanise local management to minimise impacts and enhance adaptation.

[Table Contd.

Contd. Table]

	Adaptation options	Potential activities
11	Monitoring, Control and Surveillance (MCS)	Determining the level of sustainable exploitation and other relevant information by data gathering, assessment and analysis; enforcing controls in ports and at sea; use of Global Positioning System (GPS) and Global Navigation Satellites System (GNS) applications; accessibility to satellite imagery; and use of Automatic Identification Systems (AIS) and Vessel Monitoring Systems (VMS).
12	Mainstreaming climate change	Integrating climate change adaptation into fisheries policy at the national level to ensure incorporation into broader development planning. This will also involve trade-offs, compromises, and planning with other industries affecting fisheries.
(ii)	**Research**	
13	Monitoring	Monitoring the impacts by following standard methodologies
14	Recognising opportunities	New opportunities like aquaculture-based livelihoods
15	Developing and promoting green fishing technologies	Developing and using technologies to reduce carbon emission from different activities including catching, transporting and processing of fish
(iii)	**Economics and trade**	
16	Improving value chain	Value chain from producer to consumer is weak in India, incurring huge losses. With the catches likely to decrease in future, the value realised for the same quantity of fish should be increased by better value realisation, value addition and value creation.
17	Reducing food loss and wastage	Physical loss of fish is estimated as 15% from production to consumption in India. Utilisation of the entire catch should be ensured by adopting proper measures.
18	Improving domestic marketing	Domestic fish markets and marketing system in India need to be organised and regulated. Physical facilities and infrastructure in all types of fish markets are far from satisfactory. Need to formulate a fish market policy.
19	Improving International trade	Some trade measures and barriers including subsidies can work against export of high-value processed products. Mechanisms need to be developed to bring transparency in the origin of the products by taking steps to certify the products.
(iv)	**Safety measures**	
20	Investing in safer fishing harbours and landing centres	Adequate onshore storage facilities for boats and gear can prevent loss or damage from storms and extreme events.
21	Improving safety at sea due to increased storm severity	Improved early warning and forecasting systems for severe weather events; personal and navigational safety awareness among fishers to reduce the number of accidents and increase the chances of survival if accidents occur. Steps need to be taken to improve fire safety, deck safety, use of lifesaving equipment, lighting and ventilation and make them mandatory.

[Table Contd.

Contd. Table]

	Adaptation options	Potential activities
22	Promoting disaster risk management	Disaster preparedness and protective infrastructure like "hard" options such as seawalls and flood reservoirs, or "soft" options such as buffer zones via afforestation or reforestation of mangroves
(v)	**Other measures**	
23	Considering Indigenous Technical Knowledge	What have people done and how decision-making processes worked under highly variable and extreme events? How people addressed a similar issue in the past? Can it be applied in the present context?
24	Raising awareness	In fisheries sector, awareness on the impacts of climate change needs to be raised to sensitise the stakeholders.
25	Building capacity	Capacity to monitor climate impacts and to integrate climate change issues in sustainable fisheries management need to be strengthened among researchers, officials, NGOs and fishermen.
26	Strengthening partnership	Partnerships between private, public, civil society and NGOs are vital for holistic climate change adaptation planning.
27	Financing adaptation and mitigation measures	Insurance and other innovative instruments to create effective incentives and disincentives at national level.

Fish supply

At the most fundamental level, climate change has the potential to impact on total fish production volumes available for trade, both domestic and export markets. Climate change will increase uncertainties in the supply of fish from capture fisheries. For example, the oil sardine *Sardinella longiceps* is an important food fish for the people of Kerala. Due to ENSO events, the availability of oil sardine has become erratic along the Kerala coast in the last few years. Such uncertainty will impose new challenges, making it critical for resources planning and management to take into account the greater possibility of unforeseen events, such as the increasing frequency of extreme weather events such as Ockhi cyclone.

Adaptation options across value chain

Climate change adaptation options in fisheries are generally built around production phase, but this is not sufficient as the barriers and opportunities need to be considered along seafood value chains, from capture to consumer. The impacts of climate change on the production and trade of different species will vary depending on the magnitude of the physical change brought about by climate change, the ability of fish species and ecosystems to adapt to deal with physical changes or their adaptive response to avoid these changes, and the response of fishers and communities (MacFadyen and Allison, 2009). In India, the probable consequences of climate change on fish supply chain have not yet been addressed. One of the options to address the impact of climate change is to improve the economic performance of marine fisheries. In India, post-harvest fish loss is about 15% of the fish caught due to discards and quality reduction. Moreover, 85% of the fish are consumed in fresh form without value addition. There is excellent opportunity to improve economic returns for the same amount of fish caught by improving the value chain through value realisation (for e.g., improving post-harvest handling and preservation), value addition (for e.g., promoting value added products) and value creation (for e.g., obtaining certification of selected fisheries so that the products reach high-end markets). It is important to (i) identify those activities or value chains that are at risk or under threat in some way from climate change; and (ii) analyse whether additional measures are required in order to implement the value chain successfully. These are indeed early days for predicting climate change impacts on aquatic resources and stakeholders along the seafood value chain. Linking climate change impacts and trade flows at global and national requires additional research efforts. Information on how the sector can transition to climate-ready production and post-harvest activities need to be collected (www.fao.org › in-action › globefish › fishery-information › resource-detail).

Conclusion

The impacts of climate change on marine fisheries in India are already evident. *Adapting to the impacts of climate change,* calls for a new paradigm that considers a suite of measures that would include future climate conditions. Adaptation requires actions from decision makers in the governments, private sectors, non-governmental organizations and community groups. However, current efforts are hampered by a lack of solid information about the benefits, costs, and effectiveness of various adaptation options, climate information on regional and local scales, and lack of coordination. *Adapting to the impacts of climate change* in marine fisheries calls for a national adaptation strategy that provides needed technical and scientific resources, guidance across jurisdictions and support of scientific research to expand knowledge of impacts and adaptation.

References

CMFRI, 2018. Annual Report 2017-18. Central Marine Fisheries Research Institute, Kochi, 304 pp.

CMFRI, 2019. Annual Report 2018-19. Central Marine Fisheries Research Institute, Kochi, 308 pp.

DoF, 2018. Report of the Expert Committee for Revalidation of the Potential Yield of Fishery Resources in the Indian EEZ. Department of Fisheries, Government of India, New Delhi, 104 pp.

IPCC, 2014. Climate Change 2014: Impacts, Adaptation, and Vulnerability. Part A: Global and Sectoral Aspects. Contribution of Working Group II to the Fifth Assessment Report of the Intergovernmental Panel on Climate Change. Cambridge University Press, Cambridge, United Kingdom and New York, NY, USA, 1132 pp.

MacFadyen, G. and Allison, E. 2009. Climate Change, Fisheries, Trade and Competitiveness: Understanding Impacts and Formulating Responses for Commonwealth Small States. Prepared for the Commonwealth Secretariat, 103 pp.

Shelton, C., 2014. Climate change adaptation in fisheries and aquaculture – compilation of initial examples. FAO Fisheries and Aquaculture Circular, 1088: 34 pp.

Vannuccini, S., Kavallari, A., Bellù, L.G., Müller, M. and Wisser, D., 2018. Understanding the impacts of climate change for fisheries and aquaculture: global and regional supply and demand trends and prospects. In: Impacts of climate change on fisheries and aquaculture: synthesis of current knowledge, adaptation and mitigation options. FAO Fisheries and Aquaculture Technical Paper, 627: 41-62.

Vivekanandan, E., 2011. Marine fisheries policy brief-3; climate change and Indian marine fisheries. CMFRI Special Publication, 105: 97 pp.

Vivekanandan, E., Hermes, R. and Brown, D., 2019. An ecosystem based approach to the assessment and governance of the Bay of Bengal Large Marine Ecosystem. Deep-Sea Research, 163 Part II: 87–95.

Vivekanandan, E., Hermes, R. and O'Brien, C., 2016. Climate change effects in the Bay of Bengal Large Marine Ecosystem. Environ. Dev., 17(1): 46–56.

Vivekanandan, E., Srinath, M. and Kuriakose, S., 2005. Fishing the food web along the Indian coast. Fish. Res., 72: 241-252.

Zacharia, P.U, Dineshbabu, A.P., Thomas, S., Kizhakudan, S.J., Vivekanandan, E., Pillai, S.L., Sivadas, M., Ghosh, S., Ganga, U., Rajesh, K.M., Nair, R.J., Najmudeen, T.M., Koya, M., Chellappan, A., Dash, G., Divipala, I., Akhilesh, K.V., Muktha, M. and Dash, S.S., 2016. Relative vulnerability assessment of Indian marine fishes to climate change using impact and adaptation attributes. CMFRI Special Publication 125, 192 p.

Impact of Climate Change on Hydrological Cycle, Ecosystem, Fisheries and Food Security, Pages: 261–264
Edited by: B. Madhusoodana Kurup, M.R. Boopendranath, M. Harikrishnan and A.V. Shibu

CHAPTER - 22

Fishery of Gobiid Fish *Oxyurichthys tentacularis* (Valenciennes, 1837) inhabiting Ashtamudi Lake, Kerala

Remya Mohan[1*], M. Harikrishnan[2] and E. Sherly Williams[1]

[1]*Environmental Science, Aquaculture and Fish Biotechnology Lab, Department of Zoology, Fatima Mata National College, Kollam-691001, Kerala, India*
[2]*School of Industrial Fisheries, Cochin University of Science and Technology, Kochi-682016, Kerala, India*
**E-mail: remyamohan062@gmail.com*

World over, small scale fisheries contribute considerably in ensuring food and livelihood security of rural populations. In developing countries like India, contributions from such fisheries remain poorly understood owing to lack of fundamental data which ultimately lead to managing such fisheries sustainably by conserving fishery resources, regulating exploitation and ameliorating impacts on local livelihoods (Bartley *et al.*, 2015). The state of Kerala is renowned for plentiful water resources such as rivers and estuaries. Ashtamudi Lake the second largest tropical backwater habitat on the southwest coast of India, in the Kollam district of Kerala, is renowned for its fin fish and shell fish resources which played a major role in protecting the livelihood, employment and food security for the population living in the vicinity of this backwater. More than 97 fish species belonging to 39 families which comprised of 68 species fin fishes, 5 species of crabs and 9 species each of prawns and mollusks. By this, pleanty of resources, fish and fisheries become the backbone industry for the rural fisherman residing in and around the lake (Nair *et al.*, 1983; Vimal *et al.* (2014). Along with that, in Astamudi Lake a very peculiar gobioid fishery based on the species *Oxyurichthys tentacularis* (vernacular name Koozhali) is quiet interesting because this is the only estuary in India where there is a commercial fishery for a gobiid species and it is an important income resource for local people around the Lake (Kurup and Thomas (2001). Gobiidae is one of the largest families of teleost fishes. *O. tentacularis* has a good representation in the Lake. It is a small elongated fish and can be distinguished by differences in the urogenital papilla in both sexes. It is short, blunt, and pinkish in females, while in males it is thin, pointed, pale, and smooth. In both sexes, the urogenital papillae are covered only by epithelium and not by scales. Very little has been studied on biology and stock characteristics of this species in the Lake. Remya Mohan (2018) conducted studies on the biology and population dynamics of this species while Remya Mohan *et al.* (2018) reported a detailed study on its reproductive biology. However, no concerted study has been reported on the fisheries of *O. tentacularis,* so far.

Materials and Methods

The study site selected for the present work was Ashtamudi Lake. The lake has an area of 61.42 km^2 and is located between 76^0 32' - 76^0 41'E Long. and 8^0 52' - 9^0 2'N Lat. The fishery landings of *O. tentacularis* were observed in major landing centres of Ashtamudi Lake, on fortnightly basis. The daily landings of this species were recorded from not less than 30% of fishing gears landing their catches in these landing centres. Regular catch data was collected from landing centres by observations on landing and by enquiry with the fishermen. The effort of *O. tentacularis* by different fishing gears (gill net, seine, cast net and dip net) were collected at monthly intervals from Kureeppuzha, Thekkumbhagam, Thevally and Perumon around Ashtamudi Lake during the study period, from January 2015 to December 2015.

Results and Discussion

O. tentacularis found in Ashtamudi Lake contribute sustenance fishery in the rural areas of the Lake. Temperature of the Lake ranges from 25.2 to 31.4 °C. pH of the water ranged from 7.0 to 7.2. Depth of the fishing area ranges from 1-6 m and bottom was muddy. Both men and women of the surrounding villages are engaged on fishing. Fishing activity begins in the early morning depending on the tide. In deeper areas, two persons go in dugout canoes and catch the fish. In Ashtamudi Lake, gill-nets (*Koozhali vala*) specially meant for catching *O. tentacularis* (*Koozhali*) are devised with regard to their peculiar burrowing habits. Net is usually made of PA multifilament having 22 mm and 24 mm mesh size.

Fishery

Details of catches of *O. tentacularis* from different sites are presented in Table 1. The fishery is supported by the fish of 12 -17 cm total length with the dominance of 14 cm group in all months. Maximum length and weight was observed as 17.9 cm and 18.6 g, respectively. The total fish catch from the Lake was estimated as 77.76 t. The maximum landing was observed in the month of July at the site Thekkumbhagam with 17.89 t and a minimum in the month of December at the site Thevally with 0.12 t. A close examination of the data shows that in December the catch per month was only 0.95 t. A maximum of 36.41 t was observed in July alone. In other months the fishery shows some fluctuations.

Table 1: Month-wise and station-wise landings of *O. tentacularis* in Ashtamudi Lake (tonnes)

Location	Kureepuzha	Thekkumbhagam	Thevalli	Perumon
Jan	1.94	1.21	1.65	1.39
Feb	1.37	1.12	0.53	0.89
Mar	1.41	0.9	1.2	1.06
Apr	2.3	0.9	0.74	0.91
May	1.6	0.97	0.68	1.17
Jun	1.41	2.26	0.73	0.6
Jul	7.41	17.89	5.13	5.98
Aug	1	0.53	0.7	0.52
Sep	1.27	0.49	0.48	0.74
Oct	1.15	0.85	0.62	0.57
Nov	0.54	0.79	0.67	0.54
Dec	0.16	0.13	0.12	0.54
Total	21.56	28.04	13.25	14.91
Grand total				**77.76**

O. tentacularis are caught throughout the year, and is a source of food fish during the offseason of the other commercial fisheries. It is a family oriented fishery, where men are engaged in fishing and woman in marketing. Gobiid fishes have a good market in some countries, especially in Asia. Even though the fish has an increasing demand, the fishery is marginal and there is no export market for this fish. It is usually consumed fresh and it is considered as a delicacy in the markets around Ashtamudi Lake. Increased fishery overexploitation and habitat degradation are threatening the fish stock in small-scale fisheries (Lindegren *et al.*, 2013). As an esteemed food fish, *O. tentacularis* also has an important role in the livelihood security of many fishermen living close to the Lake. In *O. tentacularis,* the spawning takes place in batches at regular intervals and the fish is a multiple spawner. The overall sex ratio was skewed towards male fishes. Male fishes were dominant in the fishery up to 175 mm (Remya mohan *et al.*, 2018). Remyamohan (2019) reported that growth overfishing and recruitment overfishing of *O. tentacularis* are not occurring in Ashtamudi Lake and overexploitation may not be the reason for dwindling catches. Extensive reviews on Ashtamudi Lake indicate that the Lake is under environmental and ecological threat mainly by pollution stress from anthropogenic sources (Razeena and Sherly, 2015). By proper management strategies, the biodiversity of the Lake can be conserved. It needs contribution of the policy makers and government institutions in the region to develop and implement suitable management programmes to improve sustainability of the gobiid fishery, in the Ashtamudi Lake.

Acknowledgement

We extend our gratitude to University Grants Commission for the research fellowship to carry out the work. We thank Dr. Vincent B. Netto, Principal, and Rev. Fr. Anil Jose, Manager, Fatima Mata National College, Kollam, Kerala, for the facilities.

References

Bartley, D.M., Graaf, G.J., Valbo, J.J. and Marmulla, G., 2015. Inland capture fisheries: status and data issues. Fish. Manage. Ecol., 22(1): 71-77.

Kurup, B.M. and Thomas, K.V., 2001. Fishery resources of the Ashtamudi estuary. Technical Report No. 14. ASR Ltd., Marine and Freshwater Consultants, Hamilton, New Zealand and Centre for Earth Sciences Studies Thiruvananthapuram, India, 58 p.

Lindegren, M., Checkley, D.M., Rouyer, T., Maccall, A.D. and Stenseth, N.C., 2013. Climate, fishing, and fluctuations of sardine and anchovy in the California. Current. Proc. Natl Acad. Sci. USA, 110: 13672-13677.

Nair, N.B., Krishnakumar, K., Nair, J.R., Abdul Azis, P.K., Dharmaraj, K. and Arunachalam, M., 1983. Ecology of Indian estuaries XI. A preliminary survey of the fishery resources of Ashtamudi estuarine system. Fish. Technol., 20: 75-83.

Razeena, K.L. and Sherly, W.E., 2015. Accumulation of heavy metals in surface water of Ashtamudi Lake, Kollam, Kerala. Nature, Environment and Pollution Technology 14(2): 434-434.

Remya Mohan, 2018. Reproductive biology and population dynamics of *Oxyurichthys tentacularis* (Valenciennes, 1837) of Ashtamudi Lake, Kollam, Kerala, Ph.D. Thesis submitted to the University of Kerala, 206 p.

Remya Mohan, Harikrishnan, M. and Sherly Williams, E., 2018. Reproductive biology of a Gobiid fish *Oxyurichthys tentacularis* (Valenciennes, 1837) inhabiting Ashtamudi Lake, South India. Journal of Applied Ichthyology 34(5): 1099-1107.

Vimal, R. V., Binushma, R., Soumya, W., Shibu, A., Lekshmi, S., Shibu, V. Y., Sruthi, S. and Tresa, R., 2014. Aquatic bioresources of Ashtamudi Lake, Ramsar site, Kerala. Journal of Aquatic Biology and Fisheries 2(1): 297-303.

Impact of Climate Change on Hydrological Cycle, Ecosystem, Fisheries and Food Security, Pages: 265–273
Edited by: B. Madhusoodana Kurup, M.R. Boopendranath, M. Harikrishnan and A.V. Shibu

CHAPTER - 23

Projecting Short-term and Long-term Total Marine Fish Production and Catch Potential in India's North-Western Coast under the Climate Change Scenario

Vinod K. Yadav*, Vidya Shree Bharti and Shrinivas Jahageerdar

ICAR-Central Institute of Fisheries Education, Panch Marg, Off Yari Road, Versova, Andheri (West), Mumbai-400061, Maharashtra, India
**E-mail: vinodkumar@cife.edu.in*

Forecasting is used to analyze the past and current behaviour to forecasts the future fish landing which intern provide aid to decision-making and in planning for the future effectively and efficiently. Autoregressive integrated moving average (ARIMA) model is the most widely used model for forecasting time series. One of the main drawbacks of this model is the presumption of linearity. Sometimes, the time series often contain non-linear components; under such condition, the ARIMA models are not adequate in modeling and forecasting. To overcome this difficulty, many parametric nonlinear models are developed to capture the nonlinear component. These parametric nonlinear models some time fails if the data generating process is highly heterogeneous, complex and nonlinear in nature. The Artificial Neural Network (ANN) is the most widely used machine learning techniques to model and forecast the complex and nonlinear data. Variability of environmental and climatic variables is an uncontrollable source in predicting the fish catch landing. The extent of influence of these variables in prediction not only depends on the magnitude of these variables but also on the distribution pattern of over the season around the year. Therefore, when forecasting is carried out for the dynamic behaviour of fish catch landing, it should be able to take advantage not only of historical data of landing of the fishery but also of the impact of various environmental and climatic variables over the different seasons of the year. In India, efforts were made by different researchers to implement different models for forecasting the marine fish catch with environmental and climatic variables (Madhavan *et al.*, 2015, Yadav *et al.*, 2016). As there is variation in the season-wise landing of total catch with variation in environmental and climatic variables, quarterly landings data of it along with quarterly average data on the mean value of exogenous (environmental and climatic variables)-Sea Surface temperature (SST), Diffuse attenuation coefficient (Kd_490), Photosynthetic active radiation (PAR) and Chlorophyll-a (Chl-a) were used for building the model and forecasting. In Modeling and forecasting, Autoregressive Integrated Moving Average with exogenous Input (ARIMAX) model has been preferred over ARIMA as it forecast more accurately (Paul 2016). Nasker *et al.*, 2017, used the ARIMAX model to study the Influence of Hydrological parameter on the abundance of Hilsa fishery in Narmada River–estuary system of India. To model the series with exogenous variables and the series contains nonlinear patterns, the artificial intelligence technique- Nonlinear Autoregressive models with exogenous input (NARX model), is commonly employed (Gao and Er, 2005).

Climate change continues to affect marine ecosystems. Marine ecosystems are not in a steady state but are affected by the environment, which varies on many spatial and temporal scales (Manjusha *et al.*, 2013). Predictions based on environment and fish catch relations have been carried out under different climate change scenarios (Brochier *et al.*, 2013). Yen *et al.* (2016) have predicted the catch potential of Skipjack tuna in the western and central Pacific Ocean under different climate change scenarios. It was found that the highest catch potential was associated with the highest greenhouse gas emissions (A2 Scenario), whereas the catch potential remained relatively stable under the scenario with lower emissions. To better understand the implications of climate change for those catches and to provide a reference for research on long-term strategies, simulated SST for the Climate Change Scenario A2 (2015–2050) obtained by the GFDL-CMIP model was used as an input for the model for estimating the total catch for a short term (2020–2025) and for a long term (2030–2050). As a benchmark, we used the average catch for the last two years (2012-2013) of fish catch data

Materials and Methods

Quarterly average data on the mean value of SST, Kd, Chl-a and PAR in the Gujarat and Maharashtra coastal area were obtained from Moderate Resolution Imaging Spectroradiometer (MODIS) level 3 standard binned images archived by the Ocean Biology Processing Group (OBPG) as an ASCII file for a period of 1998 to 2013. As complete quarter set of remotely sensed imageries value were available from the year 1998, so the quarterly landing of total catch estimate value pertaining to Gujarat and Maharashtra coastal region of India were taken from the National Marine Fisheries Data Centre (NMFDC) of Central Marine Fisheries Research Institute (CMFRI), Kochi. The year was divided into four quarters-1st (January to March), 2nd (April to June), 3rd (July to September) and 4th (October to December) as the fish landing data are available in this fashion. The datasets for the period, 1st quarter 1998 to 2nd quarter 2012 (90 % of the whole dataset, i.e., 58 points were used for training data sets and from the 3rd quarter 2012 to 4th quarter2013 (6 data set, i.e., 10%), was used as testing and validation data sets for the model. SST, Chl-a, Kd, and PAR were expressed in °C, mg/m^3, m^{-1}, and Einstein/m^2/day, respectively. Fish catch landing estimation was measured in a metric ton. Simulation SST data from the GFDL-CMIP climate model under high greenhouse gas emission climate changed scenario RCP 8.5 was collected and incorporated into the analysis of total catch potential. These data were generated from a model developed by the NOAA Geophysical Fluid Dynamics Laboratory (http://www.gfdl.noaa.gov).

Autoregressive Integrated Moving Average with Exogenous Input (ARIMAX)

The ARIMAX model is a generalization of the ARIMA model, which is capable of incorporating an exogenous input variable. The models in equation (1) with the assumption of ARIMA error and with the presence of exogenous variables, became an ARIMAX (p,d,q,k) model. In statistical form, ARIMAX model may be written as

$$Y_t = \theta_0 + \sum_{i=1}^{p} \delta_i Y_{t-i} + \sum_{j=1}^{q} \theta_j e_{t-j} + \sum_{l=1}^{k} b_l F_{t-l-1} + e_t \quad \text{... (1) (Naskar } et\ al.\text{, 2017)}$$

Where F is the exogenous variable of order k

In the present manuscript, the ARIMAX model was developed using Box and Jenkins (1976), an auto-regressive integrated moving average (ARIMA) with exogenous variables-SST, Chl-a Kd_490, and PAR. The ARIMA and ARIMAX computations have been done in R.3.6.1 software for statistical computing. The details of ARIMA and ARIMAX can be seen in Raman *et al.* (2018).

Nonlinear Autoregressive with External (Exogenous) Input (NARX) - Artificial Neural Network (ANN) model

NARX predict series y(t) given p past values of series y and another external series x(t) which can be single or multidimensional. The equation that models the NARX network behavior for time series prediction is shown as below

$$Y(t)=f(x(t-1), x(t-2)......., x(t-p),y(t-1), y(t-2),.........y(t-p))+ e(t) \quad \text{... (2)}$$

As the fish catch landing depends on different environmental and climatic parameters, hence the integration of environmental and climatic parameter data with time series fish catch could benefit the time series modeling process to provide an accurate forecast instead of a unique approach with one single value corresponding to the fish catch landing data. NARX computations have been done in R.3.6.1 software for statistical computing.

Results and Discussion

Validation of models for hold-out data

One-step-ahead forecasts from the 3rd quarter of 2012 to the 4th quarter of 2013 (the last six data points) as given by the fitted models are given in Table 1. The forecasts of these models were compared using RMSE, MAE, and average error percentage. As the developed models ARIMAX had good accuracy on hold out data, these models were used for forecasting the total catch. The predicted catch was compared with catch estimation given by CMFRI, Kochi to check the reliability of models. The predicted landing/catch values for Gujarat and Maharashtra were compared with the values reported by the ICAR-Central Marine Fisheries Research Institute (CMFRI) for 2014, 2015, and 2016 (www.cmfri.org.in/gj2014; www.cmfri.org.in/gj2015; www.cmfri.org.in/gj2016), and the results are shown in Table 2 and Table 3. The results (Table 2 and table 3) indicates that the model ARIMAX with all 4 variables, giving very close estimate of total catch for all the three year 2014, 2015 and 2016, with CMFRI release report (Note that the CMFRI data are annual aggregates, I compared the total of the four quarters as forecast to the year's actual value.

Table 1: Comparisons of the forecasting results of total catch for Gujarat coast by ARIMA, ARIMAX and NARX on last six testing data (Hold out data).

Year	Actual quarter landing/catch	ARIMA	ARIMAX	NARX
Q3_2012	114895	99399	100586	68299
Q4_2012	352986	329559	329323	337807
Q1_2013	238844	247122	227582	195852
Q2_2013	121744	172386	153868	132544
Q3_2013	98080	117280	116887	77625
Q4_2013	337994	352563	349125	293569
RMSE		25829.51	20004.46	33562.23
MAE		21935.33	18549.33	30074.46
Average Error (%)		14.84	12.12	17.62

Total catch potential using simulated SST under the A2 climate change scenario

For the short term (2020-25), the trend of the total catch, for the shorter term it is going to average increase by 99% for Gujarat coastal coast and for the Maharashtra coast for the shorter term average increase is 12.61%. Till the year 2030, there is an increase in catch percentage fluctuation in the Gujarat coast, but in the case of Maharashtra coast, there is a positive percentage change but with less rate (Fig. 1). The trend of total catch potential, for the longer term, the average increase is 84 % for Gujarat coast, but for the Maharashtra coast, overall average increase is 15.92 %. Hence the predictions for 2030 to 2050 exhibited decreases in catch potential of the total catch in Gujarat coast compare

Table 2: The comparison of the predicted value of total catch with CMFRI releases marine fish catch in the year 2014, 2015 and 2016 for Gujarat coast

Model with Input	2014		2015		2016	
	Forecasted value in a metric ton	CMFRI reported totalcatch	Forecasted value in a metric ton	CMFRI reported total catch	Forecasted value in metric ton	CMFRI reported total catch
NARX with all 4 variables	791162 (22059)	711930	913113 (22059)	721549	1035990 (22059)	774373
ARIMAX with all 4 variables	760716 (18604)		864008 (18604)		917757 (18604)	

(Value in parenthesis is Mean Absolute Error)

Table 3: The comparison of the predicted value of total catch with CMFRI releases marine fish catch in the year 2014, 2015 and 2016 for Maharashtra coast

Model with Input	2014		2015		2016	
	Forecasted value in a metric ton	CMFRI reported totalcatch	Forecasted value in a metric ton	CMFRI reported total catch	Forecasted value in metric ton	CMFRI reported total catch
NARX with all 4 variables	318590 (18663)	344648	317142 (18663)	264891	314419 (18663)	292355
ARIMAX with all 4 variables	269721 (16019)		260699 (16019)		251703 (16019)	

(Value in parenthesis is Mean Absolute Error)

to short term but in Maharashtra coast compare to short period there is an increase in the catch which is supported by GAM plot (Fig. 2 and 3).

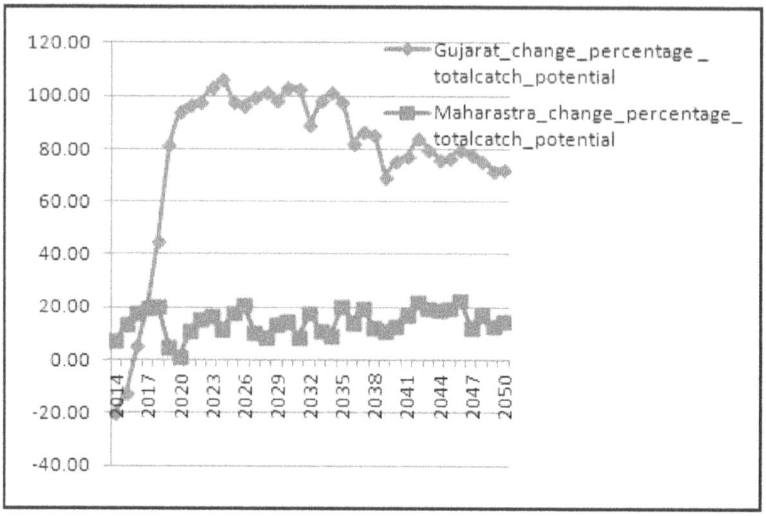

Fig. 1. Changes in total catch potential (in %) for Gujarat and Maharashtra coast

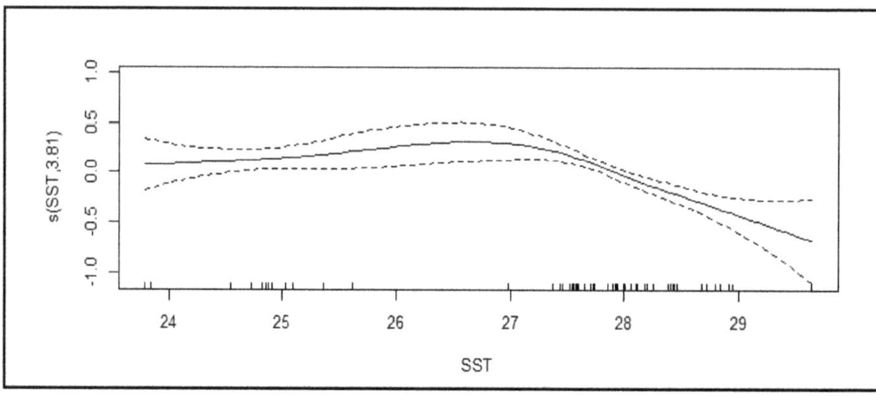

Fig. 2. Distribution of log (total catch) with respect of SST for Gujarat coast

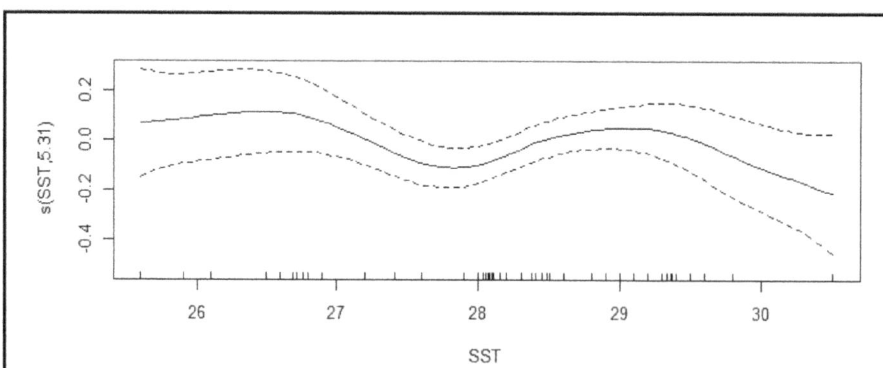

Fig. 3. Distribution of log (total catch) with respect of SST for Maharashtra coast

For the Gujarat coast, the average SST during the study period was 27.23°C but average SST during the long period was 27.49°C (Table 4), and the maximum total catch is at 27.55°C (not shown here). As average temperature during the long period was closer to the temperature of maximum catch, hence there would be chances to get a higher total catch in a long period. Also, other reason might be that an increase in average SST change might support the different activities of the other fish species that lead to increase in their catch and hence total catch. In Maharashtra coast, the average SST during the study period was 28.13°C (Table 4) but the average SST during the long period was 27.48°C and maximum total catch was at 28.15 °C (not shown here). The Average temperature during the long period has a wider difference to the temperature of maximum catch; hence, there would be chances to get a lesser total catch in the long period compared to the Gujarat coast.

Table 4: Average value of environmental variables during the study period (Year 1998 to 2013) for Gujarat and Maharashtra coast

Variables	SST	Chl-a	Kd	PAR
Gujarat	27.23 (Min-23.78, Max-29.63)	2.19 (Min-1.01, Max-3.43)	0.175 (Min-0.08, Max-0.27)	43.55 (Min-34.10, Max-54.34)
Maharashtra	28.13 (Min-25.6, max-30.5)	2.58 (Min-1.09, Max-4.43)	0.22 (Min-0.10, Max-0.30)	43.22 (Min-34.40, Max-51.30)

SST (°C), Chl-a (mg/m^3, Kd (m^{-1}) and PAR (in Einstein/m^2/day)

Various factors affect the marine ecosystems, but the SST influences the ecophysiology of vertebrates and invertebrates significantly. In this study, it was suggested that there is a certain percent of decrease/increase in total catch due to the change in SST over a period due to climate change. However, the results and analyses of change in catch potential could be improved by incorporating more remotely sensing and climate change scenario data at a finer scale, such as by downscaling. In summary, it was demonstrated that change in SST due to climate change might result in substantial changes in the catch potential of total, emphasizing the need for further research to guide the imperative development of adaptation management strategies.

Conclusion

Different methods were tested using hold-out data, and inclusion of environmental variables made the forecasts more accurate. Also, ARIMAX performed better than NARX-ANN in predicting the total catches, probably because it captures seasonal variations more accurately. The total catch was forecast for the next five years (20 quarters), and the forecast values for the first three years (2014, 2015, and 2016) were compared with actual data. The two values predicted and actual were very close at least for the first year (the forecast value was a total of four quarters whereas the actual value was aggregate for the year). With a larger data set for training, the forecasts would have been even more accurate.

Taking the average of the last two years catch data as a benchmark, for the short term (2020-25), the trend of the total catch, for the shorter term it is going to average increase by 99% for Gujarat coastal coast and for the Maharashtra coast for the shorter term average increase is 12.61%. Till the year 2030, there is an increase in catch percentage fluctuation in the Gujarat coast, but in the case of Maharashtra coast, there is a positive percentage change but with less rate. The trend of total catch potential, for the longer term, the average increase is 84 % for Gujarat coast, but for the Maharashtra coast, overall average increase is 15.92 %. However, these results and analyses can be refined by incorporating data from remote sensing and from climate change scenarios at higher resolutions. In summary, we have shown that climate change may lead to substantial changes in the quantities of the catch of Bombay duck and that further research can help in devising better management strategies to adapt to climate change.

Acknowledgment

The authors are grateful to the Director, ICAR-CIFE, Mumbai, for providing the facilities to carry the work.

References

Brochier, T., Echevin, V., Tam, J., Chaigneau, A., Goubanova, K. and Bertrand, A., 2013. Climate change scenarios experiments predict a future reduction in small pelagic fish recruitment in the Humboldt Current system. Global Change Biology, 19: 1841-1853.

CMFRI, 2014. Marine fish landings in Gujarat for the year 2014 (In tonnes) http://www.cmfri.org.in/gj2014 (Accessed on 15th October 2018).

CMFRI, 2015. Marine fish landings in Gujarat for the year 2014 (In tonnes) http://www.cmfri.org.in/gj2015 (Accessed on 15th October 2018).

CMFRI, 2016. Marine fish landings in Gujarat for the year 2014 (In tonnes) http://www.cmfri.org.in/gj2016 (Accessed on 15th October 2018).

Gao, Y. and Er, M.J., 2005. NARMAX time series model prediction: feed forward and recurrent fuzzy neural network approaches, Fuzzy Sets and Systems, 150: 331-350.

Madhavan, N., Thirumalai, V.D., Ajith, J.K. and Sravani, K., 2015. Prediction of mackerel landings using MODIS Chlorophyll-a, Pathfinder SST, and SeaWiFS PAR. Indian Journal of Natural Sciences 5(29): 4858-4871.

Manjusha, U., Jayasankar, J., Remya, R., Ambrose, T.V. and Vivekanandan, E., 2013 Influence of coastal upwelling on the fishery of small pelagic off Kerala, south-west coast of India. Indian Journal of Fisheries 60(2): 37-42.

Naskar, M., Chandra, G., Sahu, S.K. and Raman, R.K., 2017. A Modeling Framework to Quantify the Influence of Hydrology on the Abundance of a Migratory Indian Shad, the Hilsa *Tenualosa ilisha*. North American Journal of Fisheries Management 37: 1208-1219.

Paul, R.K. and Sinha, K. 2016. Forecasting crop yield: ARIMAX and NARX model. RASHI, 1(1): 77-85

Raman, R.K., Mohanty, S.K., Bhatta, K.S., Karna, S.K., Sahoo, A.K., Mohanty, B.P. and Das, B.K., 2018. The time series forecasting model for fisheries in Chilika lagoon (a Ramsar site, 1981), Odisha, India: a case study. Wetlands Ecology and Management 26: 677–687. doi.org/10.1007/s11273-018-9600-4.

Yadav, V.K., Jahageerdar, S., Ramasubramanian, V., Bharti, V.S. and Adinarayana, J. 2016. Use of different approaches to model catch per unit effort (CPUE) abundance of fish. Indian Journal of Geo-Marine Science 45(12): 1677-1687.

Yen, K.W., Su, N.J, Teemari, T., Lee, M.A. and Lu, H.J., 2016. Predicting the catch potential of Skipjack tuna in the western and Central Pacific ocean under different climate change scenarios. Journal of Marine Science and Technology 24(6): 1053-1062.

Impact of Climate Change on Hydrological Cycle, Ecosystem, Fisheries and Food Security, Pages: 275–278
Edited by: B. Madhusoodana Kurup, M.R. Boopendranath, M. Harikrishnan and A.V. Shibu

CHAPTER - 24

Length – Weight Relationship of *Oxyurichthys tentacularis*, (Valenciennes, 1837) inhabiting Ashtamudi Lake, Kerala

Remya Mohan[1*], M. Harikrishnan[2] and E. Sherly Williams[1]

[1]*Environmental Science, Aquaculture and Fish Biotechnology Lab, Department of Zoology, Fatima Mata National College, Kollam-691001, Kerala, India*
[2]*School of Industrial Fisheries, Cochin University of Science and Technology, Kochi-682016, Kerala, India*
**E-mail: remyamohan062@gmail.com*

Gobiids are small benthic fishes inhabiting a wide range of habitats (Nelson, 1994). *Oxyurichthys tentacularis*, is an inland water gobiid fish known as *Koozhali* in vernacular. As a table fish, it is economically important and palatable. Despite of its palatability and consumer appeal, little information is available on the biology of this fish (Remya Mohan, 2018). Length-weight relationship is an essential tool in fish biology, fisheries assessment and fish conservation and these provide information on the growth pattern, ontogenic changes and fish stocks (Froese *et al.*, 2011). The present study is the first work describing length-weight relationship of *O. tentacularis*.

Materials and Methods

For the present study samples were collected from Ashtamudi Lake (76^0 32' - 76^0 41'E Long. and 8^0 52' - 9^0 2'N Lat.). It is a palm-shaped extensive water body with eight prominent arms (area of 61.42 km^2). Monthly fish samples were collected from study sites namely, Kureeppuzha, Thekkumbhagam, Thevally, Perumon, Kanjirodu and Kumbalam around Ashtamudi Lake, by different fishing gears like gill net, seine, cast net and dip net. A total of 3698 samples comprising 2077 male and 1621 female fish were collected during the study period from July 2013 to June 2015. All the samples were analysed in fresh condition. The total length (from tip of the snout to tip of upper caudal lobe) of individual fish was measured to the nearest 1 mm and the individual weight was recorded to an accuracy of 0.001 g using an electronic balance. The method suggested by Le Cren (1951) was followed to compute the length and weight relationship. Thus, the length - weight relationship can be expressed by the formula:

$$W = a\ L^b$$

Results and Discussion

The mathematical relationship between total length and weight of male and female fish obtained by logarithmic regression equations are as follows (Fig. 1 and 2.):

$$\text{Log W (male)} = 1.481 \text{ Log TL} - 2.077$$

$$\text{Log W (female)} = 1.117 \text{ Log TL} - 1.293$$

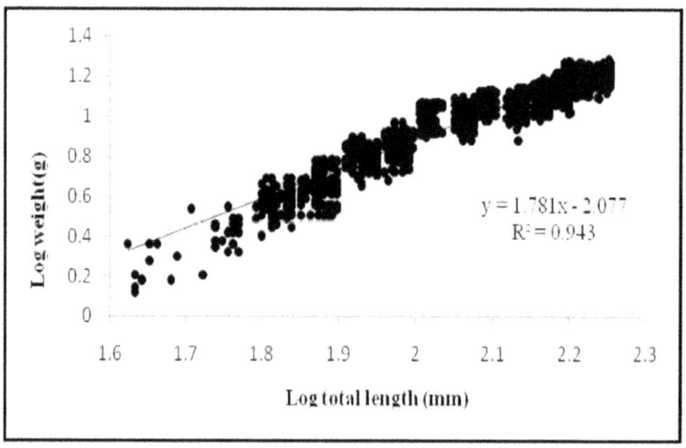

Fig. 1. Regression of log total length (mm) on log weight (g) of male *O. tentacularis* inhabiting Ashtamudi Lake

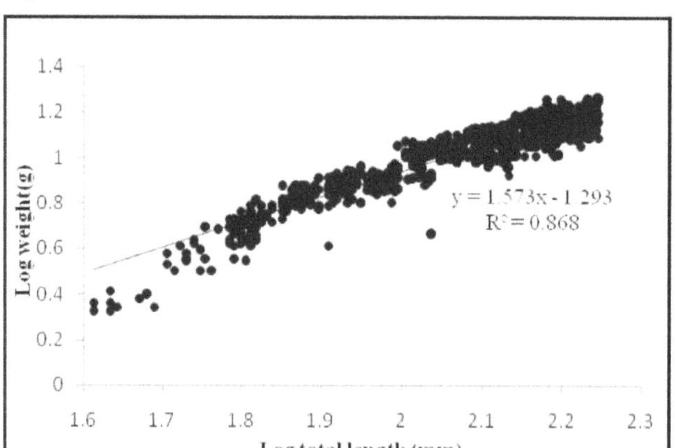

Fig. 2. Regression of log total length (mm) on log weight (g) of female *O. tentacularis* inhabiting Ashtamudi Lake

In Male and female r^2 values were recorded as 0.943 and 0.868 respectively. In male fish the b value was recorded as 1.481 and in females it was 1.117. The error in the regression coefficients was minimum in the case of both sexes. The results of the analysis of covariance (ANCOVA) have shown that the values of the slope ('b') for males and females exhibited a significant difference (F value = 6.34, significant at 1%). The results indicate a negative allometric growth.

Thus length-weight analysis is more useful for comparing life history of fish from different localities. Generally, if a fish grows isometrically, 'b' value must be equivalent to 3 and it will retain its body shape and its specific gravity will remain unchanged during their life span. But under natural condition the 'b' value can be affected by external (habitat, temperature and food availability) and internal factors (sex, fish activities, health, maturity, seasonal growth rates and also food habits) (Wootton, 1990). If the physico-chemical parameters of the lake are unfavourable, the length and weight of the fish will be negatively affected (Froese, 2006). The present study showed that the 'b' value in male fish was 1.481 and 1.117 in female fish. The results of the analysis of covariance (ANCOVA) showed a significant difference. As the 'b' value of *O. tentacularis*, was significantly lower than the isometric threshold suggesting that this goby was slim in body shape as the fish grew as defined by Froese and Binohlan (2000). The present result of *O. tentacularis* (negative allometry), is supported by the findings, in *Periophthalmus barbarus*, b=2.73 (King and Udo, 1996); and in *G. giuris* of Atrai River (Joadder, 2009). However, in some studies certain gobiids also shows a positive allometric growth, for instance, in *G. giuris* 'b' value has been reported as 3.18 (Kaur and Rawal, 2015). Similar to the present results, in a deep sea fish *Alepocephalus indicus*, the 'b' value has been reported as 1.974 and it might be due to the ecological parameters at the depth at which these species occupy (Thomas *et al.*, 2003).

The present study site, Ashtamudi Lake is under severe threat due to several reasons. The main factors responsible for weakening of the lake are pollution and encroachment. Pollution through, sewage, household waste, intensive coconut husk retting, toxic heavy metals from industries, oil hydrocarbons from fishing trawlers, tourism activities, fertilizers and pesticides through surface runoff are the major threats reported in the Lake (Kerry, 2001; Razeena and Sherly, 2014; Surya and Mophin, 2017). The pollution in Ashtamudi Lake is a serious threat to the aquatic organisms inhabiting the lake. Pollution load destroys the aquatic invertebrates which prove vital for the survival of many fishes. The negative growth pattern of the fish *O. tentacularis* in the present study may be due to the poor environmental conditions prevalent in the Ashtamudi Lake ecosystem, especially the high level of pollution. Deviation from isometric growth is commonly observed in most of the aquatic organisms which changes shape as they grow.

Acknowledgement

We extend our gratitude to University Grants Commission for the research fellowship to carry out the work. We thank Dr. Vincent B. Netto, Principal, and Rev. Fr. Anil Jose, Manager, Fatima Mata National College, Kollam, Kerala, for the facilities.

References

Froese, R. and Binohlan, C., 2000. Empirical relationships to estimate asymptotic length, length at first maturity and length at maximum yield per recruit in fishes, with a simple method to evaluate length frequency data. Journal of Fish Biology 56: 758-773.

Froese, R., Tsikliras, A. C. and Stergiou, K. I., 2011. Editorial note on weight-length relations of fishes. Acta Ichthyologica Et Piscatoria 41: 261-263.

Joadder, A. R., 2009. Length-weight relationship and condition factor (K_n) of gobi, *Glossogobius giuris* (Hamilton) from Atrai River in the Northern part of Bangladesh. J. Fish. Int., 4(1): 1-4.

Kaur, C.V., Rawal, Y.K., 2013. Length-Weight Relationship in *Glossogobius giuris* (Ham.) from Sukhna Lake. International Journal of Science and Research 4 (9): 27-31.

Kerry, B., 2001. Ashtamudi management plan - Tourism concept, 22 p.

King, P.R and Udo, M.T., 1996. Length-weight relationship of the mudskipper *P. barbarus* in the Imo River Estuary, Nigeria. NAGA, World Fish Center Quarterly 19(2): 24-27.

Le Cren, C.D., 1951. The length-weight relationship and seasonal cycle in gonad weights and condition in the perch (*Perca fluviatilis*). J. Anim. Ecol., 20: 201-219.

Razeena, K.L. and Sherly, W.E., 2014a. Bioaccumulation of Heavy metals in an estuarine fish *Liza parsia* of Ashtamudi lake-Southwest coast of Kerala, India. The International Journal of Science and Technology 2(3): 168-171.

Remya Mohan, 2018. Reproductive biology and population dynamics of *Oxyurichthys tentacularis* (valenciennes, 1837) of Ashtamudi Lake, Kollam, Kerala, Ph.D. Thesis, University of Kerala, 206 p.

Remya Mohan, Harikrishnan, M. and Sherly Williams, E., 2018. Reproductive biology of a Gobiid fish *Oxyurichthys tentacularis* (Valenciennes, 1837) inhabiting Ashtamudi Lake, S. India. Journal of Applied Ichthyology 34(5): 1099-1107.

Surya, L.P. and Mophin, K.K., 2017. Comparative assessment of water quality of Ashtamudi Lake through WQ. International Journal for Research in Technological Studies 4 (6): 23-29.

Thomas, J., Venu, S. and Kurup, B.M., 2003. Length-weight relationship of some deep-sea fish inhabiting the continental slope beyond 250 m depth along the West Coast of India. NAGA, World Fish Center Quarterly 26(2): 17-21.

Wootton, R.J., 1990. Ecology of Teleost Fishes: Fish and Fisheries Series, Chapman and Hall Ltd., 404 p.

V

Changing Biodiversity Scenarios

Impact of Climate Change on Hydrological Cycle, Ecosystem, Fisheries and Food Security, Pages: 281–299
Edited by: B. Madhusoodana Kurup, M.R. Boopendranath, M. Harikrishnan and A.V. Shibu

CHAPTER - 25

250 Years of Marine Biodiversity Scenarios in India – What Will Persist?

K. Venkataraman

National Centre for Sustainable Coastal Management, Anna University Campus, Chennai-600025, India
E-mail: venkyzsi56@gmail.com

The Indian subcontinent is surrounded by three seas and one ocean *viz.,* Arabian Sea in the west, Bay of Bengal and Andaman Sea in the east and Indian Ocean in the south. Coastal and marine ecosystems of India play a vital role in India's blue economy by virtue of their resources, productive habitats and rich biodiversity. India has a long coastline of 7,516 km of which, the mainland accounts for 5,422 km, Lakshadweep coast extends 132 km and Andaman and Nicobar Islands have coastline of 1,962 km. The coastal area is assuming greater importance in recent years, owing to increasing human population, urbanization and accelerated developmental activities. Nearly 250+ million people live within a distance of 50 km from the coast. The coastal regions are thus, an important place of human activity and the coastal ecosystems are now extremely disturbed and very much threatened. Current approaches to the management goals of coastal resources were not accomplished and the coastal environments and resources are being rapidly degraded and eroded in India.

The continental regions and the offshore islands and a very wide range of coastal ecosystems such as estuaries, lagoons, mangroves, backwaters, salt marshes, rocky coasts, sandy stretches and coral reefs, are characterized by unique biotic and abiotic properties and processes. A network of 14 major, 44 medium and numerous minor rivers together with their tributaries drain in to the Arabian Sea, Bay of Bengal and Indian Ocean. Other than these mainland coasts, there is Lakshadweep in the south Arabian Sea and Andaman and Nicobar in the eastern Bay of Bengal. The West Coast is generally exposed to heavy surf and rocky shores and headlands. The East Coast is generally shelving with beaches, lagoons, deltas and marshes. Conserving the marine biodiversity what we have today is hampered by lack of capacity to mitigate the climate change effects including awareness, outreach as well as our ability to predict the climate change impacts, what would survive in India and lack of data relating to changes in biodiversity of those environments. The present speech provides a comprehension into the biodiversity in India with a question on what will remain for future.

Material and Methods

The present synthesis of what is known of coastal and marine biodiversity in India is based mainly on historical, systematic accounts, records and reports of the Zoological Survey of India as well as other research organizations such as Central Marine Fisheries Research Institute, National Institute of Oceanography, Bombay Natural History Society, Central Salt and Marine Chemicals Research Institute; Fishery Survey of India, Geological Survey of India, Indian Institute of Science, M.S. Swaminathan Research Foundation and Wild Life Institute of India.

Results and Discussion

The inventory of marine fauna and flora through published literatures on coastal and marine biodiversity of India indicates that a total of 21,663 species of faunal and floral communities have been reported from seas around India (Venkataraman & Raghunathan, 2015). The data reveals that India contributes 6.44% of marine biodiversity in global scenario (Table 1 and 2). According to the legislation of Government of India, about 885 marine faunal species belonging to eight Phyla viz., Poriferea, Coelcntrata, Arthropoda, Mollusca, Echinodermata, Pisces, Reptilia and Mammalia are protected under the legislation (Table 3) of different categories (Schedule I, II, III, IV) as their natural population is dwindling. The taxonomy and inventory of many of the minor groups particularly invertebrates especially sponges, octocorals, ctenophores, polychaetes, sipunculates, polyclads, opisthobranchs, meiofauna, isopods, amphipods, rotifers, nemerteans, endoprocta, ectoprocta, loricifera, acanthocephala, anomura, nemertina, gastotricha, gnathostomulida, kinorhyncha, echiura, bryozoa, tardigrada, pogonophora, and tunicates remain a challenge to specialists and as a result these taxa continue to be inadequately known from Indian seas (Venkataraman & Wafar, 2005). However, considerable knowledge on the taxonomy of groups such as seaweeds, seagrasses, mangroves, hard corals, crustaceans, molluscs, echinoderms, fishes, reptiles and marine mammals are available in India. Moreover, most of the data on marine biodiversity have been collected from surveys up to the maximum depth of 200 m in the continental shelf. The gaps in knowledge extend to several smaller taxa and to large parts of the shelf and deep sea ecosystems, including seamounts (Wafer *et al.*, 2011). The data provided in this paper warrants the continued taxonomic research in least studied/unknown groups in light of current threats to marine biodiversity. We may of course never know the full extent of biodiversity in any of the world's oceans and the rate at which we increase our understanding is likely to be lowest in seas around India. The impacts of climate change will alter coastal marine ecosystems affecting the range of species and their ecology at a rate faster than it is recorded (Keesing and Irvine, 2005).

Table 1: Summary of marine biodiversity studies in India.

Sl. No.	Taxonomic Group	Period	Survey Region	Habitat	Institutions involved	Remarks
1.	Micro algae (Diatom)	1930- till date	East and west coasts, Madras Coast	Coastal and Offshore waters	University of Madras Andhra University, BSI, Annamalai University	Studies going on till today.
2.	Macro Algae (Seaweeds)	1940- till date	East and west coasts, Andaman and Nicobar Islands (A&N).	Intertidal rocky shore of west and east coasts, Coral reef areas in GOMBR, GOK, and A&N islands	BSI, FSI, ZSI, NIO, CSMCRI, CMFRI, NCSCM	Studies going on till today. Large amount of harvesting going on for commercial purposes.
3.	Sea grasses	1980 – till date	Gulf of Mannar (GOM), Andaman and Nicobar (A&N) Islands and Gulf of Kachchh (GOK).	Near shore area	Botanical Survey of India, Annmalai University, NCSCM	Studies incomplete. And going on till today. Associated fauna not worked out
4.	Mangroves	1980 – till date	East and west coasts, Sunderbans and Andaman and Nicobar Islands	Littoral Forest	BSI, FSI, ZSINIO, Annamalai University, MSSRF, NCSCM	Studies going on till today. Associated fauna is still to be done. Conservation efforts are on.
5.	Foraminifera	1798- till date	East and west coasts, Andaman and Nicobar Islands, Lakshadweep	Benthic and pelagic	Indian museum, GSI, ZSI, IISc, Agarkar Institute	National Institute of Oceanography and a very few universities studying. Very little study conducted and very few scientists working
6.	Porifera (Sponges)	1894 – till date	East and west coasts, Andaman and Nicobar Islands	Intertidal, coral reef areas	Indian Museum, ZSI, CMFRI.	Very few scientists working.

[Table Contd.

Contd. Table]

Sl. No.	Taxonomic Group	Period	Survey Region	Habitat	Institutions	Remarks
7.	Cnidaria (All groups)	1907- till date	Gulf of Mannar, Palk Bay, Gulf of Kachchh, Andaman and Nicobar Islands, Lakshadweep	Coral reef	Indian Museum, ZSI, CMFRI.	Very few scientists working. Capacity building needed for underwater studies. Status survey required. Conservation efforts are on
8.	Hydrozoa	1915- 1965	East coast	Chilka lagoon	Indian Museum, ZSI	Study incomplete. Very few scientists working. Capacity building needed for underwater studies
9.	Siphonopora	1927- 1985	East and west coasts, Madras coast	Nearshore waters	ZSI	Study incomplete. Very few scientists working. Capacity building needed for underwater studies.
10.	Anthozoa Scleractinia	1847- till date	Gulf of Mannar, Palk Bay, Gulf of Kachchh, Andaman and Nicobar Islands, Lakshadweep	Coral reef	Asiatic Society of Bengal, Indian Museum, ZSI, CMFRI, NCSCM	Study incomplete. Very few scientists working. Capacity building needed for underwater studies. Status survey required. conservation efforts are on
11.	Gorgonians	1903- till date	Gulf of Mannar, Palk Bay, Kerala coast, Andaman and Nicobar Islands	Coral reef	Indian Museum, ZSI	Very few scientists working. Capacity building needed for underwater studies. Status survey required. conservation efforts are on

[Table Contd.

Contd. Table]

Sl. No.	Taxonomic Group	Period	Survey Region	Habitat	Institutions	Remarks
12.	Scyphozoa Scyphome-dusae	1905-1983	Entire Indian Seas including Lakshadweep, Andaman and Nicobar Islands, Madras coast	Open ocean and near shore waters	Indian Museum, ZSI	Study incomplete.
13.	Ctenophora	1915-1939	Indian seas including Andaman and Nicobar Islands	Open ocean and nearshore waters	Indian Museum, ZSI	Study incomplete. Very few scientists working. Capacity building needed for underwater studies. Status survey required.
14.	Actinaria	1907-till date	East and west coasts of India especially from West Bengal (Port Canning), Orissa (Chilka lake), Tamil Nadu (Adyar backwaters and Gulf of Mannar), Kerala (Cochin backwaters and Ashtamudi), GoK, Maharashtra (Mumbai, Malvan), Goa, northern Karnataka and A&N Islands.	Intertidal rocky shores in west coast, sandy shores in east coast and coral reefs in A&N Islands	Indian Museum, ZSI, NIO.	Study incomplete. Very few scientists working.
15.	Polyclads	1902-till date	Andaman and Nicobar Islands, Lakshadweep and Gulf of Mannar	Coral reef	Indian Museum, ZSI.	Study incomplete. Very few scientists working.

[Table Contd.

Contd. Table]

Sl. No.	Taxonomic Group	Period	Survey Region	Habitat	Institutions	Remarks
16.	Archiannelida	1944-1968	East coast (Madras coast and Waltair coast)	Intertidal and nearshore waters	University of Madras, Andhra University	Study incomplete.
17.	Polychaeta	1921-1963	East coast (Chilka lake, Andhra Coast, Madras coast, Gulf of Mannar)	Intertidal and nearshore waters	ZSI, University of Madras, Andhra University, NIO	Study incomplete.
18.	Oligochaeta	1983-till date	Andaman and Nicobar Islands	Intertidal and nearshore waters	ZSI	Study incomplete. No recent studies
19.	Sipuncula	1903-1979	Andaman and Nicobar Islands, Lakshadweep Islands, Gulf of Mannar and Gulf of Kachchh	Intertidal and coral reefs	Indian Museum, ZSI	Study incomplete. Very few scientists working. Capacity building needed
20.	Echiura	1915-1964	East coast - Chilka lagoon, GoM, West Bengal and Orissa), GoK, Gulf of Kambath, Kerala coast), Lakshadweep, A&N Islands.	Intertidal and reef areas	Indian Museum, ZSI	Study incomplete. Very few scientists working. Capacity building needed
21.	Chaetognatha	1936-till date	Coastal and marine areas of India and Indian Ocean Region	Pelagic ecosystem	NIO, CMFRI, ZSI, Andhra University, Madras University, NCSCM	Study incomplete. Very few scientists working. Capacity building needed
22.	Tardigrada	1998 – till date	Intertidal and interstitial areas of Tamil Nadu coast	Free living	ZSI	Study incomplete

[Table Contd.

Contd. Table]

Sl. No.	Taxonomic Group	Period	Survey Region	Habitat	Institutions	Remarks
23.	Copepoda	1924– till date	Entire Indian coastal and marine eco-systems of India	Pelagic environment of nearshore and oceanic waters	ZSI, NIO, CMFRI, NIOT	Considerable studies have been made
24.	Parasitic Copepods	1985	Coastal and marine and interstitial waters in general	Pelagic	Kerala University	Study incomplete
25.	Ostracoda	1975 – till date	West coast of India	Pelagic environment of nearshore and oceanic waters	NIO, CMFRI,	Study incomplete.
26.	Branchiura	1983- till date	Andaman and Nicobar Islands	Intertidal and nearshore waters	ZSI	Study incomplete. No recent studies
27.	Cirrepedia	1894- till date	East coast of India	Intertidal rocky shores	Indian Museum, ZSI, University of Madras, Annamalai University	Study incomplete
28.	Mysidacea	1922- till date	East and west coasts	Pelagic ecosystem	NIO, CMFRI, ZSI.	Study incomplete
29.	Cumacea	1904- till date	East coast	Pelagic ecosystem	Indian Museum, ZSI, University of Madras	Study incomplete
30.	Tanaidacea	1923	East coast – Chilka lake	Lagoon ecosystem	ZSI	Study incomplete
31.	Isopoda	1924 - 1968	A&N Islands, Delta Ganges, Gulf of Mannar and Bombay	Pelagic waters	ZSI	Study incomplete
32.	Amphipoda	1888- till date	GoM, Travancore, Cochin, Madras and West Bengal coasts	Pelagic waters	Indian Museum, ZSI	Study incomplete

[Table Contd.

Contd. Table]

Sl. No.	Taxonomic Group	Period	Survey Region	Habitat	Institutions	Remarks
33.	Euphausiacea	1891- till date	East and west coast, Lakshadweep, Andaman and Nicobar Islands	Pelagic waters	ZSI, NIO, CMFRI.	Study incomplete
34.	Stomatopoda	1913- till date	East and west coasts	Pelagic and demersal waters	Indian Museum, ZSI, NIO, CMFRI, Annamalai Univ.	Study incomplete
35.	Decapoda Macrura	1901 - 1906	East and west coasts	Pelagic and demersal waters	Indian Museum, ZSI	Study incomplete. No recent studies on this group
36.	Brachyura	1834- till date	East and west coasts	Coastal waters	Indian Museum, ZSI, CMFRI, Annamalai University	Considerable studies are made
37.	Anomura	1972 – till date	East and west coasts	Coastal waters, estuaries	ZSI, Annamalai University	Study incomplete
38.	Mollusca	1784- till date	Throughout the Indian coast especially GoM, Palk Bay, Gulf of Kachchh, A&N Islands, Lakshadweep Archipelago	Intertidal to nearshore waters and deep sea	ZSI, NIO, CMFRI, Madras Government. Museum, BNHS, NCSCM and all Universities dealing with marine studies	Considerable studies are made
39.	Opisthobran-chia	1980- till date	Throughout the Indian coast especially GoM, Palk Bay, A&N Islands, Lakshadweep Archipelago	Intertidal to nearshore waters	ZSI, BNHS.	Study incomplete
40.	Bryozoa	1970- till date	East and west coasts, GOM.	Intertidal to deep sea	Cochin University of Science and Technology, ZSI	Study incomplete

[Table Contd.

Contd. Table]

Sl. No.	Taxonomic Group	Period	Survey Region	Habitat	Institutions	Remarks
41.	Ectoprocta	1908-1915	East coast	Coastal waters	ZSI,	Study incomplete
42.	Echinodermata	1743- till date	Throughout the Indian coast especially GOM, Palk Bay, GOK, A&N Islands, Lakshadweep Archipelago	Coastal and nearshore waters	ZSI, CMFRI NCSCM	Considerable studies are made in nearshore waters. Knowledge on deep water echino-derms is lacking
43.	Hemichordata	1994- till date	GOM and Palk Bay	Coastal and nearshore waters	ZSI,	Study incomplete
44.	Protochordata		GOM and Palk Bay	Coastal and nearshore waters	ZSI,	Study incomplete
45.	Urochordata	1986- till date	East and west coast of India, GOM, Palk Bay A&N Islands	Coastal and nearshore waters	ZSI, M. S. Univ., Annamalai Univ.	Study incomplete
46.	Pisces	1875- till date	Throughout Indian Seas	Coastal, nearshore and offshore waters	ZSI, FSI, CMFRI, NIO and all the Universities dealing with marine sciences	Considerable studies are made
47.	Reptiles	1931- till date	Throughout the Indian coast especially GOM, Palk Bay, GOK, Andaman and Nicobar Islands, Lakshadweep Archipelago and other turtle nesting sites	Coastal and nearshore waters	ZSI, IISc, WII	Sporadic studies are available. Comprehensive study is lacking
48.	Seabirds	1985- till date	Throughout the Indian coast especially GOM, Palk Bay, GOK	Coastal and nearshore waters	BNHS, ZSI,	Study incomplete

[Table Contd.

Contd. Table]

Sl. No.	Taxonomic Group	Period	Survey Region	Habitat	Institutions	Remarks
49.	Mammals	1852- till date	Entire Indian coasts including Gulf of Mannar, Andaman and Nicobar Islands	Coastal, nearshore and offshore waters	CMFRI, ZSI, NIO.	Sporadic studies are available. Comprehensive study is lacking

Acronyms: BNHS – Bombay Natural History Society; BSI – Botanical Survey of India; CMFRI – Central Marine Fisheries Research Institute; CSMCRI – Central Salt and Marine Chemicals Research Institute; FSI – Fishery Survey of India, GSI – Geological Survey of India, IISc – Indian Institute of Science, MSSRF – M.S. Swaminathan Research Foundation; NCSCN – National Centre for Sustainable coastal Management; NIO – National Institute of Oceanography; WII – Wildlife Institute of India; ZSI – Zoological Survey of India

Table 2: Marine biodiversity of India in comparison with global marine biodiversity

No.	Taxon/Group	Species in India	Species in World*
1.	Bacteria	4,800	
2.	Cyanophyta	936	1,000
3.	Chlorophyta	2,500	
4.	Phaeophyta	1,600	
5.	Rodhophyta	6,200	
6.	Seagrass	14	60
7.	Mangroves	69	100
8.	Other Protoctista	23,000	
9.	Bacillariopyta (Diatoms)	200	5000
10.	Protozoa	2577	
11.	Dinophyceae (Dinoflagellates)	90	
12.	Euglenophyta	250	
13.	Chrysophyceae	500	
14.	Dinomastigota	4,000	
15.	Radiolaria	550	
16.	Porista	532	
17.	Foraminifera	500	10,000
18.	Porifera	549	8838
19.	Cnidaria	1445	11552
20.	Actinaria	61	1000
21.	Hydrozoa	212	
22.	Scyphozoa	37	200
23.	Cubozoa	5	
24.	Anthozoa	980	2135
25.	Siphonophora	89	
26.	Ctenophora	19	199
27.	Platyhelminthes (Polyclads)	44	15000
28.	Arthropoda Crustacea	2394	44950
29.	Copepoda	540	14000
30.	Ostracoda	120	8000
31.	Branchiura	5	
32.	Cirrepedia	104	
33.	Malacostraca	474	
34.	Mysidacea	75	780
35.	Cumacea	71	1300
36.	Tanidacea	1	900
37.	Isopoda	33	

[Table Contd.

Contd. Table]

No.	Taxon/Group	Species in India	Species in World*
38.	Amphipoda	132	
39.	*Euphausiacea*	23	
40.	Stomataopoda	*139*	
41.	Decapoda	*986*	
42.	Macrura	*110*	
43.	Brachyura	*705*	
44.	Anomura	*162*	*1069*
45.	Nemertina	5	1230
46.	Gnathostomulida	4	97
47.	Rhombozoa	?	82
48.	Orthonectida	?	24
49.	Gastotricha	38	400
50.	Rotifera	3	50
51.	Ciliata	21	
52.	Kinorhyncha	4	130
53.	Loricifera	39	
54.	Acanthocephala	229	600
55.	Cycliphora	1	
56.	Entoprocta	10	170
57.	Nematoda	191	12,000
58.	Nematomorpha	5	
59.	Ectoprocta	10	5700
60.	Phornida	3	11
61.	Brachiopoda	8	391
62.	Mollusca (Except Opisthobranchs)	3400	52525
63.	Opisthobranchia	353	4000
64.	Bryozoa	200	5649
65.	Priapulida	8	
66.	Siponcula	38	147
67.	Echiura	43	176
68.	Annelida Polychaeta	883	12,000
69.	Achianellida	20	90
70.	Tardigrada	10	214
71.	Chelicerata (Pycnogonida)	36	2,267
72.	Crustacea	2394	44,950
73.	Pogonophora	148	
74.	Echinodermata	779	7,550
75.	Chaetognatha	32	121

[Table Contd.

Contd. Table]

No.	Taxon/Group	Species in India	Species in World*
76.	Protochordata	464	3061
77.	Hemichordata	14	139
78.	Urochordata	6	30
79.	Cephalopchordata	14	130
80.	Pisces	*2629*	*16475*
81.	Reptiles	*37*	*10450*
82.	Birds	*187*	*10357*
83.	Mammalia	*26*	*110*
84.	Fungi	*500*	
85.	Total marine fauna	20,444	291627
86.	Total marine flora and fauna	21,663	336447

Table 3: Protected marine faunal species in India

Marine Faunal Group	Number of Protected species	Wild Life Protection Act 1972
Porifera		
Calcareous sponges (All species)	10	Schedule III
Coelenterata		
Reef building corals (All scleractnians)	609	Schedule I
Black corals (All antipatharians)	8	Schedule I
Organ pipe coral (Tubipora misica)	1	Schedule I
Fire coral (All Millepora species)	5	Schedule I
Sea Fan (All Gorgonians)	86	Schedule I
Arthropoda		
Robber Crab (Crustacea)	1	Schedule I
Horseshoe crab (Merostomata)	2	Schedule IV
Mollusca		
Gastropoda	20	Schedule I,IV
Bivalvia	4	Schedule I,IV
Echinodermata		
Sea cucumber (All Holothurians)	163	Schedule I
Fishes		
Elasmobranchs (Shark and Rays)	10	Schedule I
Sea horse (All Syngnathidians)	23	Schedule I
Giant Grouper	1	Schedule I
Reptiles		
Marine Turtles	5	Schedule I
Saltwater crocodile	1	Schedule I
Mammals		
Marine Mammals	26	Schedule I, II
Total	**974**	

The six East India Companies (British, French, Germany, Portuguese, Dutch and Spanish) were much interested in identifying commodities that could be profitably exported to west and find substitutes for drugs. The scientific community in the west was excited about the chances that the vast landmass of India offered for scientific studies. This had led to many scientific expeditions in the seas around India *viz.* The British Challenger Expedition (1872-76), The German Expedition Valdivia (1898-99), The Dutch Expedition Siboga (1899-1900), The H.M.S. Investigator, John Murray Expedition (1930-34), Dana (1928-30), Galathea (1950-52), Albatross (1950-52) and the International Indian Ocean Expedition (1959-65). The numerous surveys (Vema, Argo, Horizon, Pioneer, Chain, Vega, Anton Brunn, Discovery, Challenger II, Vityaz, Meteor, Diamantina etc.,) conducted by marine scientists conducted by surgeon naturalists J. Armstrong Marine Survey of India's First Surgeon Naturalist (1874-1881), Lt. Col. A.W. Alcock Surgeon Naturalist (1888-1892),: Lt. Col. R.B.S. Sewell (1910-1926) etc. of several countries has culminated in 250 years of marine biodiversity scenarios in India.

Marine Survey of India

The Hydrographic Department of the Indian Navy derives its origin from the charting activities of the British East India Company, way back in the17th century. John and Samuel Thornton, Hydrographers in the East India Company compiled the first chart and Sailing Directions for the Indian Ocean in 1703. After thc dissolution of the East India Company, the Indian Marine Survey Department was established at Calcutta in 1874, which became a part of the Royal Indian Marine Survey in 1882. Marine Survey of India was merged with Zoological Survey of India in 1925. Meanwhile, the Hydrographic Surveying Service which in the beginning of 1950 was known as the Marine Survey of India and was headed by a hydrographic surveyor designated as *'Surveyor-in-Charge'* was changed to Naval Hydrographic Office on August 15, 1954.

Asiatic Society of Bengal

Sir William Jones in 1784 founded the Asiatic Society of Bengal or, as it was then called, 'Asiatic Society', whose area of investigation formed the geographical limits of Asia. Significant investigations were undertaken in the country in the last quarter of the 18th century. The Asiatic Society of Bengal was the only organization of importance in India at that time. Its zoological collections had a vital bearing on the Indian Museum with the precious heritage collections in the zoological galleries. The Asiatic Society started collecting zoological and geological specimens since 1796. The increasing collections impelled the Society to have its own building and establish a museum subsequently, which was finally achieved by setting up a museum in 1814. Dr. Nathaniel Wallich was the first Superintendent of the "Museum of the Asiatic Society" and, as well, the in-charge of the Natural History collections, including Geological and Zoological (marine) specimens, exhibits and artifacts.

In 1796, two years after Sir William Jones' death, the Royal Society announced their intention of establishing a natural history museum. In 1841 Edward Blyth (1810–1873) was brought from England as the curator. After him, John Anderson took over as the Curator in 1865. Further, in 1916, the anthropological and zoological sections of the Museum were established into Zoological Survey of India, a step of administrative significance rather than scientific.

Zoological Survey of India

While Indian Museum Authority-the Board of Trustees of the Museum is the custodian of the Zoological Galleries of the Indian Museum, the Zoological Survey of India is the guardian of the innumerable collection of zoological exhibits and artifacts in the museum. The history and progress of Zoological survey of India (ZSI) is an integral chapter in the historical development of the Indian Museum itself. The history of ZSI begins from the days of the Asiatic Society of Bengal, founded by Sir William Jones in 1784. ZSI's establishment was in fact a fulfillment of the dream of Sir William Jones, the founder of the Asiatic Society of Bengal. It was only a natural consequence of the events that the Asiatic Society should become a stronghold of the development of zoological studies. The Asiatic Society of Bengal was the mother institution not only to the Indian Museum (1875) but also to the institutions like the Zoological Survey of India and the Geological Survey of India. Although the Zoological Survey of India was officially born only in 1916, its genesis in reality had taken place in 1875 with the opening of the Indian Museum because it existed and developed as it were in embryo for many long years before it was officially born. The new museum on its inception comprised only three sections namely the Zoological, the Archaeological and the Geological. By the dedicated activity and care of the Curators of the Asiatic society of Bengal and the Superintendents of the Indian Museum, viz., McClelland, Blyth, John Anderson, James Woodmason, Alfred William Alcock and finally Thomas Nelson Annandale and his colleagues, the museum became richly endowed with a magnificent collection of animals, especially of the larger vertebrate groups. Further, additions of marine fauna to the valuable collections came through during several political and military expeditions.

Dr. Thomas Nelson Annandale, who joined the Indian museum as Deputy Superintendent in 1904, and later as the Superintendent in 1907, after strenuous years-old-struggle, achieved his aim in establishing the Zoological Survey of India in 1916. The Survey was to act *inter alia* as the guardian of the 'Standard Zoological Collections of the Indian Empire, and look after and maintain the Zoological Galleries of the Indian Museum. The Survey acquired the zoological collections which were more than a century old from the former museum of the Asiatic Society of Bengal and the Zoological Section of the Indian museum (1814-1875) in Calcutta. Zoological Survey of India thus became the custodian of the collections of zoological specimens of the Indian subcontinent, stocking and safeguarding the collections from India, Sri Lanka, Myanmar, Pakistan, Bangladesh, Nepal and Bhutan.

Marine Collections in Indian Museum

Science of preservation and maintenance of Natural History Collection today is no less prone to criticism and animosities. However, the historical perspective of the Natural History collections of major museums like Madras Museum, Chennai and Indian Museum, Kolkata is often a refreshing experience. As regards the heritage collection of the Madras Museum and Indian Museum, we are much indebted to the British naturalists (many Surgeon Naturalists), archaeologists, bio-geographers and the like. The bewildering array of heritage collections of marine exhibits and artifacts what we are seeing today in the Zoological Galleries of the Indian Museum are the product of the relentless pursuit of hobby of exploration and collection by the British pioneers and surgeon naturalists in the formative years of the Museum. They paved the way for a later professional class that proved efficient, but seemed to dedication and grit of the earlier pioneers.

Marine Protected Areas of India

Among the Asian countries, India is perhaps the only one that has a long record of inventories of coastal and marine biodiversity dating back to at least two centuries. Four MPAs are located along the east coast of India. There are four MPAs covering three separate sites along the west coast of India in the Western Indian Ocean zone. There are three MNP in the East Bay of Bengal, Mahatma Gandhi Marine National Park and Rani Jhansi Marine National Park (MNP) located in the Andaman Islands and Gulf of Mannar Marine National Park located in the south east mainland coast. There are two MNPs in the Northern Bay of Bengal zone (Sundarban National Park and Bhitar Kanika Wildlife sanctuary in India. In India protected areas along coast cover 2.4 % of the country (two national parks and 17 other protected areas) and island protected areas cover 0.3 % of the country (six national parks and 100 other protected areas), mainly in the Andaman Islands. To protect the nature as well as the marine biodiversity of the wild, 700 areas are declared as protected areas till now in India (including 103 National Parks, 528 sanctuaries, four community reserves and 65 conservation reserves and 17 Biosphere Reserves).

Protected Marine Faunal Species

According to the series of updated notifications issued under the Wildlife (Protection) Act 1972, by Ministry of Environment, Forests and Climate Change (MOEFCC), Government of India, so far 974 species (Table 3) of marine fauna belonging to eight Phyla Viz. Poriferea, Coelentrata, Arthropoda, Mollusca, Echinodermata, Pisces, Reptilia and Mammalia are well protected under the legislation of different categories (Schedule I, II, III, IV) as their natural population is in a depleting mode.

International and regional initiatives relating to MPAs

India is party to the World Heritage Convention. The only marine World Heritage Site in Asia is the Sundarbans National Park in India. India is also a party to the Ramsar Convention. Chilka Lake is a marine Ramsar site and three more sites including Point Calimere, Khijadia and Gulf of Kachchh have been proposed.

UNESCO Man and the Biosphere Programme

India has a national MAB Committee. No marine biosphere reserve has been designated but a number of them have been proposed for India, including the Gulf of Kachchh, Gulf of Mannar, Great Nicobar, Chilka Lake and the Sundarbans.

UNEP Regional Seas Programme

A Regional Sea Programme was started in 1983 for the south Asian Seas Region covering Pakistan, India, Bangladesh, Sri Lanka and the Maldives. But there is no Convention or Action Plan yet.

Gaps in the Current Systems

Policy

So far, the national committee on Mangroves and coral reefs had priorities more towards conservation of mangroves than coral reefs. MoEFCC started giving emphasis to coral reef in the past decade. Coral being a wild animal is subject of Ministry of Environment Forests and Climate Change and Wildlife Department and coral related fish is a subject of Ministry of Agriculture, Department of Fisheries, these two agencies function for the same ecosystem, one towards restricting access and the other towards increasing access by issuing fish catch. More coordination with other Ministries is essential for reducing the duplication of projects and funds for the same work.

Enforcement Agencies

Wildlife officers who manage marine parks are trained foresters and not marine zoologists and hence have little expertise or understanding of the fragility and value of marine biodiversity and protected species. Since Wild Life managers are posted for a specific tenure, officers get transferred after receiving training on management and monitoring of marine parks. The management also lacks staff trained in monitoring and management of marine biodiversity. They do not have sufficient facilities to chase poachers. Other agencies such as the Department of Tourism, Coast Guard, and Customs Department have a limited knowledge and awareness and status of marine ecosystems and their value. They also lack in trained staff for identifying endangered species that are protected by law and need constant training/orientation.

Research Institutions

Institutions such as the NIO, NCSCM, CMFRI, ICMAM and ZSI have a clear mandate to carry out research in marine biodiversity but they are hampered by some weaknesses. They do not have trained researchers to carry out taxonomy work using SCUBA diving. There is little or no interaction between fisheries biologists and marine specialists. Ministry of Environment, Forests and Climate Change funded biophysical monitoring surveys of sea grass, mangroves and coral reefs and are being carried out without conducting socio-economic surveys and hence, the data cannot be utilized for management purposes. Not much work is done on the community involvement or on the creation of awareness for conservation of marine biodiversity, either directly by the MoEFCC or by the research institutes. Coral reef research in these institutes is based on the dedication of individual scientist and once the scientist is transferred or retires the work does not get carried forward.

Non-Governmental Organizations

The greatest weakness in the NGOs is that their work is project based and the continuation of coral reef research depends on the ability to generate funding and trained manpower. Though in India the NGO movement has evolved, and Government agencies are now paying attention to the work and recommendations of the NGOs, they still have a very limited role in coral reef conservation.

Conclusion

In conclusion, it is evident that comprehensive taxonomic coverage of the marine biota of the entire region remains a monumental task, beyond the capacity of existing local taxonomic expertise. Thus, to gain an appreciable knowledge on the patterns of diversity in the region, it is necessary to identify the indicator species in response to unpredictable climate change. It would be quite appropriate if we made a careful placement of systematic studies rather than the present system of haphazard and opportunistic description of new species as and when they are discovered as remarked by Griffiths (2005). Only a few decades ago we started realizing the importance of marine life. The pressure on natural habitats associated with increasing population and economic growth will lead to the loss of biological diversity. Recognition of the scale of problem, the nature of the underlying causes and the limited resources available to counteract powerful destructive trends and ameliorating the same will definitely be a best way of conserving the biological diversity of the Marine Ecosystems of India. Let us hope it is not too late.

References

Griffiths, C.L., 2005. Coastal marine biodiversity in East Africa. Indian Journal of Marine Sciences, 33: 35-41

Keesing, J. and Irvine, T., 2005. Coastal biodiversity in the Indian Ocean: the known, the unknown and the unknowable. Indian Journal of Marine Sciences, 34(1): 11-26.

Venkataraman, K. and Wafar, M., 2005. Coastal and marine biodiversity of India. I. J. Mar. Sci., 34(1): 57-75.

Venkataraman, K., and C. Raghunathan, 2015. Marine Biodiversity in India, Status and Threats, p.1-47. (in Lesser Known Marine Animals of India 550 pp.) (Published by Director, Zool. Surv. India, Kolkata)

Impact of Climate Change on Hydrological Cycle, Ecosystem, Fisheries and Food Security, Pages: 301–319
Edited by: B. Madhusoodana Kurup, M.R. Boopendranath, M. Harikrishnan and A.V. Shibu

CHAPTER - 26

Fish Biodiversity of Western Ghats in the Context of Climate Change and Strategies for Conservation

K.V. Radhakrishnan[1]* and B. Madhusoodana Kurup[2]

[1]*Dept. of Fisheries Resource Management, College of Fisheries, Central Agricultural University (I), Agartala, Tripura-799210, India*

[2]*Founder Vice Chancellor, Kerala University of Fisheries and Ocean Studies, Kochi-682506, India; Former Vice Chancellor, Shri Venkaeshwara University, Gajraula-244236, U.P., India*

* *E-mail: krishnaradh76@gmail.com*

Climate change is the most important challenge before mankind and has serious implications for future water availability, food production, ecosystems health and well-being and even the survival of human race. Huge data collected, collated and compiled by UNFCC and IPCC indicate that the globally averaged effect since 1750 has been one of warming and by 2100 the average temperature of the world will rise by 1.8°C to 4.0°C (IPCC, 2007). Sea level is expected to rise by 18 to 59 cm by the end of 21st century. Other major changes projected are acidification of the oceans, reduced snow cover and sea ice, more frequent heat waves and heavy rainfall events, more intense tropical cyclones and slower oceanic currents. The warming shall continue over centuries and may lead to a complete meltdown of the Greenland ice-sheet increasing global sea level by about 7m (IPCC, 2014). Recent studies showed that the summer monsoon rainfall during 1901–2012 exhibited a weakening trend over parts of South Asia (Roxy *et al.*, 2015). Reduction in rainfall is linked to the rapid warming of the Indian Ocean, especially its western part, during the past century. There is a threefold increase in widespread extreme rain events over central India during 1950–2015 (Roxy *et al.*, 2015). The regional impact of global climate change shall be different in type and magnitude in the different regions, however, more severe impact is predicted in the tropical areas (mostly developing and poor countries) including India. The result of this warming which is already being felt in many parts of the world shall be heat waves, increased frequency of heavy precipitation, increased frequency of tropical cyclones, water and food scarcity, increased abundance of Malaria, dengue and other diseases, severe loss of biodiversity which will have serious social, economic, political and environmental consequences.

Sea-level rise, changes in precipitation patterns, and increased frequencies of extreme events etc. will have their own impact on biological diversity and is in fact threats to biodiversity all around the world by disrupting delicate relationships within ecosystems. Loss of biodiversity seems to be implicit in most global warming scenarios. According to the IPCC (IPCC, 2001), the range and abundance of plants and animals could change dramatically under changing climatic conditions, and some species are likely to be unable to adapt or migrate to new locations. Most plants and animals can tolerate only a narrow range of ambient temperature. If the temperature varies significantly from this range, normal physiological functioning breaks down.

Climate Change in India

The impact of climate change on developing nation is highly significant. Asia will be particularly vulnerable to the climate change, because of its concentration of major population concentrated at low elevations. It is reported that five most vulnerable countries with large populations are China, India, Bangladesh, Vietnam and Indonesia (Sundaresan, 2011). According to the majority of IPCC's Climate models, rainfall over India will become more and more erratic in nature with a plausible decrease in overall rainfall and increase in extreme weather events (Christensen *et al.*, 2013). The implications of these potential effects range from changes in ocean chemistry, forecasted sea level rise, melt of glaciers, human health and the displacement of coastal people. India has responded to the challenge of climate change in 2008 by framing the National Action Plan on Climate Change (NAPCC). It envisaged sustainable development goal coupled with long-term "climate change co-benefits". Further, the document laid out roadmap of the goals through eight national missions focusing on major thematic areas like agriculture, water, forestry, energy and urban planning. In furthering its objectives, the states were asked to prepare State Action Plans on Climate Change (SAPCC) keeping the perspective of creating an implementation framework for the NAPCC and its alignment with specific development priorities by duly considering impacts, vulnerabilities and adaptation needs. As such very few studies have been carried out on climate change in India, most of the information is available from the reports of Indian Meteorological Department (IMD), other studies available on this aspect are of Kriplani *et al.* (1996), Lal (2003), IMD (2002-2006) and most recently by Rajendran *et al.* (2012), Krishnan *et al.* (2012), Roxy *et al.* (2015), Abhilash and Krishnakumar (2017), Abhilash *et al.* (2018) and Varikoden *et al.* (2018).

Western Ghats - the Biodiversity Cradle of India

The Western Ghats, also known as Sahyadri (Benevolent Mountains), are a mountain range that covers an area of 140,000 square kilometres in a stretch of 1,600 kilometres parallel to the western coast of the Indian Peninsula, traversing the states of Kerala, Tamil Nadu, Karnataka, Goa, Maharashtra and Gujarat (Molur *et al.*, 2011). The Western Ghats is recognised as a UNESCO World Heritage Site and is one of the eight "hottest hot-spots" of biological diversity in the world and sometimes called the Great Escarpment of India (Myers *et al.*, 2000). The region contains a large proportion of the country's flora and fauna, many of which are only found in India (Molur *et al.*, 2011). The Western Ghat mountains influence Indian monsoon weather patterns by intercepting the rain-laden monsoon winds that sweep in from the south-west during late summer (Varikoden *et al.*, 2018). The range runs north to south along the western edge of the Deccan Plateau, and separates the plateau from a narrow coastal plain, called Konkan, along the Arabian Sea. Thirty-nine areas in the Western Ghats, including national parks, wildlife sanctuaries and reserve forests, were designated as world heritage sites in 2012 - twenty in Kerala, ten in Karnataka, five in Tamil Nadu and four in Maharashtra (Molur *et al.*, 2011). There are over 7,402 species of flowering plants, 1,814 species of non-flowering plants, 139 mammal species, 508 bird species, 179 amphibian species, 6,000 insects species and 290 freshwater fish species; it is likely that many undiscovered species in the Western Ghats. At least 325 globally threatened species occur in the Western Ghats (Myers *et al.*, 2000; Dahanukar *et al.*, 2004; Nayar *et al.*, 2014).

The Western Ghats form one of the four watersheds of India, feeding the perennial rivers originating from these mountains, including Godavari, Kaveri, Krishna, Thamiraparani and Tungabhadra. The majority of streams draining the Western Ghats join these rivers, and carry a large volume of water during the monsoon months. These rivers flow to the east due to the gradient of the land and drain out into the Bay of Bengal. Periyar, Bharathappuzha, Pamba, Netravati, Sharavathi, Mandovi and Zuari rivers flow westwards towards the Western Ghats, draining into the Arabian Sea, and are fast-moving, owing to the steeper gradient. Heavy precipitation does occur in the surrounding plains; changes in the direction and pace of the wind do affect the average rainfall and the wettest places might vary. However, Maharashtra and the northern part of Western Ghats in Karnataka on average receive heavier rainfall than Kerala and the southern part of Western Ghats in Karnataka. Rivers from Western Ghats drain almost 40% of Indian drainage. Therefore, it is essential to understand the impacts of the climate change on water resources in Western Ghats.

The Western Ghats Biodiversity Hotspot is endowed with high levels of endemism for freshwater taxa such as crabs (92%), amphibians (87%), shrimps (69%) and fish (59%) with majority of the endemic species localised at the southern region, within the state of Kerala (Kurup *et al.*, 2004). The unique annual and seasonal distribution of rainfall across Kerala and along the Western Ghats region in particular is mainly accountable for the exceptional repository of variety of Flora and Fauna. The region has 13 single location endemic species (Raghavan, 2019), restricted to peculiar microhabitats characterized by specific flow, velocity, depth, entrenchment ratio, light intensity etc. (Kurup *et al.*, 2004).

Habitat fragmentation is considered to be responsible for change in local biodiversity of the river, in recent years there are increasing reports of pollution causing massive fish kill, soil erosion and siltation, and agriculture expansion in the catchment area. River valley projects together with deforestation, mining, and changing agriculture patterns. Activities are responsible for degradation of riparian habitats. Developmental activities, mainly construction of dams and weirs, water level fluctuations, industrial and urban pollution, and introduction of exotic species, sand mining, bank soil extraction are major threats responsible for rapid decline in natural riparian habitats and local biodiversity.

Climate Change in Western Ghats

Change in the pattern of precipitation

Several recent studies indicated significant weakening of the large-scale Asian summer monsoon circulation, especially in the middle to upper atmosphere (Sooraj and Terray, 2014) Future regional climate model projections of summer monsoon also suggests further weakening of monsoon circulation and rain fall over peninsular India Abhilash *et al.* (2018). The Western Ghats mountain chain acts as the boundary to the elevated Deccan Plateau and interacts with the southwest monsoon circulation forming exceptionally varied climate patterns over the peninsular region (Gunnell, 1997). The onset of southwest monsoon also occurs at this region, and thus Western Ghats plays a key role in regulating Indian climate through regional climate modulations (Varikoden *et al.*, 2018). Rajendran *et al.* (2012) observed a substantial reduction in rainfall and a significant decreasing trend in seasonal mean rainfall in Western Ghats region particularly over the last half century. Based on the simulations of high resolution regional climate models, Krishnan *et al.* (2012) have projected reduced orographic precipitation especially over the Southern parts of the Western Ghats falling in Kerala and Karnataka, and increased rainfall over the interior regions of the Indian subcontinent. Most recently, Varikoden *et al.* (2018) applying Daily gridded IMD (India Meteorological Department) rainfall data for a period from 1901 to 2015 during the summer monsoon period found that the Western Ghats plays a key role in regulating Indian climate through regional climate modulations substantiating the earlier findings and projections. The Indian Network for Climate Change Assessment (INCCA, 2010) also predicts that the northern portion of the west coast, consisting of areas of Gujarat and Maharashtra, will show an increase in precipitation varies from 4% to over 25% for the 2030s, however, areas of Karnataka and Kerala will show a marginal decrease up to 4%. The reduction in water yield for Karnataka and Kerala is up to about 10%. Gujarat and Maharashtra areas see an increase in water yield, and the magnitude is up to about 50%. These results have important societal implications in the wake of concerns over the preservation of ecologically fragile Western Ghats region (Gadgil *et al.*, 2011; Kasturirangan *et al.*, 2013). The decreased precipitation over the southern regions of the Western Ghats could lead to a water stress in those places. In the northern regions, possibilities of heavy precipitation events are high, in the wake of increased wind speeds and higher moisture content of the air in a warmer climate. The South Asia is already identified as a region with high flood risk under projected warming scenarios (Hirabayashi *et al.*, 2013). The projected changes in the precipitation may induce changes in the hydrological regimes especially increase in evapo-transpiration and increased runoff (NATCOM, 2012).

Increasing temperature

In addition to rainfall, Sandeep and Ajayamohan (2015) observed a drying (wet) trend in the western coast of India during the last three decades. The authors found that the land surface experiences a faster rate of heating compared to the Oceans, as the atmosphere warms under the influence of enhanced Green House Gases (GHGs). The land-sea temperature contrast over Indian monsoon region will amplify in a global warming scenario. The INCCA Assessment (INCCA, 2010) predicts that the annual temperatures are likely to increase to 26.8°– 27.5°C in the Western Ghats region by 2030s. The rise in temperature with respect to the 1970s will be between 1.7° C and 1.8° C. The minimum temperatures may rise by 2.0° to 4.5° C, with minimum increase in those parts of Karnataka that lie in the Western Ghats. Within the region bordering the state of Kerala, the maximum temperature is likely to rise by 1°-3° C.

Increasing sediment yield in water bodies

The INCCA Assessment (INCCA, 2010) predicts that the west coast region also shows a considerable increase in the sediment yield for majority of the areas. Even those areas that are expected to receive less precipitation show an increase in sediment yield of up to 25%. The increase in sediment yield in these areas can possibly be explained due to an increase in the intensity of precipitation. This will have major impacts on water resource projects.

Increasing flood and draught events

Flood Analysis According the INCCA Assessment (INCCA, 2010), all the regions show an increase in the flooding varying between 10 to over 30% of the existing magnitudes. This has a very severe implication for the existing infrastructure such as dams for the areas and shall require appropriate adaptation and dam safety and operation measures to be taken up. It is also seen that there is an increase in the moderate drought development for Krishna, Pennar, and Cauvery basins, which have either predicted decrease in precipitation or have enhanced level of evapo-transpiration. The maximum water withdrawal takes place from Godavari and Krishna river basins in Western Ghats in all the years. The west coast region shows a general reduction in Evapo-transpiration (ET), which varies from a very nominal value to about 5% for the 2030s scenario. Areas of Gujarat and Maharashtra, which had shown an increase in precipitation, still show a reduction in ET perhaps because of high intensity of the rainfall. The climate change is expected to increase threats to biodiversity and causes large scale depletion in stock or species losses in Western Ghats. Changes in phenology are expected to occur for many species (NATCOM, 2012).

Fish Biodiversity in Western Ghats and Potential Impacts of Climate Change

As of 2011, 290 species of freshwater fishes belonging to 11 orders, 33 families and 106 genera are recognized from the Western Ghats (Dahanukar *et al.*, 2011). Several new species have been described from the region from 2000 onwards (e.g., *Pseudolaguvia austrina*, Radhakrishnan *et al.*, 2010; Dario urops, Britz *et al.*, 2012). The Western Ghats also has a rich endemic fish fauna of 189 species, belonging to seven orders, 23 families and 69 genera. Twelve genera, *Betadevario*, *Dayella*, *Horabagrus*, *Horalabiosa, Hypselobarbus*, *Indoreonectes*, *Lepidopygopsis*, *Longischistura*, *Mesonoemacheilus*, *Parapsilorhynchus*, *Rohtee* and *Travancoria*, are endemic to the Western Ghats (Dahanukar *et al.*, 2011).There is a higher fish richness in the southern part of the Western Ghats than in the northern (Molur *et al.*, 2011), and the highest is in the Chalakudy River, which alone holds 98 species (Raghavan *et al.*, 2008). Other rivers with high species numbers include the Periyar, Bharatapuzha, Pamba and Chaliyar, as well as upstream tributaries of the Kaveri, Pambar, Bhavani and Krishna rivers (Molur *et al.*, 2011). Cypriniformes (178 species) is the most species rich order followed by Siluriformes (50 species), while the species rich families are Cyprinidae (72 species); Balitoridae (34 species); Bagridae (19 species); and Sisoridae (12 species) (Dahanukar *et al.*, 2011). The region is home to several brilliantly coloured ornamental fishes like the Denison barb (or red line torpedo), melon barb, several species of Dawkinsia, zebra loach, *Horabagrus* and pufferfishes. The rivers are also home to Osteobrama bakeri, and larger species such as the Malabar snakehead (Beniger *et al.*, 2011) and different species of mahseer. A few are adapted to subterranean life, including swamp eels (Gobi *et al.*, 2002), and the catfish Horaglanis and Kryptoglanis (Vincent and Thomas, 2011). According to the IUCN, 97 freshwater fish species from the Western Ghats were considered threatened in 2011, including 12 critically endangered, 54 endangered and 31 vulnerable (Dahanukar *et al.*, 2011). An additional 26 species from the region are considered data deficient (their status is unclear at present). The primary threats are from habitat loss, but also from overexploitation and introduced species (Dahanukar *et al.*, 2011).

Since there is no comprehensive study so far been taken up focusing on the impact of climate change on the fish biodiversity in any of the water bodies of Western Ghats region, some potential impacts based on the observation from survey and sampling of fishes conducted by the authors in different water bodies of Western Ghats in Kerala during 2000-2005 period and consulting recent literature on fish diversity and biogeographic distribution of fishes in Western Ghats is briefed below. In addition, the information and knowledge generated by thoroughly going through the literatures on the impact of climate change on the fish diversity globally and India, in particular, is also incorporated to derive plausible conclusions.

Reduction in Species Diversity

Climate change will induce tropical river systems to have faster water loss to the atmosphere (Jacobs, 1993) and tropical lakes inundated during flood stages may become isolated from the main channel during the dry season (Val and Almeida-Val, 1995). In high latitude rivers, the expected increase in precipitation will result in an earlier, smaller flood pulse (ACIA, 2004) and later, these stream systems may experience lower minimum flows (Nijssen *et al.*, 2001). Smaller flood stages would decrease the allochthonous input of material in to the aquatic system. In addition, a smaller inundated floodplain would also translate in to reduced spawning habitat. Floods or exceptionally large seasonal flows can displace adults and displace or injure juveniles and larvae (Harvey, 1987), and droughts or prolonged dry spells can cause population declines, reduced abundance, and altered species composition (Matthews and Marsh-Matthews, 2003). Increased evapo-transpiration could reduce the volume and quality of available fish habitat during the dry season. Western Ghats region is a biodiversity hotspot, encompassing a lot of diversified fishes and their microhabitats. The varying pattern of precipitation and subsequent flood and draughts will negatively impact the fishes, leading to their large scale reduction in diversity.

Impact on Endemic and Localised Fishes

Endemic species, species in fragmented habitats, or those in east-west oriented systems are naturally prone to extinction (Hanski,1994; Angermeier, 1995; Ashley *et al.*, 2007). Therefore, another possible effect of climate change is the loss of biodiversity through the extinction of specialized or endemic fish species. Fishes living in lotic systems have adapted to a specific set of hydrologic parameters; changes in these conditions can result in extirpation of locally specialized species (Cross and Moss, 1987). There are 189 endemic species recorded in the aquatic systems of Western Ghats, including several point endemics (Radhakrishnan and Kurup, 2010; Raghavan *et al.*, 2008. The drastic effect of climate change of climate change thus will have a stronger impact on endemic fish diversity of this region.

Impact on Threatened Fishes

With a temperature increase of only 1-2° C, disturbances such as deforestation, pollution, and overfishing would likely pose a larger and more immediate threat to fish populations

(Abell and Allan, 2002; Dudgeon, 2003). There are 103 fishes in Western Ghats coming under threatened category (CR-12, EN-54, VU-31, NT-6) (Dahanukar *et al.*, 2011) of which 99 are endemic. The distribution of threatened species (Figure 3.6) shows the Periyar, Chalakkudy and Pambar rivers at the Southern Western Ghats have the highest number of threatened species. The climate change will aggravate the rate of endangering the fish population already under threat.

Proliferation of Exotic Fishes

Fishes living in lotic systems have adapted to a specific set of hydrologic parameters; changes in these conditions can result in increased success of invasive species (Ross *et al.*, 2001). Changes in variability could select for generalist species or those with the ability to rapidly colonize degraded habitats and possibly lead to a loss of locally adapted ones (Poff *et al.*, 2001). Under a changing climate, alien species would expand in to regions in which they previously could not survive and reproduce (Walther *et al.*, 2009). Invasive species have already been identified as another major threat affecting the native fish biodiversity in Western Ghats (Knight 2010). Thirteen species of exotic fish, including *Clarias gariepinus*, *Cyprinus carpio*, *Oncorhyncus mykiss*, *Pangasianodon hypophthalmus*, *Oreochromis niloticus*, *O. mossambicus*, *Osphronemus goramy*, *Pterygoplichthys multiradiatus*, *Piaractus brachypomus*, *Trichopodus trichopterus*, *Xiphophorus maculatus*, *Poecilia reticulata* and *Gambusia affinis*, are recorded in river systems of the Kerala part of the Western Ghats, an area showing both highest number of endemics and threatened species (Dahanukar *et al.*, 2011).

Change in Distribution Pattern of Fishes

Because increased temperatures can have sub-lethal effects on tropical fish, it is likely that tropical species will experience a pole-ward expansion in distribution as the thermal isolines are shifted (Mulholland *et al.*, 1997). Our own sampling surveys in different river systems in Kerala part of Western Ghats also observed that the cyprinid minnows predominantly seen at the middle and lower stretches became proliferated even at the upper stretches of rivers and downstream fishes like *Glossogobius giuris*, *Xenentodon cancila* showed extended distribution to upper stretches. Vaas *et al.* (2009), Sarkar *et al.* (2012) and Das *et al.* (2013) also reported that in Ganga River System, a number of fish species which were earlier never reported in the upper stretch and were predominantly only available in the lower and middle stretches were recorded from the upper cold-water stretch.

Change in Breeding and Recruitment Pattern of Fish

Climate change may affect the fish diversity in lotic systems by reducing their recruitment (Mion *et al.*, 1998; Modde *et al.*, 2001) rate. Because the aerobic metabolic rates of most cold-blooded aquatic organisms increase with temperature, an increase in temperature both decreases the DO supply (through reduced saturation concentrations relative to air) and increases the biological oxygen demand (BOD) (Kalff, 2000). Decreased levels of dissolved oxygen can reduce the growth and reproductive success of individual fishes and prevent them from dealing as effectively with other environmental challenges.

Reduction in Growth Rate

If climate change results in reduced levels of dissolved oxygen, it is possible that individual fish could be smaller, less fecund, and less likely to survive. Higher temperatures and a fixed ration would result in decreased growth rates by increasing in metabolism. Temperature changes will have an impact on the suitability of species for a given location. In temperate areas increasing temperatures could bring the advantages of faster growth rates and longer growing seasons (Vaas *et al.*, 2009). McCauley and Beitinger (1992) observed higher growth rate for Channel catfish with every 1°C rise in temperature up to 30°C.

Change in predator-prey relationship

Though prey fish could be more susceptible to capture in an increasing temperature scenario, harsh abiotic conditions in isolated habitats could also lead to increased fish death and reduced survival of these fishes (Meisner, 1992). Vaas *et al.* (2009), Sarkar *et al.* (2012) and Das *et al.* (2013) observed that the predator prey ratio in the middle stretch in the river Ganga had declined from1:4.2 to 1:1.4 in the last three decades and one of the reasons might be Climate change. Western Ghats is rich in different prey fish species; hence similar effect of change in predator-prey relationship could be anticipated in this region as well.

Increased Susceptibility to Poisoning and Diseases

The toxicity of common pollutants (e.g., and heavy metals) to fish generally increases at higher temperatures (Murty, 1986). Climate change may alter host-parasite dynamics by changing transmission opportunities and host susceptibility (Marcogliese, 2001). Changing global temperatures will also affect fish susceptibility to disease because the immune functions of fish is compromised in the presence of stressors, including crowding, high temperatures, and osmotic stress. Most of the hillstream fishes in Western Ghats are highly susceptible to disease and death due to chemical pollution. Recent among them are in Nilgiris, Mysore and Kodagu regions (Pandey and Das, 2002), Valparai in the Anamalai Hills (Raghavan *et al.*, 2008), Ashambu hills ((Abraham *et al.*, 2011) etc. Under the scenario of changing climate, the susceptibility to disease and toxicity of pollutants may further increase in Western Ghats.

Deterioration of Fish Habitats

In lentic systems, increased temperatures will likely result in a general increase in lake trophic status (Allan *et al.*, 2005). Lower stream flows could increase water residence times and reduce flushing of nutrients from lake systems, thereby increasing trophic status (Bertahas *et al.*, 2006). When coupled with the input of anthropogenic pollutants, climate change can accelerate the eutrophication process (Klapper, 1991; Adrian *et al.*, 1995) or delay recovery from anthropogenic eutrophication (Nicholls, 1998; Jankowski *et al.*, 2006). An increase in global temperatures will strengthen stratification in larger lakes because increased heating of the epilimnion (Gerdaux, 1998). Lake mixing is essential for the movement of oxygen to the hypolimnion and nutrients to the epilimnion. More pronounced and longer-lasting stratification will reduce the amount of oxygen exchange to the hypolimnion from the oxygen-rich epilimnion. When the oxygen demand in the hypolimnion exceeds the supply, hypoxic or anoxic conditions will occur.

Specific Case of Southern Western Ghats

In western Ghats, the highest species richness is found in the river drainages of southern Western Ghats including the Periyar, Chalakkudy, Bharatapuzha, Pamba, Chaliyar and upstream tributaries of Cauvery, Pambar, Moyar and Bhavani rivers (Dahanukar *et al.*, 2011). Of the total 39 protected areas in Western Ghats (national parks, wildlife sanctuaries and reserve forests), 30 are in southern region. The highest level of endemism (between 77-102 species per sub-basin) is also found in the west flowing rivers Chaliyar, Bharatapuzha, Chalakkudy, Periyar, and Pamba in the Kerala part of the Western Ghats (Dahanukar *et al.*, 2011). In the World Bank technical report, streams of southern Western Ghats, especially in the state of Kerala have been reported as one of the few sites in the world showing exceptional biodiversity and high degree of endemism of freshwater fishes (Kottelat and Whitten, 1996). The Kerala part of the Western Ghats region is the most richest in endemic species and is the most diverse area in the entire Western Ghats area (Kurup *et al.*, 2004). The water bodies in this region encompass different species of endemic fishes including those restricted to single locations such as *Lepidopygopsis typus, Crossocheilus periyarensis*, *Horabagrus nigricollaris*, *Horaglanis alikunhii*, *H. krishnai*, *Mesonoemacheilus pambarensis*, *Travancoria elongata*, *T. jonesi* and *Dayella malabarica.* However, a lot of recent studies (Rajendran *et al.*, 2012; Krishnan et. Al., 2012; Abhilash and Krishnakumar, 2017; Abhilash *et al.*, 2018; Varikoden *et al.*, 2018) indicated reduced orographic precipitation especially over the Southern parts of the Western Ghats falling in Kerala and Karnataka, and increased rainfall over the Northern region as well as interior regions of the Indian subcontinent. Moreover, Sandeep and Ajayamohan (2015) observed a drying trend in the southern part of the western coast of India in the last three decades. Abhilash and Krishnakumar (2017) and Abhilash *et al.* (2018) also found that monsoon rainfall decreasing over south Kerala during recent decades, while the post-monsoon rainfall increasing whereas occurrences of longer time scale droughts are becoming more frequent in the recent decades. These authors also observes significantly increasing trend in the severity of heat waves during hot weather seasons over Kerala region, suggesting that the future climate of Kerala would be warmer (Abhilash and Krishnakumar, 2017). Change in precipitation pattern and warming climate may lead to the increase in the frequency and severity of extreme rainfall events (Abhilash *et al.*, 2018). Recently, the state witnessed unprecedented flood in the year 2018 which resulted in impacts to the morphology of many rivers. A survey conducted immediately after flood event (Raghavan, 2019) observed heavy impact on the critical microhabitats located in the middle and lower reaches of the major river systems. Large numbers of alien fish species were recorded in the major rivers and associated backwaters of Kerala after flood, many of which were encountered for the first time in the natural waters of Kerala (Raghavan, 2019). The floods have devastated the pristine riverine ecosystems and many endemic, threatened fish species are feared to have large scale reduction in population. Specific mitigation, adaptation and conservation measures is the need of the hour in order to cope up with the climate induced hazards and degrading biodiversity including fishes, in southern Western Ghats.

Knowledge Gaps and Research Needs

How can the states surrounding Western Ghats respond more effectively to climate change? Do we have the necessary knowledge to deal with the risks from climate change? How do we perceive global warming and climate change and its implications? Do we understand the linkages between the expected changes under climate change and their impact on fish biodiversity? Many such questions need to be tackled while we deal with the current impacts and prepare for an indecisive future. Considerable number of climate studies have been carried out in India. However, there has been no comprehensive study taken up specifically addressing the impact of climate change on biodiversity in any geographical region including Western Ghats. As result, most of the current information and studies could be termed as qualitative assessments and analysis. Attempts are made by modelling change in vegetation types by climate change, but not further on to specific impacts on ecosystems, ecological functions species and biodiversity. Moreover, climate change is too often seen by those responsible for biological research, conservation and even policy makers as a future problem that is currently insignificant in comparison to more immediate pressures such as pollution and deforestation. This results in the development of national conservation strategies which have not taken future climate change challenges, into account. There is an urgent need to conduct cross-sector, cross-boundary, and trans-disciplinary research on climate change and its impacts on fish biodiversity in this region and analyse the movement of climate sensitive and invasive species considering altitudinal and latitudinal aspects, protected area coverage and effectiveness to enable vulnerability assessments for effective risk reduction on our biodiversity, including fishes, under climate change. Delay in initiating and adopting resilience building measures would make the adaptation difficult and costly. It is important to practice the doctrine of sustainable development by emphasizing on resource conservation and recognizing the limits of carrying capacity of natural ecosystems for minimizing the risks under climate change.

Mitigation, Adaptation and Conservation

- There is urgent need to protect the aquatic ecosystems and their biodiversity in the Western Ghats. The conservation of these habitats is very much and directly linked to the livelihood security of the people in this region. In this regard, the River Regulation Zone (RRZ) legislation (2009) should be given wide publicity among fishermen, people or concerned government officials to realise the importance of protecting our natural resources and the ways and means to it. The directions of NAPCC and SAPCC should be strictly enforced at regional level.
- The fragile aquatic ecosystems of Western Ghats are already under stress by the ever increasing anthropogenic interventions such as construction of dams and barrages, domestic and industrial pollution, river diversions, abstraction of water and sand, agricultural encroachment and other habitat destruction activities along with the over exploitation of fish resources. Studies shows that with climate change and the subsequent warming, these disturbances would likely pose a larger and more immediate threat to fish populations. Hence it is important to take steps for reducing anthropogenic stresses on fish populations by bringing strict legislation and thus make fish diversity more resilient to climate change.
- Endemic species, species in fragmented habitats and already threatened species are more prone to further reduction in population under climate change. Western Ghats, especially the southern region encompass a large number of endemic and/or threatened fishes. Such species should be delineated and conservation plans such as standardising captive breeding technologies, habitat restoration and rehabilitation, gene banks, stock enhancement methods in natural habitats including ranching etc. could be taken up.
- The decreased precipitation over the southern regions of the Western Ghats could lead to a water stress in those places. In the northern regions, possibilities of heavy precipitation events are high. Water demand in coming years by urban and industrial sectors would increase; substantial increases in withdrawals for irrigation are expected; a growing human population will increase water demand more than will climate warming. An integrated water resource management approach should be planned and executed for judicious utilisation of water by different stake holders.
- Apparently there is no comprehensive disaster management plan or an early warning system against climate induced hazards in Western Ghats. Hence the magnitude of flood or draught events become further amplified leading to heavy loss ecosystem

services. It is important to implement and early warning system in order to cope with the climate induced hazards.

- The Ministry of Environment, Forests & Climate Change should develop an action plan for determining vulnerability and adaptation of ecological systems to climate change in Western Ghats. There is a need to develop guidelines and criteria for conservation & management measures/activities that will promote ecosystem resilience to climate change and the prescribed procedures in the manual should be incorporated in to the development planning in this region. Such a manual can be the prime tool for assisting conservationists, planners and stakeholders in their work for make adapting ecosystems and their biodiversity to the changing climate.
- Traditional fishermen have a lot of devices and techniques to cope up with the climate induced hazards and alleviate their impacts. It is very important to document such knowledge and augment research on the use of traditional knowledge in climate change, local indicators as well as local adaptation strategies.
- Considering the inherent uncertainties involved in the climate projections over Western Ghats, there is an urgent need for regional cooperation, efficient linkage between government, non-governmental and research institutions working in the field of climate change. Effective capacity building at regional and local level should be taken up and promoted to assimilate the already existing knowledge on climate change impact on various sectors like water resource, biodiversity, etc.
- Cost-effective energy efficiency and energy conservation measures can simultaneously advance development and climate change objectives. The use of bio-fuels should be encouraged, provided such fuels are not produced at the expense of food crops. Energy efficiency must be improved in the industrial sector. Promotion of renewable energy measures like wind, solar and hydropower help to reduce carbon emissions. Replacing diesel by CNG or by solar and electric vehicles can address public health concerns arising from air pollution. Green methods of urban waste disposal systems result in cutting down emissions of methane.
- Conservation of freshwater ecosystems and fish will only be fulfilled through a co-management approach, involving the communities who are directly linked with them for their livelihood. Hence, it is very important to empower the people by capacity building and mass awareness programme.

Conclusion

It is proven without doubt and with very strong confidence by IPCC that India is going to be severely hit by climate change. Freshwater resources and biodiversity are the two pillars on which our survival and sustenance is dependent. The Western Ghats is the main watershed of peninsular India from which all the principal rivers mainly Godavari, Krishna and Cauvery emerges. The Ghats also contribute over 48% of the freshwater fish biodiversity in India. The rich and diverse fish fauna in these places comprise several rare and endangered species, many of which are potentially important either as cultivable or as ornamental fish species. However, the aquatic habitats and fish in this globally recognized biodiversity hotspot is going to be seriously affected by climate change. The adverse impact of climate on freshwater systems shall aggravate the impacts of other stresses, such as population growth, economic activities, land use and urbanization and possibly a major bulk of the species shall be endangered in next 50 years. Implications of such a major loss of biodiversity are perhaps not estimated yet, but certainly the impact on our economy, and well-being and sustainability of our ecosystems shall be very high. Hence, it is very important to formulate mitigation, adaptation and conservation plans for climate risk management in concordance with NAPCC and SAPCC and take steps to reduce the vulnerability associated with climate risks and thereby increase resilience, for sustainable utilisation of fish diversity and fishery resources of Western Ghats in the long run.

References

Abell, R. and Allan, J.D., 2002. Riparian shade and stream temperatures in an agricultural catchment, Michigan, USA. Int. Ver. Theor. Angew. Limnol., Verh. 28: 232-237. doi.org/10.1080/ 03680770.2001.11902578.

Abhilash, S. and Krishnakumar, E.K., 2017. Future Climate of Kerala - Warm and Dry? Proceedings on the Workshop on Climate Variability in Kerala: Climate Change Perspectives, organized by Institute of Climate Change Research, Kerala.

Abhilash, S., Krishnakumar, E.K., Vijaykumar, P., Sahai, A.K., Chakrapani, B. and Gopinath, G., 2018. Changing Characteristics of Droughts over Kerala, India: Inter-Annual Variability and Trend. *Asia-Pac. J.* Atmos. Sci., 55: 1-17.

Abraham, R.K., Kelkar, N. and Kumar, A.B., 2011. Freshwater fish fauna of the Ashambu Hills landscape, Southern Western Ghats, India, with notes on some range extensions. J. Threat. Taxa. 3, 1585-1593.

ACIA, 2004. Impacts of a warming Arctic-Arctic climate impact assessment. Cambridge University Press, Cambridge, UK.

Adrian, R., Deneke, R., Mischke, U., Stellmacher, R. and Lederer, P., 1995. A long-term study of the Heiligensee (1975-1992). Evidence for effects of climatic change on the dynamics of eutrophied lake ecosystems. Arch. Hydrobiol., 133: 315-337.

Allan, J.D., Palmer, M. and Poff, N.L., 2005. Climate change and freshwater ecosystems. In: Climate Change and Biodiversity (Lovejoy, T.E., Hannah, L.,Eds), pp 274–295, Yale University Press, New Haven, CT.

Angermeier, P.L., 1995. Ecological attributes of extinction prone species: loss of freshwater fishes of Virginia. Conserv. Biol., 9: 143-158. doi.org/10.1046/j.1523 1739.1995. 09010143.x.

Ashley, D.F., Christopher, A., Myrick, L.J.H., 2007. Potential impacts of global climate change on freshwater fisheries. Rev. Fish. Biol. Fisheries 17: 581-613.

Benziger, A., Philip, S., Raghavan, R., Anvar Ali, P.H., Sukumaran, M., Tharian, J.C., Dahanukar, N., Baby, F., Peter, R., Devi, K.R., Radhakrishnan, K.V., Haniffa, M.A., Britz, R. and Antunes, A., 2011. Unravelling a 146 years old taxonomic puzzle: validation of malabar snakehead, species - status and its relevance for channid systematics and evolution. *PLoS One,* 6: e21272. doi.org/10.1371/journal.pone.0021272.

Bertahas, I., Dimitriou, E., Karaouzas, I., Laschou, S. and Zacharias, I., 2006. Climate change and agricultural pollution effects on the trophic status of a Mediterranean lake. Acta Hydrochim. Hydrobiol., 34: 349-359. doi.org/10.1002/aheh.200500637.

Britz, R., Ali, A. And Philip, S., 2012. *Dario urops*, a new species of badid fish from the Western Ghats, southern India (Teleostei: Percomorpha: Badidae). *Zootaxa* 3348: 63-68.

Cross, F.B. and Moss, R.E., 1987. Historic changes in fish communities and aquatic habitats in plains streams of Kansas. In: Community and evolutionary ecology of North American stream fishes (Matthews, W.J., Heins, D.C., Eds.), pp. 155-165, University of Oklahoma Press, Norman, Oklahoma.

Dahanukar, N., Raghavan, R., Ali, A., Abraham, A., Shaji, C.P., 2011. The status and distribution of freshwater fishes of the Western Ghats. In: The Status of Freshwater Biodiversity in the Western Ghats, India (Molur, S., Smith, K.G., Daniel, B.G., Darwall, W.R.T., Eds.), pp 21–48, International Union for Conservation of Nature (IUCN) Gland, Switzerland and Zoo Outreach Organization (ZOO), Coimbatore, India.

Dahanukar, N., Raut, R. and Bhat, A., 2004. Distribution, endemism and threat status of freshwater fishes in the Western Ghats of India. *J. Biogeogr., 31: 123-136.* doi:10. 1046/j.0305-0270.2003.01016.x.

Das, M.K., Sharma, A.P., Sahu, S.K., Srivastava, P.K., Rej, A., 2013. Impacts and vulnerability of inland fisheries to climate change in the Ganga River system in India. Aquat. Ecosyst. Health Manag., 16: 415-424.

Dudgeon, D., 2003. The contribution of scientific information to the conservation and management of freshwater biodiversity in tropical Asia. In. Aquat. Biodivers. pp. 295-314.

Gadgil, M., Krishnan, B.J., Ganeshaiah, K.N., Vijayan, V.S., Borges, R., Sukumar, R., Noronha L., Nayak, S.V., Subramaniam, D.K., Varma, R.V., Gautam, S.P., Navalgund, R.R. and Subrahmanyam, G.V., 2011. Report of the Western Ghats Ecology Expert Panel Part 1, New Delhi.

Gerdaux, D., 1998. Fluctuations in lake fisheries and global warming. In: Management of lakes and reservoirs during global climate change (George, D.G., Ed.), pp 263–272, Kluwer Academic Publishers, Dodrecht, The Netherlands.

Gobi, K.C., 2002. A new synbranchid fish, Monopterus digressus from Kerala, Peninsular India. Rec. Zool. Surv. India. 100: 137–143.

Gunnell, Y., 1997. Relief and climate in South Asia: the influence of the Western Ghats on the current climate pattern of peninsular India. Int. J. Climatol., 17: 1169-1182.

Hanski, I., 1994. A practical model of metapopulation dynamics. J. Anim. Ecol., 63: 151-162.

Harvey, B.C., 1987. Susceptibility of young-of-the-year fishes to downstream displacement by flooding. Trans. Am. Fisher. Soc. 116: 851-855.

Hirabayashi, Y., Mahendran, R., Koirala, S., Konoshima, L., Yamazaki, D., Watanabe, S., Kim, H., and Kanae, S., 2013. Global flood risk under climate change. Nat. Clim. Change 3: 816-821. doi:10.1038/nclimate1911.

IMD, 2002 to 2006. South-west monsoons of 2002, 2003, 2004, 2005 - end of session reports. Indian meteorological Department, Pune, India.

INCCA, 2010. Climate change and India: a 4x4 assessment – a sectoral and regional analysis for 2030s. http://moef.nic.in/downloads/public-information/fin-rpt-incca.pdf.

IPCC, 2001. Climate Change 2001: Impacts, Adaptation and Vulnerability, Contribution of Working Group II to the Third Assessment Report of the Intergovernmental Panel on Climate Change (McCarthy, J.J., Canziani, O.F., Leary, N.A., Dokken, D.J., White, K.S., Eds.). Cambridge University Press, Cambridge, UK.

IPCC, 2014. Climate Change 2014: Synthesis Report. Contribution of Working Groups I, II and III to the Fifth Assessment Report of the Intergovernmental Panel on Climate Change (Core Writing Team, Pachauri, R.K. and Meyer, L.A., Eds.). IPCC, Geneva, Switzerland.

IPCC, 2017. IPCC Fifth Assessment Report (AR5) Observed Climate Change Impacts Database, Version 2.01. Palisades, NY: NASA Socioeconomic Data and Applications Center (SEDAC). doi.org/10.7927/H4FT8J0X.

Jacobs, J.W., 1993. International River Basin Development and Climatic Change: The Lower Mekong of Southeast Asia. Ph.D. Dissertation, University of Colorado.

Jankowski, T., Livingstone, D.M., Bührer, H., Forster, R. and Niederhauser, P., 2006. Consequences of the 2003 European heat wave for lake temperature profiles, thermal stability, and hypolimnetic oxygen depletion: Implications for a warmer world. Limnol. Oceanogr., 51: 815-819.

Kalff, J., 2000. Limnology. Prentice Hall, Upper Saddle River, New Jersey.

Kasturirangan, K., Babu, C.R., Mauskar, J.M., Chopra, K., Kishwan, J., Shanar, D., Narain, S., Roy, P.S., Tyagi, A., Chandrasekharan, I., 2013. Report of the High Level Working Group on Western Ghats, Vol. 1., New Delhi.

Klapper, H., 1991. Control of eutrophication in Inland waters. Ellis Horwood Ltd., West Sussex, UK.

Knight, J.D.M., 2010. Invasive ornamental fish: a potential threat to aquatic biodiversity in peninsular India. J. Threat. Taxa. 2: 700-704.

Kottelat, M. and Whitten., T. 1996. Freshwater biodiversity in Asia with special reference to fish. World Bank Technical Paper 343.

Krishnan, R., Sabin, T.P., Ayantika, D.C., Kitoh, A., Sugi, M., Murakami, H., Turner, A.G., Slingo, J.M., Rajendran, K., 2012. Will the South Asian monsoon overturning circulation stabilize any further? Clim. Dyn., 40: 187-211.

Kurup, B.M., Radhakrishnan, K.V., Manojkumar, T.G., 2004. Biodiversity status of fishes inhabiting rivers of Kerala (S. India) with special reference to endemism, threats and conservation measures. In: Proceedings of 2nd Large Rivers Symposium. (Welcome, R.L., Petr, T., Eds.), pp. 163-182, Mekong River Commission and Food and Agricultural Organization.

Lal, M., 2003. Global climate change: India's monsoon and its variability. Journal of Environmental Studies and Policy 6: 1-34.

Marcogliese, D.J., 2001. Implications of climate change for parasitism of animals in the aquatic environment. Can. J. Zool., 79: 1331-1352.

Meisner, J.D. 1992 Assessing potential effects of global climate change on tropical freshwater fishes. Geojournal 28: 21-27.

Mion, J.B., Stein, R.A. and Marschall, E.A., 1998. River discharge drives survival of larval walleye. Ecol. Appl., 8: 88-103.

Modde, T., Muth, R.T. and Haines, G.B., 2001. Floodplain wetland suitability, access, and potential use by juvenile razorback suckers in the middle Green River, Utah. Trans. Am. Fisher. Soc., 130: 1095-1105.

Molur, M., Smith, K.G., Daniel, B.A., Darwall, W.R.T., 2011. The status and distribution of freshwater biodiversity in the Western Ghats, India. IUCN Red List of Threatened Species – Regional Assessment.

Mulholland, P.J., Best, G.R., Coutant, C.C., Hornberger, G.M., Meyer, J.L., Robinson, P.J., Stenberg, J.R., Turner, R.E., Vera-herrera, Francisco and Wetzel, R.G., 1997. Effects of climate change on freshwater ecosystems of the south-eastern United States and the Gulf Coast of Mexico. Hydrol. Process., 11: 949-970.

Murty, A.S., 1986. Toxicity of Pesticides to Fish. I. CRC Press, Boca Raton, FL.

Myers, N., Mittermeier, R.A., Mittermeier, C.G., Fonseca, G.A.B.D., Kent, J., 2000. Biodiversity Hotspots for Conservation Priorities. Nature 403: 853-858.

NATCOM, 2012. National communication report submitted by government of India to UNFCCC. http://unfccc.int/resource/docs/natc/indnc2.pdf

Nayar, T.S., Rasiya Beegam, A., Sibi, M., 2014. Flowering Plants of the Western Ghats, India (2 Volumes). Thiruvananthapuram, India: Jawaharlal Nehru Tropical Botanic Garden and Research Institute. 1700 p.

Nicholls, K.H., 1998. El Nino, ice cover, and Great Lakes phosphorus: implications for climate warming. Limnol. Oceanogr., 43: 715-719.

Nijssen, B., O'Donnell, G.M., Hamlet, A.F. and Lettenmaier, D.P., 2001. Hydrologic sensitivity of global rivers to climate change. Clim. Change 50: 143-175.

Pandey, A.K and Das, P. 2002. Ichthyobiodiversity conservation for sustainable production, In: Riverine and Reservoir Fisheries of India. (Boopendranath, M.R., Meenakumari, B., Joseph, J., Sankar, T.V. Pravin, P. and Edwin, L., Eds.), pp. 70-90, Society of Fisheries Technologists (India), Cochin.

Poff, N.L., Angermeier, P.L., Cooper, S.D., Lake, P.S., Fausch, K.D., Winemiller, K.O., Mertes, L.A.K., Oswood, M.W., Reynolds, J., Rahel, F.J., 2001. Fish diversity in streams and rivers. In: Global Biodiversity in a Changing Environment: Scenarios for the 21st Century (Chapin, F.S.I., Sala, O.E., Huber-Sannwald, E., Eds.), pp. 351–368, Springer, New York.

Radhakrishnan, K.V. and Kurup, B.M., 2010. Ichthyodiversity of Periyar Tiger Reserve, Kerala, India. J. Threat. Taxa., 2: 1192-1198.

Radhakrishnan, K.V., Sureshkumar, S., Ng, H.H., 2010. *Pseudolaguvia austrina*, a new species of sisorid catfish (Osteichthyes: Siluriformes) from Peninsular India. Ichthyol. Explor. Fres., 21: 377-383.

Raghavan, R., 2019. Impact of 2018 Kerala floods on aquatic biodiversity with special reference to single location endemic species. Report submitted to the Kerala State Biodiversity Board, 21 p.

Raghavan, R., Prasad, P., Ali, A. and Pereira, B. 2008. Fish fauna of Chalakudy River, part of Western Ghats biodiversity hotspot, Kerala, India: patterns of distribution, threats and conservation needs. Biodiversity and Conservation 17: 3119-3131.

Rajendran, K., Kitoh, A., Shrinivasan, J., Mizuta, R. and Krishnan, R., 2012. Monsoon circulation interaction with Western Ghats orography under climate change. Theor. Appl. Climatol., 110: 555–571.

Ross, R.M., Lellis, W.A., Bennett, R.M. and Johnson, C.S., 2001. Landscape determinants of non indigenous fish invasions. Biol. Invasions 3: 347-361.

Roxy, M., Ritika, K., Terray, P., Murtugudde, R., Ashok, K. and Goswami, B.N., 2015. Drying of Indian subcontinent by rapid Indian Ocean warming and a weakening land-sea thermal gradient. Nat. Commun., 6: 7423. doi.org/10.1038/ncomms8423.

Sandeep, S. and Ajayamohan, R.S., 2015. Poleward shift in Indian summer monsoon low level jetstream under global warming. Clim. Dyn., 45: 337-351.

Sarkar, U.K., Pathak, A.K., Sinha, R.K., Sivakumar, K., Pandian, A.K., Pandey, A., Dubey, V.K. and Lakra, W.S., 2012. Freshwater fish biodiversity in the River Ganga (India): changing pattern, threats and conservation perspectives. Rev. Fish. Biol. Fisher., 22: 251-272.

Sooraj, K.P., Terray, P., Mujumdar, M., 2014. Global warming and the weakening of the Asian summer Monsoon circulation: assessments from the CMIP5 models. Clim. Dyn., 45: 1-20.

Sundaresan, J. and Patel, L.K., 2011. Climate change impact - a novel, initiative for Kerala. Indian J. Mar. Sci., 40: 483-486.

Val, A.L. and de Almeida-Val, V.M.F., 1995. The Amazon ichthyofauna. In: Fishes of the Amazon and their Environment, pp. 28-69, Springer, Berlin, Heidelberg.

Varikoden, H., Revadekar, J.V., Kuttippurath, J. and Babu, C.A., 2018. Contrasting trends in southwest monsoon rainfall over the Western Ghats region of India. Clim. Dyn., 52: 4557-4566.

Vass, K.K., Das, M.K., Srivastava, P.K. and Dey, S., 2009. Assessing the impact of climate change on inland fisheries in River Ganga and its plains in India. Aquat. Ecosyst. *Health.* Manag., 12: 138-151.

Vincent, M. and Thomas, J. April 2011. *Kryptoglanis shajii*, an enigmatic subterranean-spring catfish (Siluriformes, Incertae sedis) from Kerala, India. Ichthyol. Res. 58, 161–165. doi:10.1007/s10228-011-0206-6.

Walther, G.R., Roques, A., Hulme, P.E., Sykes, M.T., Pysek, P., Kühn, I., Zobel, M., Bacher, S., Botta-Dukát, Z., Bugmann, H., Czúcz, B., Dauber, J., Hickler, T., Jarosík, V., Kenis, M., Klotz, S., Minchin, D., Moora, M., Nentwig, W., Ott, J., Panov, V. E., Reineking, B., Robinet, C., Semenchenko, V., Solarz, W., Thuiller, W., Vilà, M., Vohland, K. and Settele, J., 2009. Alien species in a warmer world: risks and opportunities. Trends Ecol. Evol., 24: 686-693.

Impact of Climate Change on Hydrological Cycle, Ecosystem, Fisheries and Food Security, Pages: 321–340
Edited by: B. Madhusoodana Kurup, M.R. Boopendranath, M. Harikrishnan and A.V. Shibu

CHAPTER - 27

Impact of Climate Change on Fish Biodiversity of North-East India and Conservation Strategies

Pramod Kumar Pandey[1*] and K.V. Radhakrishnan[2]

[1] *College of Fisheries, Central Agricultural University, Lembucherra, Agartala, Tripura-799210, India*
[2] *Department of Fisheries Resource Management, College of Fisheries, Central Agricultural University, Lembucherra, Agartala, Tripura-799210, India*
**E-mail: pkpandey_in@yahoo.co.uk*

Climate change is one of the biggest environmental threats facing the world, potentially impacting food production and security, sustained water supply, biodiversity and natural ecosystems, human health and settlements. Changes in the climate of the earth have become evident both on global and regional scales in the past few decades. The most notable and significant changes associated with climate change are the gradual rise of global mean temperatures and a gradual increase in atmospheric green house gases, both of which have been aptly synthesized and documented (IPCC, 2013). Climate change is projected to impact by increasing pressure on all livelihoods and food supplies, including those in the fisheries and aquaculture sector. Food fish production, as is the case in all other primary production sectors, is expected to be influenced and or impacted to varying degree by climate change in different parts of the world (De Silva and Soto, 2009). An analysis of Global climate showed warming of 0.89 (0.69 to 1.08)°C over the period 1901–2012 which is mainly attributed to anthropogenic activities (IPCC, 2013). The findings indicate that warming is more pronounced than expected. The impact would be particularly severe in the tropical areas, which mainly consist of developing countries, including India (Sathaye *et al.*, 2006). Increasing temperature trends of the order of 0.60°C during last 112 years (IMD 2012) and increase in heavy rainfall events and decrease in low and medium rainfall events (Goswami *et al.*, 2006) over India have been observed. Changes in rainfall and temperatures have also been reported by Dash *et al.* (2009), MoEF (2010), Kothawale *et al.* (2010) and others.

Climate Change and its Impact on Freshwater Ecosystems and Fish

Most models on Global climate change indicate that the mountain snow packs are likely to decline in the west, which would bring adverse impact on fish populations, hydropower, water recreation and water availability for agricultural, industrial and residential use. Partial loss of ice sheets on polar land could imply meters of sea level rise, major changes in coastlines and inundation of low-lying areas, with greatest effects in river deltas and low-lying islands. Such changes are projected to occur over millennial time scales, but more rapid sea level rise on century time scales cannot be excluded (ICMOD, 2013).

Impact of Climate Change on Ecosystems

All aquatic ecosystems, including that of freshwater are influenced by climate change (Parry *et al.*, 2007). Rising temperatures, hydrologic intensification and sea-level rise are all expected to exert impacts on freshwater ecosystems across the spectrum of fish individuals, populations, species and communities (Heino *et al.*, 2016). Fishes living in lotic systems have adapted to a specific set of hydrologic parameters; changes in these conditions can result in increased success of invasive species (Ross *et al.*, 2001; Anneear*et al.*, 2004), reduced recruitment (Mion *et al.*, 1998; Modde *et al.*, 2001), and extirpation of locally specialized species (Cross and Moss, 1987). In high latitude rivers, the expected increase in precipitation and decrease in snowpack will result in an earlier, smaller flood pulse (ACIA, 2004). Without high spring flows, these stream systems may experience lower minimum flows (Nijssen *et al.*, 2001). This could negatively affect populations of economically important fishes. Tropical lakes inundated during flood stages may become isolated from the main channel during the dry season (Val and Almeida-Val, 1995). Tropical river systems may well experience faster water loss to the atmosphere (Jacobs, 1992). Increased evapotranspiration could reduce the volume and quality of available fish habitat during the dry season. Though prey fish could be more susceptible to capture, harsh abiotic conditions in isolated habitats could also lead to increased fish death and reduced survival of these fishes (Meisner, 1992). High flows would bring an increase in catastrophic flooding. Smaller flood stages would decrease the allochthonous input of material in to the aquatic system. In addition, a smaller inundated floodplain would also translate in to reduced spawning habitat. It is not well studied whether climate change has resulted in the decrease of the seasonal pulse in tropical systems. Floods or exceptionally large seasonal flows can displace adults and displace or injure juveniles and larvae (Harvey, 1987), and droughts or prolonged dry spells can cause population declines, reduced abundance, and altered species composition (Matthews and Marsh-Matthews, 2003). As a result, changes in variability could select for generalist species or those with the ability to rapidly colonize degraded habitats and possibly lead to a loss of locally adapted ones (Poff *et al.*, 2001). Lower dry-season flows in tropical and temperate rivers may also present problems with respect to seawater intrusion. This change in flows would be further exacerbated by a small rise in sea level (Roessig *et al.*, 2004). In lentic systems, increased temperatures will likely result in a general increase in lake trophic status (Allan *et al.*, 2005). Lower stream flows could increase water residence times and reduce flushing of nutrients from lake systems, thereby increasing trophic status (Bertahas *et al.*, 2006). When coupled with the input of anthropogenic pollutants, climate change can accelerate the eutrophication process (Klapper,1991; Adrian *et al.*, 1995) or delay recovery from anthropogenic eutrophication (Nicholls, 1998; Jankowski *et al.*, 2006). An increase in global temperatures will strengthen stratification in larger lakes because increased heating of the epilimnion (Gerdaux, 1998). Lake mixing is essential for the movement of oxygen to the hypolimnion and nutrients to the epilimnion. Longer periods of stratification create favorable conditions for blue-green algae, which are inedible to most species of zooplankton that planktivorous fishes eat (George *et al.*, 1990; Kangur *et al.*, 2002), so such a shift in phytoplankton composition can negatively affect fisheries productivity. In addition, some of blue-green algae (e.g., Anabaena) produce alkaloids that are toxic to fish (de Souza *et al.*, 1998) or their prey items (Bucka, 1998). This could lead to increased accumulation of toxins such as microcystins in the flesh of edible fish. More pronounced and longer-lasting stratification will reduce the amount of oxygen exchange to the hypolimnion from the oxygen-rich epilimnion. When the oxygen demand in the hypolimnion exceeds the supply, hypoxic or anoxic conditions will occur.

Impact of Climate Change on Fish

Because biochemical reaction rates vary as a function of body temperature, increasing global temperatures can affect individual fish by altering physiological functions such as thermal tolerance, growth, metabolism, food consumption, reproductive success, and the ability to maintain internal homeostasis in the face of a variable external environment (Fry, 1971). Though tropical fishes can currently endure these temperatures (Milstein *et al.*, 2000), a slight (1–2^{0}C) increase in regional temperatures may cause the daily temperature maxima to exceed these limits, particularly for populations that currently exist in thermally marginal habitats (Roessig *et al.*, 2004). Because increased temperatures can have sub-lethal effects on tropical fish, it is likely that tropical species will experience a pole-ward expansion in distribution as the thermal isolines are shifted (Mulholland *et al.*, 1997). Alien species would expand in to regions in which they previously could not survive and reproduce (Walther *et al.*, 2009). Because the aerobic metabolic rates of most cold-blooded aquatic organisms increase with temperature, an increase in temperature both decreases the DO supply (through reduced saturation concentrations relative to air) and increases the biological oxygen demand (BOD) (Kalff, 2000). Decreased levels of dissolved oxygen can reduce the growth and reproductive success of individual fishes and prevent them from dealing as effectively with other environmental challenges. Therefore, if climate change results in reduced levels of dissolved oxygen, it is possible that individual fish could be smaller, less fecund, and less likely to survive. Higher temperatures and a fixed ration would result in decreased growth rates by increasing in metabolism. The effects of altered flow regimes will be more detrimental than a change in temperature (Poff *et al.*, 2001). With an ambient temperature increase of only 1–2^{0}C, disturbances such as deforestation, pollution, and overfishing would likely pose a larger and more immediate threat to fish populations (Abell and Allan, 2002; Dudgeon, 2003). The toxicity of common pollutants (e.g., and heavy metals) to fish generally increases at higher temperatures (Murty, 1986).Climate change may alter host-parasite dynamics by changing transmission opportunities and host susceptibility (Marcogliese, 2001). Changing global temperatures will also affect fish susceptibility to disease because the immune functions of fish is compromised in the presence of stressors, including crowding, high temperatures, and osmotic stress. Endemic species or those with limited ranges are naturally prone to extinction (Brown and Kodric-Brown, 1977; Stacey and Taper, 1992; Hanski,1994; Angermeier, 1995). Therefore, another possible effect of climate change is the loss of biodiversity through the extinction of specialized or endemic fish species.

Climate Change in India and Impact on Fish Biodiversity

In recent years the climate is showing perceptible changes in the Indian subcontinent, where the average temperature is on the rise over the last few decades. The observed changes include an increase in air temperature, regional monsoon variation, frequent droughts and regional increase in severe storm incidences in coastal states and Himalayan glacier recession (Vass *et al.*, 2009). Climate change modelling studies for India show that the Indian sub-continent is likely to experiencea warming of over 3–5^{0}C and significant changes (increases and decreases) in flood and drought frequency and intensity. World Bank's World Development Report points out that degraded ecosystems and natural resources in South Asia are more vulnerable to climate change. Further, the poorest people are the most vulnerable to adverse impacts of climate change because they often reside in high-exposure areas and also have low adaptive capacity to cope with climate risks.

Studies on impact of Climate change on fish biodiversity of India are relatively poor. However, studies conducted in these lines in River Ganga are notable. Das *et al.* (2013) observed that the percentage of total rainfall in the peak breeding period (May-August) declined by 7% while it increased by 4% in the post-breeding period of Indian Major Carps (IMC) when resorption of eggs of their sets in. This shift and decrease in the rainfall pattern compounded by the abstraction of water in the river Ganga has resulted in the alteration of the required flow and turbidity during the breeding season, causing failure in breeding and consequent recruitment of young ones of IMC in the River Ganga. An adequate flood level during the monsoon months (June-September) is required to inundate the floodplains of the River Ganga where the majority of the Gangetic carps breed. This phenomenon has a close relationship with the spawn availability in river

(Natarajan, 1989). Similar declines in availability of Asian carp spawn as a result of breeding failure due to inadequate water discharge and velocity was recorded in Yangtse River in China (Duan *et al.*, 2009).

From analysis of 30 years' time series data on river Ganga and water bodies in the plains, Vass *et al.* (2009) reported that the annual mean minimum water temperature in the upper colder stretch of River Ganga at Haridwar during the period 1980–2009 increased by 0.99°C. As a result, the stretch of River Ganga around Haridwar has become a more congenial habitat for warm water fishes of the middle stretch of the river. The study conducted from 2005 to 2009 found several species of fish in the upper cooler stretch of the River Ganga from Deoprayag to Haridwar that had not been recorded previously from this stretch but previously found in the middle warmer stretch of the river (Menon, 1954). These are the warm water fish species: *Glossogobius giuris, Puntius ticto, Xenentodon cancila* and *Mystus vittatus*. Thus a shift in geographic distribution of the fishes has occurred into the upper colder stretch of the Ganga River. Moreover, the predator-prey ratio in the middle stretch of the river has been reported to be declined from 1:4.2 to 1:1.4 in the last three decades.

Sarkar *et al.* (2012) observed that in recent years the IMCs maturing and spawning as early as March, making it possible to breed them twice a year. Thus, there is an extended breeding activity as compared to a couple of decades ago (Dey *et al.*, 2007), which appears to be a positive impact of the climate change regime. The fish spawn availability index declined from an average of 1529 ml during (1965–1969) to an average of 568 ml in recent years (2005 to 2009) (Natarajan, 1989; CIFRI Annual Report, 1965–2009). It also showed a continuing decline of IMC spawn with decreasing percentage of major carp spawn (46% in 1965–1969 to 10% in 2005–2009) whereas other fish spawn increased from 54% in 1965–1969 to 90% in 2005–2009 in the total spawn collection. The low fish production of the major fish groups in the river Ganges is believed to be the recruitment failure of the young ones due to degradation (decreased runoff, changes in flow, turbidity) of the natural spawning habitat and climate change. Sarkar *et al.* (2012) also observed range extensions of several fish species including *Mastacembelus armatus* and *Cyprinus carpio var. specularis* to the upper stretch (between Tehri and Rishikesh) and *Glossogobius giuris*, *Macrognathus aral*, *Sperata aor*, *Clupisoma garua*, *Puntius sarana* and *Ompok pabda* at Haridwar stretch indicating a perceptible shift in distribution pattern of fishes. Species like *Glyptothorax brevipinnis* and *G. telchitta*, common inhabitants of upland waters were also recorded in the middle stretch of river Ganga during pre-monsoon periods confirming the range extension of these species towards downstream. On the other hand, the distribution range of *Schizothorax* spp, a typical coldwater fish was shrunken towards the upstream of Ganga River, which could be considered as a warranting situation due to temperature increase.

Impact of Climate Change in Eastern Himalayas

The Himalayas has shown consistent warming trends during the past 100 years (Yao *et al.* 2006). Global warming is accelerating the melting of glaciers in the Himalayas, which are melting faster than the global average (Bajracharya *et al.*, 2007); the snow line will rise and potentially some glaciers could disappear or stabilise at a much reduced mass (IPCC, 2007). In the short term, this means an increase in meltwater and in the longer term, a reduction. The Eastern Himalayas traversing through five countries (India, Myanmar, Bhutan, China and Nepal) and encompass parts of three of the 34 global Biodiversity Hotspots: 39 per cent of the Himalayan, 8 percent of the Indo-Burma, and 13 per cent of the Mountains of Southwest China Hotspot; taking in 25 ecoregions of the world (WWF 2006). Eastern Himalayas is particularly vulnerable to climate change due to its ecological fragility and economic marginality. In this region, a substantial proportion of the annual precipitation falls as snow. With rising temperatures, areas covered by permafrost and glaciers are decreasing in much of the region and the snowmelt begins earlier and winter is shorter. Whereas snow masses have acted as a natural form of storage, releasing moisture slowly into the rivers, water is now increasingly only available at the time of precipitation. Spatial analysis of water balance components indicates that stream flows will increase significantly across most of the Eastern Himalayas in response to precipitation and temperature changes, though the effects will vary spatially. Studies also suggest that the seasonal temperature fluctuations arc changing in both timing and extent and the rate of change of temperature with altitude is becoming less (ie, higher altitude areas are warming faster than lower areas so the difference in temperature between them is becoming less) (ICMOD, 2013). Projections of ICMOD (2013) also suggest an increase in winter, pre-monsoon, monsoon, post-monsoon, and annual precipitation in the region, in the tune of 18 per cent by the middle of the century, and by 13 to 34 per cent by the end of the century. Higher monsoon precipitation is also projected at higher altitudes, and lower at lower altitudes. The changes in precipitation pattern result in changes in the hydrological regime, an increase in hazard frequency and intensity, and impacts aquatic biodiversity. It is clear that the hydrological systems of the Eastern Himalayas are very sensitive to climate change, variability, and extremes, both on a seasonal basis and over longer time periods. There is also evidence of noticeable increases in the intensity and frequency of many extreme weather events in the region such as heat waves, tropical cyclones, prolonged dry spells, intense rainfall, snow avalanches, thunderstorms, and severe dust storms (Cruz *et al.*, 2007).

Climate Change in North-Eastern India

The North-Eastern (NE) India comprises of the states of Arunachal Pradesh, Assam, Manipur, Nagaland, Mizoram, Sikkim and Tripura and stretches between 21°50' and 29° 34' N latitude and 85° 34' and 97° 50' E longitude (Fig. 1). The diversity within the freshwater ecosystems in this region, forming part of the Himalayas and Indo-Burma are both highly diverse and of great regional importance in terms of the livelihood and the economic importance of the people living around it (Goswami *et al.*, 2012). The region has inaccessible terrain, fragility, marginality, excessive sloping land with rolling topography, rich and unique ethnicity and socio-ecological set up. The NE region receives high rainfall and therefore clothed with diverse and dense vegetation. Beyond the transitional pre-glacial region with increasing altitude, there appears the greater Himalayan region devoid of significant vegetative cover. The altitudinal pattern of the NE region varies from place to place. The entire geographical area of NE is criss-crossed by numerous rivers belonging to the three major drainage systems: the Barak-Brahmaputra, the Kaladan, and the Chindwin (Vishwanath, 2017). The varied physiological features and altitudinal differences gives rise to varied types of climate ranging from near tropical to temperate and alpine. The annual precipitation in the region is received mainly from south-west monsoon from middle of May and continue till October. On an average, the NE region receives about 2450 mm of rainfall. The region shows great variation in temperature regime too, varies from 15°C to 32°C in summer and 0 to 26°C in winter. The NE region is very rich in biodiversity and harbors the largest number of endemics and species then anywhere in the country. The distribution of various species has been, again, restricted due to topographical features, deep valleys, slopes and river system.

The NE region is expected to be highly prone to the consequences to climate change because of its geo-ecological fragility, strategic location vis-à-vis the Eastern Himalayas and international borders, its trans-boundary river basins and its inherent socio-economic instabilities. Environmental security and sustainability of the region are and will be greatly challenged by these impacts. A study conducted based on hydrological model SWAT by Ravindranath *et al.* (2011) found that the trend in precipitation in the NE region exhibits considerable spatial variability with respect to the predictions for near term. The northern part of the NE shows a reduction in precipitation varying from 3% in the northwestern portion to about 12% in the northeastern portion. In the remaining part of the NE, there is an increase in precipitation varying from 0% to as much as 25% in the central portion. Majority of the NE region, except for some parts of Mizoram, Tripura, Manipur and Assam, show an increase in the evapotranspiration during the near-term scenario. Even those parts of Arunachal Pradesh that showed a decrease in the precipitation exhibited an increase in evapotranspiration, all leading to increased water availability and thus lesser water vulnerability. This can only be explained by the higher temperatures that will enhance the evaporative force. However, the increase in evapotranspiration ranges from a small fraction to about 20%. The reduction in evapotranspiration in the southern portion is only marginal. The trend in the water yield in the NE region is similar to that in precipitation. The areas that have shown less increase in precipitation show correspondingly low water yield. The reduction in water yield for Arunachal Pradesh is up to about 20% and the increase in the water yield in Assam and Manipur areas is up to about 40%.

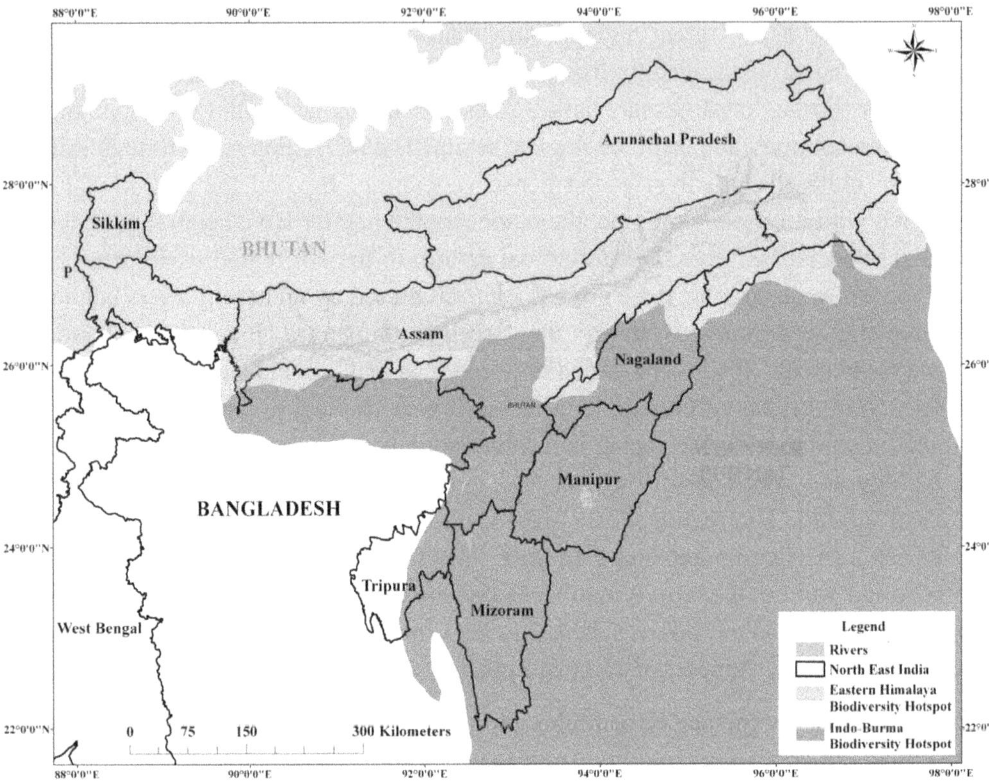

Fig. 1. The north eastern states (8 states) bearing both Himalayan and Indo-Burma biodiversity hotspots zones (Source: Goswami *et al.*, 2012)

Ravindranath *et al.* (2011) also observed that, in accordance with the change in pattern of precipitation, both intensity of floods and drought severity are likely to increase in many parts of the North East region. There will be a general increase in flood magnitude of the Barak basin compared to Brahmaputra basin in future. State-wise analysis shows that in Tripura, Mizoram, Manipur, parts of Meghalaya and Nagaland, the flood magnitude is likely to increase by about 25% in the future compared to the present. Arunachal Pradesh, Assam, Sikkim and parts of Meghalaya are likely to experience floods of lower magnitude (about 5-10% less) in future compared to the present. The numbers of drought weeks during monsoon months shows an increasing trend in Arunachal Pradesh, parts of Assam, Meghalaya, Mizoram, Tripura and Manipur, to the tune of about 25% increase in future. A few districts in Assam, Nagaland, Meghalaya and Mizoram show improvement in drought situation during the onset of monsoon. Many parts of the Brahmaputra basin show a tendency of extreme soil moisture stress during monsoon months, which is likely to lead to moderate to extreme drought condition.

Potential Impacts of Climate Change on the Fish Biodiversity of NE India

The NE India has rich freshwater fish diversity. The diversity is attributed to the past geological history, i.e., the collision of Indian, Burmese, and Chinese plates, and the Himalayan orogeny which played an important role in the speciation and evolution of groups inhabiting mountain streams (Kottelat, 1989). A total of 318 fish species under 113 genera and 36 families were so far recorded from NE region of which 229 species are in the Brahmaputra drainage, 103 in the Chindwin, and 27 in the Kaladan. Of these, Twenty-seven species are endemic to the Brahmaputra basin, as many as 41 species of fish are endemic to the Chindwin drainage and about 20 endemics are expected to occur in Kaladan basin. However, 48 fish species of this list are already categorized as threatened (CR-1, EN-13, and VU-34) due to various anthropogenic and other reasons (Vishwanath, 2017) of which Climate change is recognized as a major factor.

Since there is no comprehensive study so far been taken up focusing on the exact impact of Climate change on the fish biodiversity in any of the water bodies of NE region, some potential impacts based on the observation from survey and sampling of fishes conducted by the authors in different water bodies of NE India for the past 7 years (2012-2019) as part of COE-FAB Project (Centre of Excellence on Fisheries and Aquaculture) is briefed below. In addition, the information and knowledge generated by thoroughly going through the literature on the impact of climate change on the fish diversity of Ganga River in the Northern-central India is also incorporated to derive plausible conclusions.

Breeding and recruitment of fish

There observed a continuing decline of IMC spawn with decreasing percentage of major carp spawn and increase in other fish spawn in Ganga River (Vaas *et al.*, 2009; Das *et al.*, 2013). Further, the shift and decrease in the rainfall pattern compounded by the abstraction of water caused alteration of the required flow and turbidity during the breeding season and resulted in the failure in breeding and consequent recruitment of young ones of IMC in the River Ganga. The changing pattern of precipitation as detailed by Ravindranath *et al.* (2011) along with the continuing habitat alteration activities will reduce the breeding and spawning success and result in recruitment failure of major carp species in the rivers of NE region as well.

Biogeographic distribution of fishes

Surveys revealed that a number of fish species which were earlier never reported in the upper stretch of the river Ganga and were predominantly only available in the lower and middle Ganga in 1950s are now recorded from the upper cold-water stretch (Vaas *et al.*, 2009; Sarkar *et al.*, 2012; Das *et al.*, 2013). Our own sampling surveys in different NE river systems also observed that the cyprinid minnows predominantly seen at the middle and lower stretches became proliferated even at the upper stretches of rivers and downstream fishes like *Glossogobius giuris*, *Xenentodon cancila* showed extended distribution to upper stretches.

Predator-Prey Relationship

Changes in the thermal structure of an aquatic ecosystem impact on the prey densities and distribution. If the same number of prey is distributed across large volume of the habitat, prey densities encountered by a predator would be reduced. Reduced prey densities would reduce the predator encounter rate with prey, which would reduce predator consumption rate. The predator prey ratio in the middle stretch in the river Ganga has found declined from1:4.2to1:1.4in the last three decades, possibly one of the reasons might be Climate change (Vaas *et al.*, 2009; Sarkar *et al.*, 2012; Das *et al.*, 2013). Given the high fish diversity in NE India, similar effect of an alteration in predator-prey relationship is anticipated.

Qualitative changes in distribution and abundance of plankton

The total number of genera of phytoplankton has found declined in Ganga River (Vaas *et al.*, 2009). There was a shift in plankton diversity as well. Temperature is an important environment factor affecting the distributions of diatoms (Ward and Whipple, 1959). Genera such as *Tetracycles* and *Amphycampus* usually found in cold mountainous regions and the genera *Diatoma* and *Ceratoneus* encountered in cool temperate regions than in warmer areas of Ganga were declined. Increased water temperature might have resulted in depletion of the stenothermal phytoplankton genera mentioned above. Similar impact may also be affected to different plankton communities of the Himalayan states like Sikkim and Arunachal Pradesh in NE India as well as both these states have relatively colder climate compared to the downside states.

Growth of the fish

Temperature changes will have an impact on the suitability of species for a given location. In temperate areas increasing temperatures could bring the advantages of faster growth rates and longer growing seasons (Vaas *et al.*, 2009). McCauley and Beitinger (1992) observed higher growth rate for Channel catfish with every 1^0C rise in temperature up to 30^0C. Similar results were also obtained for Indian Major Carps up to 33^0C (Vaas *et al.*, 2009). It may be possible that the growth rate of many coldwater fishes in NE India could also be enhanced by increasing temperature. However, focused research may be needed to substantiate this.

Fish production

Consequent to the breeding and recruitment failure, the average fish landings in the upper Ganga river have come down from 119.35 t in 1959 to 78.15 t in 2004. The decline of the Indian major carp (IMC) fisheries in river Ganga has been the major cause for reduced fish production in the river. Increasing water temperature in the aquatic bodies of NE India will also cause changes in the breeding and recruitment success of fishes in them; finally result in low fish production.

Distribution and abundance of coldwater fishes

Coldwater fishes are important ecological indicators for climate change as they are very sensitive to changes in water temperature and other environmental conditions. The North eastern Himalayan coldwater fish species may be at the highest risk of global warming as many of them are endangered. In many areas, they are already living at the upper end of their thermal range and ecological models demonstrate that there would be significant losses of temperate fish species as climate change may lead to a reduction of fish habitat and diversity (Mohseni *et al.*, 2003). Eaton and Scheller (1996) opined that increasing temperatures in North American streams could result in a decrease in coldwater fish habitat of nearly 50% and Mulholland *et al.* (1997) were of the opinion that increase in temperature by 3.88^{0}C can drastically reduce the range of brook trout (*S. fontinalis*) in the southeastern United States. Meisner (1990) based on a study in Humber and Rouge river systems in Canada indicated that a 4.18^{0}C increase in summer air temperature would reduce thermally suitable brook trout habitats by 42% and 30% respectively. Sarkar *et al.* (2012) observed shrunken distribution range of indigenous snow trouts (*Schizothorax* spp.) towards the upstream of Ganga River, possibly due to the temperature increase at the upper stretches of the river. The NE Region of India is well-known for its rich coldwater fish resources including snowtrouts, mahseers, catfishes, balitorids and other cyprinids. The global warming and consequent increasing temperature of aquatic resources of NE India at the Himalayan plains will pressurize the coldwater fishes to migrate to further higher latitudes. These fish would also compete for space in their "new" ecosystems, possibly at the expense of native species. The northern migration of predatory species could also have a deleterious effect on native prey fish.

Mitigation, Adaptation and Strategies for Conservation of Fish Diversity of NE Region in Response to Climate Change

In order to cop up with the looming Climate change scenario in the NE region, the following measures are suggested.

Mitigation of wetland habitats

Wetlands are considered to be natural reserves for the fish. It is important to maintain the hydrology, incorporating ecosystem/integrated approach for wetland management, de-siltation of wetlands, opening and widening of linkage channels, provision of appropriate number of deep pools in lotic water bodies, prevention of additional anthropogenic stresses, control of exotics, improvisation of fishing tools, ranching, sensitization among stakeholders etc. Besides, proper scientific appraisal of the climate change impacts to reduce risks and adapt to climate-driven changes in wetlands (Sarkar *et al.*, 2018).

In situ and Ex situ conservation

Measures like in situ conservation, habitat fingerprinting, ex-situ conservation and developing live gene banks could be implemented to conserve the native fish diversity. In this light, in Northern India, observations were made in the water bodies of the selected wildlife sanctuaries in order to conceptualize the need and approach for developing freshwater aquatic sanctuaries (FAS) within the protected area network (Sarkar *et al.*, 2018).

Adopt 'Responsible Fisheries' practices

To build resilience to the effects of climate change and derive sustainable benefits, fisheries and aquaculture managers need to adopt and adhere to best practices such as those described in the FAO 'Code of Conduct for Responsible Fisheries', reducing overfishing and rebuilding fish stocks. These practices need to be integrated more effectively with the management of river basins, watersheds and coastal zones. Fisheries and aquaculture need to be blended into National Climate Change Adaptation Strategies. In absence of careful planning, aquatic ecosystems, fisheries and aquaculture can potentially suffer as a result of adaptation measures applied by other sectors such as increased use of dams and hydro power in catchments with high rainfall, or the construction of artificial coastal defenses or marine wind farms (Mohanty *et al.*, 2010).

Standardize tools for stress monitoring

There is need to standardize tools suitable for monitoring stress resulting from global warming and climate change impacts, in the aquatic animals from both capture fisheries and aquaculture systems (Mohanty *et al.*, 2010).

Adopt Ecosystem based approaches

In freshwater systems, ecosystem health and productivity is linked to water quality and flow and the health of wetlands. Ecosystem-based approaches to fisheries and coastal zone management are highly beneficial as such approaches recognize the need for people to use the ecosystem for their food security and livelihoods while enabling these valuable natural assets to adapt to the effects of climate change, and to reduce the threats from other environmental stresses (Hoegh-Guldberg *et al.*, 2007; Mohanty *et al.*, 2010).

The measures that can be adopted in aquaculture scenario are briefed below.

Water stress

The availability of water will be a major constraint for fish biodiversity in the future. The reduced water in natural aquatic systems will further result in decrease in water quality, increased diseases, and altered and reduced freshwater supply to aquaculture. Operational costs of maintaining aquaculture ponds will increase along with conflict with other water users. Reduced production capacity will occur. Aquaculture should be integrated with other farming systems for many fold use and reuse of water areas (Vaas *et al.* 2009). The resource of waste water and degraded water should be extensively used with modified aquaculture practices.

Increasing water temperature

The experience in Ganga River showed that the IMCs cultured maintain their normal metabolic activities up to 34°C. Beyond this temperature growth and food conversion is affected. Since low oxygen is an issue with enhanced temperature, substitution with low oxygen tolerant species like catfishes in the culture system could be taken up (Vaas *et al.,* 2009).

Floods

Heavy floods will cause damage to the infrastructure facilities of aquaculture farms. There may be loss of fish stock and introduction of fish predators and disease too. To reduce the loss of fish due to floods, provision for continuous supply of seed from fish hatcheries must be assured (Vaas *et al.,* 2009). Management practices should emphasize harvesting of fish at smaller sizes and selection of fish species that require short culture periods with minimum expenditure on inputs.

Drought

Draught will be resulting in reduced water availability to the culture farms and in reduced water quality as well. As a result, there will be loss in fish production. To mitigate this, smaller ponds that retain water for 2-4 months can be used for fish culture with species like catfish with appropriate management practices.

Cyclones and storms

Inundation, flooding and salinity changes occur in the fish culture facilities resulting in the loss of fish/prawn stock and introduction of predators (Vaas *et al.*, 2009). Damage to the facilities occurs with heavy financial loss. An early warning system for weather events is essential to cope up with Cyclone induced damages.

Knowledge Gaps and Research Needs

Little is known in detail about the climatic characteristics of North-Eastern India both because of the paucity of observations and because insufficient theoretical attention has been given to the complex interaction of spatial scales in weather and climate phenomena. There is a considerable gap in our knowledge of the natural resources in the monitoring, documentation or to check the status of biodiversity in the region. Despite various projections and observed changes, the region also lacks adequate scientific evidences to understand the impact of climate change on fish biodiversity and also different aspects of human wellbeing. This makes the region in predicting future scenarios more uncertain for vulnerability assessment and risk management. Moreover, across the entire region, the limited research that is available in NE region only focus on the adverse impacts of climate change and overlooks the adaptation mechanisms and the potential new opportunities. There is no consolidated handbook of proven biodiversity conscrvation tcchniques, or climate adaptation techniques, targeted at this region. There also appears to be a lack of human and institutional set ups and policy imperatives to tackle climate change issues.

The following are some of the most important policy research needs identified:

- Conduct a detailed inventory of wetlands in NE India and generate quantitative data on their ecological characteristics, functions, and services, and assess the effects of temperature and enhanced CO_2 levels on growth, carbon sequestration, and methane emission in different types of wetland.
- Conduct cross-sector, cross-boundary, and trans-disciplinary research on climate change and its impacts on fish biodiversity in this region and analyse the movement of climate sensitive and invasive species considering altitudinal and latitudinal aspects, protected area coverage and effectiveness, adaptability of biodiversity entities, and impact on aquatic productivity.
- Prepare a consolidated handbook of proven biodiversity conservation techniques, or climate adaptation techniques, targeted at this region. In future, design and develop programmes in the context of the vulnerability situations of the NE region based on the results of the assessment carried out by research as mentioned earlier.
- Strengthen and review policies to make them more sensitive to the interaction in processes and the linkages between the consequences of biodiversity and climate change.

Conclusions

The NE India is a part of two biodiversity hotspots of the world (Eastern Himalayas and Indo-Burma) and shares the highest species richness and endemicity along with Western Ghats and Peninsular India, with respect to freshwater fishes. However, by virtue of its strategic location in Eastern Himalayas, the effect of climate change will be felt hard and even, will be hardest, since many of the aquatic ecosystems in this region are already subjected to various anthropogenic stresses. One of the prime requirements to sustain the ability of freshwater ecosystems to support fish population and biodiversity is to maintain adequate water and its flow. However, in the climate-warming scenario, it is expected that low water current in rivers would be more severe along with a general water stress. Water demand in coming years by urban and industrial sectors would increase; substantial increases in withdrawals for irrigation are expected; a growing human population will increase water demand more than will climate warming. The adaptive capacity under such situation will depend on the abilities to implement integrated water resource management approaches. Unfortunately, the existing policies and development programmes have not yet recognized climate change and its potential impacts as factors that can significantly jeopardise the environmental and socioeconomic systems in the region. It is time for the national and regional planners to reconcile with the reality of climate change and take appropriate action for its mitigation. Reliable assessment and mapping of vulnerability and risk and promoting appropriate adaptation strategies are also equally important to ensure mitigation of the water and climate induced vulnerability on the fish biodiversity in NE region. Immediate steps should be initiated by state Governments in NE India through appropriate policies and practices in water, forest, environment and disaster management to make environmental and climate concerns inbuilt in all development activities. Besides, proper scientific appraisal of the climate change impacts, capacity building of the vulnerable population to reduce risks and adapt to climate-driven changes in natural ecosystems and human societies are the urgencies of the hour.

References

Abell, R. and Allan, J.D., 2002. Riparian shade and stream temperatures in an agricultural catchment, Michigan, USA. Int. Ver. Theor. Angew. Limnol., Verh. 28: 232-237.https://doi.org/ 10.1080/03680770.2001.11902578.

Adrian, R., Deneke, R., Mischke, U., Stellmacher, R. and Lederer, P., 1995. A long-term study of the Heiligensee (1975-1992). Evidence for effects of climatic change on the dynamics of eutrophied lake ecosystems. Arch. Hydrobiol. 133: 315-337. http://pascal-francis.inist.fr/vibad/ index.php?action=getRecordDetail&idt=3583922.

Allan, J.D., Palmer, M. and Poff, N.L., 2005. Climate change and freshwater ecosystems. In:Climate change and biodiversity (Lovejoy, T.E., Hannah, L.,Eds), pp 274–295, Yale University Press, New Haven CT.

Angermeier, P.L., 1995. Ecological attributes of extinction prone species: loss of freshwater fishes of Virginia. Conserv. Biol. 9: 143-158. https://doi.org/10.1046/j.1523 1739.1995.09010143.x

Arthington, A.H., Dulvy, N.K., Gladstone, W. and Winfield, I.J., 2016. Fish conservation in freshwater and marine realms: status, threats and management. Aquat. Conserv., 26: 838-857.

Assessment, A.C.I., 2004. Impacts of a warming Arctic-Arctic climate impact assessment. ISBN 0521617782. Cambridge, UK: Cambridge University Press, December 2004., 144 p.

Bajracharya, S.R., Mool, P.K. and Shrestha, B.R., 2007. Impact of climate change on Himalayan glaciers and glacial lakes: case studies on GLOF and associated hazards in Nepal and Bhutan. International Centre for Integrated Mountain Development (ICIMOD).

Bertahas, I., Dimitriou, E., Karaouzas, I., Laschou, S. and Zacharias, I., 2006. Climate change and agricultural pollution effects on the trophic status of a Mediterranean lake. Acta Hydrochim. Hydrobiol. 34: 349-359. https://doi.org/10.1002/aheh.200500637

Bhatt, J.P., Manish, K., Mehta, R. and Pandit, M.K., 2016. Assessing potential conservation and restoration areas of freshwater fish fauna in the Indian river basins. J. Environ. Manage. 57: 1098-1111.

Brown, J.H. and Kodric-Brown, A., 1977. Turnover rates in insular biogeography: effect of immigration on extinction. Ecology 58: 445-449. https://doi.org/10.2307/1935620

Bucka, H., 1998. The mass invasion of several blue-green algae in two drinking-water supply reservoirs in southern Poland. In:Management of lakes and reservoirs during global climate change (George, D.G., Ed.), pp 145–151, Kluwer Academic Publishers, Dodrecht, The Netherlands.

Cross, F.B. and Moss, R.E., 1987. Historic changes in fish communities and aquatic habitats in plains streams of Kansas. In:Community and evolutionary ecology of North American stream fishes (Matthews, W.J., Heins, D.C., Eds.), pp 155–165, University of Oklahoma Press, Norman, OK.

Cruz, R.V., Harasawa, H., Lal, M., Wu, S., Anokhin, Y., Punsalmaa, B., Honda, Y., Jafari, M., Li, C. and Huu, N.N., 2007. Climate change 2007: Impacts, adaptation and vulnerability. In: Contribution of Working Group II to the Fourth Assessment Report of the Intergovernmental Panel on Climate Change (Parry, M.L., Canziani, O.F., Palutikof, J.P., van der Linden, P.J., Hanson, C.E., Eds), pp 469-506, Cambridge University Press, Cambridge.

Das, A., Ghosh, P.K., Choudhury, B.U., Patel, D.P., Munda, G.C., Ngachan, S.V. and Chowdhury, P., 2009. Climate change in North East India: recent facts and events–worry for agricultural management. ISPRS Archives XXXVIII-8. pp. 32-37, In W3 Workshop Proceedings: Impact of Climate Change on Agriculture.

Das, M.K., 2009. Impact of recent changes in weather on Inland Fisheries in India. In: Global climate change and Indian agriculture, Case studies from the ICAR Network Project, pp.102, Indian Council of Agricultural Research, New Delhi.

Das, M.K., Sharma, A.P., Sahu, S.K., Srivastava, P.K. and Rej, A., 2013. Impacts and vulnerability of inland fisheries to climate change in the Ganga River system in India. Aquat. Ecosyst. *Health* Manag. 16: 415-424.

De Silva, S.S. and Soto, D., 2009. Climate change and aquaculture: potential impacts, adaptation and mitigation. In: Climate change implications for fisheries and aquaculture: overview of current scientific knowledge (Cochrane, K., Young, C., De Soto D. and Bahri, T., Eds.), pp. 151-212, FAO Fisheries and Aquaculture Technical Paper. No. 530, FAO, Rome.

Dey, S., Srivastava, P.K., Maji, S., Das, M.K., Mukhopadhyaya, M.K. and Saha, P.K., 2007. Impact of climate change on the breeding of Indian major carps in West Bengal. J. Inland Fish. Soc. India, 39: 26-34.

Dudgeon, D., 2003. The contribution of scientific information to the conservation and management of freshwater biodiversity in tropical Asia. In Aquat. Biodivers, pp. 295-314, Springer, Dordrecht.

Eaton, J.G. and Scheller, R.M., 1996. Effects of climate warming on fish thermal habitat in streams of the United States. Limnol. Oceanogr. 41: 1109-1115.

Ficke, A.D., Myrick, C.A. and Hansen, L.J., 2007. Potential impacts of global climate change on freshwater fisheries. Rev. Fish. Biol. Fisher. 17: 581-613.

Fry, F.E.J., 1971. The effect of environmental factors on the physiology of fish. In: Fish Physiology, Environmental Relations and Behavior (Hoar, W.S. and Randall, D.J., Eds.), pp 1–98, Academic Press, New York.

George, D.G., Hewitt, D.P., Lund, J.W.G. and Smyly, W.J.P., 1990. The relative effects of enrichment and climate change on the long term dynamics of Daphnia in Esthwaite Water, Cumbria. Freshw. Biol.23, 55-70.

Gerdaux, D., 1998. Fluctuations in lake fisheries and global warming. In: Management of Lakes and Reservoirs during Global Climate Change (George, D.G., Ed.), pp 263–272, Kluwer Academic Publishers, Dodrecht, The Netherlands.

Hanski, I., 1994. A practical model of metapopulation dynamics. J. Anim. Ecol. 63: 151–162.

Harvey, B.C., 1987. Susceptibility of young-of-the-year fishes to downstream displacement by flooding. Trans. Am. Fisher. Soc. 116: 851–855.

Hassol, S., 2004. Impacts of a warming Arctic - Arctic climate impact assessment. Cambridge University Press.

Heino, J., Erkinaro, J., Huusko, A. and Luoto, M., 2016. Climate change effects on freshwater fishes, conservation and management, pp 76-106, Cambridge University Press: Cambridge.

Hoegh-Guldberg, O., Mumby, P.J., Hooten, A.J., Steneck, R.S., Greenfield, P., Gomez, E., Harvell, C.D., Sale, P.F., Edwards, A.J., Caldeira, K. and Knowlton, N., 2007. Coral reefs under rapid climate change and ocean acidification. Sci. 318: 1737-1742.

Jacobs, J.W., 1993. International River Basin Development and Climatic Change: The Lower Mekong of Southeast Asia. PhD Dissertation, University of Colorado.

Jankowski, T., Livingstone, D.M., Bührer, H., Forster, R. and Niederhauser, P., 2006. Consequences of the 2003 European heat wave for lake temperature profiles, thermal stability, and hypolimnetic oxygen depletion: Implications for a warmer world. Limnol. Oceanogr. 51: 815-819.

Kalff, J., 2000. Limnology. Prentice Hall, Upper Saddle River, New Jersey.

Kangur, A., Kangur, P. and Pihu, E., 2002. Long-term trends in the fish communities of Lakes Peipsi and Võrtsjärv (Estonia). Aquat. Ecosyst. Health. Manag., 5: 379-389.

Klapper, H., 1991. Control of eutrophication in Inland waters. Ellis Horwood Ltd., West Sussex, UK.

Matthews, W.J. and Marsh Matthews, E., 2003. Effects of drought on fish across axes of space, time and ecological complexity. *Freshw. Biol.*, 48: 1232-1253.

McCauley, R. and Beitinger, T., 1992. Predicted effects of climate warming on the commercial culture of the channel catfish, *Ictalurus punctatus*. GeoJournal. 28,61-66.

Meisner, J.D. and Shuter, B.J., 1992. Assessing potential effects of global climate change on tropical freshwater fishes. Geo Journal 28: 21-27.

Meisner, J.D., 1990. Effect of climatic warming on the southern margins of the native range of brook trout, *Salvelinus fontinalis*. Can. J. Fish. Aquat. Sci., 47: 1065-1070.

Milstein, A., Zoran, M., Peretz, Y. and Joseph, D., 2000. Low temperature tolerance of pacu, *Piaractusmesopotamicus*. Environ. Biol. Fishes, 58: 455-460.

Mion, J.B., Stein, R.A. and Marschall, E.A., 1998. River discharge drives survival of larval walleye. Ecol. Appl., 8: 88-103.

Modde, T., Muth, R.T. and Haines, G.B., 2001. Floodplain wetland suitability, access, and potential use by juvenile razorback suckers in the middle Green River, Utah. Trans. Am. Fisher. Soc., 130: 1095-1105.

Mohanty, B., Mohanty, S., Sahoo, J. and Sharma, A., 2010. Climate Change: Impacts on Fisheries and Aquaculture, Climate Change and Variability, Suzanne Simard (Eds.), ISBN: 978-953-307-144-2. , http://www. intechopen. com/books/climate-change-and-variability/climate-change impacts-on-fisheries-and-aquaculture.

Mohseni, O., Stefan, H.G. and Eaton, J.G., 2003. Global warming and potential changes in fish habitat in U.S. streams. Clim. Change, 59: 389-409.

Mulholland, P.J., Best, G.R., Coutant, C.C., Hornberger, G.M., Meyer, J.L., Robinson, P.J., Stenberg, J.R., Turner, R.E., Vera-Herrera, F. and Wetzel, R.G., 1997: Effects of climate change on freshwater ecosystems of the southeastern United States and the gulf coast of Mexico. Hydrological Processes, 11,: 949–970.

Murty, A.S.,1986a. Toxicity of Pesticides to Fish I, CRC Press, Boca Raton, FL.

Murty, A.S.,1986b. Toxicity of Pesticides to Fish II,CRC Press, Boca Raton, FL.

Natarajan, A.V., 1989. Environmental impact of Ganga basin development on gene-pool and fisheries of the Ganga River system. In: Proceedings of the International large river symposium (Odge D.P.D., Ed.), Can. Spec. Publ. Fish. Aquat. Sci. 106.

Nicholls, K.H., 1998. El Nino, ice cover, and Great Lakes phosphorus: implications for climate warming. Limnol.Oceanogr. 43: 715–719

Nijssen, B., O'Donnell, G.M., Hamlet, A.F. and Lettenmaier, D.P., 2001. Hydrologic sensitivity of global rivers to climate change. Clim. Change 50, 143–175.

Poff, N.L., Angermeier, P.L., Cooper, S.D., Lake, P.S., Fausch, K.D., Winemiller, K.O., Mertes, L.A.K., Oswood, M.W., Reynolds, J., Rahel, F.J., 2001. Fish diversity in streams and rivers. In: Global Biodiversity in a Changing Environment: Scenarios for the 21st Century (Chapin, F.S.I., Sala, O.E. and Huber-Sannwald, E., Eds.), pp 351–368, Springer, New York.

Ravindranath, N.H., Rao, S., Sharma, N., Nair, M., Gopalakrishnan, R., Rao, A.S., Malaviya, S., Tiwari, R., Sagadevan, A., Munsi, M. and Krishna, N., 2011. Climate change vulnerability profiles for North-East India, Curr. Sci., 101(3): 384-394.

Sarkar, U.K. and Borah, B.C., 2018. Flood plain wetland fisheries of India: with special reference to impact of climate change. Wetl. Ecol. Manag., 26: 1-15.

Sarkar, U.K., Pathak, A.K., Sinha, R.K., Sivakumar, K., Pandian, A.K., Pandey, A., Dubey, V.K. and Lakra, W.S., 2012. Freshwater fish biodiversity in the River Ganga (India): changing pattern, threats and conservation perspectives. Rev. Fish. Biol. Fisher., 22: 251-272.

Sharma, A.P., Joshi, K.D., Naskar, M. and Das, M.K., 2015. Impact of climate variation on breeding of major fish species in inland water. Bulletin No. 185. Report number: NICRA/Policy/2015-16/1, ICAR - Central Inland Fisheries Research Institute, Barrackpore.

Sharma, E., Chettri, N., Tse-Ring, K., Shrestha, A.B., Jing, F., Mool, P. and Eriksson, M., 2009. Climate change impacts and vulnerability in the Eastern Himalayas. ICIMOD, Kathmandu, Nepal. ISBN: 978 92 9115 134 9

Singh, A.K., 2015. Advances in Indian coldwater fisheries and aquaculture. J. Fish. Sci., 9(3): 48-54

Solomon, S., 2007. The physical science basis: Contribution of Working Group I to the fourth assessment report of the Intergovernmental Panel on Climate Change. Intergovernmental Panel on Climate Change (IPCC), Climate Change 2007, 996.

Val, A.L. and de Almeida-Val, V.M.F., 1995. The Amazon ichthyofauna. In: Fishes of the Amazon and their Environment, pp. 28-69, Springer, Berlin, Heidelberg.

Vass, K.K., Das, M.K., Srivastava, P.K. and Dey, S., 2009. Assessing the impact of climate change on inland fisheries in River Ganga and its plains in India. Aquat. Ecosyst. Health. Manag., 12: 138-151.

Walther, G.R., Roques, A., Hulme, P.E., Sykes, M.T., Pysek, P., Kühn, I., Zobel, M., Bacher, S., Botta-Dukát, Z., Bugmann, H., Czúcz, B., Dauber, J., Hickler, T., Jarosík, V., Kenis, M., Klotz, S., Minchin, D., Moora, M., Nentwig, W., Ott, J., Panov, V. E., Reineking, B., Robinet, C., Semenchenko, V., Solarz, W., Thuiller, W., Vilà, M., Vohland, K. and Settele, J., 2009. Alien species in a warmer world: risks and opportunities. Trends Ecol. Evol., 24(12): 686-693.

Ward, H.B. and Whipple, G.C. Fresh-water biology. Second Edition. Edmondson, W.T. (Ed.). pages. John Wiley & Sons, Inc., New York, 1248 p.

Yao, T., Guo, X., Thompson, L., Duan, K., Wang, N., Pu, J., Xu, B., Yang, X. and Sun, W., 2006. ä 18 O record and temperature change over the past 100 years in ice cores on the Tibetan Plateau. Sci. China: Ser. D Earth Sci., 49: 1-9.

Impact of Climate Change on Hydrological Cycle, Ecosystem, Fisheries and Food Security, Pages: 341–351
Edited by: B. Madhusoodana Kurup, M.R. Boopendranath, M. Harikrishnan and A.V. Shibu

CHAPTER - 28

Taxonomic and Phylogenetic Indices of Species Assemblages in the Deep-Sea Demersal Ichthyofauna of the South-eastern Arabian Sea: Simple Measures of Diversity for Marine Ecosystem-Based Management

Sileesh Mullasseri[1*] and B. Madhusoodana Kurup[2]

[1]*School of Ocean Science and Technology, Kerala University of Fisheries and Ocean Studies (KUFOS), Panangad, Kochi-682506, Kerala, India*
[2] *School of Industrial Fisheries, Cochin University of Science and Technology (CUSAT), Lake Side Campus, Kochi-682 016, Kerala, India*
[*]*E-mail: sileeshm@gmail.com.*

The marine fisheries sector plays a significant role in the economic development of the nation by providing nutritional security and employment opportunities. The marine fish production contributes 90% catch from the coastal area up to 50 m depth (Balachandran *et al.*, 1996) Considering biodiversity aspects, most new marine taxa are likely to be found in the deep slopes and deep reefs (Eschmeyer *et al.*, 2010). Therefore, the assessment of diversity is promising for the development of fisheries. There are many studies reported on diversity of deep sea fishes of the Arabian sea (Kurup and Samuel, 1987; Venu and Kurup, 2002a), as well as distribution, biology, etc. (Kurup and Samuel, 1987; Kurup *et al.*, 2010; Kurup *et al.*, 2005; Venu and Kurup, 2002a; 2002b; 2006). The quantitative analysis of diversity is scarce in the deep water fishes of the southwest coast of India except a few studies conducted to measure classical diversity indices (Hashim, 2012; Sileesh *et al.*, 2017; Sudhakar *et al.*, 2013), from the region.

There are about 200 nations who have signed up the Convention for Biological Diversity, which was aimed at reducing the rate of biodiversity loss by 2010. It is very important to measure biodiversity to understand the rate of loss, pattern of loss, etc. There are a number of biodiversity measures such as Shannon index, and Simpson's index (Patil and Taillie, 1982). However, taxonomic diversity generally accounts for species composition (Magurran, 1988), and abundance and phylogenetic diversity is taken care of by the evolutionary history of the species in the community (Cavender-Bares *et al.*, 2009). There has been no previous report on the analysis of taxonomic and phylogenetic diversity indices of the deep sea ichthyofauna of the southwest coast of India. In this paper, we have analyzed taxonomic and phylogenetic diversity indices of deep sea fishes and compared the two depth zones of the mesopelagic realm, which is ~200 m and ~1000 m depths. These quantitative measurements will serve as a benchmark reference points for future investigation in the study area.

Materials and Methods

Samples were collected from onboard FORV Sagar Sampada (71.51m LOA, IMO: 8300080, Ministry of Earth Sciences, Government of India), from the exclusive fishery cruises conducted in the deep waters of Southeastern Arabian Sea, during 2014-2017. The study area and the trawling stations are marked in Fig. 1. High-Speed Demersal Trawl (HSDT, Designed by ICAR-CIFT), with a modification known as Crustacean Version (II CV), was deployed in the present study. This trawl had a head rope length of 38 m, footrope of 44.5 m and codend with stretched mesh size of 30 mm, gradually increasing to 130 mm in the front trawl sections. Samples were identified onboard using standard taxonomic keys and electronic databases (Alcock, 1889; Bianchi, 1991; Froese and Pauly, 2000; Smith and Heemstra, 1986). The depth of sampling ranged from 180 to 1010 m in the Arabian Sea, The stations with the average depth 180 m to 230 m were considered as ~200 m stations and the depth ranges from 970 m to 1010 m considered as ~ 1000 m stations. We have also taken a few environmental parameters using Sea-Bird SBE 9, and Software Version Seasave V7.22.

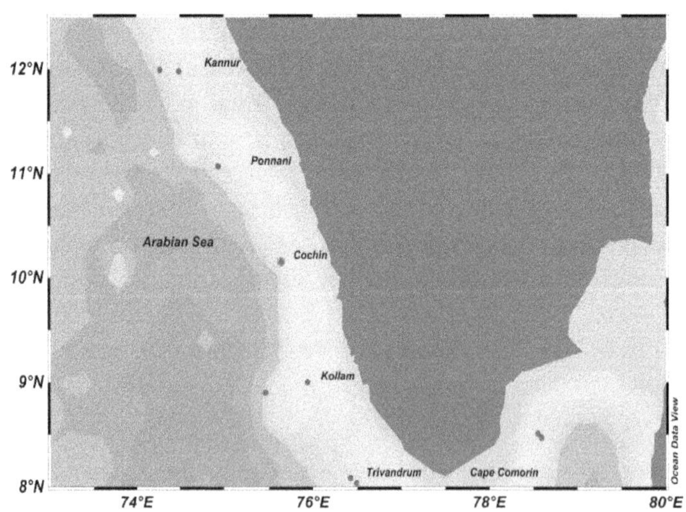

Fig. 1. Map showing trawling stations in Southeastern Arabian Sea

Statistical analysis

The analysis of the diversity indices and community structure were made using PRIMER Version 6 for Windows (Clarke and Gorley, 2006; Clarke and Warwick, 2001).

Taxonomic diversity index

The new biodiversity indices viz., taxonomic diversity (") and taxonomic distinctness ("*) (Warwick and Clarke, 1995), are useful in capturing not only the distribution of abundances amongst species but also the taxonomic relatedness of the species in each sample. The taxonomic distinctness divided based on presence/absence data into two types is namely (i) average taxonomic distinctness ("+) and (ii) variation in taxonomic distinctness ("+). Average taxonomic distinctness index (Ä+) and variation in taxonomic distinctness (ë+), studied graphically by employing the funnel method. Combined ë+ and Ä+, represented by ellipse plot.

Phylogenetic diversity

Phylogenetic diversity (PD) is a biodiversity measure that takes account of phylogenetic relationships between taxa. Thus provide a better measurement for conservation than measuring richness. (Rodrigues & Gaston, 2002), therefore we have calculated phylogenetic diversity (PD) and average phylogenetic diversity.

Results and Discussion

In the trawling experiments conducted, we have encountered 79 deep sea fish species from all the ten trawls. The environmental parameters such as pressure, temperature, and salinity Oxygen, water density turbidity, and conductivity are shown in table 1. The pressure in the ~200m depth zone ranges from 141.84 dB to 201.23dB, and that for ~1000m depth zone is in the range of 871.80 dB to 1008.16, usually pressure increases with depth. The temperature at the bottom at the time of operation for 200m depth zone ranged from 13.10 0C to 20.010C and that of 1000m depth zone is 6.850C to 8.530C. The salinity in the 200m depth zone ranged from 35.08 PSU to 35.12 PSU and that of 1000m depth zone was 35.05 - 35.05 PSU. The oxygen in the 200m depth zone ranged from 0.22 to 0.66 ML/L and for 1000m was 0.43 to 0.84. the water density in the 200m depth zone ranged from 24.83 to 26.11 Kg/m3 and for 1000m depth zone it was27.32 - 27.41 Kg/m3. The conductivity of the 200m depth zone ranged from 4.13 to 4.81 S/m and for 1000 m depth zone it was 3.55 - 3.72 S/m.

The principal component analysis of abundance and environment data are plotted in Fig. 2. The variable depth, pressure, Oxygen, and density in the positive axis (PC1), as well as salinity, conductivity, turbidity and temperature (PC2) are in the negative axis. The diversity in the 100 m depth zone is in the positive axis of the plot. PC1 is responsible for the 72.2% variation and PC2 is 19.8% variation. The taxonomic diversity (“) for the ten stations (Fig. 3), on an average for the 200 m depth zone, is 60.15, lowest 31.39, and highest 78.36. Taxonomic diversity for 1000 m depth zone is 76.01(average), lowest 55.05 and highest is 87.54. Taxonomic distance (“*) of deep-sea fishes from the ten stations of the Arabian Sea is shown in Fig. 4. The value of Taxonomic distance for the 200 m depth zone is 92.68 (average), the lowest is 85.81, and the highest 99.74. The average of Taxonomic distance (“*) for the 1000 m depth zone is 94.36, and the lowest 85.38, the highest 97.73. Average Taxonomic Distance (“+)is shown in Fig. 5, showing an average value for the 200m depth range is 94.73, and the lowest 89.74, highest is 98.22. For the 1000m depth zone, the average is 94.08 and the lowest is 94.08 and the highest value is 97.83. Total taxonomic distance (s”+) in Fig. 6, the average value for 200 m depth is 1639, the lowest is 948.5 and the highest is 2455.56. The average value of TTD for 1000 m depth is 2317.25, the lowest is 2069.84 and the highest is 2738. The variation in the taxonomic distance (•+) of the deep-sea fishes from the stations of the Arabian Sea is depicted in Fig. 7. The average value for 200m depth is 159.12, the lowest is 70.91 and the highest is 265.18. The average for 1000m depth is 193.54, the lowest value is 115.66 and the highest is 255.42. The phylogenetic distance (s§+) for the deep-sea fishes from the stations of the Arabian Sea is depicted in Fig. 8. The average value for 200m depth is 1433.33, the lowest value is 866.7 and the highest value is 2166.67. The average value for 1000m depth is 1866.67, the lowest value is 1700 and the highest is 2266.67. The average Phylogenetic distance (§+) of deep-sea fishes from the ten stations of the Arabian Sea Is shown in Fig. 9. The average value for 200 m depth is 79.31, with the lowest is 76.92, and the highest is 86.67. The average value for the 1000m depth is 81.27, the lowest is 73.9 and the highest is 86.66.

Table 1: The environmental parameters of each trawling stations at the Arabian Sea

Stations	Depth (m)	Pressure (dB)	Temperature (^{0}C)	Salinity (PSU)	Oxygen (ML/L)	Density (kg/m^3]	Turbidity [NTU]	Conductivity [S/m]
CAPE	200	194.19	14.79	35.11	0.22	26.11	0.32	4.29
Trivandrum	200	201.23	14.91	35.10	0.27	26.07	0.27	4.30
Kollam	200	201.23	13.10	35.12	0.44	26.47	0.30	4.13
Cochin	200	201.22	15.16	35.08	0.22	26.00	0.25	4.33
Kannur	200	141.84	20.01	35.08	0.66	24.83	0.48	4.81
Tuticorin	1000	1008.16	7.38	35.05	0.68	27.42	0.29	3.61
Trivandrum	1000	871.80	8.53	35.14	0.43	27.32	0.29	3.72
Kollam	1000	1008.09	7.13	35.00	0.79	27.42	0.28	3.59
Ponnani	1000	958.56	8.09	35.04	0.58	27.31	0.28	3.68
Kannur	1000	1008.08	6.85	34.94	0.84	27.41	0.30	3.55

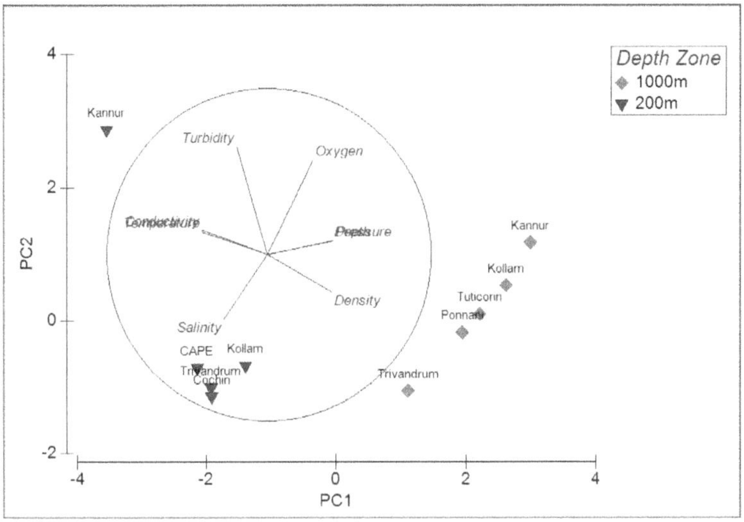

Fig. 2. Principal components analysis of the Environment data and abundance of the two extreme depth zones of the mesopelagic zone in the Arabian Sea

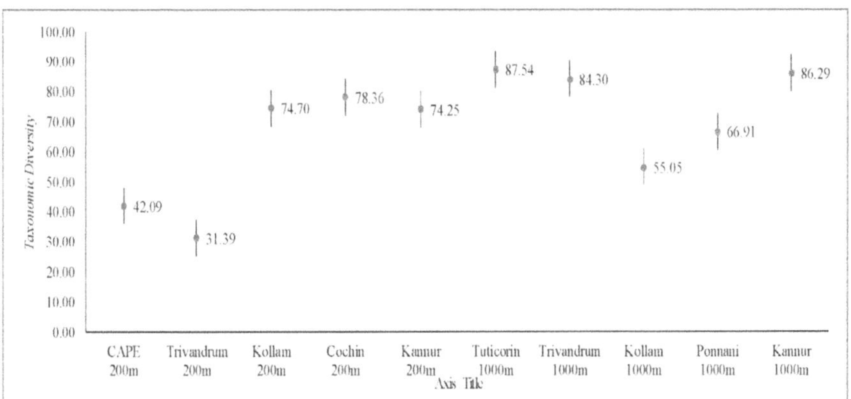

Fig. 3. Taxonomic diversity (Δ) of deep-sea fishes from the ten stations of Arabian Sea

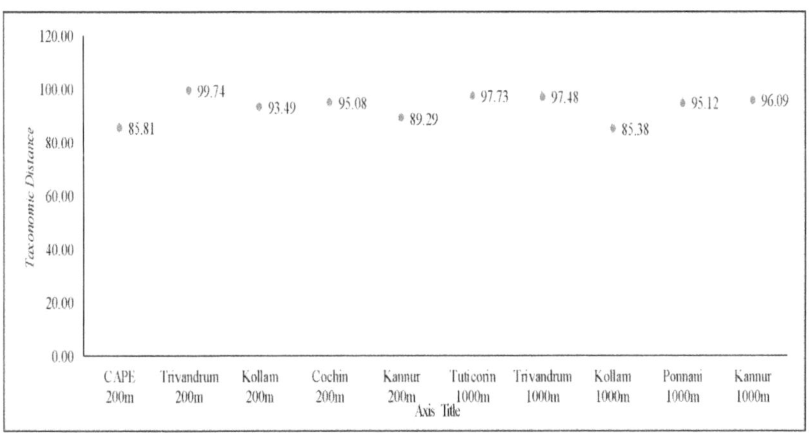

Fig. 4. Taxonomic distance (Δ*) of deep-sea fishes from the ten stations of Arabian Sea

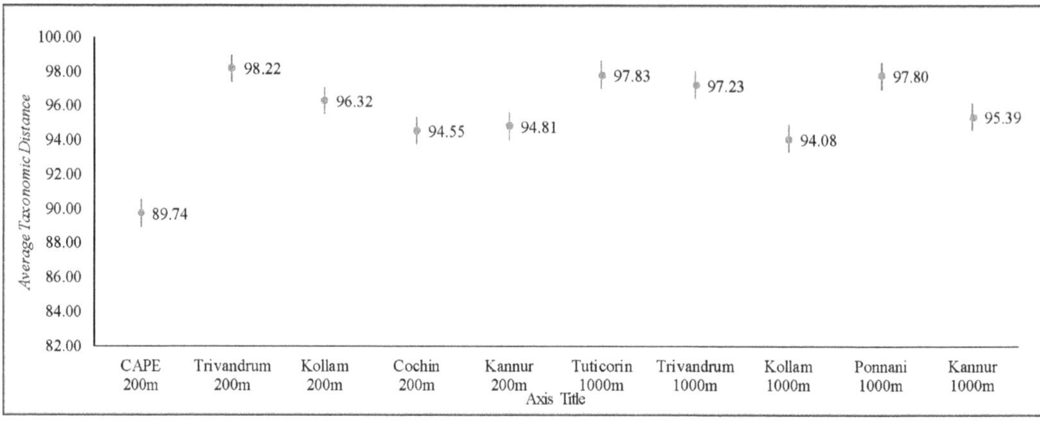

Fig. 5. Average Taxonomic Distance (Δ+) of deep-sea fishes from the ten stations of Arabian Sea

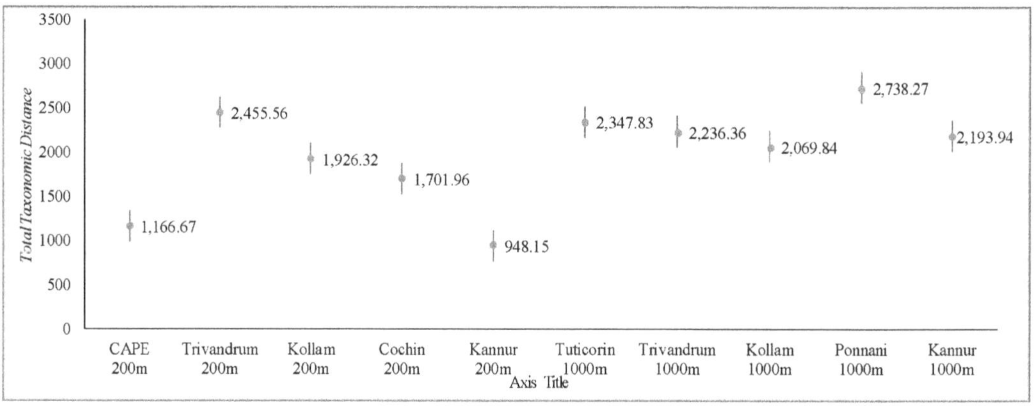

Fig. 6. Total Taxonomic Distance of deep-sea fishes from the ten stations of Arabian Sea

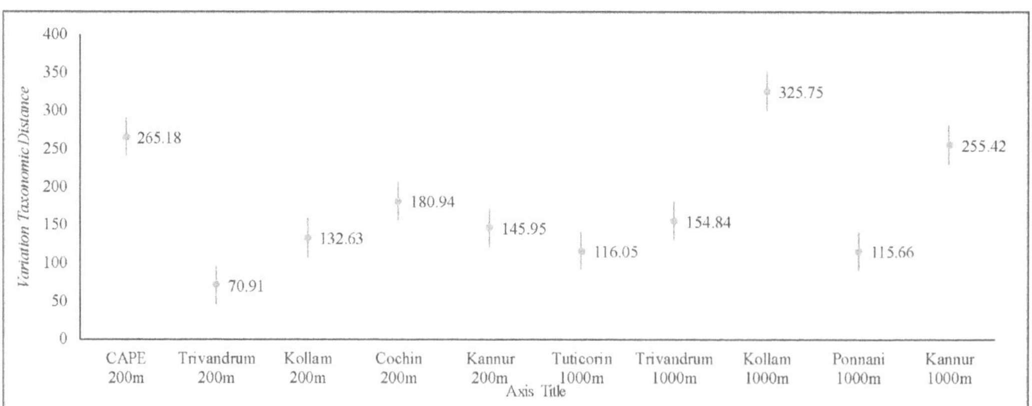

Fig. 7. Variation in the taxonomic distance (Δ+) of deep-sea fishes from the ten stations of Arabian Sea

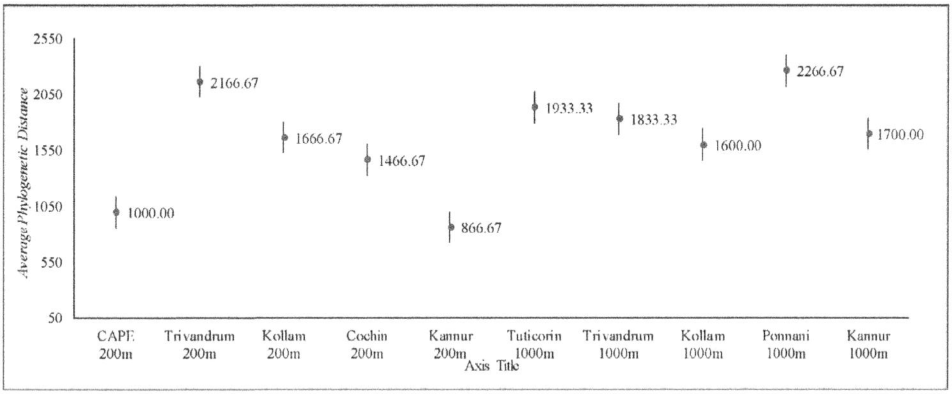

Fig. 8. Phylogenetic distance (s§+) of deep-sea fishes from the ten stations of Arabian Sea

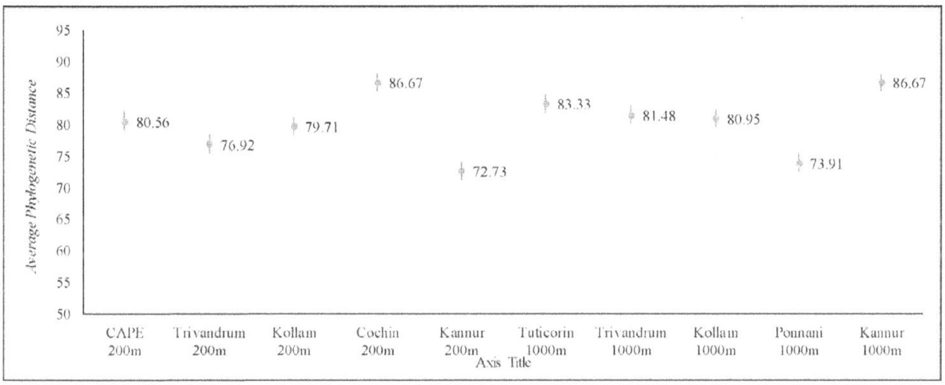

Fig. 9. Average Phylogenetic distance (§+) of deep-sea fishes from the ten stations of Arabian Sea

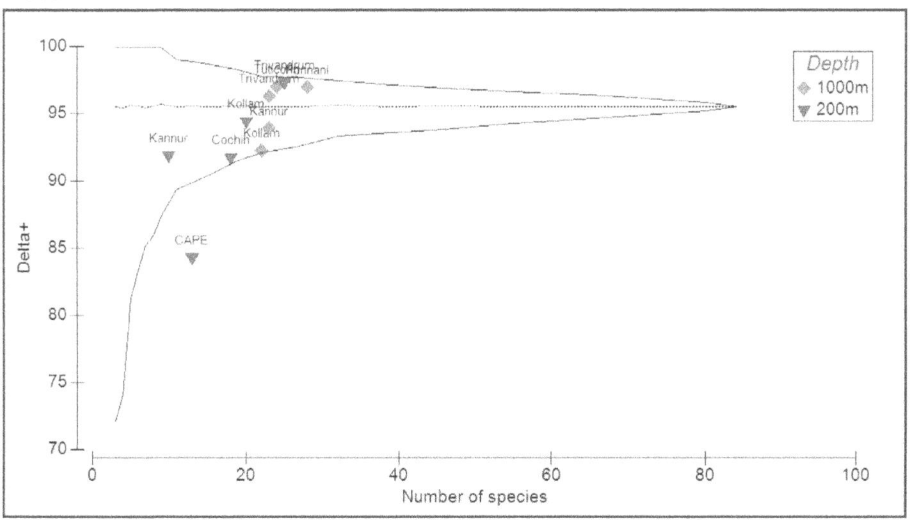

Fig. 10. Funnel plot for simulated average taxonomic distinctness (Δ+) (Thin lines indicate limits within which 95% of simulated values of Delta+ indices lie, whereas central thick lines indicate mean Delta+ for the full inventory of deep-sea fishes of the southwest coast of India, Dots represent the true values of Delta+ indices for the studied sites with species richness.)

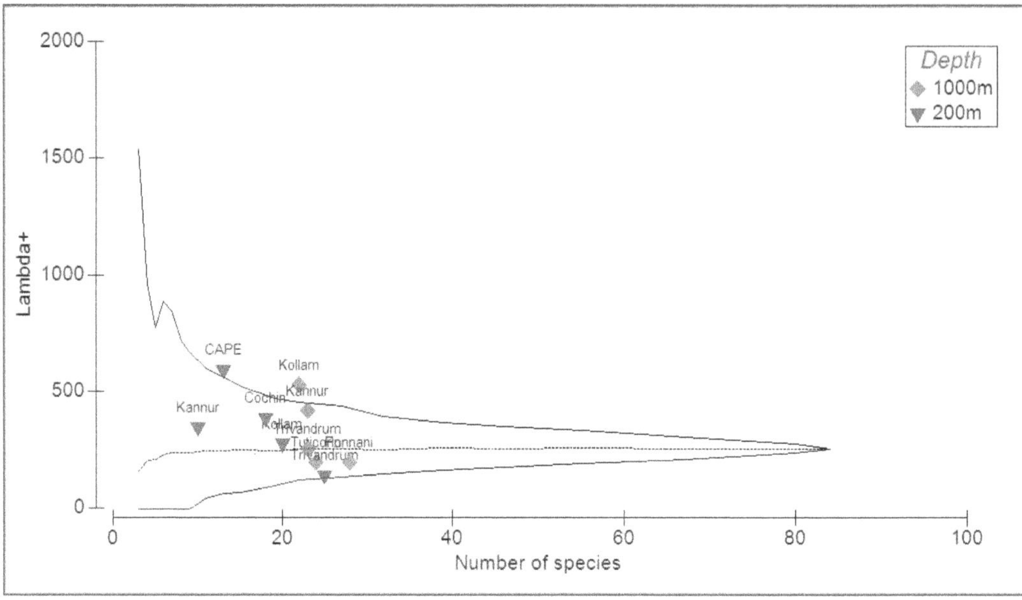

Fig. 11 Funnel plot for simulated average variation in taxonomic distinctness (•+) (Thin lines indicate limits within which 95% of simulated values of •+ indices lie, whereas central thick lines indicate mean •+ for the full inventory of deep-sea fishes of the southwest coast of India, Dots represent the true values of •+ indices for the studied sites with species richness.)

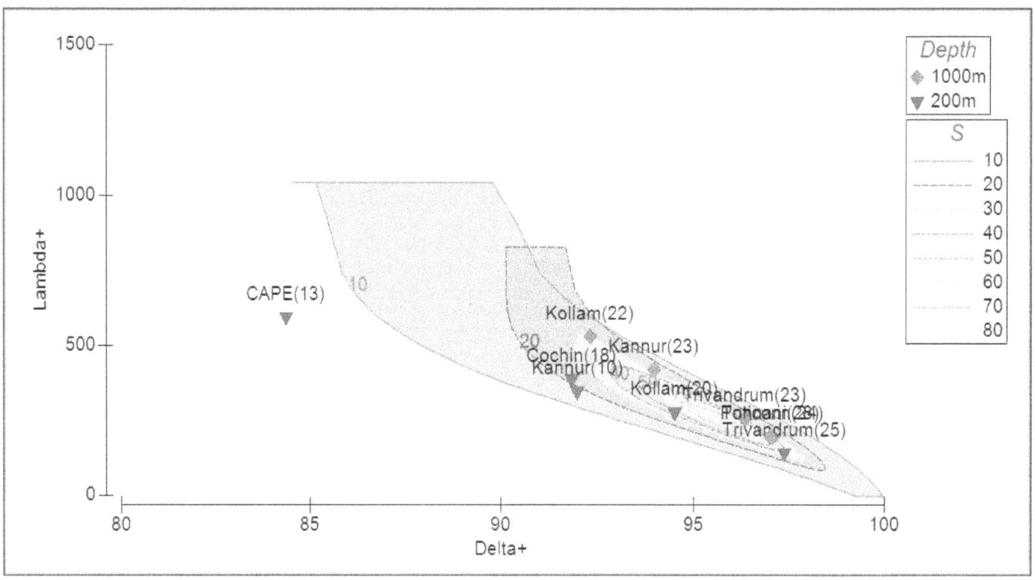

Fig. 12. Scatter plot of simulated VarTD, AvTD pairs, with superimposed probability contours. (95% of simulated values expected to fall within the outer envelope there is a statistically significant deviation between depth ranges in the diversity of fishes of deep waters of the southwest coast of India.)

In this study, 79 species of deep sea fishes were identified from the ten trawls in the two depth zones. However, the present study focus on taxonomic and phylogenetic indices of the species assemblages of deep sea ichthyofauna of the southwest coast of India. In the case of classical diversity indices, the available report that says that the diversity of deep-sea Ichthyofauna has shown as an increasing trend with depth up to 900 m (Sudhakar *et al.*, 2013) from the southwest coast of India. Moreover, an increase in diversity with increasing depth also reported from the Gulf of Mexico (Bianchi, 1991), Mediterranean Sea (Moranta *et al.*, 1998) and North Atlantic Ocean (Farina *et al.*, 1997; Merrett *et al.*, 1991).

There is no previous report on taxonomic and phylogenetic diversity indices for the deep sea fishes of the southwest coast of India. While comparing the two depth zone (200 m and 1000 m) all the taxonomic indices are showing a higher value for 1000 m depths, which is in full conformity with that of Sudhakar *et al.* (2013). Taxonomic diversity and phylogenetic diversity indices are more useful for spatial-temporal comparison of diversities because these indices have a negligible influence on sampling bias; the measures are based on presence/absence of species when the taxonomic hierarchy is ignored and the taxonomic diversity is similar like classical diversity indices such as Simpson diversity (Clarke and Warwick, 1998). In the funnel plot for simulated average taxonomic distinctness (^{3}%+), the 1000 m depths are showing higher diversity both in simulation and sampling stations have a deviation from the normal distribution, i.e. all the stations do not fall in 95% confidence interval (Fig. 10 and 11; Fig. 12). For the variation in taxonomic distinctness (•+), the 1000 m depths are showing higher diversity in simulation and sampling stations have a deviation from the normal distribution. The measured values produced in the study will serve as benchmark data for future studies on deep sea fish biodiversity along the southwest coast of India. These results will also serve as valuable information for ecosystem-based conservation and management strategies for the region.

Acknowledgments

We would like to thank the Centre for Marine Living Resources and Ecology (Ministry of Earth Sciences, Government of India) for the funding and for providing facilities and the Fisheries Oceanographic Research Vessel Sagar Sampada for fishing operations.

References

Alcock, A.W., 1889. Natural history notes from H. M. Indian marine survey steamer, Investigator. No. 13. On the bathybial fishes of the Bay of Bengal and neighbouring waters, obtained during the seasons 1885-1889. Annals and Magazine of Natural History (Series 6), 4(24): 450-461.

Balachandran, K., Menon, N.G., and Pillai, N.G.K., 1996. Scope for the exploitation and management of non-conventional fish resources from the distant waters of Indian EEZ. Proceedings of the Fourth Indian Fisheries Forum: 405–409.

Bianchi, G., 1991. Demersal assemblages of the continental shelf and slope edge between the Gulf of Tehuantepec (Mexico) and the Gulf of Papagayo. Marine Ecology Progress Series, 73: 121-140.

Cavender-Bares, J., Kozak, K.H., Fine, P.V.A. and Kembel, S.W., 2009. The merging of community ecology and phylogenetic biology 12(7): 693-715. doi: 10.1111/j.1461-0248.2009.01314.x.

Clarke, K.R. and Gorley, R.N., 2006. PRIMER v6: User Manual/Tutorial. PRIMER-E. 192.

Clarke, K.R. and Warwick, R.M., 1998. A taxonomic distinctness index and its statistical properties. Journal of Applied Ecology 35: 523-531.

Clarke, K.R. and Warwick, R.M., 2001. Change in Marine Communities: An Approach to Statistical Analysis and Interpretation. 2nd Edition, PRIMER-E, Ltd., Plymouth Marine Laboratory, Plymouth.

Eschmeyer, W.N., Fricke, R., Fong, J.D. And Polack, D.A., 2010. Marine fish diversity: history of knowledge and discovery (Pisces). Zootaxa 50: 19-50. doi: 10.11646/zootaxa.2525.1.2.

Farina, A.C., Fereire, J. and Gonzalez-Gurriran, E., 1997. Demersal fish assemblages in the Galician continental shelf and upper slope (NW Spain): spatial structure and long-term changes. Estuarine Coastal Shelf Science 44: 435-454.

Froese, R. and Pauly, D., 2000. FishBase. World Wide Web electronic publication. Retrieved October 18, 2019, from The International Center for Living Aquatic Resources.

Hashim, M., 2012. Distribution diversity and biology of deep-sea fishes in the Indian EEZ. Ph.D. Thesis, Cochin University of Science and Technology, Cochin.

Kurup, B.M. and Samuel, C., 1987. Ecology and fish distribution pattern of a tropical estuary. Proceedings of the National Seminar on Estuarine Management, Trivandrum: 339-349.

Kurup, B.M., Cubelio, S.S., Joseph, G. and Venu, S., 2010. New record of the deep-sea fish *Neobythites multistriatus* (Teleostei: Ophidiidae) from the south-west coast of India. Marine Biodiversity Records 3(117): 1-4. doi: 10.1017/S1755267210001016.

Kurup, B.M., Jiji, T. And Venu, S., 2005. Distribution and biology of Chlorophthalmus bicornis Norman, beyond 250m depth off the south west coast in the Indian EEZ. Journal of Marine Biological Association of India, 47(1): 57- 62.

Magurran, A.E., 1988. Ecological Diversity and Its Measurement: Princeton University Press.

Merrett, N.R., Gordon, J.D.M., Stehmann, M. and Haedrich, R.L., 1991. Deep demersal fish assemblage structure in the porcupine seabight (eastern North Atlantic): slope sampling by three different trawls compared. Journal of the Marine Biological Association of the United Kingdom 71: 329-358.

Moranta, J., Steianescu, C., Massuti, E., Morales-Nin, B. and Lloris, D., 1998. Fish community structure and depth related trends on the continental slope of the Balearic Islands (Algerian basin, western Mediterranean). Marine Ecology Progress Series 171: 247–259.

Patil, G.P. and Taillie, C., 1982. Diversity as a Concept and its Measurement. Journal of the American Statistical Association 77(379): 548-561. doi: 10.1080/01621459.1982.10477845.

Rodrigues, A.S.L. and Gaston, K.J., 2002. Maximising phylogenetic diversity in the selection of networks of conservation areas. Biological Conservation 105: 103-111.

Sileesh, M.S., Alphi, K., Harish, K.C. and Viji, V., 2017. Species assemblages and community structure of deep-sea demersal ichthyofauna of the South-eastern Arabian Sea (SEAS). Journal of the Marine Biological Association of the United Kingdom 98(7): 1775-1781. doi: 10.1017/S0025315417001151.

Smith, M.M. and Heemstra, P.C.,1986. Smith's Sea Fishes: Springer Verlag Publishers.

Sudhakar, G.V.S., Sreedhar, U. And Meenakumari, B., 2013. Abundance, bathymetric distribution and diversity of deep sea demersal fin fish resources along the South-West Coast of India. Indian Journal of Fisheries Research 60: 1-6.

Venu, S. and Kurup, B.M., 2002a. Distribution and abundance of deepsea fishes along the West coast of India. Fishery Technology 39: 20-26.

Venu, S. and Kurup, B.M., 2002b. Distribution and biology of the deep sea fish *Psenopsis cyanea* (Alcock) inhabiting continental dope of the west coast of India. Journal of Marine Biological Association of India 44(1&2): 176-186.

Venu, S. and Kurup, B.M., 2006. Distribution and biology of deepsea fishes Neoepinnula orientalis Gilchrist and von Bonde 1924 and Psenes squamiceps Lloyd, 1909 from west coast of Indian EEZ. Journal of Marine Biological Association of India 48: 24-28.

Warwick, R.M. and Clarke, K.R., 1995. New biodiversity measures reveal a decrease in taxonomic distinctness with increasing stress. Marine Ecology Progress Series 129: 301-305.

Impact of Climate Change on Hydrological Cycle, Ecosystem, Fisheries and Food Security, Pages: 353–358
Edited by: B. Madhusoodana Kurup, M.R. Boopendranath, M. Harikrishnan and A.V. Shibu

CHAPTER - 29

Marine Faunal Species Inventory for the Indian EEZ: Necessity and Significance

Vinay P. Padate*, Rufus Thella, Rajeeshkumar, M.P., Aiswarya Gopal, Usha Parameswaran, Aswini Nayak, Sudhanshu Dixit, S. Kumaralingam, K. Chandrasekar, Johnny Konjarla and N. Saravanane

Centre for Marine Living Resources & Ecology, Ministry of Earth Sciences, Government of India, Atal Bhavan, LNG Road, Puthuvype, Kochi, Kerala-682508, India
**E-mail: vinaypadate@gmail.com*

Species inventories are potential tools to "inform and catalyze action for biodiversity conservation and policy change, critical to protecting the natural resources we need to survive" (IUCN, 2019). India with approximately 7,500 km of coastline and 2.02 million km^2 of Exclusive Economic Zone (Wadhawan, 2016) supports a rich assemblage of marine fauna across 24 animal phyla. The marine faunal diversity of Indian waters has been assessed several times in the recent past (Venkataraman & Wafar, 2005; Wafar *et al.*, 2011; Venkataraman & Raghunathan, 2015a, 2015b; Dash, 2018), with variable estimates (Fig. 1). This may be attributed to

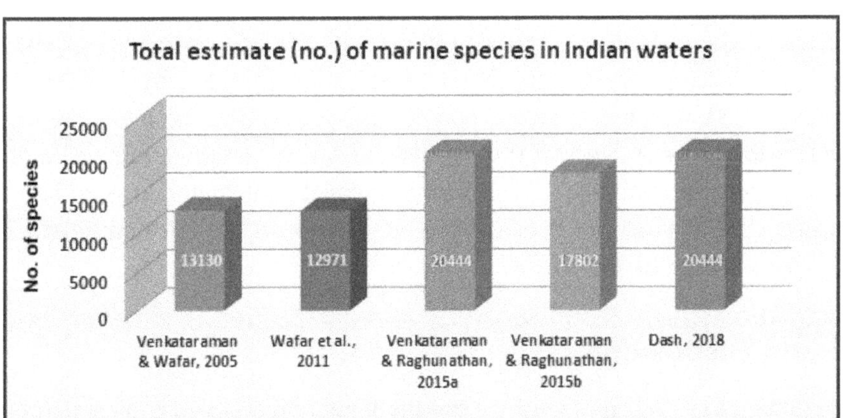

Fig. 1. Total estimate (No.) of marine species in Indian waters (based on published literature).

(i) **Taxonomic status of species** — recent taxonomic revisions using phylogenetic, morphometric and morphological analyses have resulted in discovery of cryptic species across animal phyla, and relegation of older binomial nomenclature to synonymy. It is assumed that these estimates might not have accounted for the changes arising out of these developments.

(ii) **Inconsistencies in species enumeration** — for example, different publications in the same year (Venkataraman & Raghunathan, 2015a, 2015b) have reported widely different estimates. Such inconsistencies may arise out of gleaning data from uncurated check-lists.

Costello *et al.* (2013) emphasized the issue of species checklists being "compromised" due to lack of curation resulting in "inheriting past misuse of names", "use of same name for dissimilar species in different locations", "use of differing names for the same species in different regions", or "combination of these."

The ambiguities in the marine faunal species reported from the Indian waters, coupled with the uncertainty about the genuineness of some of the existing faunal check-lists, necessitated a re-assessment of the marine faunal diversity of Indian waters based on collation of species records from the above literature and verification of the taxonomic status of these species using the World Register of Marine Species (WoRMS) database (WoRMS Editorial Board, 2017). The present study attempted to compile a checklist, and verify the validity, of marine species from the Indian waters based on published records. Secondly, the paper discusses the discrepancies of previous estimates of marine species diversity in this region.

Materials and Methods

This exercise was undertaken by the Centre for Marine Living Resources and Ecology, Kochi and involved a team of project scientists and other project staff. Each participant was allotted an animal phylum / several phyla to compile data on species reported from Indian waters in Microsoft Excel software along with the reference source. The validity of the data was assessed using the "Match taxa" tool provided by WoRMS database (Horton *et al.*, 2018), which notifies the taxonomic status of the taxon concerned, assigns a unique numeric identifier (AphiaID), presents taxonomic classification, and indicates its environment (wherever available). Data from individual contributors was collated and re-processed using the R software package provided by WoRMS (Horton *et al.*, 2018) to corroborate the validity of the species.

Results and Discussion

The present attempt at compilation of the marine faunal diversity of Indian waters yielded 13,250 species constituting Version 1.0 of the Check-list of Marine fauna of Indian waters. Validation of the taxonomic nomenclature with the WoRMS database yielded 12,890 taxa (species and sub-species) with valid AphiaID. Out of these, 12,535 taxa were categorized as exclusively marine, 202 taxa as admixture of brackish, freshwater and terrestrial, and 30 taxa without details of "environment" category (Fig. 2a). Subsequently, 123 taxa without valid AphiaID were categorized into 5 "taxonomic status" categories provided in the WoRMS database (Fig. 2b). The remaining 360 taxa were not recognized by the database (bearing no AphiaID) (Fig. 2b).

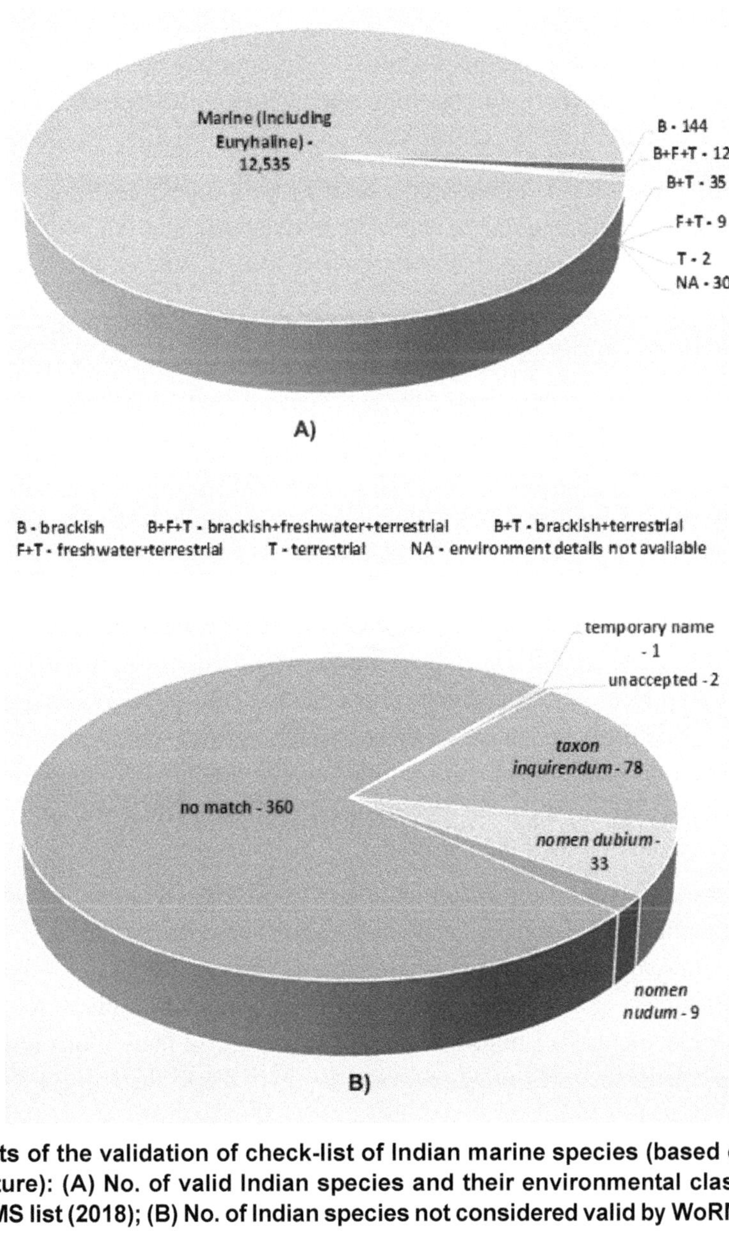

Fig. 2. Results of the validation of check-list of Indian marine species (based on published literature): (A) No. of valid Indian species and their environmental classification by WoRMS list (2018); (B) No. of Indian species not considered valid by WoRMS list (2018).

The Check-list compilation exercise, although enormous in scale, was facilitated by availability of taxonomic literature from open access journals hosted by their respective publishers, as well as classical taxonomic works hosted by the Biodiversity Heritage Library (Gwinn & Rinaldo, 2009), AToL Decapoda database (Crandall & Feldmann, 2010), and the Digital Archive-Publications of Zoological Survey of India (2018). These websites have consolidated widely scattered natural history journals, monographs and books on a single platform, and provided free access to internet users in general, thereby ensuring rapid exchange of the most updated taxonomic information. Moreover, the advent of computer applications such as Microsoft Excel has facilitated easy and rapid compilation, processing, comparison and presentation of data. In comparison, prior to the era of the desktop computer and internet access, obtaining taxonomic literature as well as data management and collation were cumbersome and time-consuming processes.

However, the compilation process encountered two major obstacles, availability of published literature and ambiguous taxonomic nomenclature, both of which are discussed below in detail.

(i) **Availability of published literature**: the survey of substantial amount of observational papers and a few checklists is pending due to unavailability of reprints of some of the recognized taxonomic works, and restricted access to articles in subscription-based online taxonomic journals.

(ii) **Ambiguous taxonomic nomenclature**: the presence of ambiguous binomial/ trinomial nomenclature in the published literature delayed the compilation process owing to the time spent in verifying the validity of the concerned taxon, and uncertainty over retaining such names in the check-list. Ambiguous taxonomic nomenclature is a result of (a) erroneous spellings or typographical errors; (b) existence of multiple synonyms of a species in different taxonomic studies (literature sources); (c) presence of secondary homonyms (resulting from transferring a species to a new genus) (Remsen, 2016), and the failure to distinguish between these, which resulted in erroneous data entry; (d) nomenclature that is either unavailable (*nomen nudum*), of uncertain application (*nomen dubium*), belonging to an incompletely defined taxon (*taxon inquirendum*), or unaccepted by the WoRMS database.

All of the above factors could result in an artificial inflation in the number of species, thereby jeopardizing the estimation of alpha-biodiversity (Dayrat, 2005). In view of the limitations encountered during the compilation of Version 1.0 of the Check-list, it is proposed to undertake rigorous search of all available taxonomic literature, keep abreast of the latest developments in animal taxonomy, and exercise meticulous attention to the use of taxonomic nomenclature during the compilation of the forthcoming version (Version 2.0) of the Check-list.

Acknowledgements

The authors are grateful to the Director, Centre for Marine Living Resources and Ecology, Ministry of Earth Sciences, Government of India, Kochi, Kerala.

References

Costello, M.J., Bouchet, P., Boxshall, G., Fauchald, K., Gordon, D., Hoeksema, B.W., Poore, G.C.B., van Soest, R.W.M., Stohr, S., Walter, T.C., Vanhoorne, B., Decock, W., Appeltans, W., 2013. Global Coordination and Standardisation in Marine Biodiversity through the World Register of Marine Species (WoRMS) and Related Databases. PLoS One 8(1), e51629. http://dx.doi.org/ 10.1371/journal.pone.0051629

Crandall, K.A., Feldmann, R.M., 2010. AToL Decapoda — The decapod crustacean tree of life. In Why study the tree of life? — The scientists speak (ed. Mikkelsen, P.). Am. Paleontol. 18(3), 11 p.

Dash, S., 2018. Marine faunal diversity. Curr. Sci. 115(2): 199–200.

Dayrat, B., 2005. Towards integrative taxonomy. Biol. J. Linnean Soc., 85: 407–415.

Digital Archive-Publications of Zoological Survey of India, 2018. http://faunaofindia.nic.in/index.php (accessed on 14 July 2018)

Gwinn, N.E. and Rinaldo, C.A., 2009. The Biodiversity Heritage Library: Sharing biodiversity with the world. IFLA J. 35 (1), 25–34. https://doi.org/10.1177/0340035208102032

IUCN, 2019. The IUCN Red List of Threatened Species. Ver. 2019-3. http://www.iucnredlist.org. (Accessed on 10 December 2019)

Remsen, D., 2016. The use and limits of scientific names in biological informatics. In: Anchoring Biodiversity Information: From Sherborn to the 21st Century and Beyond, Michel, E. (Ed.). ZooKeys 550: pp. 207–223. https://doi.org/10.3897/ zookeys.550.9546

Venkataraman, K. and Raghunathan, C., 2015a. Marine Biodiversity in India: status and threats. In: Marine Faunal Diversity in India: Taxonomy, Ecology and Conservation (Venkataraman, K. and Sivaperuman, C., Eds.), pp. 1–47, Academic Press, London, San Diego. https://doi.org/10.1016/C2014-0-00821-0

Venkataraman, K. and Raghunathan, C., 2015b. Coastal and Marine Biodiversity of India. In: Marine Faunal Diversity in India (Venkataraman, K. and Sivaperuman, C., Eds), pp. 303–348, Elsevier, Amsterdam. https://doi.org/10.1016/B978-0-12-801948-1.00019-7

Venkataraman, K. and Wafar, M., 2005. Coastal and marine biodiversity of India. Indian J. Mar. Sci., 34(1): 76–87.

Wadhawan, S.K., 2016. Marine Geology and Coastal Surveys for Seabed Mapping and Mineral Resource Evaluation in India – An Overview. Proc. Indian Natn. Sci. Acad. 82(3): 675–683. https://doi.org/ 10.16943/ptinsa/2016/48476

Wafar, M., Venkataraman, K., Ingole, B., Ajmal Khan, S. and Loka Bharathi, P., 2011. State of Knowledge of Coastal and Marine Biodiversity of Indian Ocean Countries. PLoS One 6(1), e14613. https://doi.org/10.1371/journal.pone.0014613

WoRMS Editorial Board, World Register of Marine Species, 2018. http://www.marinespecies.org (Accessed on 14 July 2018)

Impact of Climate Change on Hydrological Cycle, Ecosystem, Fisheries and Food Security, Pages: 359–364
Edited by: B. Madhusoodana Kurup, M.R. Boopendranath, M. Harikrishnan and A.V. Shibu

CHAPTER - 30

Widespread Invasion of an Alien Mussel in Ashtamudi Lake, a Ramsar Site in Kerala, India

P.R. Divya[1]*, T.T. Ajith Kumar[1], Kuldeep Kumar Lal[2]

[1]*PMFGR Centre, ICAR National Bureau of Fish Genetic Resources, Kochi-682 018, India*
[2]*ICAR-National Bureau of Fish Genetic Resources, Canal Ring Road, P.O. Dilkusha, Lucknow-226 002, India*
**E-mail: divyanbfgr@gmail.com*

Invasive Marine Species (IMS), also known as introduced marine pests, transported via fouling of ship hulls (Farrapeira *et al.*, 2010) and ballast water exchange (Spinuzzi *et al.*, 2013) pose threats to marine environments the world over. Marine bio-invasion has been considered one of the greatest threats challenging the health of the oceans. Over 90% of the world's cargo is transported by sea and eventually it leads to the trans-oceanic movement of nearly 10 billion tonnes of ballast water annually. This is a prime vector for the introduction of living organisms, including pathogens, into alien regions, usually along the coasts of various countries. The Ashtamudi lake, one of the Ramsar sites in Kerala, is a tropical backwater habitat along the southwest coast of India, covering an area of about 32 km^2. Lying between latitudes 80°52' and 80°60' N and longitudes 76°30' and 76°40' E, Ashtamudi faces several threats including soil erosion due to deforestation from the wetland catchment, pollution and saline intrusion. The mangrove ecosystem in the lake has also been considerably disturbed by the construction of tourist resorts, hotels and godowns. Ashtamudi forms a natural bed for Asian green mussel *Perna viridis* and yellow clam *Paphia malabarica.*

P. viridis is found in the coastal waters of the Indo-Pacific region and is considered a candidate species for mariculture in India. Extraordinary growth performance, natural abundance, adaptability to new environments and fairly simple culture techniques make mussels of the genus *Perna* an ideal shellfish crop in many Asian countries (Vakily, 1989). In Kerala, naturally occurring populations of *P. viridis* are recorded from backwaters of

Kollam, Cochin, Calicut and Kannur. This green mussel is an important fishery resource in the lower stretches of Ashtamudi with the production estimated at 2800 t (Gomathi *et al.*, 2017). In India, the technology for farming the green mussel, *Perna viridis*, was developed during 1970s and was subsequently tested for feasibility at various locations along the country's southeast and southwest coasts by ICAR Central Marine Fisheries Research Institute (ICAR-CMFRI), Kochi. Ashtamudi forms natural beds for green mussel and MSC (Marine Stewardship Council) certified yellow clam. MSC certification was done to ensure that this valuable resource is not overfished and its ecosystem is protected. Thereby occurrence of the invasive alien species in this ground has to be looked into with caution and the present study aims to explore and report the extended distribution and invasion of *M. strigata* in the natural beds of yellow clam and green mussel in Ashtamudi Lake.

Materials and Methods

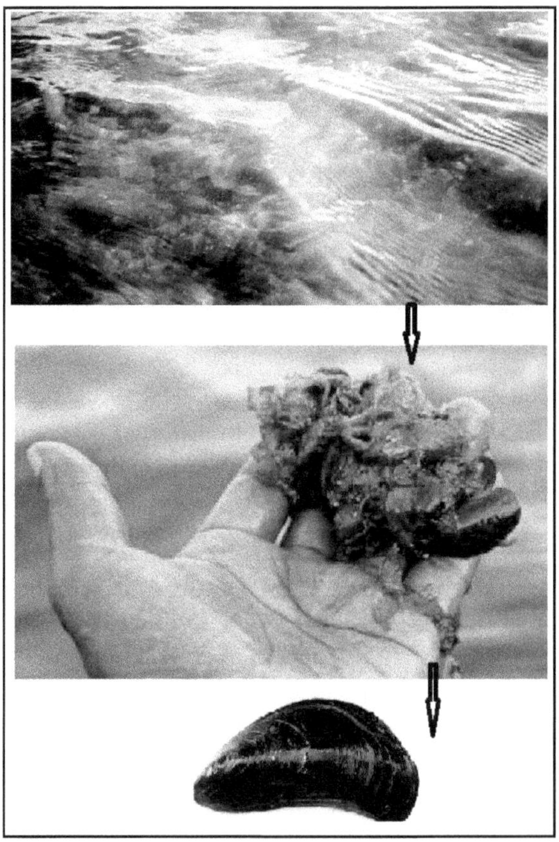

Fig. 1. Alien species *Mytella strigata* in the natural beds of *Perna viridis*, in Ashtamudi Lake.

As part of regular exploration in Ramsar sites in Kerala, explorations were conducted in Ashtamudi. We observed widespread distribution of an alien species of mussel in the natural beds of *P. viridis and P. malabarica* (Fig. 1) in Ashtamudi lake. Samples were brought to lab and morphological examinations were made. Mantle tissues were collected and preserved in 90% alcohol. Total genomic DNA was extracted from the mantle tissue using DNeasy Blood and Tissue Kit (Qiagen, Germany). The partial sequence of mitochondrial COI gene was amplified using corresponding primers (Folmer *et al.*, 1994). The reaction was carried out in a 25µl reaction volume, with components of 1x Taq buffer, 2.0 mM $MgCl_2$, 0.5µm each primer, 200 µM of each dNTP, 0.75 U Taq DNA polymerase and 25ng of template DNA. The PCR conditions were maintained as initial denaturation at 95°C 4 min, followed by 35 cycles of 94°C for 30s, 50°C for 30s, 72°C for 45s for denaturation, annealing and extension, with a final extension at 72°C for 10 min. The PCR products were sequenced bidirectionally using ABI 3730 capillary sequencer, following manufacturer instructions. The sequences generated in our study were compared to sequences on GenBank to determine the identity of the species. Temperature and salinity of the sampling sites were also noted at the time of collection.

Results and Discussion

Morphological and molecular anaysis suggested that the alien species is *Mytella strigata.* DNA barcoding results from eight individuals (GenBank accession numbers: Accession No MN531546- MN531553) matched with the sequences attributed to *Mytella charruana/ Mytella strigata* with an average match of 99.9-100%. Lim *et al.* (2018), Tay *et al.* (2019) and Sanpanich and Wells (2019) used the name *Mytella strigata* for this mussel, based on a detailed examination of 19th century publication (Lim *et al.*, 2018). The World Registry of Marine Species still uses *M. charruana* (Mollusca Base, 2019).

This alien invasive biofouling mussel *M. strigata* (Hanley, 1843) (Mollusca: Mytilidae) was recently recorded for the first time in India from Vembanad lake, a Ramsar site in Kerala (Jayachandran 2019). In this report, we record the extended distribution of *M. strigata* in Ashtamudi lake, which is approximately 150 km away from Vembanad lake. The egg and larvae introduced in our waters through ballast waters may be capable of settlement only in two weeks post fertilization time (Tay *et al.*, 2018). Migration of the planktonic larvae happening due to coastal currents in this two weeks' time, may promote the rapid extension of this species in our waters. To add, *M. strigata* is described as primarily marine, it is capable of tolerating a wide range of salinities from 2 to 40 (Yuan *et al.* 2009) and a wide range of temperatures (6–36°C) (Brodsky *et al.*, 2011). Coupled with their proven tendency to spread quickly, these fouling species have great potential to survive and establish populations in a non-native environment. The temperatures recorded at the collection sites during day time (6 AM - 6 PM) in Ashtamudi ranged from 23 to 25.5°C and salinity ranged from 20 to 24 PPT. Given the adaptability of mytilids to estuarine and coastal conditions and their high reproductive rates in eutrophic conditions, it is highly likely that *Mytella* will be established and be an invasive in Ashtamudi lake. This alien species are occupying the similar grounds of commercially important *P. viridis* species in our waters and the local fishermen were really concerned. Introduced marine pests are posing threats to marine environments. As we know, IMS can dominate their new environment, outcompete native species, threaten fisheries and aquaculture and may introduce new diseases. *M. strigata* is considered as an invasive alien species in World Register of Introduced marine species (Wo RiMS) database too.

M. strigata is native to the Caribbean coast of South America. Recently, this species has been reported as invasive, especially from Southeast Asian Countries. *M. strigata* was first reported as an alien species in the Philippines in 2014 (Mediodia *et al.*, 2017; Vallejo *et al.* 2017) and in Singapore in March 2016(Lim *et al.*, 2018). They observed that densities of *M. strigata* was very high, exceeding 10,000 individuals 100 cm$^{"2}$, leading to the exclusion of the native *Perna viridis*. *M. strigata* is now reported as an invasive species in Manila Bay, Philippines (Vallejo *et al.*, 2017). This mussel has been reported from Thailand as well (Sanpanich and Wells, 2019). *M. strigata* might have been introduced through ballast water (Farrapeira *et al.*, 2010) and fouling, directly from its native provinces, or spread from other South East Asian countries *viz.,* Singapore, Philippines or Gulf of Thailand, where there are already established populations.

M. strigata occupies the same ecological niche as *P. viridis* and this non indigenous species may affect the populations of the indigenous *P. viridis*. The situation is alarming and, a detailed observation about the origin and distribution of this species in our natural waters is called for. According to local fishermen, *M. strigata* has been observed in these waters for the past two years, but the abundance has increased this year, leading to low production of *P. viridis*. *M. strigata* is not preferred by the fishermen due to its low meat content and shell value and they are considering it as a pest. The mitochondrial COI gene sequences of *M. strigata* from our waters exhibit haplotypes exactly matching those reported from Florida, USA and Singapore (NCBI Accession EU917168 and MG736075 respectively). Vembanad and Ashtamudi lake open to the Arabian Sea, where the ballast waters of ships anchoring in Cochin port are discharged. The International Container Trans-shipment Terminal (ICTT), part of the Cochin Port is the largest container trans-shipment facility in India and handled about 32.02 MMT of shipping in 2018-19. These facilities may be providing a platform for the introduction of above-mentioned alien species. Perhaps, the physical and biological parameters occurring in our coastal waters will be favouring the establishment and spread of these species. With known invasiveness of the species elsewhere, we recommend more in-depth studies on the reproduction, ecology, community dynamics, population genetics and the phylogenetics of *M. strigata* in Ashtamudi lake and other parts of India. We observed a low number of mtDNA haplotypes among the introduced population, perhaps due to founder effect. Again, an in-depth study is required to confirm the same.

As per National Biodiversity Authority of India's report, *Mytiloposis sallei* (Recluz, 1849) and *Perna perna* (Linnaeus, 1758) are the other alien marine molluscan species reported from Indian waters though ballast waters (Sandilyan *et al.*, 2018). The Biological Diversity Act, 2002 and other laws for preserving endangered marine flora and fauna seldom find application in the control of routine ballast discharges. The massive destructive power of biological pollution through ships as a vector needs to be addressed and successfully controlled only by passing a comprehensive law on ballast pollution.

Acknowledgment

The authors acknowledge the Director, ICAR-NBFGR for providing guidance and support for carrying out the work. Kerala State Council for Science, Technology and Environment, Govt. of Kerala is also acknowledged for funding the YSA Scheme Project.

References

Brodsky, S., Walters, L., Schneider, K. and Hoffman, E., 2011. Cold Temperature Effects on Byssal Thread Formation Production by the Native Mussel *Geukensia demissa* versus the Non-native Mussel *Mytella charruana*. The University of Central Florida Undergraduate Research Journal 5: 15-24.

Farrapeira C.M.R., Ferreira GFdA., Tenório DdO., 2010. Intra-regional transportation of a tugboat fouling community between the ports of Recife and Natal, northeast Brazil. Braz. J. Oceanogr. 58: 1–14.

Folmer, O., Black, M., Wr, H., Lutz, R., Vrijenhoek, R., 1994. DNA primers for amplification of mitochondrial Cytochrome C oxidase subunit I from diverse metazoan invertebrates. Mol. Mar. Biol. Biotechnol. 3: 294-299.

Gomathi, P., Anil, M.K., Raheem, P.K. and Raju, B., 2017. Large scale farming of green mussel in Ashtamudi lake, Kerala. Mar. Fish. Infor. Serv., T & E Ser. 231: 15-16.

Jayachandran, P.R., Aneesh, B.P., Oliver, P.G., Philomina, J., Jima, M., Harikrishnan, K., Bijoy Nandan, S., 2019. First record of the alien invasive biofouling mussel *Mytella strigata* (Hanley, 1843) (Mollusca: Mytilidae) from Indian waters. BioInvasions Records 8(4): 828-837.

Kripa, V. And Mohammed, K.S., 2008. Green mussel, *Perna viridis*, Farming in Kerala, India - Technology diffusion process and socioeconomic impacts. Journal of the World Aquaculture Society 39: 612-624.

Lim, J.Y., Tay, T.S., Lim, C.S., Lee, S.S.C., Teo, S.L.M. and Tan, K.S., 2018. *Mytella strigata* (Bivalvia: Mytilidae): an alien mussel recently introduced to Singapore and spreading rapidly. Molluscan Res., 38: 170–186. https://doi.org/10.1080/13235818.2018.1423858

Mediodia, D., Santander - de Leon, Sheila M., Añasco, N. and Baylon, C., 2017. Shell morphology and anatomy of the Philippine Charru mussel *Mytella charruana* (d'Orbigny 1842). Asian Fisheries Science 30: 185-194.

Mollusca Base, 2019. *Mytella charruana* (d'Orbigny, 1842). Accessed through: World Register of Marine Species at: http://www.marinespecies.org/aphia.php?p=taxdetails&id=397146 on 2019-10-03.

Sandilyan, S., Meenakumari, B., Babu, C.R. and Rupam Mandal, 2018. Invasive alien species of India, National Biodiversity Authority, Chennai. http://nbaindia.org/uploaded/pdf/Iaslist.pdf

Sanpanich, K., Wells, F.E., 2019. *Mytella strigata* (Hanley, 1843) emerging as an invasive marine threat in Southeast Asia. Bioinvasions Rec., 8, 2, 343–356, https://doi.org/10.3391/ bir.2019.8.2.16

Spinuzzi, S., Schneider, K.R., Walters, L.J., Yuan W.S., Hoffman, E.A., 2013. Tracking the distribution of non-native marine invertebrates (*Mytella charruana, Perna viridis* and *Megabalanus coccopoma*) along the south-eastern USA. Mar. Biodivers. Rec., 6, e55, https://doi.org/10.1017/s1755267213000316.

Tay, T.S., Bin, Q.G., Serina Lee, S.C., Lim, C.S., Tan, K.S. and Teo S.L.M., 2018. Larval development of the invasive charru mussel, *Mytella strigata (*Bivalvia: Mytilidae). Invertebrate Reproduction and Development, 62(4): 248-256. doi: 10.1080/07924259.2018.1505671.

Vakily, J.M., 1989. The biology and culture of mussels of the Genus *Perna*. International Centre for Aquatic Living Resources Management, Manila, Philippines, 63 p.

Vallejo Benjamin Jr., Jeniffer Conejar-Espedido, Leanna Manubag, Kevin Carlo C. Artiaga, Amor M. Damatac II, Ivan Christian V.J. Imperial, Tyrll Adolf B. Itong, Ian Kendrich Fontanilla, Ernelea P. Cao, 2017. First record of the charru mussel *Mytella charruana* d'Orbignyi, 1846 (Bivalvia: Mytilidae) from Manila Bay, Luzon, Philippines. Bio Inv. Rec., 6.

Yuan, W., Walters, L., Hoffman, E., Schneider, K., 2009. Limits on survival: examining salinity tolerance in the non-native charru mussel, *Mytella charruana*. Integrative and Comp. Biol., 49, (Supplement 1): e330.

Impact of Climate Change on Hydrological Cycle, Ecosystem, Fisheries and Food Security, Pages: 365–377
Edited by: B. Madhusoodana Kurup, M.R. Boopendranath, M. Harikrishnan and A.V. Shibu

CHAPTER - 31

Assessing the Drivers of Forest Degradation in Peechi-Vazhani Wildlife Sanctuary, Western Ghats

Shalu George*, Manjusha, K., Mukesh Lal Das, Nadirsha P.S. Nawab, S. Anbazhagi and Muthukumar Muthuchamy

Department of Environmental Science, Central University of Kerala, Tejaswini Hills, Kasaragod-671316, Kerala, India
**E-mail: shalu.george.963@gmail.com*

Forest degradation increases the possibility of declining the relevant functions expecting from a natural forest system (ITTO, 2002). The changes associated with degradation are exacerbated by disturbances, which may be linked with natural or human-induced or a combination of both (FAO, 2009). Virtually all the forest ecosystem have been degraded to a higher extent through timber extraction, logging, fuel wood collection, fire, grazing, invasive species, fragmentation, deforestation, overexploitation, and climate change; which brings drastic loss of productivity and biodiversity of the forest (Hosonuma *et al.*, 2012; Morris, 2010). Studies revealed that tropical forest has been drastically degraded, which dramatically accelerated the CO_2 emission levels rather than acting as a carbon sink (Ahrends *et al.*, 2010).

Land surveys are still not appropriate for determining the specific factors of the degradation, long term assessment is needed to be performed to monitor the accurate changes of vegetation and climatic fluctuations. Satellite sensor data is more satisfactory for identifying long term changes related to the degradation process in larger areas. Satellite-based NDVI (Normalized Difference Vegetation Index) and LST (Land Surface Temperature) in proportion with time, provide useful information about the actual conditions of an area. Understanding the temporal changes is important to understand the climate and human-assisted issues in a better way; which is significant to develop adaptation strategies and policies to prevent the situation worsening (Hosonuma *et al.*, 2012; Mertz *et al.*, 2009).

Studies are rather limited to identify the forest degradation factors in protected areas of Western Ghats. Hence this study is focused on (i) Identify the factors contributing to forest degradation in Peechi- Vazhani WLS, a protected area of Southern Western Ghats, Kerala through fieldwork; (ii) Determine NDVI and LST variations throughout 1988 to 2018 using Arc GIS and Landsat dataset; (iii) Find out the relationship with NDVI and LST.

Materials and Methods

Study area

Peechi-Vazhani Wildlife Sanctuary (WLS) is located at Trissur district of Kerala state, lies in between 10° 28' - 10° 40' N latitude to 76° 18' - 76° 28'E longitude with a total extent of 125 sq. km (Figure 1). Peechi-Vazhani WLS is formed in 1958 and it is the second oldest sanctuary in Kerala. Nearly eighty percent of the sanctuary area is covered by the moist deciduous forest. The evergreen and semi-evergreen forest patches are confined to the high altitude areas and catchments of Peechi and Vazhani reservoirs. Sanctuary is divided into six sections namely, Velllani, Vazhani, Olakara-Ponmudi, Vaniyampara, Peechi, and Mannamangalam. Plantation of *Tectona grandis* (teak) and *Bombax ceiba* (Indian bombax) is practiced in a considerable portion of the forest area.

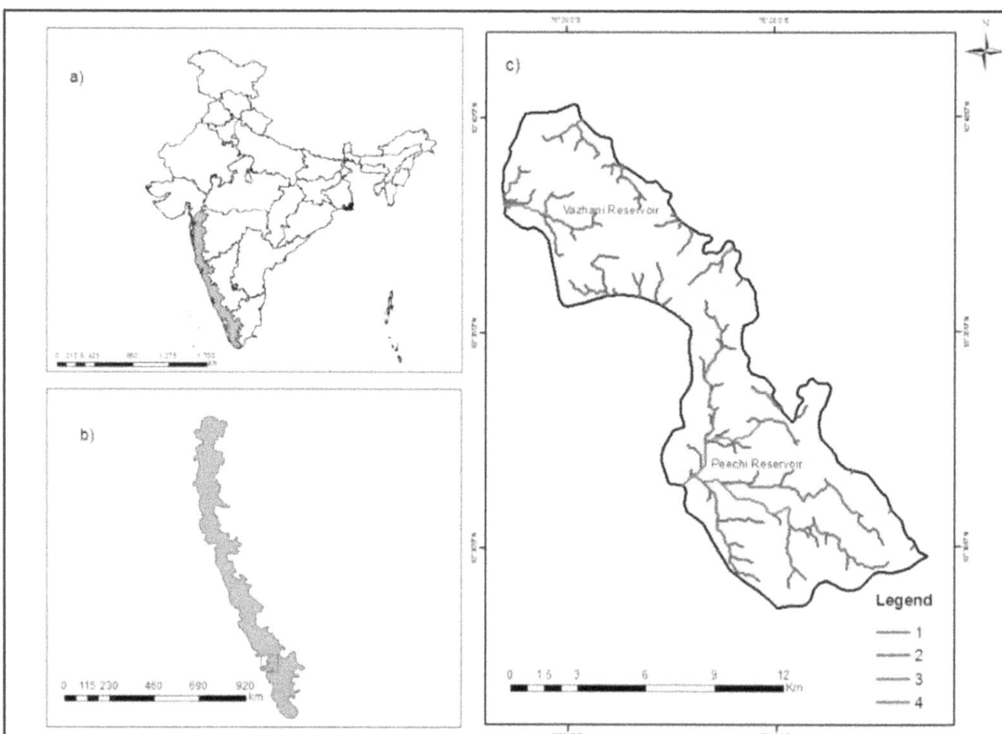

Fig. 1. (a) Map of India; (b) Map of Western Ghats highlighting the target area; (c) Peechi-Vazhani WLS drainage system

Methods

The schematic representation of the procedure is shown in Fig. 2.

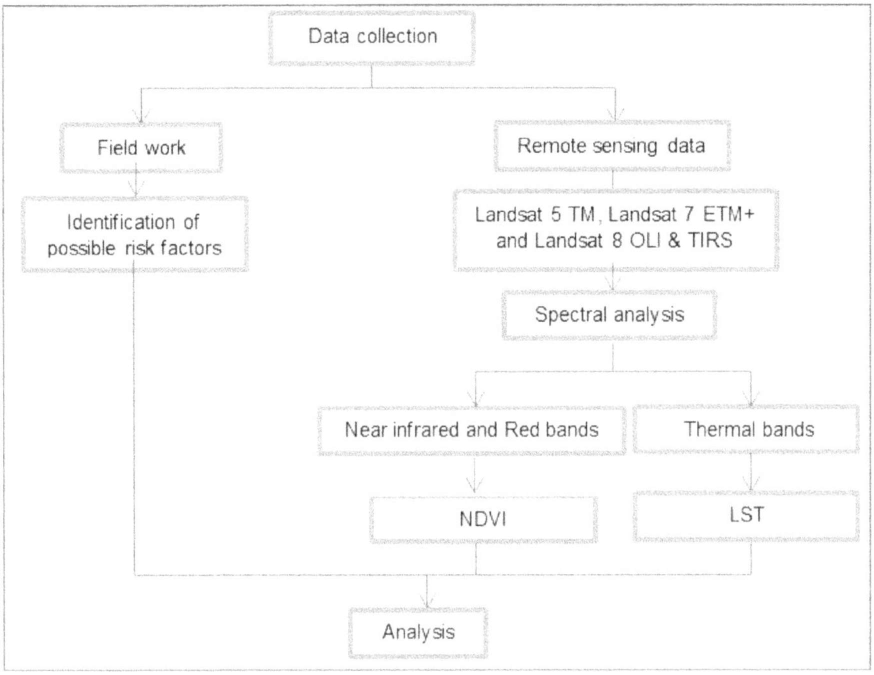

Fig. 2. Schematic representation of the research procedure

Identification of risk factors through field survey

The potential risk factors that result in the degradation of forest area in Peechi- Vazhani WLS were studied through field-work. These fieldworks were carried out in September 2019 in plantation areas, moist deciduous, semi-evergreen and evergreen forest widely distributed in six sections of Peechi-Vazhani WLS. Field observations regarding the forest plantations, invasive species, exploitation of species and the other threats were noted down. Apart from the field observations, information collected from the tribal people and concerned staff of the forest division.

Satellite image acquisition and processing

The cloud-free Landsat 5 Thematic Mapper (TM), Landsat 7 Enhanced Thematic Mapper (ETM+), Landsat 8 Operational Land Imager (OLI) and Thermal Infrared Sensor (TIRS) datasets for 1988, 2002 and 2018 respectively were downloaded from the USGS Earth explorer portal. Details of Landsat data are provided in Table 1.

Table 1: Details of Landsat data collection

Date of Image	Satellite/ Sensor	Reference System/ Path/ Row	Target output
19.1.1988	Landsat 5 TM	WRS/ 144/ 53	NDVI/ LST
18.2.2002	Landsat 7 ETM+	WRS/ 144/ 53	NDVI/ LST
21.1.2018	Landsat 8 OLI & TIRS	WRS/ 144/ 53	NDVI/ LST

Normalized Difference Vegetation Index (NDVI)

Vegetation indices are used to analyze the greenness of the target based on the reflectance pattern (Gandhi *et al.*, 2015). NDVI is calculated using equation 1.

$$\text{NDVI}=\frac{\text{NIR-RED}}{\text{NIR+RED}} \qquad \dots (1)$$

NDVI values are ranging from -1 to +1. Whereas, higher positive values are indicating the presence of good vegetation and values closer to 0 or -1 corresponds to low vegetation or absence of vegetation or water bodies.

Land Surface Temperature (LST)

For deriving LST from satellite data the following steps were followed, (i) conversion of band data to Spectral Radiance (L_{λ}) using the radiance rescaling factors provided in metadata file; (ii) conversion of Spectral Radiance to Satellite Brightness temperature in Kelvin and Kelvin to Celsius using the thermal constants; (iii) land surface emissivity determination from the NDVI.

Conversion of band data to Spectral Radiance (L_{λ}): Band data obtained from Landsat were converted to Spectral Radiance (L_{λ}) using Equation (2)

$$L_{\lambda} = M_L Q_{cal} + A_L \qquad \dots (2)$$

Where, L_{λ} = Spectral radiance (W/ (m^2*ster*μm)); M_L = Band specific multiplicative rescaling factor; Q_{cal} = Quantized and calibrated standard product pixel values (DN); A_L = Band specific additive rescaling factor

Conversion of Spectral Radiance to Satellite Brightness Temperature (T_B): Spectral Radiance were converted to satellite brightness temperature using the thermal constants provided in the metadata file. The conversion formula used for the study is given in equation (3)

$$T_B = \frac{K2}{\ln\left(\frac{K1}{L_l}+1\right)} \qquad \dots (3)$$

Where, T_B = effective at satellite temperature in K; L_λ = spectral radiance in W/m^2 ster ìm; K_1 and K_2 = pre- launch Calibration constants.

Conversion of Kelvin to Celsius: To convert the temperature from Kelvin to Celsius, the following equation was used,

$$T_B = T_B - 273.15 \quad \ldots (4)$$

Proportion of Vegetation (Pv): Pv is the proportion of vegetation which was obtained using the following equation (5)

$$Pv = ((NDVI - NDVI_{min})/(NDVI_{max} - NDVI_{min}))^2 \quad \ldots (5)$$

Whereas, $NDVI_{min}$ is minimum value of NDVI and $NDVI_{max}$ is the maximum value of NDVI.

Lands surface emissivity (ε): Land surface emissivity was calculated using the equation (6)

$$\varepsilon = m\ Pv + n;\ \text{with } m= \varepsilon_v - \varepsilon_s - (1 - \varepsilon_s)\ F\varepsilon_v \text{ and } n = \varepsilon_s + (1 - \varepsilon_s)F\varepsilon_v \quad \ldots (6)$$

Where ε_v= emissivity of vegetation; ε_s = emissivity of soil; F = shape factor

By substituting the value of soil and vegetation emissivity and shape factor got the final expression as:

$$\varepsilon = 0.004\ Pv + 0.986 \quad \ldots (7)$$

This emissivity value incorporated into the final land surface temperature equation.

Land Surface Temperature i.e ground temperature: Finally LST was calculated using the formula (8)

$$LST = T_B/1 + w * (T_B/p) * Ln(\varepsilon) \quad \ldots (8)$$

Where, T_B = Satellite Brightness Temperature; w = Wavelength of emitted radiance; p = h*c/s ($1.438*10^{-2}$ m K); ε= land surface emissivity; h= Planck's constant ($6.626* 10^{-34}\ m^2\ kg / s$); s = Boltsmann constant ($1.38* 10^{-23}\ m^2\ kg\ s^{-2}\ K^{-1}$); c = Velocity of Light ($2.998* 10^8$ m/s).

Correlation between NDVI and LST

For correlation analysis, 1000 points were selected from one NDVI-LST map by the multipoint value selection method in Arc GIS. Similarly, points were selected for 1988, 2002 and 2018 and plotted in excel and examined the correlation between two sets of data.

Results and Discussion

Identification of risk factors

The field observation in Peechi-Vazhani WLS shows that the forest ecosystem is in a stage of degradation. The major disturbances observed here are logging, fuelwood collection, grazing, cutting of trees, extraction of medicinal plants, forest plantation, and dependence on Non-Timber Forest Products (NTFPs). In many areas of the sanctuary *Tectona grandis-Bombax ceiba* plantation was practiced especially in the Vellani and Vaniyampara sections. Vaniyampara forest section is disturbed with plantation activities and grazing pressure. This study noticed the above plant species and *Macaranga peltata* in the fringes of the semi-evergreen forest and surviving well with other forest species. This sanctuary is a known resource of many medicinal plants including *Wrightia tinctoria, Schleichera oleosa, Terminalia bellirica, Adina cordifolia, Myristica beddomei, Knema attenuata* and *Strychnos nux-vomica.* Most of the species are collected in a relatively high quantity because of their commercial wilderness, placed some species are at risk of extinction. Among those species, both *Myristica beddomei* and *Knema attenuata* are endemics to the Western Ghats and also belongs to the IUCN red list category. Weedy growth of the invasive species *Lantana camara* and *chromolaena odorata* are predominant and higher densities of *Helicteres isora* are also noted. The major NTFPs collected from this sanctuary is gum from the tree species *Canarium strictum* (Black dammar or Karutha Kunthirikkam), honey and fruits of *Myristica beddomei* and *Knema attenuata* (condiment/ spice/ medicinal). The way of extracting the gum from the bark of *Canarium strictum* affects their natural growth (drilling the bark in deep); extraction is happening even in the core zone of higher altitude semi-evergreen forest areas. Peechi-Vazhani WLS is prone to forest fires during summer. Both the anthropogenic and climate factors are responsible for these forest fires. Cases reported that fire started as a result of setting the fire to pacify the bees for honey collection. Study observed that disturbances are lesser in core zone areas; which indicates that forest degradation has a higher relationship with human accessibility.

Normalized Difference Vegetation Index (NDVI)

The NDVI maps were prepared from Landsat 5 TM, Landsat 7 ETM+, Landsat 8 OLI & TIRS for 1988, 2002 and 2018 and results were presented in Fig. 3. Peechi- Vazhani WLS experienced a substantial decrease in vegetation cover from 1988 to 2002. NDVI value of 1988 showed a good vegetation index as 0.005 to 0.685 and a major portion of the area shows dense canopy cover (>0.4) (Fig. 3a). These higher values show a declining trend in the succeeding year of 2002 (Fig.3b). This observation was supported by the other research works carried out in the study area. A remote sensing data used study revealed that 29% of the forests in Peechi-Vazhani WLS were in a highly and moderately degraded state (Mammen, 2000). Related to Peechi and Vazhani dam construction work during the years the 1950s and 1960s, people have encroached in this WLS, as a result, it caused an increase in population density and deforestation rate. Because of this population pressure, socio-economic dependence mainly for fuelwood, fodder, grazing, and non-timber forest products (medicinal products, green manure, food, etc.) are also higher (Anitha & Muraleedharan, 2002). Teak and Eucalyptus plantation are carried out in different parts of the sanctuary and the analysis showed soils in the plantation were degraded (Geetha & Balagopalan, 2005). Paul (2013) found that forests in these areas are secondary moist deciduous climax forests; which indicates the intense human disturbances during the past period. Athira *et al.*, (2017) stated that a significant level of forest loss has occurred in Peechi- Vazhani WLS from 1985 (101.87 km^2) to 1995 (95.78 km2). Anitha & Muraleedharan (2002) conducted a study during 2000-2002, they reported that the availability of resources in this sanctuary making it an open access area. They are continuing with the findings a) Peechi-Vazhani WLS facing higher threats from grazing, extraction of medicinal plants, harvesting of non-timber forest products and over-exploitation of species mainly *Terminalia paniculata, Xylia xylocarpa and grewia tiliifolia;* b) In an around 1,400 tons of firewood are annually obtained for sale and consumption. Due to these higher human interactions and changes in weather patterns, forest fire is a common phenomenon in these places. Ajin *et al.*, (2017) found that abundance of dry fuel, prevailing winds and hot and dry climate influencing the forest fire. The selected factors such as land cover, distance from the settlement, distance from the road, slope and elevation confirmed the root cause of many forest fires in this area are manmade. The prominent vegetation changes in this area negatively affected the water resources specifically the water level of Peechi and Vazhani dam reservoirs; which is considered an important catchment area of rivers and sources of drinking water.

While considering the period 2002 to 2018, The NDVI values show that there is no considerable decrease in vegetation in Peechi - Vazhani WLS after 2002 (Fig. 3c). The prominent change observed was the declining trend of dense forest. But the area shows an overall increment in the areal extent of other vegetation types (semi-evergreen forest, moist deciduous forest, grassland). It indicates that some conservation efforts have taken in the sanctuary after 2002 to attain stability in forest cover. Athira *et al.*, (2017) decadal analysis (2005-2015) underlines the point, Peechi- Vazhani WLS were unchanged during this period. Effective conservation activities and policies from the Government and Forest department succeeded to minimize the degradation rate in this protected area. The Biological Diversity Act and Biological Diversity Rules enacted in 2002 and 2004 respectively for the legal conservation and sustainable use of biodiversity. National Biodiversity Authority (NBA) constituted under the provision of this act in 2003 and also formed Biodiversity Boards for the respective states. The most significant movement from the Government authority is the legal regulation of diversion of forest land to non-forest and plantation purposes (Anon, 2019) As per the order dated 09.05.2002 any activity related to the destruction of wildlife or habitat will not allow in protected areas without the approval by

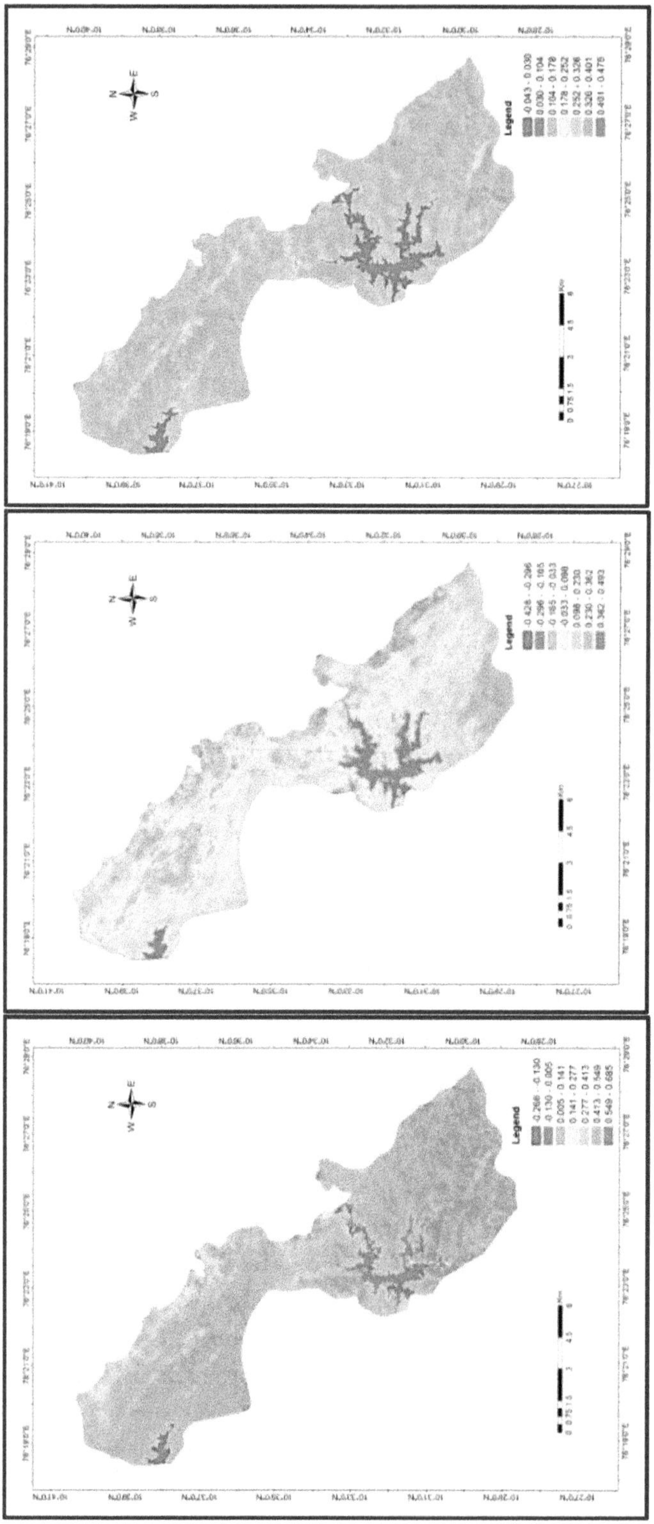

Fig. 3. NDVI map of Peechi-Vazhani WLS for (a) 1988; (b) 2002; (c) 2018

the standing committee of the National Board for Wildlife (NBWL). Also, the activities within a distance of 10 kilometers of protected areas or any project within the protected areas are required to obtain permission from NBWL (ELDF & WWF India, 2009; MoEF, 2008). It has been noted that, further *Tectona grandis (*Teak*)* - *Bombax ceiba* plantation activities are not allowing in the sanctuary; resulted in the automatic regeneration of natural forest in the plantation areas. All these strategies helped to conserve biodiversity in protected areas to a greater extent.

Land Surface Temperature (LST)

LST was obtained from Landsat 5, Landsat 7 and Landsat 8 for 1988, 2002 and 2018 and shown in Fig. 4. The analysis has shown that in 1988, the LST values range from 19.72 p C to 31.64p C (Figure 4a) and these values are changed in 2002 as 21.70p C to 34.45p C (Figure 4b). Decadal analysis (1988-2002) indicates a considerable increment in the minimum and maximum temperature in Peechi - Vazhani WLS in 14 years. These changes can be directly linked with vegetation changes occurred in this sanctuary. Many researchers mentioned the strong inverse correlation of LST with NDVI (Kumar & Shekhar, 2015; Guha *et al.*, 2018; Sruthi & Aslam, 2015). The drastic reduction of vegetation from 1988 to 2002 had witnessed a dramatic increment in overall temperature in the study area. The analysis of data revealed that the vegetated areas and water bodies record lower temperatures, while the area with less vegetation or barren areas exhibits higher temperatures. This relationship was validated and shown in Figure 5. When compared to the NDVI of 2002 with 2018 data, detected the area improved in the aspect of vegetation; as a result, LST is also decreased where the area improved their vegetation (Fig. 4c). So, conservation activities suggesting are a) afforestation with the indigenous species; b) prevention of human encroachment in protected zones; c) prevention of forest plantation with exotic tree species.

Conclusion

Forest degradation is a state where the qualities of a forest ecosystem are diminishing by a combination of factors. The word degradation does not mean only the reduction of forest cover but the definition may consider the areas where a forest lost its structure and qualities. This study attempted to assess the degradation factors of Peechi - Vazhani WLS in past and current scenarios. The NDVI and LST of forest area from 1988 to 2018 were analyzed using Landsat datasets and fieldwork. Results revealed that sanctuary shown a sharp decline in vegetation during the period 1988 to 2002. Concerning these vegetation changes the area shown increment in its local temperature. Besides these changes, the anomalous deviation of water level in dam reservoirs was observed. While considering the last decade (2002- 2018), observed an overall increment in the areal extent of vegetation and less temperature where the areas are covered with vegetation. It indicates the

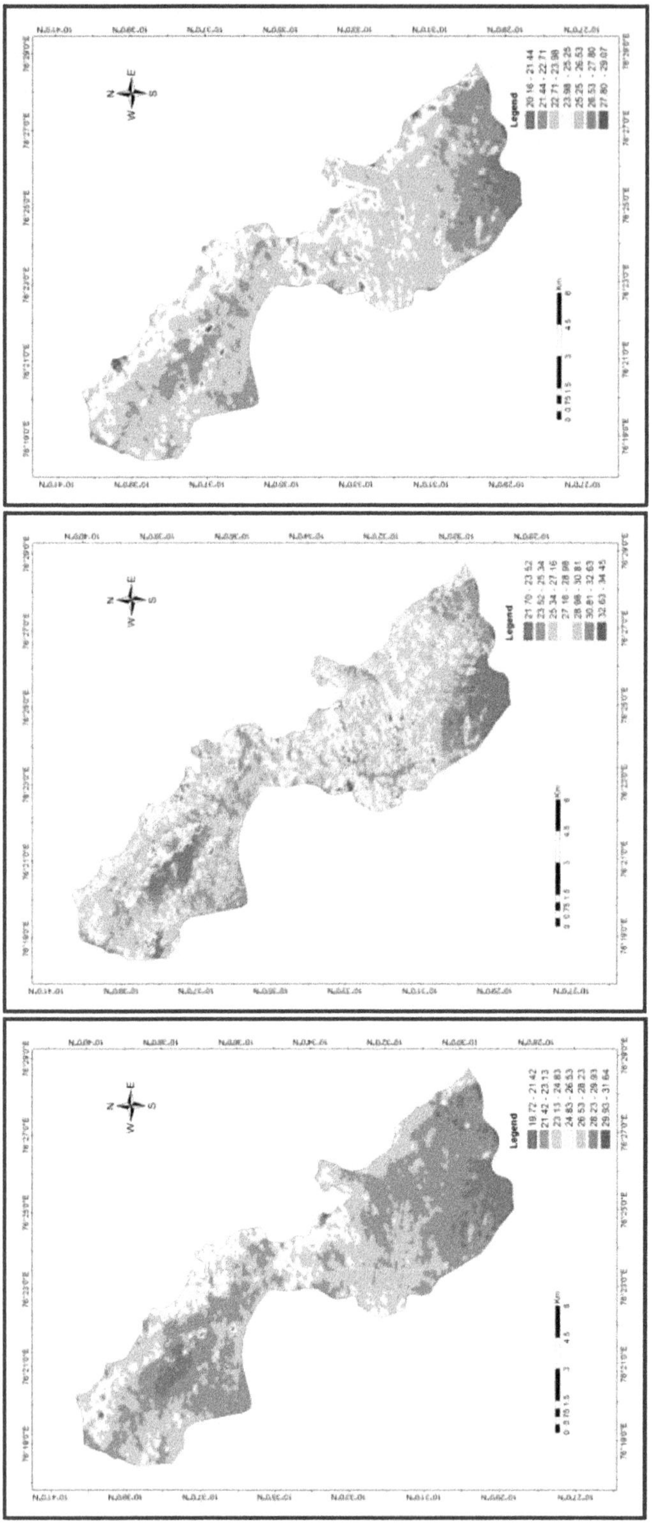

Fig.4. LST (^{0}C) map of Peechi-Vazhani WLS for (a) 1988; (b) 2002; (c) 2018

effectiveness of conservation strategies and policies adopted in protected areas. In current scenarios, some degradation activities still exist in Peechi – Vazhani WLS mainly because of its huge availability of resources and easy accessibility of the area. So, further strategies and policies are required to achieve the ultimate goal of proper conservation. Further conservation activities suggested are (a) afforestation with the indigenous species; (b) prevention of human encroachment in protected zones; and (c) prevention of forest plantation with exotic tree species, which harm biodiversity and water resources in the sanctuary.

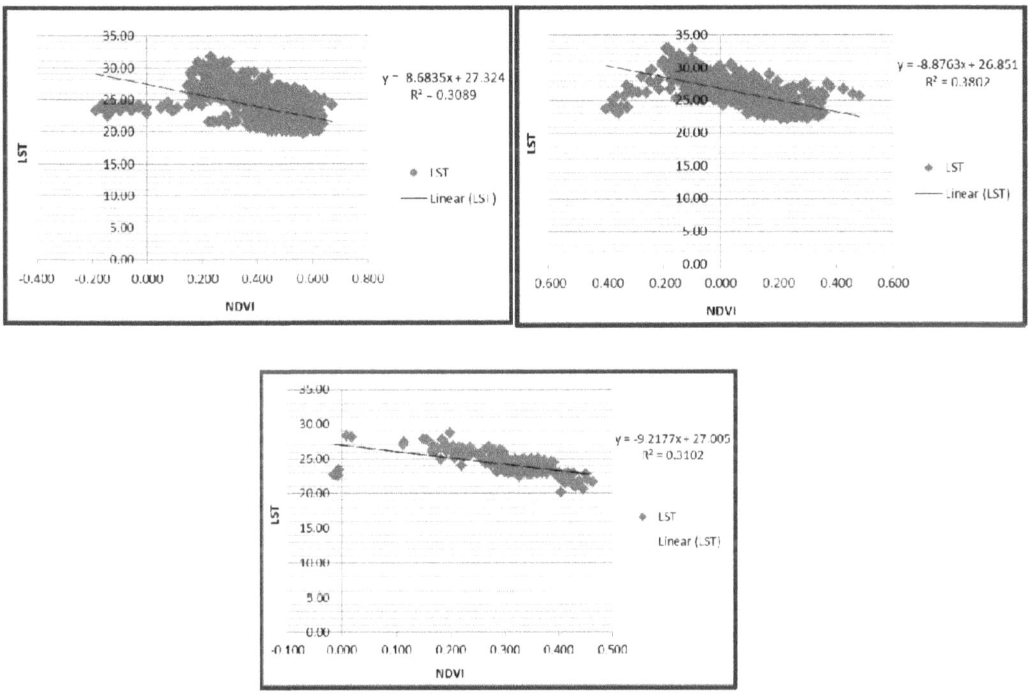

Fig. 5. Correlation Graph between LST and NDVI for (a) 1988; (b) 2002; (b) 2018

Acknowledgments

We would like to thank Dr. Sreenath Subrahmanyam, Director (Research and Innovation), Institute of BioEcoSciences, USA for providing technical discussions with us. Also, we would like to thank the Kerala Forest Department for given permission and assistance provided during the field visit. Also, we extend our thanks to SERB for providing financial assistance.

References

Ahrends, A., Burgess, N.D., Milledge, S.A., Bulling, M.T., Fisher, B., Smart, J.C., Clarke, G.P., Mhoro, B.E. and Lewis S.L., 2010. Predictable waves of sequential forest degradation and biodiversity loss spreading from an African city. Proceedings of the National Academy of Sciences, 107(33): 14556-14561.

Ajin, R.S., Loghin, A.M., Vinod, P.G. and Jacob, M.K., 2017. Mapping of forest fire risk zones in Peechi-Vazhani Wildlife Sanctuary, Thrissur, Kerala, India: A Study Using Geospatial Techniques. J. Wetlands Biodiversity, 7: 7-16.

Anitha, V. and Muraleedharan, P.K., 2002. Final Report of Project KFRI/352/2000.

Anon, 2019. Biodiversity Conservation Programmes, http://www.kerenvis.nic.in/WriteRead Data/ UserFiles/file/Biodiversity%20Conservation%20Programmes.pdf (Accessed 24 December 2019).

Athira, K., Reddy, C.S., Saranya, K.R.L., Joseph, S. and Jaishanker, R., 2017. Habitat monitoring and conservation prioritisation of protected areas in Western Ghats, Kerala, India. Environmental Monitoring and Assessment 189(6): 295.

ELDF and WWF India, 2009, Conserving Protected Areas and Wildlife: A judicial journey. New Delhi.

FAO, 2009. Towards defining degradation, FRA Working Paper 154, FAO, Rome.

Gandhi, G.M., Parthiban, S., Thummalu, N. and Christy, A., 2015. NDVI: vegetation change detection using remote sensing and GIS - a case study of Vellore District. Procedia Computer Science, 57: 1199-1210.

Geetha, T. and Balagopalan, M., 2005. Soils as influenced by teak and eucalypt plantations - a case study in Peechi-Vazhani Wildlife Sanctuary, Kerala. Journal of Soils and Crops 15(2): 264-268.

Guha, S., Govil, H., Dey, A. and Gill, N., 2018. Analytical study of land surface temperature with NDVI and NDBI using Landsat 8 OLI and TIRS data in Florence and Naples city, Italy. European Journal of Remote Sensing 51(1): 667-678.

Hosonuma, N., Herold, M., De Sy, V., De Fries, R. S., Brockhaus, M., Verchot, L., Angelsen, A., Romijn, E. and Romijn, E., 2012. An assessment of deforestation and forest degradation drivers in developing countries. Environmental Research Letters, 7(4): 4009.

ITTO, 2002. ITTO guidelines for the restoration, management and rehabilitation of degraded and secondary tropical forests. ITTO Policy Development Series No. 13. Yokohama, Japan.

Kumar, D. and Shekhar, S., 2015. Statistical analysis of land surface temperature–vegetation indexes relationship through thermal remote sensing. Ecotoxicology and Environmental Safety 121: 39-44.

Mammen, D.P., 2000. Landscape evaluation of Peechi-Vazhani Wildlife Sanctuary with special reference to Peechi region using remote sensing techniques. MSc. Dissertation (unpublished), School of Environmental Studies, Mahatma Gandhi University, Kottayam, Kerala, India.

Mertz, O., Halsnæs, K., Olesen, J.E. and Rasmussen, K., 2009. Adaptation to climate change in developing countries. Environmental Management, 43(5): 743-752.

MoEF, 2008. Minutes of the Meetings of Indian Board for Wild Life and National Board for Wild Life (1998-2009). Right to Information response from the MoEF, 14 November (F No 25-1/2005 WL-I), available on http://moef.nic.in/index.php. – (2009): Protected Areas in Country for Wildlife, Press Release. http:// www.pib.nic.in/release/release.asp?relid= 54028, (accessed 02 January 2020).

Morris, R.J., 2010. Anthropogenic impacts on tropical forest biodiversity: a network structure and ecosystem functioning perspective. Philosophical Transactions of the Royal Society B: Biological Sciences, 365(1558): 3709-3718.

Paul, C.R., 2013. Quantification of Anthropogenic Disturbances in Forest as a Function of Distance to Human Habitation - A Case Study from Peechi-Vazhani Wildlife Sanctuary. Doctoral Dissertation, College of Forestry, Vellanikkara.

Sruthi, S. and Aslam, M.M., 2015. Agricultural drought analysis using the NDVI and land surface temperature data; a case study of Raichur district. Aquatic Procedia, 4: 1258-1264.

Impact of Climate Change on Hydrological Cycle, Ecosystem, Fisheries and Food Security, Pages: 379–389
Edited by: B. Madhusoodana Kurup, M.R. Boopendranath, M. Harikrishnan and A.V. Shibu

CHAPTER - 32

Structural Analysis of Major Mangrove Forests of India

S. Sreelekshmi[1*], S. Bijoy Nandan[2] and M. Harikrishnan[1]

[1]*School of Industrial Fisheries,*
Cochin University of Science and Technology Cochin-682016, Kerala, India
[2]*Department of Marine Biology, Microbiology and Biochemistry, School of Marine Sciences,*
Cochin University of Science and Technology, Cochin-682016, Kerala, India
**E-mail: sreelekshmis87@gmail.com*

Taxonomically, mangroves constitute diverse angiosperm plants exhibiting a set of physiological adaptations (Tomlinson, 1986), protecting the coastal zone from frequent cyclones, tidal thrust, and tsunami, and provide a wide variety of goods and services used by local populace (Zhang *et al.*, 2006). Mangrove ecosystems are unique habitats of the coastal wetland systems but are facing intense pressure due to wanton destruction by humans for various developmental needs. These are highly productive ecosystems on tropical and subtropical coastlines, currently occupying approximately 140,000 sq km, globally (Giri *et al.*, 2011). While, in India, it covers an area of 4740 sq.km, which accounts for nearly three percent of world's mangrove cover, of which 2114 sq km is in Sundarbans, 615 sq km in Andamans and only 9 sq km in Kerala (FSI, 2019). Even though, several authors reported the distribution of mangrove species in these forests, statistical evidence for structural variation among different mangrove forests of India remains scanty. With this background, our objective was to analyse the variation in structural attributes among major mangrove forests of India - Andamans, Sundarbans and Kerala

Materials and Methods

The present study followed Polidoro *et al.* (2010), for species identification. Fixed area plot measurement was used for the characterization of the structural attributes of true mangroves based on the methods suggested by Cintrón and Schaffer-Novelli (1984). Quadrat method was used for the structural analysis of true mangrove flora. In each site, four transects were laid perpendicular to the waterline at 50 m interval. In each transect, five quadrats (10x10 m) were laid at 50 m interval considering mangrove diversity and accessibility. Each species in the quadrat were identified, counted and analyzed and diameter at breast height (DBH) was measured using a measuring tape. The diameter of all plants of more than 2cm DBH was measured 1.3 m above the ground. Structural variables like stem density, relative density and abundance, % frequency, relative frequency, basal area, relative basal area were calculated from the data using standard formula (Cintrón and Schaffer-Novelli 1984). The importance value index (IVI) was calculated by adding the values of relative density, relative frequency, and relative basal area.

Results and Discussion

Species composition and distribution pattern

Altogether thirty-three true mangrove species were identified from the study area of which, mangroves of Andaman islands comprises 28 true mangroves whereas Sundarbans comprises 27 and Kerala represented 18 true mangrove species (Table 1). In Andamans, *Avicennia officinalis*, *A. marina*, *Rhizophora mucronata*, *R. apiculata*, *Bruguiera gymnorhiza*, *Sonneratia alba,* and *Nypa fruticans* were the dominant species whereas *Sonneratia ovata, Aegialitis rotundifolia, Bruguiera sexangula, Acrostichum speciosum* and *Lumnitzera racemosa* were found to be rare species. However, *Sonneratia griffithii*, a "critically endangered" plant as per the IUCN Red List, was found to be a rare plant in Indian Sundarbans and *Heritiera fomes*, an endangered species, showed an occasional distribution in the study. While, *Ceriops decandra*, *Aegialitis rotundifolia* and *Phoenix paludosa*, which are "Near threatened" species, exhibited a common distribution in Indian Sundarbans in the study. In Kerala *Rhizophora mucronata*, *Avicennia officinalis and Excoecaria agallocha were found to be dominant species whereas Avicennia alba, Excoecaria indica, Ceriops tagal and Sonneratia alba were found to be rare species.*

Structural analysis

In Andamans, total densities were highest for *Rhizophora apiculata* (64935±2348 no.ha^{-1}, accounting for 19.54% of all stems) followed by *Bruguiera gymnorhiza* and *Rhizophora mucronata* (Table 1). However, *Rhizophora apiculata* (accounting for 21.24% of the stand basal area) dominated with respect to the individual basal area. Mean stem density ranged from 133-14000 ha^{-1} whereas basal area varied from 0.07-78.75 m^{2}ha^{-1}. Lowest density was contributed by *Sonneratia ovata* (0.04%) and basal area by *Aegialitis rotundifolia* (0.02%). From the Importance Value Index, it is clear that *Rhizophora apiculata* was the most important species in South Andaman (IVI 44.95), Baratang (IVI 88.49) and Middle Andaman (IVI 77.39) whereas *Avicennia officinalis* (IVI 49.16) in Mayabunder and *Rhizophora mucronata* (IVI 61.98) in Diglipur forest division. The study characterized a mixed species composition since the IVI will be 300 for monospecific mangrove stands. The dbh class 11-20 showed the highest number of stems followed by 21-30 and 31-40 class in all the sites (Fig. 1). It was also found that beyond 11-20 class the number of stems decreased with increase in DBH.

Table 1: Structural analysis of mangroves of Andamans, Sundarbans and Kerala

Sl No	Species	Andamans		Sundarbans		Kerala	
		Density	BA	Density	BA	Density	BA
1	*Acanthus ilicifolius*	6267±1043	–	907±1160	–	13846 ± 8794	–
2	*Acrostichum aureum*	28266±1427	–	80±180	–	2319 ± 2116	–
3	*Acrostichum speciosum*	533±116	–	–	–	–	–
4	*Aegialitis rotundifolia*	267±58	6.8±1.48	2347±2806	0.27±0.28	–	–
5	*Aegiceras corniculatum*	1733±294	13.97±1.94	1520±2243	0.15±0.2	145 ± 195	0.16 ±0.33
6	*Avicennia alba*	–	–	1573±2166	1.07±1.91	10 ± 32	0.08 ±0.26
7	*Avicennia marina*	19468±1477	328.26±22.08	2987±3092	2.57±4.48	1107 ± 2541	3 ±5.21
8	*Avicennia officinalis*	14666±1341	215.78±12.45	2227±2195	4.01±4.72	1373 ± 923	20.19 ±10.54
9	*Bruguiera cylindrica*	2133±301	28.8±3.02	413±967	0.21±0.43	516 ± 511	2.15 ±2.35
10	*Bruguiera gymnorhiza*	42100±3070	398.34±23.37	1480±1093	0.79±1.22	153 ± 213	0.70 ±1.06
11	*Bruguiera sexangula*	400±87	8.1±1.77	133±281	0.03±0.07	243 ± 555	0.77 ±2.18
12	*Ceriops decandra*	2399±382	34.4±3.54	3347±6204	0.71±1.1	–	
13	*Ceriops tagal*	18665±893	89.7±3.17	1040±1266	0.39±0.48	13 ± 4	0.08 ±0.27
14	*Dolichandrone spathacea*	3734±361	108.5±11.53	–	–	–	–
15	*Excoecaria agallocha*	12134±841	189.1±10	3947±3526	3.42±3.2	971 ± 784	3.66 ±4.87
16	*Excoecaria indica*	–	–	–		96 ± 203	0.06 ±0.14
17	*Heritiera fomes*	–	–	520±825	0.47±0.81	–	–
18	*Heritiera littoralis*	1200±130	108.4±11.81	–		–	–
19	*Kandelia candel*	–	–	13±42	0.004±0.01	312 ± 469	1.45 ±2.01
20	*Lumnitzera racemosa*	1733±176	31.9±4.18	160±421	0.02±0.06	148 ± 243	0.02 ±0.05
21	*Lumnitzera littorea*	3200±327	52.6±6.54	–		–	–
22	*Nypa fruticans*	10134±1465	16.6±14.71	507±1249	2.59±8.02	–	–
23	*Phoenix paludosa*	16266±1332	59.8±4.83	3680±5452	0.71±0.92	–	–
24	*Rhizophora apiculata*	64935±2348	417.05±13.75	200±269	0.14±0.19	158 ± 200	1 ±1.29

[Table Contd.

Contd. Table]

Sl No	Species	Andamans		Sundarbans		Kerala	
		Density	BA	Density	BA	Density	BA
25	*Rhizophora mucronata*	37733±1719	289.51±11.45	13±42	0.06±0.19	696 ± 609	5.82 ±5.16
26	*Scyphiphora hydrophyllacea*	4133±320	45.8±2.76	53±169	0.04±0.01	–	–
27	*Sonneratia alba*	3199±271	111.7±10.10	13±42	0.04±0.01	28 ± 206	0.50 ±1.19
28	*Sonneratia apetala*	–	–	880±2047	0.69±0.96	–	
29	*Sonneratia caseolaris*	800±127	48.9±7.7	93±252	0.24±0.64	328 ± 459	6.6 ±11.02
30	*Sonneratia griffithii*	1334±187	73.6±10.94	560±1157	0.16±0.38	–	–
31	*Sonneratia ovata*	133±30	10.7±2.33	–	–	–	–
32	*Xylocarpus granatum*	6667±579	224.13±22.82	1133±1909	1.13±2.02	–	–
33	*Xylocarpus moluccensis*	1467±196	74.6±10.56	427±705	0.27±0.47	–	–

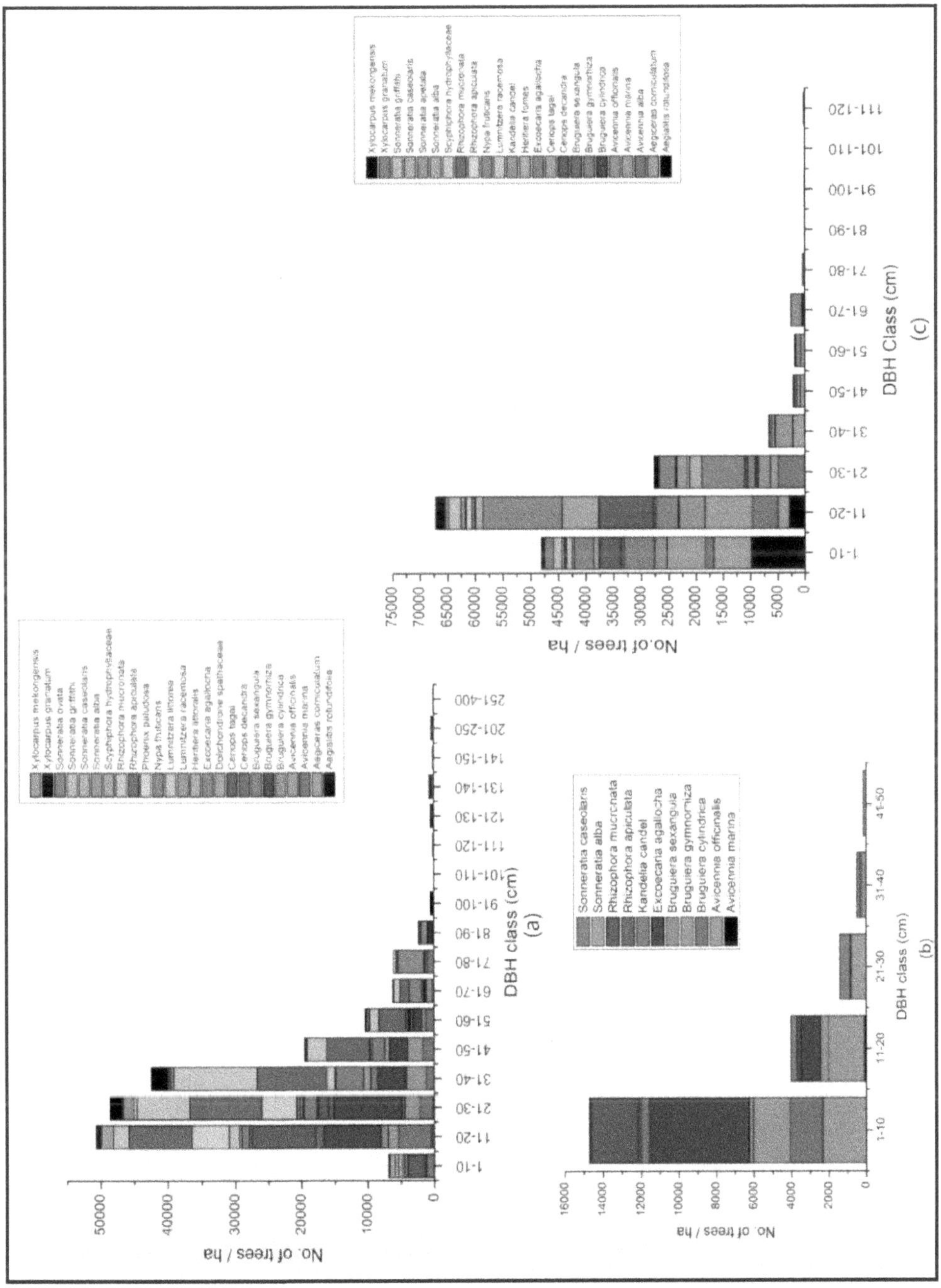

Fig. 1. Species-wise DBH class distribution of mangroves in (a) Andamans (b) Kerala (c) Sundarbans

While in Sundarbans, total densities were highest for *Excoecaria agallocha* (3947 ±3526 ha^{-1}, accounting for 13% of all stems) followed by *Phoenix paludosa* and *Ceriops decandra*. However, *Avicennia officinalis* (accounting for 21% of the stand basal area) dominated with respect to the individual basal area (Table 1). Mean stem density varied from 133 to 19333 $ha^{"1}$ and mean individual basal area 0.02-14.5 m^2 ha^{-1}. Regarding importance value index (IVI), *Avicennia marina* was the dominant species (119–9) in Indian Sundarbans followed by *Avicennia officinalis* (110–15), *Excoecaria agallocha* (106–14) and *Avicennia alba* (105–106). The DBH class 11–20 showed the highest number of stems followed by 1–10 class in all the islands (Fig. 1). It was also found that beyond 11–20 class the number of stems decreased with increase in DBH.

Structural analysis of mangroves of Kerala revealed that the mean stem density was highest for *Acanthus ilicifolius* (13846± 8794 ha^{-1}), followed by *Acrostichum aureum* and *Avicennia officinalis*. The least mean stem density was represented by *Ceriops tagal* (10±32 ha^{-1}) (*Table.1*). However, *Avicennia officinalis* showed the highest mean basal area (20.19 ±10.54 m^2 ha^{-1}) contributed up to 43.64% of the total basal area. *Avicennia officinalis* was found to be the most important species since it had the highest importance value index (63.31). Despite very low basal area (mean 1.11 m^2 ha^{-1}), *Acanthus ilicifolius* ranked second (IVI, 47.53), due to its gregarious occurrence in all the sites. Structural analyses of the mangroves of Kerala revealed the domination of the shrub, *Acanthus ilicifolius* (50.2%) and the importance of *Avicennia officinalis* in most of the habitats.

Altogether twenty-seven true mangrove species were identified from selected islands of Indian Sundarbans. This is in contrast to the studies by Naskar and Guhabakshi (1987) and Banerjee *et al.* (1989), having reported 37 obligate mangroves whereas, Chaudhuri and Choudhury (1994) reported 36 species from entire Indian Sundarbans. Mitra (2000) and Mitra and Pal (2002) documented 34 true mangrove species whereas Barik and Chowdhury (2014) reported 24 true mangrove species from the Indian Sundarbans. The variations in the number of species identified in the study could be due to the differences in study sites and species considerations (Van Nguyen *et al.*, 2013). The number of species (27) documented now is in the range of other mangroves of Indonesia (Hinrichs *et al.*, 2009). Other notable observations were that *Avicennia alba* was the most dominant species in the study as reported by Giri *et al.* (2014). *Sonneratia alba* was reported for the first time in Indian Sundarbans, but, species like *Bruguiera parviflora* and *Aglaia cucculata*, reported previously in Indian Sundarbans by Naskar and Guhabakshi (1987) were not encountered in the study.

Altogether twenty eight true mangrove species were identified from selected sites in Andaman islands. In 1923, Parkinson reported 17 species, Mall *et al.* (1985) reported 26 species, while Debnath (2004) identified 36 species. However, Goutham-Bharathi *et al.* (2014) and Kiruba-Sankar *et al.* (2018) reported only 25 mangrove species from Andaman and Nicobar islands whereas Ragavan *et al.* (2015) reported 38 mangrove species. Species such as *Cynometra iripa, Excoecaria indica, Pemphis acidula, Sonneratia lanceolata* and *Brownlowia tersa* reported by Ragavan *et al.* (2015) from Andaman islands were not encountered in the present study as well as the studies of Goutham-Bharathi *et al.* (2014). While, our study observed the presence of certain other species which were not reported recently such as *Ceriops decandra, Aegialitis rotundifolia* and *Bruguiera sexangula* in South Andaman. *C. decandra* was considered as very rare species two decades ago (Dagar *et al.,* 1991). Further Dagar and Singh (1999), Debnath (2004) and Yao *et al.* (2011) also reported the presence of this particular species in Andaman and Nicobar islands. *B. sexangula* was first reported by Singh *et al.* (1987) in Andaman and Nicobar islands and it was reported by all subsequent studies except Goutham- Bharathi *et al.* (2014) and Ragavan *et al.* (2015).

The study reported 18 true mangrove species from Kerala which is in agreement with the reports of Ramachandran *et al.* (1986) and Basha (1991; 1992). However, Anupama and Sivadasan (2004) and Vidhyasagaran and Madhusoodhanan (2014) reported 15 true mangroves belonging to 9 genera and 7 families and 49 mangrove associates. Recently, George *et al.* (2018) reported only 13 species from the mangrove wetlands of Kerala.

Mangroves are generally considered as even-aged forests, primarily developing following a disturbance or colonization of mud flats (Saenger, 2002). The stand age data of mangroves of India are unavailable, however, the structural attributes of the mangrove stands reflect an uneven-aged mangrove forest. The diameter distribution of both trees and saplings represents an uneven-aged mangrove forest (Trettin *et al.*, 2016). The density values reported from Andamans were comparable with previous reports from Andamans (Das *et al.*, 2014), Coringa (Satyanarayana *et al.*, 2009) and Kakinada bay mangroves (Satyanarayana *et al.*, 2002). While the basal area of Andaman islands was found to be higher than other mangrove systems of India and was comparable with Sibuti mangroves, Malaysia (Shah *et al.*, 2015).

The density values of Sundarbans were comparable with the reports from Coringa (Satyanarayana *et al.*, 2009) and Kakinada bay mangroves (Satyanarayana *et al.*, 2002). While, the basal area values are similar to the reports from Odisha (Upadhyay and Mishra, 2014), Lothian island, Sundarbans (Joshi and Ghose, 2003) and Indonesia (Hinrichs *et al.*, 2009). The high density of *Ceriops decandra* in Bagna was comparable to the density of *Aegiceras corniculatum* (15938 ha^{-1}) in Lothian island (Joshi and Ghose, 2014) and Harinbhanga island (16400 ha^{-1}) in the eastern Sundarbans (Matilal *et al.*, 1986). The forest structure was simple and represented mainly by small dbh trees, in all the 10 islands attributed to the low structural development depicting an expanding or evolving or rejuvenating population in Indian Sundarbans (Joshi and Ghose, 2014).

The structural attributes of the mangrove stands of Kerala also reflect an uneven-aged mangrove forest. The density values obtained from Kerala were found to be comparable with the studies of Satyanarayana *et al.*, (2002) in Coringa mangroves and Thivakaran *et al.* (2003) in Kachchh mangroves. While, basal area values were appeared similar to Andaman mangroves (Kiruba-Sankar *et al.*, 2017) and Zambezi deltaic mangroves, Brazil (Trettin *et al.*, 2016). The high density of *Acanthus ilicifolius* was comparable with the studies of Rani *et al.* (2016) in Cochin mangroves. The forest structure was simple and represented mainly by medium dbh trees in most of the study sites.

Conclusion

Indian mangroves that extend from coastal wetlands to deltaic regions are noted for their diverse and unique floral and faunal characteristics. The species composition, distribution pattern and floristic structure of Sundarbans indicate a growing and emerging forest, favourable to sustain a mature forest in the end. However, highly diverse and mature mangrove forests, with respect to species composition, dbh (diameter at breast height) and basal area, were observed in Andaman islands. But in Kerala, the vegetation appeared to be less diverse with moderate structural development. The outcome of this study on phytosociological analysis of mangroves from Indian Sundarbans, Andaman islands and Kerala will cater to our knowledge and future mangrove conservation efforts in India.

Acknowledgments

The first author is grateful for the award of UGC BSR Senior Research Fellowship. Our sincere gratitude is due to the Head, Department of Marine Biology, Microbiology & Biochemistry for the facilities for undertaking the work and also to those who assisted in our field surveys and sampling works.

References

Anupama, C. and Sivadasan, M., 2004. Mangroves of Kerala, India. Rheedea 14: 9-46.

Banerjee, L.K., Sastry, A.R.K. and Nayar, M.P., 1989. Mangroves in India, Identification Manual. Botanical Survey of India, Kolkata.

Barik, J. and Chowdhury, S., 2014. True mangrove species of Sundarbans delta, West Bengal, Eastern India, Check List 10(2): 329-334.

Basha, S.C., 1991. Distribution of mangroves in Kerala. The Indian For., 117(6): 439-448.

Chaudhuri, A.B. and Choudhury, A., 1994. Mangroves of the Sundarbans Vol. I: India. IUCN, Bangkok, Thailand, 247 p.

Cintron, G. and Schaffer-Novelli, Y.S., 1984. Methods for studying mangrove structure. In: Samuel, C.S. and Jane, G.S. (Eds.), The Mangrove Ecosystem: Research Methods, pp. 91-113, UNESCO, Paris.

Dagar, J.C. and Singh, N.T., 1999. Plant Resources of the Andaman and Nicobar Islands Vol. I. Bishen Singh and Mahendra Pal Singh, Dehradun, India., 271 p.

Dagar, J.C., Mongia, A.D. and Bandyopadhyay, A.K., 1991. Mangroves of Andaman and Nicobar Islands, Oxford and IBH Publishing Co., Pvt. Ltd, New Delhi, India., 166 p.

Debnath, H.S., 2004. Mangroves of Andaman and Nicobar Islands: Taxonomy and Ecology. Bishen Singh, Mahendra Pal Singh, Dehradun, India., 133 p.

FSI, 2019. India State of Forest Report, 2019. Forest Survey of India, Dehradun.

George, G., Krishnan, P., Mini, K.G., Salim, S.S., Ragvan, P., Tenjing, S.Y., Muruganandam, R., Dubey, S.K., Gopalakrishnan, A., Purvaja, R. and Ramesh, R., 2018. Structure and regeneration status of mangrove patches along the estuarine and coastal stretches of Kerala, India. J. For. Res. 30: 507-518. doi.org/10.1007/s11676-018-0600-2.

Giri, S., Mukhopadhyay, A., Hazra, S., Mukherjee, S., Roy, D., Ghosh, S., Ghosh, T. and Mitra, D., 2014. A study on abundance and distribution of mangrove species in Indian Sundarban using remote sensing technique. J. Coast. Conserv., 18: 359-367. doi:10.1007/s11852-014-0322-3.

Goutham-Bharathi, M.P., Dam Roy, S., Krishnan, P., Kaliyamoorthy, M. and Immanuel, T., 2014. Species diversity and distribution of mangroves in Andaman and Nicobar Islands, India. Botanica Marina 57: 421-432.

Hinrichs, S., Nordhaus, I. and Geist, S.J., 2009. Status, diversity and distribution patterns of mangrove vegetation in the Segara Anakan lagoon, Java, Indonesia. Reg. Environ. Change., 9: 275-289.

Joshi, H. and Ghose, M., 2003. Forest structure and species distribution along soil salinity and pH gradient in mangrove swamps of the Sundarbans. Tropical Ecology 44: 197-206.

Joshi, H. and Ghose, M., 2014. Community structure, species diversity and aboveground biomass of the Sundarbans mangrove swamps. Tropical Ecology 55(3): 283-303

Kiruba-Sankar, R., Krishnan, P., Dam Roy, S., Raymond, J.J.A., Goutham-Bharathi, M.P., Lohith Kumar, K., Ragavan, P., Kaliyamoorthy, M., Muruganandam, R., Rajakumari, S., Purvaja, R. and Ramesh, R. 2018. Structural complexity and tree species composition of mangrove forests of the Andaman islands, India. Journal of coastal conservation 2018 22(2): 217-234. doi 10.1007/s11852-017-0588-3.

Mall, L.P., Singh, V.P., Garge, A. And Pathak, S.M., 1985. A new approach towards the mangrove forests flora of Andaman Island. Indian Forester 3: 290-300.

Matilal, S., Mukherjee, B.B., Chatterjee, N. and Gupta, M.D., 1986. Studies on soil and vegetation of mangrove forests of Sundarbans. Indian Journal of Marine Sciences 15: 181-184.

Mitra, A. and Pal, S., 2002. The oscillating mangrove ecosystem and the Indian Sundarbans, (S. Banerjee and F. Tampal, Eds.), WWF-India-WBSO, New Delhi.

Mitra, A., 2000. The Northeast coast of the Bay of Bengal and deltaic Sundarbans. In: Sheppard, C. (Ed.) Seas at the Millennium – An Environmental Evaluation, pp. 143-157, Pergamon, Amsterdam.

Naskar, K.R. and Guha-Bakshi, D.N., 1987. Mangrove Swamps of the Sundarbans: An Ecological Perspective. Naya Prokash, Kolkata, India.

Parkinson, C.E., 1923. A forest flora of the Andaman Islands. Indian Government Central Press, Shimla, India., 325 p.

Polidoro, B.A., Carpenter, K.E., Collins, L., Duke, N.C., Ellison, A.M., Ellision, J.C., Farnsworth, E.J., Fernando, E.S., Kathiresan, K., Koedam, N.E., Livingstone, S.R., Miyagi, T., Moore, G.E., Nam, V.N., Ong, J.E., Primavera, J.H., Salmo, S.G., Sanciangco, J.C., Sukardjo, S., Wang, Y., Yong, J.W.H., 2010. The loss of species: mangrove extinction risk and geographic areas of global concern. Plos One 5: 1-10.

Ragavan, P., Saxena, A., Mohan, P.M., Ravichandran, K., Jayaraj, R.S.C. and Saravanan, S., 2015. Diversity, distribution and vegetative structure of mangroves of the Andaman and Nicobar Islands, India, Journal of Coastal Conservation 19: 417-443.

Ramachandran, K.K., Mohanan, C.N., Balasubramonian, G., Johnson, K. and Jessy T., 1986. The mangrove ecosystem of Kerala, its mapping, inventory and some environmental aspects. A project sponsored by the State Committee on Science, Technology and Environment.

Rani, V., Sreelekshmi, S., Asha, C.V. and Bijoy Nandan, S., 2016. Forest structure and community analysis of Cochin mangroves, South-West coast of India, Proceedings of the Indian National Science Academy. India, Sect. B Biological Sciences. 88: 111-119. doi 10.1007/s40011-016-0738-7.

Saenger, P., 2002. Mangrove Ecology, Silviculture and Conservation. Kluwer Academic Publishers, Dordrecht.

Satyanarayana, B., Raman, A.V., Dehairs, F., Kalavati, C. and Chandramohan, P., 2002. Mangrove floristics and zonation pattern of Coringa, Kakinada Bay, East coast of India. Wetlands Ecology and Management 10: 25-39.

Satyanarayana, B., Raman, A.V., Mohd-Lokman, H., Dehairs, F., Sharma, V.S. and Farid, D.G., 2009. Multivariate methods distinguishing mangrove community structure of Coringa in the Godavari Delta, east coast of India. Aquat. Ecosyst. Health., 12: 401-408.

Shah, K., Mustafa Kamal, A.H., Rosli, Z., Hakeem, K.R. and Hoque, M.M., 2015. Composition and diversity of plants in Sibuti mangrove forest, Sarawak, Malaysia. For. Sci. Technol., 12: 70-76.

Thivakaran, G.A., Saravanakumar, A., Serebiah, J.S., Sunderraj, W. and Vijayakumar, V., 2003. Vegetation structure of Kachchh mangroves, Gujarat, Northwest coast of India. Indian Journal of Marine Sciences 32(1): 37-44.

Tomlinson, P.B., 1986. The Botany of Mangroves. Cambridge University Press, New York.

Trettin, C.C., Stringer, C. E., Zarnoch, S.J., 2016. Composition, biomass and structure of mangroves within the Zambezi River Delta. Wetl Ecol Manag 24:173–186

Upadhyay, V.P. and Mishra, P.K., 2014. An ecological analysis of mangroves ecosystem of Odisha on the Eastern coast of India. Proc. Indian Natl. Sci. Acad., 80: 647-661.

Van Nguyen, T., Le Nhu, H., Lin, S.M., Steen, F. and De Clerck, O., 2013. Checklist of the marine macroalgae of Vietnam. Bot. Mar., 56: 207-227.

Vidyasagaran, K. and Madhusoodanan, V.K., 2014. Distribution and plant diversity of mangroves in the west coast of Kerala, India. Journal of Biodiversity and Environmental Sciences 4(5): 38-45.

Yao, Y., Bera, S., Naskar, K., Liao, W. and Li, C., 2011. A comparative study of mangrove floras in China and India. Forestry Studies in China 13: 173-182.

Zhang, Y., Wang, W., Wu, Q., Fang, B. and Lin, P., 2006. The growth of *Kandelia candel* seedlings in mangrove habitats of the Zhang jiang estuary in Fujian, China. Acta Ecologica Sinica 26: 1648-1656.

VI

Food and Nutritional Insecurity

Impact of Climate Change on Hydrological Cycle, Ecosystem, Fisheries and Food Security, Pages: 393–398
Edited by: B. Madhusoodana Kurup, M.R. Boopendranath, M. Harikrishnan and A.V. Shibu

CHAPTER - 33

Food and Nutritional Insecurity in the Perspective of Climate Change

K. Gopakumar

Professor of Eminence,
Kerala University of Fisheries and Ocean Studies, Panangad P.O., Kochi-682506, Kerala, India
Email: kgopa@yahoo.com

World Food Summit 1996, Rome Declaration on World Food Security defined that "*Food security exists when all the people, all the times have physical and economic access to sufficient, safe and nutritious food to meet their dietary needs and food preferences for an active and healthy life*" (FAO 1996). According to reputed Indian economist Bhattacharya (2001) the above definition emphasizes several critical dimensions of food security such as (i) Supply of desired food in sufficient quantity; (ii) right to access; and (iii) stability in supply which in a globalized scenario will include access to global market.

FAO made an assessment and estimated that about 34 million people in developed countries and 790 million in developing countries suffer from malnutrition (1995-97). In developing countries about 75% among the suffering group are in Asia, Pacific and South-Sahara Africa. In 2006, it was reported by World Bank that globally the number of people who are overweight has surpassed the number who are undernourished. The world had more than one billion people who were overweight, and an estimated 800 million who were undernourished. Worldwide around 852 million people are chronically hungry due to extreme poverty, while up to 2 billion people lack food security intermittently due to varying degrees of poverty. 17,000 children die of hunger and malnutrition related diseases every day, which equals 6 million children who die of hunger every year.

Until the time of Napoleon, there were less than 1 billion people on Earth at any one time. Since the Second World War, we have been adding a billion people to the global population every 12-15 years. Our population is more than double today what it was in 1970. The Indian agricultural production system faces the daunting task of having to feed

17.5 % of the global population with only 2.45 of land and 4% of water resources at its disposal (NAAS, 2013). The situation in India has shown drastic fall in percentage of people living below extreme poverty in 2018 according to a recent report of India News, based on World Data Lab. This is due to excellent policy implementation by Central and state governments by way of supplying free ration to people below poverty line and other financial assistance. It is now expected that if the present trend continues in 2030 only 3 million people will be living below extreme poverty line in India. This is indeed a great achievement considering the size of Indian population and high birth rate.

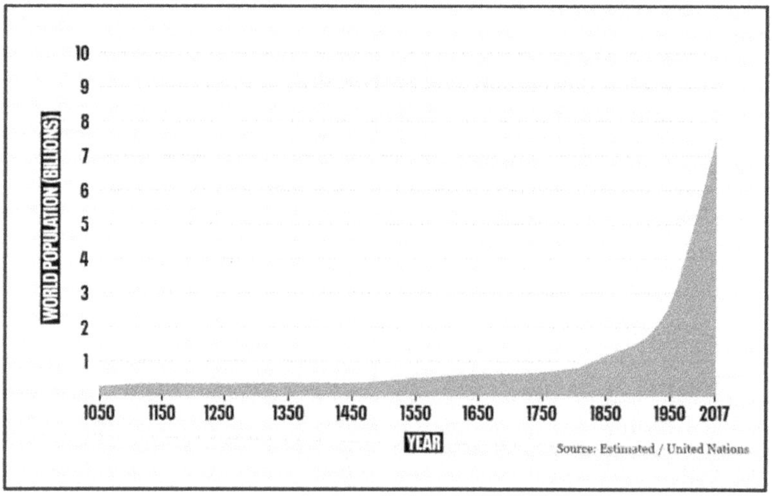

Fig. 1. Human population growth (1050-2017) (Source: UN)

Factors affecting food security today may include

- **Global water crisis** - Water table reserves are falling in many countries (including Northern China, the US and India) due to widespread over-pumping and irrigation.
- **Climate change** - Rising global temperatures are beginning to have a ripple effect on crop yields, forest resources, water supplies and altering the balance of nature.
- **Land degradation** - Intensive farming leads to a vicious cycle of exhaustion of soil fertility and decline of agricultural yields.
- **Greedy land deals** - Corporations and Governments buying rights to millions of acres of agricultural land in developing countries to secure their own long-term food supplies.

Scientists say that the impacts of climate change - higher temperatures, extreme weather, drought, increasing levels of carbon dioxide and sea level rise - threaten to decrease the quantity and jeopardize the quality of our food supplies. Climate change affects food security in complex ways. It impacts crops, livestock, forestry, fisheries and aquaculture, and can cause grave social and economic consequences in the form of reduced incomes, eroded livelihoods, trade disruption and adverse health impacts.

Impact of Climate Change on Agriculture

Climate change and its variability are emerging as the major challenges influencing the performance of Indian agriculture (NAAS, 2013). Climate change has got profound impact on agriculture both directly and indirectly. The effect of climate change on agriculture will depend on the magnitude of the climate change on a global basis. Failure on crop production will effect decreased supply of fodder to cattle and limit the use of cereals for production of both animal and fish feed. Reduction of water resources due to draught will make less water for drinking and domestic uses. Reduction or lowering of water in lakes, rivers and ponds will reduce fish farming. Climate Resilient Agriculture (CRA) encompassing adaptation and mitigation strategies and the effective use of bio-diversity at all levels-genes, species and ecosystem- is thus an essential pre- requisite for sustainable development in face of changing climate (NAAS 2013).

Impact of Climate Change on Fisheries

Climate change is modifying fish distribution and the productivity of marine and freshwater species. This has impacts on the sustainability of fisheries and aquaculture, on the livelihoods of the communities that depend on fisheries, and on the ability of the oceans to capture and store carbon. It has both positive and negative impacts.

Positive impacts may include (i) Lower natural winter mortality; (ii) Faster growth rate among fish in higher altitudes; (iii) Opportunities for regions like Andaman and Nicobar Islands where agriculture is not economically viable and chances for aquaculture in saline waters; and (iv) certain areas are predicted to witness abundance in some species.

Negative impacts may include: (i) temperature increase in sea water resulted in bleaching and death of coral reefs; (ii) Migration of small pelagic fish like mackerel and oil sardine from their existing home grounds; (iii) Smaller fish is likely to dominate high value big fish like tuna and butter fish; (iv) Changes in planktonic species composition; and (v) Changes in ecosystem structure and function.

Despite these uncertainties and potential negative impacts of climate change on fisheries there are opportunities to reduce the vulnerability to climate related impacts. As the first step, projections on fish distribution and abundance catches need to be developed for planning better management adaptations (Vivekanandan, 2011). National Academy of Agriculture Sciences, New Delhi in the policy paper 65, on Climate Resilient Agriculture in India (NAAS, 2013) suggested following measures to reduce negative impact on fisheries due to climate change: (i) Evaluating the adaptive capacity of important fish groups; (ii) Identifying adaptive fishing and post-harvest practices to sustain fish production and quality; (iii) Supporting energy efficient fishing craft and gear; (iv) Cultivating aquatic algae, which have positive response to climate change for food and pharmaceutical purposes and for production of biodiesel; (v) Increasing climate literacy among the fish farming communities; (iv) Organizing weather watch groups; (v) Establishing effective coast protection structures; and (vi) Evolving decision support systems for fisheries production systems.

Global Fisheries

Fish is an important food. It is also an important source for animal protein and omega 3 fatty acids. One of the best studies reported so far on the impact of climate change on marine fisheries is by Teixeira *et al.* (2014) on Portuguese fisheries. According to their study, the present worrying situation of the world fisheries is the result of the interaction of different factors such as climatic variations, natural oscillations, and technological innovation, but the greatest challenges are overfishing and climate change. The species richness of the Portuguese coast is generally higher than that found in northern Europe and similar to that of the Mediterranean, since many species have their southern or northern distribution limits along the Portuguese coast. The zoogeographic importance of this latitudinal area has long been recognized, representing the transition between north-eastern Atlantic warm-temperate and cold-temperate regions, which makes the Portuguese coast an area of great sensitivity to the detection of climate change. Official landing data for commercial species that have the Portuguese coast as their distribution limit were analysed together with sea surface temperature for the period of 1927–2011. In general, landings of species with affinity for temperate waters presented a decreasing trend, whereas species with affinity for subtropical/tropical waters showed an increasing trend. These variations were associated with variation in temperature: lower landings of temperate species occurred mainly in warm years, whereas landings of subtropical/tropical species presented an opposite trend, with higher landings in warm years. Preparing for the impacts of climate change is crucial for the sustainability of fisheries. Management should take into consideration information about environmental factors that affect species distribution and abundance (Teixeira *et al.*, 2014).

Heath *et al.* (2012) in their study on recent and projected future changes in the temperature and chemistry of marine waters around the UK and Ireland reported that changes in sea water temperature are having, and will in the future have, effects on the phenology, productivity and distribution of marine fish and shellfish. During warm periods, southern species have tended to become more prominent and northern species less abundant.

Indian Fisheries

The oil sardine fishery is one which has shown visible impact of climate change on landings. Indian oil sardine is distributed on the entire west coast of India from Gujarat to Kerala, and also on Tamil Nadu, Pondicherry, Andhra Pradesh and Orissa in the Indian east coast, but the highest abundance is observed off Kerala and Karnataka coasts. The stock structure of the oil sardine in India has been the subject of different studies in the last years (Boyce, 2009; Remya *et al.,* 2015; Sukumaran *et al.,* 2016).

After witnessing a record low catch of 45,000 tonnes in 2016, oil sardine seems to be showing signs of recovery with the landings in 2017 at 1.27 lakh tonnes. It is pointed out that oil sardine production was at a record high in the country as well as in Kerala with a production of 720,270 tonnes and 399,786 tonnes, respectively, in 2012 (Source: CMFRI, Kochi). The decline in oil sardine abundance was attributed to the retarded growth in the species and total failure of spawning in 2015-16 following warming up of coastal waters around peninsular India with focal point along Kerala coast. The warming up was prelude to the onset of El Nino, which gradually created an environment unfavorable to oil sardines (Source: CMFRI).

Barange *et al.* (2018) conducted a detailed study on the vulnerability of marine fisheries to climate change and existing and potential responses to adapt to the changes are examined in more detail for 13 different marine regions covering a range of ecological, social and economic conditions. It is concluded that adaptations to climate change must be undertaken within the multifaceted context of fisheries, with any additional measures or actions to address climate change complementing overall governance for sustainable use. It is recognized that some of these measures will require institutional adaptation. The report also indicates that short-term climate change impacts on aquaculture can include losses of production and infrastructure arising from extreme events such as floods, increased risks of diseases, parasites and harmful algal blooms. Long-term impacts can include reduced availability of wild seed as well as reduced precipitation leading to increasing competition for freshwater. The report also indicates that the interaction between ecosystem changes and management responses is crucial to minimize the threats and maximize the opportunities emerging from climate change. Production changes are partly a result of expected shifts in the distribution of species, which are likely to cause conflicts between users, both within and between countries. Reduced availability of fish can develop unemployment to marine fishers, reduce availability of coastal people and also threaten food security of the nation.

References

Barange, M., Bahri, T., Beveridge, M.C.M., Cochrane, K.L., Funge-Smith, S. and Poulain, F. (Eds.), 2018. Impacts of climate change on fisheries and aquaculture: synthesis of current knowledge, adaptation and mitigation options. FAO Fisheries and Aquaculture Technical Paper No. 627. FAO, Rome, 628 pp.

Bhattacharya, B. 2001. Food Security, NAP and WTO. Yojana, January, Vol. 45(1): 16-20

Boyce, M.S. 2009. Oil sardine (*Sardinella longiceps*) off the Malabar Coast: density dependence and environmental effects. Fisheries Oceanography, 18(5): 359-370

FAO 1996. Rome Declaration on World Food Security. Report of the World Food Summit,13-17 November 1996, WFS 96/REP, FAO, Rome.

Heath, M.R. , Neat, F.C., Pinnegar, J.K., Reid, D.G., Sims, D.W. and Wright, P.J. 2012. Review of climate change impacts on marine fish and shellfish around the UK and Ireland. Aquatic Conservation: Marine and Freshwater Ecosystem 22(3): 337-367

NAAS, 2013. Climate Resilient Agriculture in India. Policy Paper No. 65, National Academy of Agricultural Sciences, New Delhi: 20 p.

Remya, R. and Vivekanandan, E. and Sreekanth, G.B. and Ambrose, T.V. and Nair, P.G. and Manjusha, U. and Thomas, S. and Mohamed, K.S. (2015) Stock structure analysis of oil sardine *Sardinella longiceps* (Valenciennes, 1847) from southeast and southwest coasts of India. Journal of the Marine Biological Association of India, 57 (1): 14-20

Sukumaran, S., Gopalakrishnan, A. Sebastian, W., Vijayagopal, P., Nandakumar Rao, S., Raju, N., Ismail, S., Abdussamad, E.M., Asokan, P.K., Said Koya K.P. and Rohit, P. 2016. Morphological divergence in Indian oil sardine, *Sardinella longiceps* Valenciennes, 1847 – Does it imply adaptive variation? Journal of Applied Ichthyology, 32(4): 706-711

Teixeira, C.M., Leitγo, F., Gamito, R. and Cabral, H. 2014. Trends in landings of fish species potentially affected by climate change in Portuguese fisheries. Regional Environmental Change 14: 657-669.

Impact of Climate Change on Hydrological Cycle, Ecosystem, Fisheries and Food Security, Pages: 399–405
Edited by: B. Madhusoodana Kurup, M.R. Boopendranath, M. Harikrishnan and A.V. Shibu

CHAPTER - 34

Export Oriented Aquaculture for Seafood Exports from India

K.S. Srinivas

The Marine Products Export Development Authority, MPEDA House, P.B.No. 4272, Panampilly Avenue, Panampilly Nagar P.O., Kochi-682036, Kerala, India
E-mail: chairman@mpeda.gov.in

There is steady growth of seafood exports from the country from the sixties. But this growth has been predominant during the nineties, the decade of the new millennium and the present decade. The seafood exports has been touching an all-time high, both in terms of quantity and value almost every succeeding year, during the last decade, barring the exception of 2013-14 (Fig. 1). One of the main factors for the same is the stupendous growth of export oriented aquaculture.

Export oriented aquaculture production continues to increase year over the year in the country. Shrimp aquaculture production from the country mainly comprising of two species of shrimps such as *Penaeus monodon* and *Litopenaeus vannamei* and one species of freshwater prawn (*Macrobrachium rosenbergii*). The major culture areas are in the states of West Bengal, Odisha, Andhra Pradesh, Tamil Nadu, Kerala, Karnataka, Maharashtra and Gujarat.

Indian Seafood Export Scenario During 2018-19

India has exported 14,37,445 tonnes of seafood worth US$6.73 billion (46,589 crores) during the financial year 2018-19. This is from an export quantity of 6,78,436 tonnes ten years back. The Fig. 2 shows the Indian Seafood exports during the last ten years in terms of quantity and Fig. 3 in terms of value. The USA continued to be the major importer of Indian seafood as in the previous year with a share of 34.84% in terms of USD and imported 2, 81,913 tonnes of seafood.

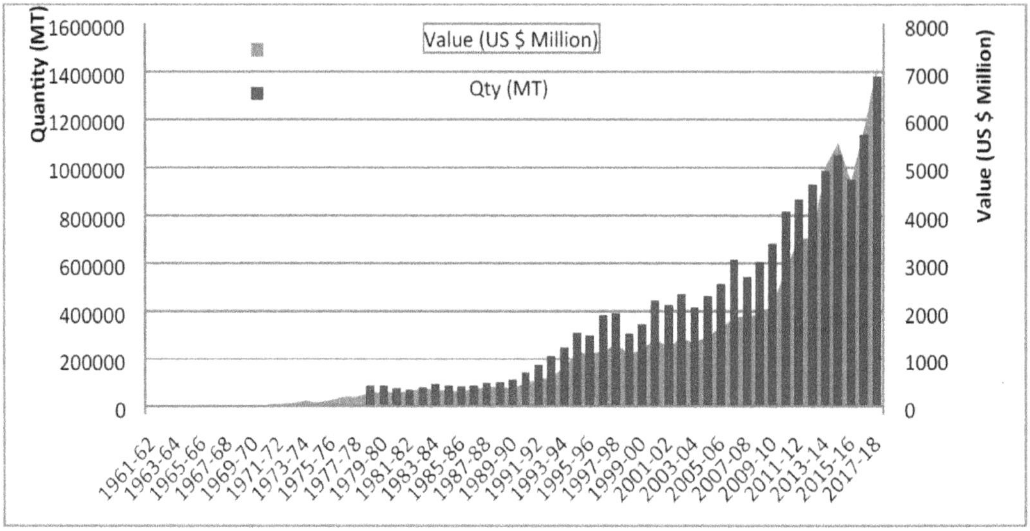

Fig. 1. Seafood exports from 1960s

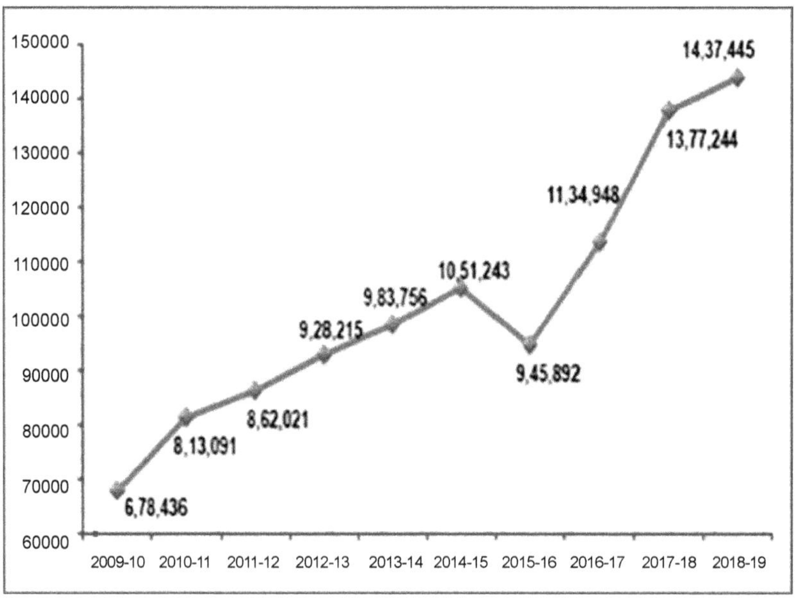

Fig. 2. Indian Seafood exports during the last 10 years (tonnes)

South East Asia remains as the second largest market accounting for a share of 22.76 % in USD terms followed by European Union, China, Japan, Middle East and other countries. The major South East Asian markets for Indian marine products are Vietnam with a percentage share of 65.67 in US$ terms, followed by Thailand, Taiwan, Malaysia, Singapore, South Korea and other countries. Vietnam alone imported 237114 tonnes of Indian sea foods.

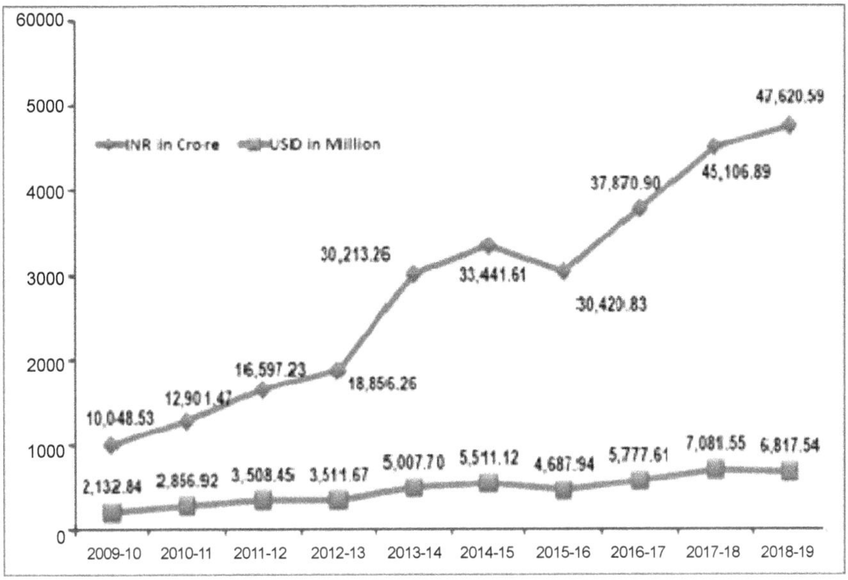

Fig. 3. Indian Seafood exports during the last 10 years (value)

Frozen shrimp is the major item of export and it contributes 44.10% in quantity and 68.52%in value of the total USD earnings. Frozen shrimp export has increased by 8.51% in terms of quantity and 3.02% in rupee value. The overall export of shrimp during 2018-19 was 6,14,145 tonnes worth USD 4,610.59 million. USA is the largest market imported 2,58,551 tonnes of frozen shrimp followed by South East Asia (1,16,061 tonnes), EuropeanUnion(70,367 tonnes),China(66,929 tonnes), Japan (36,649 tonnes), Middle East countries (28,505 tonnes) and other countries (37,084 tonnes). Frozen fish is the second largest export item, contribute share of 24.34% in quantity and 10.39% in USD earnings. Export of frozen fish has shown a positive growth of 5.19 % in terms of rupee value. Frozen squid, chilled items, frozen cuttlefish, dried items, and live items also contribute to exports of seafood. The Indian marine seafood products were exported through 31 different sea/air/ land ports in the country. Vizag, Kochi, Kolkata, JNP, Pipavav and Krishnapatanam are the major ports which handled the marine cargo.

Impact of Export Oriented Aquaculture Production on Marine Product Exports

The advent of export oriented aquaculture in the country has had its direct positive impact on the seafood exports. Shrimp from the capture fisheries was contributing to the major share of the seafood exports in the seventies and the eighties. Since the nineties, and during the first decades of the new millennium, though all seafood commodities showed increase in exports, the frozen fish had emerged as the one contributing the largest share in terms of quantity. However, the exponential increase of shrimp production through aquaculture during the second decade of the new millennium saw frozen shrimp surging ahead of frozen fish from 2014-15 onwards (Fig. 4). The correlation between the increase in value of seafood exports and increase in value of shrimp exports is of very high degree, underlining the importance of shrimp in our seafood exports.

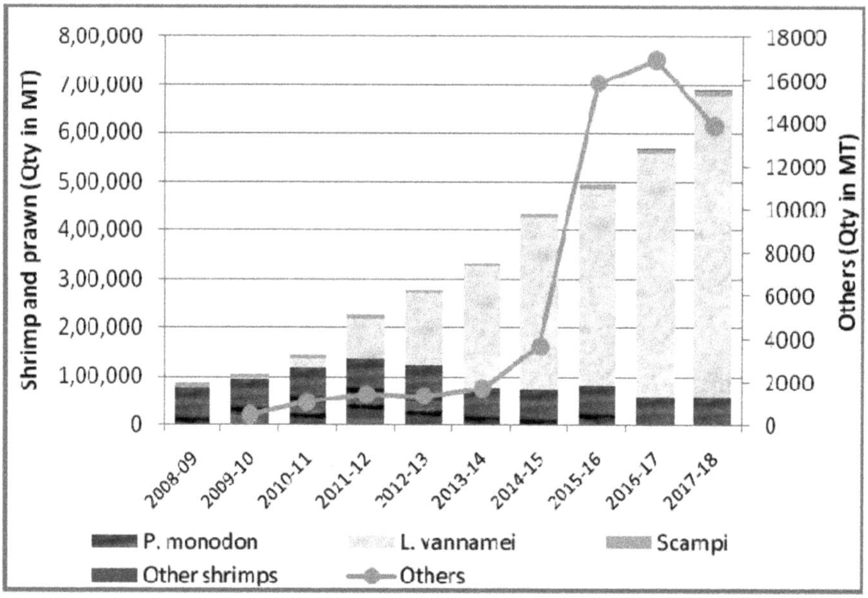

Fig. 4. Changes in the pattern of production of export oriented species

Export Oriented Aquaculture Production During 2018-19

Export oriented Aquaculture Production during 2018-19 was of 6,99,350 tonnes. Since the introduction of *Litopenaeus vannamei* for commercial aquaculture production in India, cultured shrimp production increases year by year. The shrimp production comprising *L. vannamei* and Black Tiger shrimp was about 6,80,000 tonnes, during 2018-19 (Table 1).

Table 1: Species wise aquaculture production during 2018-19

Species	Production (tonne) 2018-19
L . vannamei	6,18,678
P. monodon	54,901
Other shrimps	1,344
Scampi	7,222
Total	**6,82,145**

The Pacific White shrimp (*Litopenaeus vannamei*) continued to contribute the lion's share of India's cultured shrimp and prawn production. The share of *L. vannamei* in total shrimp and prawn production of 2018-19 was 90.70% of the total shrimp aquaculture production in the country (Fig. 5). The total production of the species for the year 2018-

19 was 6, 18,678 tonnes from an area of 75,494 ha. Andhra Pradesh tops in both production and productivity followed by Gujarat. Average productivity showed an increase to 8.2 tonne/ha/yr from the previous year 6.66 tonne/ha/yr. The reason for higher yield could be due to the farm stocking at higher rate and harvesting smaller size shrimp. The tiger shrimp production during the 2018-19 was 54,902 tonnes from an area of 57,691 ha. Tiger shrimp production trend decreased marginally by about 0.81% when compared with the year 2017-18. West Bengal contributes about 87.14% of the total tiger shrimp production of the country. The total scampi production during the year reported was 7,222 tonnes, showing a decrease of about 27.66% compared to the previous year. West Bengal was the major producer of Scampi.

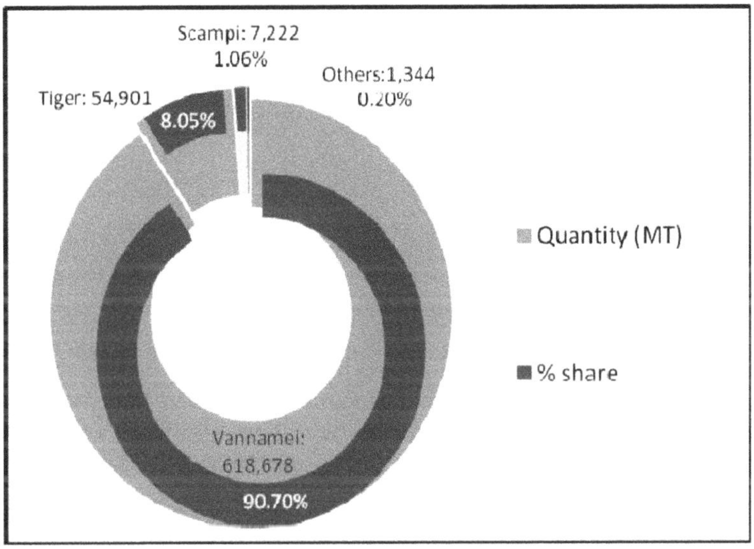

Fig. 5. Composition of shrimp species in aquaculture production 2018-19.

Potential for diversified species

Crab, Sea bass, and Tilapia (GIFT) besides Pangassius and small amounts of Cobia and Pompano were also cultured as other diversified species during 2018-19. In the case of Cobia and Pompano, only trial productions and demonstration programs were reported. During the year 2018-19, the estimated production of diversified species was 29,929 tonnes, which included 1,671 tonnes of Mangrove Crab, 3,952 tonnes of Seabass, 8,910 tonnes of Tilapia and 15,396 tonnes of Pangassius. The production of diversified species of fish and shellfish for export is still in the early stages. Large scale development can be possible with the involvement of corporate bodies who can invest in setting up integrated units of exportable varieties of fish and shellfish including hatcheries, farms and processing plants.

Role of MPEDA in Export Oriented Aquaculture Development

One of the main objectives of MPEDA, the nodal agency to promote export oriented aquaculture in the country is to develop responsible aquaculture practices in the country through better management practices (BMPs). The responsible aquaculture means production of pathogen free, safe aquaculture produces without using any harmful antibiotics and pharmacologically active substances and also safeguarding the environment through sustainable aquaculture methods. Field level Demonstration programmes are being carried out by the fields offices to demonstrate the farmers about the new technologies like Re-circulated Aquaculture System, Biofloc, etc., and diversified species (Sea bass, Crab, Cobia, etc.). The National Residue Control Plan (NRCP) is formulated for monitoring the presence of residues of veterinary drugs and environmental contaminants in aquaculture products, to ensure an overall monitoring of the aquaculture products at different stages of production to guarantee safe products from farm to fork. MPEDA has implemented a scheme to enroll all the aquaculture farms in the country . Samples are being collected from all the enrolled farms 15 days prior to the date of harvest to test the presence of antibiotics/chemicals/ pharmacologically active substances in ELISA Labs established by MPEDA throughout the country. The Central Government has formulated a new agri-export policy and aquaculture is being given special thrust under the Pradhan Mantri Matsya Sampada Yojana. In view of this, MPEDA has formulated State wise Action Plan augmenting fish production and fish exports. The hinterland States are also being focused now for expanding the production base. MPEDA has developed responsible aquaculture practices in the country through better management practices (BMP's) for export production. Apart from the promotional activities directly conducted by MPEDA, the societies established under the organization are also extending assistance to the aquaculture sector. The Rajiv Gandhi Centre for Aquaculture (RGCA), under the umbrella of MPEDA is engaged in developing commercial technologies for aquaculture diversification programmes in India. RGCA is also operating the only Aquatic Quarantine Facility (AQF) in India, for ensuring the disease free supply of seeds to Indian shrimp farmers. The National Centre for Sustainable Aquaculture (NaCSA) is another outreach organization established by MPEDA for empowerment of small scale shrimp farmers in the country and to ensure the adoption of BMPs in sustainable operation of aquaculture farms. In order to ensure the quality of aquaculture produce exported from the country, MPEDA has set up designated ELISA labs in various parts of the coast to help the farmers to conduct Pre Harvest Tests (PHT). Further, the National Residue Control Plan (NRCP) for monitoring the presence of residues of veterinary drugs and environmental contaminants in aquaculture products is implemented by MPEDA, to ensure an overall monitoring of the aquaculture products with at different stages of production to guarantee safe products from farm to fork. The competitive global seafood market is moving towards certified seafood, to ensure traceability and accountability. In this connection, MPEDA is soon launching an online certification platform, for enabling the Indian shrimp farmers to adopt certification at a reasonable cost. The aquaculture farms and hatcheries can avail this opportunity. With all these measures, it is hoped that India will surge ahead in seafood exports, and be able to achieve the projected target of US$ 17 billion by the year 2030.

Impact of Climate Change on Hydrological Cycle, Ecosystem, Fisheries and Food Security, Pages: 407–423
Edited by: B. Madhusoodana Kurup, M.R. Boopendranath, M. Harikrishnan and A.V. Shibu

CHAPTER - 35

A Meta-Analysis Study on Types of Seafood Mislabelling in Canada

Lovena Mathew*

University of Guelph, 50 Stone Rd. E, Guelph, ON N1G 2W1, Canada
**E-mail: mathewl@uoguelph.ca*

Food fraud occurs when a food product is illegally sold with the intention of misleading consumers, usually for economical gain. Food crime consists of any criminal activity that affects safety or authenticity of food; which can include product mislabelling, species substitution, misbranding, counterfeiting, dilution and adulteration. Although the motivation for food fraud is usually financial by tricking consumers, it could also cause public health concerns among consumers (Reilly, 2018). A gap was noticed through the review of seafood mislabelling studies done in Canada (Table 1). None of the scientific studies on mislabelling had done a meta-analysis review on the types of seafood mislabelling based on criteria for product authenticity. Such a study would be able to recognize what kind of mislabelling is most prevalent and thereby suggest solutions for this issue at the point in the supply chain where the specific mislabelling occurs.

Materials and Methods

Data collection

We conducted a literature review using various scientific databases including but not limited to Web of Science, Primo, etc., to compile studies in Canada (Table 1) that collected seafood samples at different points in the market supply chain and barcoded DNA to identify its true identity. Seafood sample data was compiled from various studies done in Canada that reported seafood mislabelling over the past decade from 2008 to 2019 (Table 1). A total of 1784 seafood samples were used from various studies conducted in Canada from 2008 to 2019. They were collected at different points in the marketing chain such as retail, wholesale, import or export stages and restaurants (including sushi and non-sushi restaurants) (Table 2).

Table 1: List of Canadian studies that reported seafood mislabelling across various levels of the market supply chain.

	Scientific studies reporting seafood mislabelling in Canada	Year of publication	Author	Samples	Total Mislabelling rate
1	Oceana: Ottawa	2018	Oceana	98	45 (46%)
2	Survey of mislabelling across finfish supply chain reveals mislabelling both outside and within Canada	2019	Shehata *et al.*	201	56 (28%)
3	Oceana: Victoria	2018	Oceana	15	10 (1.5%)
4	Study of fish products in Metro Vancouver using DNA barcoding methods reveals fraudulent labeling	2018	Hu *et al.*	275	71 (26%)
5	Oceana: Halifax	2018	Oceana	89	34 (38%)
6	Oceana: Toronto	2018	Oceana	96	56 (58%)
7	Oceana: Vancouver	2018	Oceana	84	21 (25%)
8	DNA barcoding as a regulatory tool for seafood authentication in Canada	2018	Shehata *et al.*	328	45 (14%)
9	DNA barcoding detects market substitution in North American seafood	2008	Wong and Hanner	90	25 (28%)
10	FISH-BOL and seafood identification: Geographically dispersed case studies reveal systemic market substitution across Canada	2011	Hanner *et al.*	238	75 (32%)
11	Utilization of DNA barcoding to determine fish sample species from an Ontario seafood distributor.	Unpublished data	Phoenix Gu.	93	18 (19%)
12	Community engagement in seafood identification using DNA barcoding reveals market substitution in Canadian seafood	2015	Naaum and Hanner	292	66 (23%)
13	Sea Choice Fish Labelling survey	2017	Sea Choice	442	30 (6.8%)

Table 2: **Total amount of seafood samples collected from different distribution outlets and associated mislabelling rates reported in the Canadian seafood mislabelling studies (Table 1).**

Distribution level in the supply chain	Number of samples	Number of mislabelled samples	Rate of mislabelling
Retail	1203	194	16%
Wholesale	339	48	14%
Import/Export stage	145	26	18%
Restaurants(including sushi bars)	654	284	43%
Sushi restaurants	183	82	45%

Data categorization schema

We screened each publication for usable data on seafood mislabeling. We compiled and coded the following information when possible: (i) study, (11) content of the label, (iii) species English name (common name) reported on the CFIA fish list, (iv) species reported on CFIA fish list (Canadian Food Inspection Agency, 2019), (v) genus reported on CFIA fish list, (vi) the location in the supply chain (e.g., restaurant), (vii) the true identity according to BOLD ID testing (e.g., species common name, scientific name, genus), (viii) food safety concerns, and (ix) environmental concerns of the identified product. Fishbase was used as a taxonomic reference to cross check and find common names from these sources (i.e., for species and families) (Luque and Donlan, 2019).

The data was then filtered to isolate the total amount of mislabelled samples and categorize them separately to determine the types and rates of mislabelling (Table 3). The mislabelled species data set was also used to filter samples where species indicated on the label or BOLD identified species were missing on the fish list.

Meta-analysis

The average of estimates provided by a group of studies is closer to the solution for the issue than the estimate provided from an individual study (Luque and Donlan 2019). This model was applied to seafood data to filter out various categories and determine the types of mislabelling.

Results and Discussion

The criteria for a product to be considered authentic was if the scientific name of the species (indicated by the label) matched the BOLD identified species' scientific name. A product was considered to not be authentic if the scientific name of the species (indicated by the label) was completely different and did not match the BOLD identified species' scientific name.

Table 3: Types of mislabelling: product substitution and label non conformance. 1784 seafood samples were analysed; 61% of the samples were not mislabelled and 39% of the samples were mislabelled.

Types of mislabelling or substitution	Mislabelling or substitution of product details	Product Label (L)	Species common name on CFIA fishlist (A): CFIA ID (a)	Species Common name identified (B): BOLD ID (b)		Label conformance	Number of samples	Is the authentic (Y/N)
Product Label (non conformance)	(a): Product label **consists of species common name with additional descriptor** (location of catch, size, method of growth or processing, etc.). Product label does not indicate the identified common name.	Fogo island cod	Cod: *Gadus morhua* or *Gadus macrocephalus*	Atlantic cod: *Gadus morhua*	A≠B a=b	L≠A	21	Y
	(b): Product label **consists of species common name with additional descriptor** (location of catch, size, method of growth or processing, etc.). Product label does indicate the identified common name.	BC albacore tuna	Albacore tuna: *Thunnus alalunga*	Albacore tuna: *Thunnus alalunga*	A=B a=b	L≠A	24 152	Y
	(c): Product label **consists of species common name with additional descriptor** (usually an unacceptable market name).	Red tuna (bigeye)	Bigeye tuna: *Thunnus obesus*	Bigeye tuna: *Thunnus obesus*	A=B a=b	L≠A	6	Y

[Table Contd.

Contd. Table]

Types of mislabelling or substitution	Mislabelling or substitution of product details	Product Label (L)	Species common name on CFIA fishlist (A): CFIA ID (a)	Species Common name identified (B): BOLD ID (b)		Label conformance	Number of samples	Is the authentic (Y/N)
	(d): Product label **consists of species common name but is not completely descriptive or compliant** to the product common name that is allowed by the CFIA fish list.	Sockeye	Sockeye salmon: *Oncorhynchus nerka*	Sockeye salmon / red salmon: *Oncorhynchus nerka*	A=B a=b	L≠A	10	Y
	(e): 1. Product label **does not consist of species common name and is an unacceptable market name** for (single) product species. 2. Product label **does not consist of species common name and is an unacceptable market name** for (multiple) product species.	1. White tuna 2. Pacific/ Wild salmon	1. Tuna / Albacore tuna: *Thunnus alalunga* 2. Pacific Pink Salmon / Pink Salmon: *Oncorhynchus gorbuscha* Chum Salmon / Keta Salmon / Silverbrite Salmon: *Oncorhynchus keta* Blueback Salmon / Medium Red Coho Salmon / Coho Salmon: *Oncorhynchus kisutch*	1. Albacore tuna: *Thunnus alalunga* 2. Chum salmon: *Oncorhyncus keta*	A=B a=b	L≠A	82	Y

[Table Contd.

Contd. Table]

Types of mislabelling or substitution	Mislabelling or substitution of product details	Product Label (L)	Species common name on CFIA fishlist (A): CFIA ID (a)	Species Common name identified (B): BOLD ID (b)	Label conformance	Number of samples	Is the authentic (Y/N)
			Red Salmon / Red Sockeye Salmon / Sockeye Salmon: *Oncorhynchus nerka* White Chinook / Chinook Salmon / Pink Chinook / Chinook / King Salmon / Red Chinook / Spring Salmon: *Oncorhynchus tshawytscha*				
	(f): 1. Product label is not compliant as **product species is missing on the fishlist**. 2. Product label **does not consist of common name of product species which is missing on the fishlist.** 3. Product label is an **unacceptable market name and product species is missing on the fish list.**	1. Yellowedge grouper 2. Katsuwonus *Hyporthodus* pelamis 3. Jade perch	N/A	1. Yellowedge grouper: *flavolimbatus* 2. Skipjack tuna: *Katsuwonus pelamis* 3. Barcoo grunter: *Scortum barcoo*	B, b missing	9	Y

[Table Contd.

Contd. Table]

Types of mislabelling or substitution	Mislabelling or substitution of product details	Product Label (L)	Species common name on CFIA fishlist (A): CFIA ID (a)	Species Common name identified (B): BOLD ID (b)		Label conformance	Number of samples	Is the authentic (Y/N)
Product Label (non conformance) Product species (substituted)	(g): Product label **consists of species common name with additional descriptor** (location of catch, size, method of growth or processing, etc.). Product label does indicate the identified common name. Product species does not match the BOLD identified species.	1. North atlantic cod 2. Tilapia-ASC	1. Atlantic cod: *Gadus morhua* 2. Tilapia: *Oreochromis mossambicus / Oreochromis niloticus / Oreochromis urolepis hornorum / Sarotherodon galilaeus / Sarotherodon melanotheron*	1. Pacific cod: *Gadus macrocephalus* 2. Blue tilapia: *Oreochromis aureus*	A≠B a ≠b	L≠A	37 128	N
	(h): 1. Product label **does not consist of species common name and is an unacceptable market name** for (single) product species. Product species does not match the BOLD identified	1. Chilean seabass 2. Pacific/ Wild salmon	1. Patagonian toothfish: *Dissostichus eleginoides* 2. Pacific Pink Salmon / Pink Salmon: *Oncorhynchus gorbuscha* Chum Salmon / Keta Salmon / Silverbrite Salmon: *Oncorhynchus keta*	1.Antarctic toothfish: *Dissostichus mawsoni* 2. Salmon / Atlantic salmon: *Salmo salar*	A≠B a≠b	L≠A	76	N

[Table Contd.

Contd. Table]

Types of mislabelling or substitution	Mislabelling or substitution of product details	Product Label (L)	Species common name on CFIA fishlist (A): CFIA ID (a)	Species Common name identified (B): BOLD ID (b)		Label conformance	Number of samples	Is the authentic (Y/N)
	species. 2. Product label **does not consist of species common name and is an unacceptable market name** for (multiple) product species. Product species does not match the BOLD identified species.		Blueback Salmon / Medium Red Coho Salmon / Coho Salmon: *Oncorhynchus kisutch* Red Salmon / Red Sockeye Salmon / Sockeye Salmon: *Oncorhynchus nerka* White Chinook / Chinook Salmon / Pink Chinook / Chinook / King Salmon / Red Chinook / Spring Salmon: *Oncorhynchus tshawytscha*					
	(i): 1. Product label is not compliant as **product species is missing on the fishlist.** Product species does not match the BOLD identified species. 2. Product label and BOLD identified product indicate **species that are missing on the fishlist.**	1. White snapper 2. Black grouper	N/A	1. Seabream / Porgy / Red Porgy: *Pagrus pagrus* 2. Gag: *Mycteroperca microlepis*	1. A, a missing 2. A, a, B, b missing	L≠A	15	N

[Table Contd.

Contd. Table]

Types of mislabelling or substitution	Mislabelling or substitution of product details	Product Label (L)	Species common name on CFIA fishlist (A): CFIA ID (a)	Species Common name identified (B): BOLD ID (b)		Label conformance	Number of samples		Is the authentic (Y/N)
Product species (substituted)	(j): Labelled product **species does not match** the BOLD identified product species.	1. Pickerel 2. Atlantic cod	1 Pickerel: *Sander vitreus* 2. Atlantic cod: *Gadus morhua*	1. Pike perch: *Sander lucioperca* 2. Whiting / Hake: *Merluccius productus*	A≠B a ≠ b	L=A	317	413	N
	(k): Labelled product **species does not match** the BOLD identified product species.	Silverside: *Menidia menidia*	Silverside: *Menidia menidia*	Silverside: *Odontesthes regia*	A=B a ≠b	L=A	2		N
	(l): Labelled product species does not match the BOLD identified product species (which is **missing on the fishlist**).	1. King fish	1. King fish: *Scomberomorus cavalla*	1. Australian spotted mackerel: *Scomberomorus munroi*	A ≠B a ≠ b	L=A	94		N
Product Label (compliant) Product species (authentic)	(m): Product **species is a match but scientific name of species is outdated** in CFIA fishlist.	Alaska pollock	Alaska pollock: *Theragra chalcogramma*	Alaska pollock: *Gadus chalcogrammus*	A=B a=b	L=A	22	1091	Y

[Table Contd.

Contd. Table]

Types of mislabelling or substitution	Mislabelling or substitution of product details	Product Label (L)	Species common name on CFIA fishlist (A): CFIA ID (a)	Species Common name identified (B): BOLD ID (b)		Label conformance	Number of samples	Is the authentic (Y/N)
	(n): Product **species is a match but identified common name of species' missing** on the CFIA fishlist.	Emperor	Emperor: *Lethrinus* spp.	Spangled emperor: *Lethrinus nebulosus*	B missing	L=A	2	Y
	(o): Product **species is a match and label is compliant** to the species common name according to the CFIA fish list.	Arctic char	Arctic char: *Salvelinus alpinus*	Arctic char: *Salvelinus alpinus*	A=B a=b	L=A	1067	Y

A**: species common name on CFIA fishlista**: scientific name of species on CFIA fishlist***B**: product common name identified through a mandated fishlist***b**: scientific name of product (BOLD ID)***L**: product label name

Types of mislabelling

Table 3 highlights the types of mislabelling in seafood samples through two significant types of non-conformities: label non-conformance and product substitution. 152 samples (8.5%) demonstrated some kind of product label non-conformity, 413 samples (23%) indicated product substitution, and 128 samples (7.2%) were a combination of both of these kinds of mislabelling. 1067 samples (60%) were not subject to any kind of mislabelling.

Product label non-conformances

The first category of mislabelling consists of product label non-conformances where the seafood product label is not compliant to the common name for the species on the CFIA fish list although the product is still authentic and not replaced or substituted by another species. This accounts for 22% of mislabelling among the 693 mislabelled samples analysed in this study (Table 3).

Label includes species common name

Product label non-conformances can occur when the label includes the species common name indicated for that species on the CFIA fish list but also includes an additional descriptor (Table 3: (a), (b), (c)) or if the label consists of a species common name that is similar but not entirely compliant to the common name on the CFIA fish list (Table 3: (d)). These labels account for 41% of product label non-conformances. Product label non-conformances with additional descriptors can either indicate the identified common name of the species (b) or a different common name for the same species (a). When the label indicates a different common name for the same species (which is attributed to the single species in the CFIA fish list having multiple common names: Table 4) consumers may perceive the species to be of a different value than what it actually identifies to be. A label that states Fogo island cod may suggest to consumers that the product is Cod from Fogo island than what it identifies to be which is Atlantic cod. Although, the species' inherently authentic, this difference in understanding the product could affect the consumers' perception of the value of the product. Additional descriptors are coupled with these labels which are usually details about the location of catch, size or body part of the fish being sold, method of growth or processing, etc., (a) (b) or consist of an unacceptable or local market name for the product (c). Product label non-conformances can also occur with species common names that consist of parts of the common name on the CFIA fish list but not completely match it (d) which could eliminate important details about the type of fish and thereby confuse the consumer.

Label does not include species common name

Product label non-conformances can also occur when the label does not include the species common names and instead consists of an unacceptable market name for a single species or multiple species on the CFIA fish list (e). These labels account for 54% of product label non-conformances where the species' authentic but a market name for the product has been used instead of the acceptable species common name. Wild salmon labels (e) can provide consumers with the assumption of the fish being either King or Sockeye salmons which are usually the most expensive due to their higher fat content and firmer texture. These labels allow economically motivated mislabelling to occur whereby chum or pink salmons are sold at a higher price. Product label non-conformances also occur when the label indicates the common name, scientific name or an unacceptable market name for a species that is missing on the fish list (f).

Product substitution

The second category of mislabelling consists of product substitution where the labelled seafood product is in accordance with the species common name on the CFIA fish list but does not match the BOLD identified product, thereby indicating that the product is not authentic. This accounts for 60% of mislabelling among the 693 mislabelled samples analysed in this study (Table 3).

Product substitution can occur when the species on the fish list that is identified based on the label (common) market name does not match the BOLD identified species (j). Labels can also indicate a common name that pertains to multiple species on the fish list (k) (Table 4). The labelled species common name may also not match the BOLD identified product which is missing on the fish list (l) (Section 3.3); this could pose health threats or other dangers to the susceptible populations if the product is not allowed on the fish list for environmental or health concerns.

Product substitution and label non conformance

Product label non-conformance and species substitution were found to occur together in 18% of the 693 mislabelled samples analysed in this study. Product label non-conformance can occur when the label includes the species common name indicated for that species on the CFIA fish list but also includes an additional descriptor, or the label does not includes the species common name but instead has an unacceptable market name is a missing species on the fish list. These label non-conformities are coupled with species substitution whereby the product indicated by the label is not what it identifies to be (Table 3: (g) (h) (i)).

Authentic product and compliant product labels

Compliant product labels and authentic species accounted for 61% of the total samples analysed in the study (Table 3: (m) (n) (o)). A small subset of these samples consisted of species with scientific names that required to be updated on the CFIA fish list (m)

Table 4: **Categorization inconsistencies of the CFIA fish list where a single market name is applied to multiple species, multiple common names exist for a single species, or a single common name exists for multiple species. Scientific names for species**

Species English name (common name)	Scientific name	Scientific family name
Single common name	**Multiple scientific names**	**Multiple families**
- Snapper	*Macolor* spp. *Ocyurus* spp. *Paracaesio* spp. *Parapristipomoides* spp. *Pinjalo* spp. *Rhomboplites* spp. *Symphorichthys* spp. *Symphorus* spp. *Lutjanus* spp. *Apsilus* spp. *Hoplopagrus* spp. *Lipocheilus* spp.	Lutjaenidae Scorpaenidae
- Shark	*Sebastes ruberrimus* *Squalus acanthias* *Squalus blainville* *Centroscyllium fabricii* *Lamna nasus* *Prionace glauca* *Scyliorhinus canicula*	Squalidae Etmopteridae Lamnidae Carcharhinidae Scyliorhinidae
- Butterfish	*Peprilus medius* *Peprilus burti* *Peprilus simillimus* *Peprilus paru* *Peprilus triacanthus*	Stromatidae
- Hake	*Merluccius bilinearis* *Merlucciidae* *Merluccius gayi gayi* *Merluccius paradoxus* *Merluccius merluccius* *Merluccius gayi peruanus* *Merluccius productus*	Merlucciidae
- Bee liner / B-liner	*Rhomboplites aurorubens*	Lutjanidae
- American Smelt / Lake Smelt / Rainbow Smelt / Smelt	*Osmerus mordax*	Osmeridae
Single common name	**Multiple scientific names**	**Single family**
- Cod	*Gadus morhua* *Gadus macrocephalus*	Gadidae
- Halibut	*Hippoglossus hippoglossus* *Hippoglossus stenolepis*	Pleuronectidae

[Table Contd.

Contd. Table]

	Single common name	Scientific names not updated to current standards	Single family
-	Alaskan pollock	*Theragra chalcogramma*	Gadidae
	Single / Multiple common names	**Proper scientific names for various species that are incomplete**	**Single family**
-	Conger eel / Conger	*Conger* spp.	Congridae
-	Marlin	*Tetrapturus* spp.	Istiophoridae

(Table 4) and some samples where the product species common name was missing on the fish list (n). There was a 40% rate of non-conformance labelling and 60% product substitution among the 693 mislabelled samples in the study. Fig. 1 describes the relations between product species and its corresponding label. It is closely related to Table 3 and helps identify what category of authenticity a seafood sample can be categorized into after DNA barcoding testing.

	Product species (Authentic) a=b	Product species (Substituted) a≠b
Product label (non-conformance) L≠A	A=B a=b (80%) A≠B a=b (14%)	A≠B a≠b (100%)
Product label (compliant) L=A	A=B a=b (99%) A≠B a=b (0.2%)	A≠B a≠b (99%) A=B a≠b (0.5%)

Fig. 1. Summary of Table 3 indicating types of mislabelling and areas of label non-conformance

A fish supply chain could consist of four or more countries being involved in the harvesting/farming, processing (e.g. freezing, filleting, canning), and packaging. The packaged fish product may be sold domestically or exported to countries with labels that are based on the regulations of the country where the fish was harvested (Hu *et al.*, 2018). A lack of monitoring, surveillance and inspection can also lead to misrepresentation of seafoods.

Product label non-conformances (Table 3: (e)) where the label consists of an unacceptable market name could be intentional mislabelling to increase the marketing and purchase value of the product or unintentional mislabelling due to misidentification or ambiguous labelling regulations (Hu *et al.*, 2018).

Product substitution occurs due to various reasons such as a high demand for a limited supply, a greater profit margin, a growth in international trade of processed seafood, lack of strict seafood labelling and naming regulations, etc. (Hanner Becker, 2011). Product substitution can have negative impacts such as the loss of integrity in the fish markets resulting in economic loss for consumers and stakeholders in the fish industry, lack of consumer trust in governmental agencies, creating a market where unregulated and dangerous products are allowed while affecting the consumer's ability to make informed decisions to support sustainable fisheries (Hu *et al.*, 2018).

Retailers and consumers are also exposed to health and environmental risks. 16% (282) of the seafood samples compiled in the study indicated some type of food safety concern, primarily histamine, while 34% (614) of the seafood samples indicated environmental concerns. Public health is endangered when toxic fish species that are illegal for sale in Canada are substituted or labelled as other allowable species.

Mislabelling across the market supply chain

The highest amounts of mislabelling were found to be in restaurants (Table 2). This is the point in the supply chain that regulatory efforts should be focused at. Table 3 has indicated various instances where BOLD identified seafood products were missing on the fish list (f).

Regulatory Framework (CFIA fish list) Concerns

Species missing on CFIA fish list

The mislabelled species data set was used to filter and identify the number of samples that consisted of species missing on the CFIA fish list. Eight substituted samples consisted of labels indicating species with common names and associated scientific names missing on the fish list; e.g. White snapper *(Pagrus auratus)* or Golden pomfrct *(Xenobrama microlepis)*. Ninety-one substituted samples and 3 authentic samples consisted of BOLD identified product common names and associated scientific names (B, b) that were missing on the fish list; e.g. Spinycheek grouper (*Epinephelus diacanthus)* or Spotted goatfish: *Pseudupeneus maculates*. Nine substituted samples and 27 authentic samples consisted of BOLD identified scientific names (b) that were missing on the fish list. A major contributor to these BOLD identified missing species was the Alaskan pollock which is present on the fish list with the scientific name, *Theragra chalcogramma,* which is now outdated and requires to be updated to *Gadus chalcogrammus* (Orr and Stevenson, 2013). Eight substituted samples and 9 authentic samples consisted of the label and BOLD identified product common names and associated scientific names missing on the fish list; e.g. Yellowedge grouper *(Hyporthodus flavolimbatus),* Sawtooth barracuda (*Sphyraena putnamae*), etc. Twenty-five product labels indicated species that were missing on the CFIA fish list. Hundred and one samples indicated species with BOLD identified common names and associated scientific names that were missing on the fish list.

Conclusions

The common name and/or scientific name of the species based on the Canadian regulations must be present on the packaging to combat mislabelling and eliminate cases where naming conventions may cause mislabelling (Roebuck *et al.*, 2017). Establishing an agreed list of fish names can also help combat the issue. National governments must publish a Seafood list that links common species names to scientific nomenclature. USDA (United States Food and Drug Administration, 2018) and the CFIA (Canadian Food Inspection Agency, 2019) have published seafood lists which consist of a list of common names and scientific names that is mandatory for product compliance in commercial trade. Challenges in creating a global and harmonized fish list for international trade was discussed by the Codex Committee on Fish and Fishery Products in 2003.

A gap analysis of the species common name and scientific name entries in the fish list was conducted in a study where it was suggested that it would be useful to expand the CFIA fish list to include individual species within each genus as separate entries on the list (Shehata, 2018). Mandatory labelling requirements must be implemented for seafood products with the label clearly and accurately stating the common name and/or scientific name of the product. This would eliminate many label nonconformities and allow the consumer to make an informed decision on purchasing the product. Official food controls in monitoring and surveillance programs can be strengthened through development of DNA barcoding detection methods and stricter regulations for recall of samples that are non-compliant. Effective traceability systems must be in place through documentation and accessible information to mitigate fish fraud risks (Luque and Donlan, 2019). Fig. 1 describes an effective model on categorizing seafood samples to determine authenticity and recognizing types of mislabelling. Product substitution proved to be the most prevalent issue among food samples while restaurants are the establishments with the greatest risk for mislabelling to occur. Effective measures must be taken by regulatory programs and consumers to combat this issue.

References

Canadian Food Inspection Agency, 2019. CFIA fish list. Canadian Food Inspection fssa/fispoi/fplist/ fplist.asp?lang=e.

Hanner, R., Becker, S., Ivanova, N.V. And Steinke, D. 2011. FISH-BOL and seafood identification: Geographically dispersed case studies reveal systemic market substitution across Canada. Mitochondrial DNA 22 Suppl 1(S1): 106-22. doi: 10.3109/19401736.2011.588217

Hu, Y. , Huang, S.Y. , Hanner, R., Levin, J. and Lu, X., 2018. Study of fish products in Metro Vancouver using DNA barcoding methods reveals fraudulent labeling. Food Control 94: 38-47. doi:10.1016/j.foodcont.2018.06.023.

Luque, G. and Donlan, C., 2019. The characterization of seafood mislabelling: A global meta-analysis. Biological Conservation, 236: 556-570. https://doi.org/10.1016/j.biocon.2019.04.006

Naaum, A. and Hanner, R., 2015. Community engagement in seafood identification using DNA barcoding reveals market substitution in Canadian seafood. DNA Barcodes 3(1): 80-84. http://dx.doi.org/10.1515/dna-2015-0009.

North American seafood. Food Research International, 41(8): 828-837. doi. org/10.1016/j.foodres. 2008.07.005,

Orr, J.W. and Stevenson, D.E., 2013. Yes, Virginia, Walleye Pollock is *Gadus chalcogrammus*. www.afsc.noaa.gov/Quarterly/OND2013/divrptsRACE1.htm.

Reilly, A. 2018. Overview of food fraud in the fisheries sector, Fisheries and Aquaculture Circular No. 1165. FAO, Rome, Italy. http://www.fao.org/documents/card/en/c/I8791EN.

Roebuck, K., Turlo, C., Fuller, S. and Wallace, S., 2017. Canadians Eating in the Dark: A Report Card of International Seafood Labelling Requirements. SeaChoice. ISBN 978-1-988424-05-7.

Sea Choice, 2017. Fish Labelling survey. Sea Choice.org.

Shehata, H.R., Naaum, A., Garduno, R.A. and Hanner, R., 2018. DNA barcoding as a regulatory tool for seafood authentication in Canada. Food Control 92: 147-153. doi.org/10.1016/j.foodcont. 2018.04.045.

Wong, E.H.-K. and Hanner, R.H., 2008 DNA barcoding detects market substitution in North American seafood. Food Research International 41(8): 828-837.

Impact of Climate Change on Hydrological Cycle, Ecosystem, Fisheries and Food Security, Pages: 425–437
Edited by: B. Madhusoodana Kurup, M.R. Boopendranath, M. Harikrishnan and A.V. Shibu

CHAPTER - 36

State of Biocultural Diversity, Climate Change and Ethnic Meat and Fish Products of Northeast India: A Review

Ranendra Kumar Majumdar

Dept. of Fish Processing Technology
College of Fisheries (CAU-I), Lembucherra, Tripura-799210, India
E-mail: drrkmcof@gmail.com

The north-eastern region of India (NEI), lie deep in the lap of easternmost Himalayan hills, harbour a cauldron of different people and cultures. This region comprises of States like Arunachal Pradesh, Nagaland, Manipur, Tripura, Mizoram, Meghalaya, Assam and Sikkim, and connected to rest of India by merely 20 km of wide land (Fig. 1). It shares over 2,000 km of border with Bhutan, China, Myanmar and Bangladesh. The NEI's total geographical area is 2,55,168 sq km with population of 40 million (as per 2011 census) which contributes about 3.1% of the country's population. People of all religions like Hinduism, Islam, Buddhism and Christianity live here. This part of India is home to about 166 numbers of ethnic groups, each with their own distinct culture, which gives them a unique cultural identity. Their existence can be traced back to the pre-historic times, and can be largely associated with the ethnic groups of Indo-Mongoloids, Tibeto-Burmese and proto-Austrioloids which represent the Asio-Austric culture in Indian hilly regions. From the ethnic and linguistic angles, ethnic populace of NEI provide a cultural bridge between India on one side and Southeast Asia, China, Inner Asia and Burma on the other.

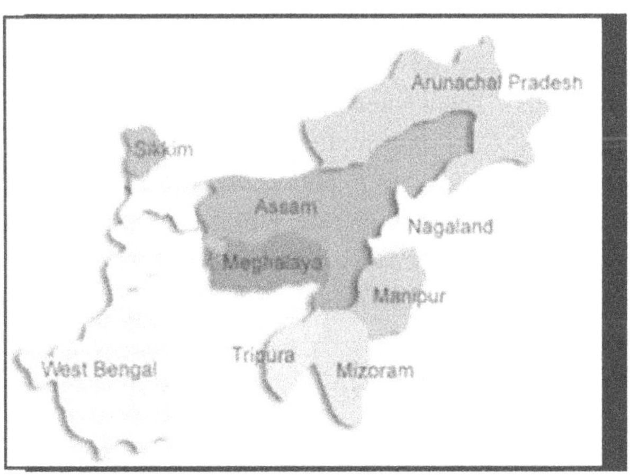

Fig. 1. North-eastern region of India (NEI)

Northeast India is a Treasure of Indigenous Knowledge

The region has a great treasure of cultural, social and linguistic variability, conserved by tribal people. Further, the region is famous for indigenous knowledge systems pertaining to agriculture, medicine, food and natural resources management. Indigenous people are accustomed to live and survive with the forest and follow the so called 'jhum' cultivation culture, which ensure a range of ethnic foods rich in nutrition and compatible to culture and ethnicity of tribes. Traditional agricultural practices, a high degree of forest dependency and use of ethnic foods and medicines indicates NEI a rich heritage of culturally-embedded traditional knowledge (Singh & Sureja, 2006). These not only protect biodiversity and ecology (Ramakrishnan *et al.*, 1996, 2002; Singh & Sureja, 2006), but also ensure to sustain the cultural diversity of NEI (Singh, 2004; Singh & Srivastava, 2009). Several indigenous practices, beliefs and knowledge built-up concerning food, medicines, hunting, fishing, agriculture, home gardening, handicrafts and other skills within a group of people across generations of inhabiting close contract with nature are known as traditional knowledge (Turner & Berkes, 2006). It evolves due to adaptation to local environmental and social circumstances, which is transmitted through generation without much improvement (Berkes, 2009; Berkes & Berkes, 2009).

Biocultural Diversity of NEI

Biocultural diversity encompasses the variability of biological species and ecosystems, and the uniqueness of cultural groups who interact with these resources (Berkes & Berkes; 2009, Singh & Srivastava, 2009). The NEI is considered one of the most bioculturally diverse sections of India (Yumnam, 2008; Singh & Srivastava, 2009). Traditional knowledge and biocultural diversity are intermingled with each other and is considered to be essential components to ensure sustainable development of communities inhabiting mountain ecosystems (Braton, 1989; Agrawal & Gibson, 1999). It is important to assess the value of biocultural diversity and associated traditional knowledge for their conservation and Intellectual Property Rights (IPR). The significance of conserving traditional knowledge and biocultural diversity is now widely recognised as well as of growing concern (Pilgrim *et al.*, 2007). The women in this region are major stakeholders and custodians of knowledge, conserving food and medicinal plants in both jhum land and home gardens. However, contribution of women in conserving biocultural diversity in the region is rarely recognised at policy level. Moreover, the policies and technologies formulated by the government on resource conservation are usually implemented without due recognition and participation of local people and their indigenous institutions.

Climate of Northeast India

The climate is predominantly humid sub-tropical with hot summers and severe monsoons and mild winters. Northeast India lies close to the tropics and to a large extent displays the character of tropical climate, especially in the valleys. There are three distinct seasons in the area, winter, summer and rainy season. Although the rainy season, as in the rest of India, coincides with summer months. The main source of rain being the southwest monsoon and June is the rainiest month. However, a climatic contrast exists between the valleys and the mountainous region. According to Dikshit & Dikshit (2013), in the plains the summer temperatures vary between 30 and 33°C, while the hills experience a mean maximum and minimum summer temperature of around 20°C and 15°C respectively. In the month of January, the temperature in the valley region of Assam is around 16°C, while the mountainous region of Arunachal Pradesh and Nagaland drift around a maximum of 14°C and a minimum of sub-zero temperature. In this region, heavy snowfall is witnessed only in the elevated areas of Arunachal Pradesh, like West Kameng and Tawang areas. Minimum average rainfall in any part of NEI is above 1000 mm. Cherrapunji, situated on the top of the scarp, receives a mean annual rainfall of 11,465 mm. The average rainfall of Brahmaputra valley is around 2,000 mm with local variations. Guwahati receives only 1,717 mm of rain. About 90 % of the rain is received during the southwest summer monsoon, and June is by far the rainiest month. The hilly areas of the region receive 2,000–3,000 mm of rain, though places like Kohima in Nagaland and Imphal in Manipur, because of their being in the shadow of the mountains, receive less than 2,000 mm of rains.

State of Climate Change in NEI

The apprehended high expectancy of the consequences of climate change in the northeast Indian region of India is due to its geo-ecological fragility, strategic location vis-a-vis the eastern Himalayan landscape and international borders, its trans-boundary river basins and its inherent socio-economic instabilities (Das *et al.*, 2009). For a long time, the NEI was a white spot on the map of climate change data, compared with other, better-researched regions in India. Initial scientific research and field observations confirm that the region is suffering from the impacts of climate change already. Due to its unique location and topography it has distinct precipitation and drainage patterns. From March to May, thunderstorms contribute about 20% of annual rainfall. From June to September, monsoon rains supply another 70%. The monsoon season is marked by frequent floods, as melting Himalayan snow and torrential rains feed the majestic Brahmaputra River. Moreover, because of climate change, rainfall is becoming more unpredictable and erratic. A significant decreasing trend of summer monsoon rainfall over this region at an approximate rate of 11 mm per decade during the last century has been reported (Mirza *et al.*, 1998).

Indian scholars from the Indian Institute of Science have recently compiled, verified and aggregated existing data to evaluate climate change of NEI (Ravindranath *et al.*, 2011). The compiled data show that current climate trends will continue for the years 2021 to 2050 in India's northeastern states. Analysis of macro-data supplemented by an on-the-ground assessment of local conditions suggests that there can be no more doubt that precipitation and temperature patterns are changing. Moreover, the region has warmed significantly during the last decade. The rise in surface air temperature has been primarily due to rapid increases of both maximum and minimum temperatures, with minimum temperatures rising faster – a trend which is found in other studies in the Himalayas. The climate forecasts indicate that these trends will be exacerbated in the future. Average temperatures are projected to increase by about 1.7°C in almost all the districts of the northeast. In some districts, they are likely to go up by more than 2°C. The rainfall is projected to increase in 57 of the 78 districts, with some districts expected to experience almost 25% more rainfall. The number of extreme rainfall events per year is also expected to increase by 26%, and obviously there will be more floods in this region.

Food security is one of the leading concerns associated with climate change. Climate change affects food security in complex ways. It is a threat to livestock production, forestry, fisheries and aquaculture, and can cause grave social and economic consequences in the form of reduced incomes, eroded livelihoods, trade disruption and adverse health impacts. However, the net impact of climate change depends not only on the extent of the climatic shock but also on the underlying vulnerabilities. Strategic programmes to challenge the climate change requires formulation of appropriate long term adaptation approaches and mitigation options for the livestock sector. Besides these, it is also important to plan different activities for the promotion of diversified income sources for vulnerable communities.

Food Habits of Populace of NEI

Northeast India is known for its diverse food culture comprising fermented and non-fermented ethnic foods and alcoholic beverages (Sathe & Mandal, 2016). From time immemorial, most of the people of the region bear their own methods of fermenting food materials for the purpose of preservation and taste enhancement. Indigenous fermented foods contribute a major share in the diet of the ethnic tribes in the Himalayan belt of India, and is considered as the oldest and most economic methods for development of a diversity of aromas, flavors, and textures, as well as food preservation and biological enrichment of food product by the manipulation of different microbial population (Sekar & Mariappan, 2007). More than 250 different types of fermented foods are region and even community-specific and locally available materials such as milk, vegetable, bamboo, soybean, meat, fish, and cereal are commonly fermented (Das & Deka, 2012; Tamang *et al.*, 2012). Processing of food not only contributes to the food security by minimizing the wastage, but also conserve the ethnic indigenous technology through increasing their availability and market value (Rolle & Satin, 2002).

The people of northeast India have strong preference for food of animal origin especially for meat. With more than 90% of the population being non-vegetarian against the national average of 70%, there is heavy demand for meat and other animal products including fish. The strong preference of NEI people for foods of animal origin might be due to their relationship with the Mongoloid culture of southeast Asian countries. It is said that the Mongoloid people need more iron in their blood so as to support normal respiratory process at high altitude and meat is a good source of iron. Therefore, meat consumption both from conventional and non-conventional animals is a tradition for tribal people of NEI. However, there is a large gap between the demand and supply of meat and meat products and thus the NEI states import huge quantity of fresh and processed animal foods from outside the region. Therefore, to meet the demand for meat, varieties of non-conventional meats like Mithun, Yak, Deer, Rabbit, Dog, Jackal, Monkey, Tiger, Beer, Elephant, Rhinoceros, Snakes, Mongoose, Frog, Tortoise etc. are also consumed regularly, although wild animals are not permitted to kill under Wild Life Protection Act. The NEI shares about 5.8% of the country's domesticated animals apart from rare varieties of wild species of animals.

Northeast sector of India is land locked and enriched with many freshwater species of fish and is considered to be one of the hot spots of freshwater fish biodiversity in the world. About 180 to 266 species of freshwater fish including a considerable amount of ornamental fish have been reported in the NEI by several authors. However, present availability of fish is around 9.18 kg/person/year (considering 95% fish eating population) as compared to the national average of 13.13 kg/person/year (considering 56% fish eating population as per Planning Commission Report). The annual average fish production in the region is estimated to be 0.34 MMT from both capture and culture fisheries with nearly 72% of production coming from Assam alone.

Traditional Fish Products of Northeast India

Northeast region is bestowed with many fermented fish products such as shidal, ngari, hentaak, lonailish, tungtap etc. Fermented fish contributes a regular menu especially in the diets of tribal people (Table 1). Unlike salt fermented fish products of southeast Asia, the salt-free fermented fish products of the northeast indicates the fact that the technology of salt-free fermentation originated long before the people started using salt (sodium chloride). Later on salt was used for fermentation as it was observed in case of lonailish. Smoked fish products are mostly popular in Manipur state and to lesser extent in certain tribes of Meghalaya, Nagaland and Mizoram states of NEI. Though there are several reports on smoke curing of fish in India and abroad, the smoking of fish in Manipur is unique in nature. Preparation of such fish products is simple but most of the critical steps are optimized by experiences for getting a yield of different qualities. Till today traditional practices are followed due to lack of proper scientific intervention and standardization of methods.

The fish fermentation technology in the northeastern states evolved by compulsion of people (Majumdar *et al.*, 2015). The northeastern states of India, being the highest rainfall area of the world, do not provide a congenial environment for simple sun drying of fish. People used to preserve fish for use in lean period by drying under sun. Hence, such drying used to be prolonged due to high humid atmosphere and frequent rainfall particularly during the peak fishing seasons (i.e. from May to September). Northeast, being abundant in low lying areas where accumulation of water during rainy season offers an ideal habitat for the breeding of weed fishes such as *Puntius* spp., 'Darkina' (*Esomus danricus*), 'Mola' (*Amblypharyngodon mola*). The clever fishermen, therefore, was in search of a method through which they could preserve the heavy catches of such less valued weed fishes for consumption and sale in the dry seasons (November to April) when there was scarcity of raw fresh fish in the market. Moreover, due to non-availability of ice and good road communications, these perishable produce could not have been transported to the distant markets where they could get a higher price which they never used to earn in the village markets.

Traditional Meat Products of Northeast India

From time immemorial, a number of indigenous fermented meat products has been developed in the region for their increased nutritional value and long term edibility (Table 2). Traditionally, excess meat is processed and preserved by drying or by drying and smoking or by salting and drying with local herbs or by fermentation. Some indigenous herbs, leaves of plants, roots, seeds, liquid vegetable extracts, spices and oils are also used for preservation of some meat products. There is a people's claim that certain herbal products used in meat processing and preservation has lypolytic, proteolytic, cholesterolytic and hepatoprotective properties, and therefore, due to heavy consumption of meat and meat products, few people in the region are prone to heart problems.

Nutritional Properties of the Ethnic Fish and Meat Products of Northeast India

Indigenous fish and meat products of NEI are more palatable, nutritious, and better keeping qual-ity at room temperature. The products usually are free from preservatives and artificial colouring agents, which could create attractions for the health-conscious consumers. The techniques employed for processing and the special ingredients added are partially responsible for imparting the characteristic flavour to the products which make them unique. The most common tech-niques used in such traditional products are fermenta-tion, smoking, and drying. Generation of acids in the fermentation process is responsible for the enhancement of nutritional value and shelf life of the products by reducing pH (Lueck, 1980). Organic acids are primarily generated in the products by lactic acid bacteria such as *Lactobacillus*, *Streptococcus, Pediococcus, Leuconostoc, Lactococcus*, and *Enterococcus* (Gilliland, 1985). These bacteria can metabolize several saccharides of fish and meat into lac-tic acid, alcohols, lipids, and some amino acids

Table 1: Some traditional fish products of NEI with their colloquial term, origin and type

Name of the product	Origin	Type and brief description
Shidal (local name *seedal*, *sepaa*, *hidal*, *verma*)	Erstwhile East Pakistan. Now in Tripura and Assam	Salt-free fermented *Puntius* spp. or *Setipinna phasa*. Fermentation is done is earthen container for a period of 3-5 months. Long shelf life at ambient temperature
Ngari	Manipur	Similar to Shidal with little difference in processing
Hentaak	Manipur	Fermented fish paste prepared from dry fish *Puntius* spp, *Amblypharyngodan mola*, *Esomus danricus* etc. along with vegetable like *Colocasia* spp.
Tungtap	Khasi & Jaintia communities, Meghalaya	Fermented fish paste made with fish like *Puntius*, *Danio* sp. etc. along with chilies, onion, mint leaves, turmeric powder and salt
Numsing	Missing tribe of Assam	Dried and smoked small fish species like *Puntius*, *A. mola*, *Channa*, *Tengra* etc., mixed with spices followed by fermentation in air tight bamboo container
Dang-pui-thu	Mizoram	Prepared from sundried small prawn, crab and fish species, which are ground to paste and fermented in earthen *container* for 2-4 months
Lonailish	Tripura, Assam	Sliced Hilsa (*Tenualosa ilisha*) fermented in saturated brine kept in tin container for 3-5 months
Napti	Tripura	Fermented fish paste from small variety of fish
Ngamu-leirou	Manipur	Smoked *channa* fish
Ukabi-leirou	Manipur	Smoked *Anabus* fish
Nga-phak	Manipur	Smoked common carp in butterfly style
Pengba-ayaiba	Manipur	Smoked *Pengba*
Ngapai-ayiba	Manipur	Smoked *Notopterus* fish
Khazing-ayiba	Manipur	Smoked prawn
Kharang & Kha Pyndong	Meghalaya	Smoked bigger varieties of fish (Chocolate Mahseer, Rohu, Mrigal, Grass carp etc.)
Ayaiba	Manipur, Mizoram	Fish Smoked fish; pickle
Bordia	Assam, Manipur, Mizoram	Salted and dried small varieties of fish
Chucha	Assam	Sun-dried fish product
Gnuchi	Sikkim	Smoked *Schizothorax* spp., *Labeo* sp.
Karati	Assam, Meghalaya	Dried, salted *Gudusia chapra*
Lashim	Assam, Meghalaya	Dried, salted *Cirrhinus reba*
Sidra	Sikkim	Dried *Puntius sarana*
Suka ko maacha	Sikkim	Dried and smoked *Schizothorax* spp.

Table 2: Some traditional meat products of NEI with their colloquial term, origin and type

Name of the product	Origin	Type and brief description
Adin	Arunachal Pradesh	Smoked, dried pork, beef or mithun meat with fermented bamboo shoot
Ar sa ret	Mizoram	Dry smoked chicken
Ashi kioki	Naga tribes, Assam	Thinly sliced dried pork taken with fermented bamboo shoot /soybean
Arjia	Western part of Himalaya & Sikkim	Soft brown sausage type mutton product
Bai	Mizoram	Pork with spinach and bamboo shoot taken as curry with rice
Bagjinam	Sema Naga of Nagaland	Fermented pork
Cheu	Deuri community, Assam	Semi-Boiled pork pieces smeared with turmeric, red chilli and salt woven in bamboo sticks and roasted over fire (Charcoal)
Chilu	Sikkim, Ladak and Bhutan	Fat of yak/Beef/Lamb stored in empty stomach of sheep used as edible oil
Dawn sa ret	Mizoram	Dry smoked beef
Doh klong	Meghalaya	Small pieces of boiled pork with local spices- fried in oil & hot served
Doh khleik	Meghalaya	Pig brain and head meat with bamboo shoot and spices and fried in oil
Doh kpu	Meghalaya	Finely minced pork-mixed with onion, garlic, green chillies and local spices, fried in oil
Doh snam	Meghalaya	Blood is filled in intestine with spices, cooked for half an hour
Doh Snier	Meghalaya	Cleaned intestine filled with molten fat, moist cooked, spices and turmeric
Doh pheret	Meghalaya	Small cut pieces from edible and inedible organs cooked in a pot for long time with spices
Doh tyrkhong	Meghalaya	Smoked-dried pork/beef
Eg-adin banum	Mising community of Assam	Pork pieces mild cooked- marinated with local spices and condiments-woven in bamboo sticks-roasted over fire until turns into golden brown
Honohein/ Mongong grain	Dimasa tribe of Karbi angling	Dried pork/buffalo meat
Kheuri	Sikkim	Meat product prepared from Yak or Beef chopped intestine in empty stomach of sheep
Kargyong	Sikkim, Arunachal Pradesh, Darjeeling	Soft or hard brownish sausage like product prepared from yak/Beef/Pork

[Table Contd.

Contd. Table]

Name of the product	Origin	Type and brief description
Momo	All the states of North Eastern Region	Flour with water-dough, small balls are rolled, filled with pre fried minced meat with spices, steam cooked
Ngam phoat	Arunachal Pradesh	Dry smoked meat with salt and turmeric paste
Ngam toongpak	Arunachal Pradesh	Small pieces of pork with bamboo shoot kept inside a tightly covered hollow bamboo
Noau soum	Assam	Pork with boiled rice in sealed bamboo container and allowed fermentation
Pakku	Tripura	Mutton curry with rice
Sathu	North cachar hills of Assam	Pieces of pork fat-half boiled –kept inside the sathu (Water gourd) container -kept near the fire for 4-5 days
Saphak	North cachar hills of Assam	Boiled fats of pork-kept in air tight bottle and used in curry preparation
Suka ko maso	Sikkim	Smoked-dried mutton/buffalo meat
Sun/Yen akangha	Manipur	Dried beef/chicken
Sun/Yen ayaiba	Manipur	Smoked beef/chicken
Satchu	Sikkim, Arunachal Pradesh	Smoked-dried yak meat or beef
Tambe-Akom	Misimi tribe of Arunachal Pradesh	Small pork pieces mixed with ginger, garlic and salt paste stuffed into bamboo hollow container-burnt in fire
Yoo-Aso	Apatani tribes of Arunachal Pradesh	Pork back fat with skin treated with a herbal preparation called pila (extract of ash of millet and a local fern)
Zadoh snam	Meghalaya	Fresh beef/pork blood mixed with half cooked rice and fried in oil for 5-6 minutes with local spices

resulting improvement of the nutritional value. In certain traditional products, these bacteria simulta-neously reduce nitrate and nitrite to nitric oxide, the nitric oxide is combined with myoglobin protein of meat which gives cured colour.

The technique of smoke processing is used to develop some most popular traditional meat products. Smoked fish and meat products are of great demand by the food lovers in the region. Smoke generates innumer-able antimicrobial and aroma compounds such as phe-nols and carbonyls and organic acids such as formic, acetic, propionic, butyric, and isobutyric acids and retards the growth of microbes such as bacteria, moulds, and yeast on the surface of the meat besides adding aroma compounds (Ockerman & Basu, 2008). Smoke also adds antioxidative properties of the fish and meat products alongwith improvement of organoleptic taste (Leistner, 1995). Nevertheless, the duration, temperature and source of smoke are responsible for colour, texture and flavour of the smoked meat products.

Indigenous fermented meat and fish products of NEI contain a diverse group of prebiotic compounds that attract and stimulate the growth of probiotics. Basic understanding about the relationship between food, beneficial microorganisms, and health of the human being is important to improve the quality of food as well as prevention of several diseases (Das *et al.*, 2016). Furthermore, concern and awareness among the consumers about the type and quality and the raw materials fish and meat used for preparation of various types of fermented products and their safety is continually increasing. Although there are a number of challenges in the processing and production of various fermented food products, it is clear that the traditional knowledge of the natives of the northeastern states of India in preparation of different types of fermented fish and meat items offers an enormous opportunity for the development of various food industries related to fermented products, quality improvements, and enhanced wider-scale and even cross border marketing.

Constraints in the Traditional Meat Products

In spite of many nutritional advantages, the traditional fish and meat products have a number of shortcomings which act as constraints for commercialization at a broader level. These include:

- Lack of quality control and hygiene in the production units and marketing points
- Lack of standardization of most of the indigenous techniques
- Lack of intervention by the FSSAI
- Lack of good management practices (GMPs)
- Lack of adequate scientific intervention
- Lack of adequate logistics for scaling up to large production
- Lack of knowledge on proper packaging materials and transport system
- Lack of institutional support mechanism
- Lack of branding and trademark in manufactured products
- Lack of market networking
- Lack of availability and accessibility of meat processing equipment
- Lack of training for skill development to create entrepreneurs
- Lack of managerial and marketing skills for entrepreneurship development

Conclusion

Cost of the traditional food production units are rural, seasonal, labor intensive, informal, and capital deficient. Commonly, traditional meat and fish products are sold and consumed in the areas where they are produced. The methods of processing were developed in homes and improvements were based on the observations of the practitioners. The processes are normally handed down from generation to generation. There is little interest in knowing the role of microorganisms and the physical and chemical changes that occur in the products. What is recognized are changes in colour, odour, and taste that result from modifications of the process or variations in the ingredients or conditions. Most processes are conducted on a trial-and-error basis with little quality control. Product quality primarily depends on the experience of the processor.

Northeast India is immensely rich in traditional cuisine based on meat and fish since time immemorial. Although presently the level of production is area or even ethnic group specific and scientific intervention in the technology is still lacking to a great extent. Some of such ethnic food products have already been reached to the stage of extinction and some are endangered. Loss of these ethnic meat delicacies can be prevented only by increasing its availability and market value. There is urgent need of scientific intervention toward the optimization of preparation methods, validation, popularization and commercialization of these traditional food products. There is also a need to upgrade the overall quality of the traditional meat products by improving the hygiene parameters and packaging procedures. Hygienic production with proper labelling and packaging of the products can reach these to the export market as well. Branding the products properly highlighting their nutritional benefits, functional significance, and regional ethnicity will increase their visibility and penetration level in the market. Scaling up of these ethnic technologies with mechanization of the production system in compliance with Food Safety and Standards Authority of India is essential for creation of employment opportunities for the youths as well as generation of revenue for the region (Govindasamy *et al.*, 2018). Proper showcasing of these products would explore the possibilities to boost food tourism in the region. There is extensive potential of the products to promote at global level for keeping the ethnicity instinct.

References

Agrawal, A. And Gibson, C.C., 1999. Enchantment and disenchantment: the role of community in natural resource conservation. World Development, 27(4): 629–649.

Berkes, F. and Berkes, M.K., 2009. Ecological complexity, fuzzy logic, and holism in indigenous knowledge. Futures. 41: 6 12.

Berkes, F., 2009. Evolution of co-management: role of knowledge generation, bridging organizations and social learning. J. Env. Manage. 90: 1692–1702.

Braton, M., 1989. The politics of government-NGO relation in Africa. World Development. 17(4): 569–587.

Das, A., Ghosh, P.K., Choudhury, B.U., Patel, D.P., Munda, G.C., Ngachan, S.V. and Chowdhury, P., 2009. Climate change in North East India: recent facts and events–

worry for agricultural management. ISPRS Archives XXXVIII-8. In W3 Workshop Proceedings: Impact of Climate Change on Agriculture (pp. 32-37).

Das, A.J. and Deka, C., 2012. Fermented foods and beverages of Northeast India. Int. Food Res. J. 19: 377–392.

Das, G., Patra, J. K., Singdevsachan, S.K., Gouda, S. and Shin, H.S., 2016. Diversity of traditional and fermented foods of the Seven Sister states of India and their nutritional and nutraceutical potential: a review. Front. Life Sci. 9(4): 292-312.

Dikshit, K.R. and Dikshit, J.K., 2014. Weather and climate of Northeast India. In. North-East India: Land, People and Economy, Advances in Asian Human-Environmental Research, pp: 149-173, Springer, Dordrecht.

Gilliland, S.E., 1985. Bacterial starter cultures for foods. CRC Press, Inc., Boca Raton, Florida

Govindasamy, K., Banerjee, B.B., Milton, A.A.P, Katiyar, R. and Meitei, S., 2018. Meat-based ethnic delicacies of Meghalaya state in Eastern Himalaya: preparation methods and significance, J. Ethnic Foods 5: 267-271.

Leistner, L., 1995. Stable and safe fermented sausages world-wide, Fermented meats. Blackie Academic and Professional, London.

Lueck, F., 1980. Antimicrobial food additives. Springer- Verlag, New York

Majumdar, R.K., Roy, D., Bejjanki, S. and Bhaskar, N., (2015). Chemical and microbial properties of shidal, a traditional fermented fish of Northeast India. J. Food Sci. Technol. 53(1), 401-410.

Mirza, M.M.Q., Warrick, R.A., Ericksen, N.J., Kenny, G.J.,1998. Trends and persistence in precipitation in the Ganges, Brahmaputra and Meghna river basins. Hydrol. Sci. J. 43(6): 845-858.

Ockerman, H.W. and Basu, L., 2008. Fermented meat products: Production and Consumption. In: Handbook of Fermented Meat and Poultry (Toldra, F., Ed.), pp 47-68, Blackwell Publishing Ltd., Oxford, UK.

Pilgrim, S.E., Smith, D. and Pretty, J., 2007. A cross-regional quantitative assessment of the factors affecting ecoliteracy: policy and practice implications. Ecol. Applic.17(6): 1742–1751.

Ramakrishnan, P.S., Das, A.K., Saxena, K.G., 1996. Conserving biodiversity for sustainable development. Indian National Science Academy, New Delhi, 246 p.

Ramakrishnan, P.S., Rai, R.K., Katwal, R.P.S. and Mehndiratta, M. (Eds.), 2002. Traditional Ecological Knowledge for Managing Biosphere Reserves in South and Central Asia. UNESCO and Oxford IBH., New Delhi. 536 p.

Ravindranath, N.H., Rao, S., Sharma, N., Nair, N., Gopalakrishnan, R., Rao, A.S., Malaviya, S., Tiwari, R., Sagadevan, A., Munsi, M., Krishna, N. and Bala, G., 2011. Climate change vulnerability profiles for North East India. Cur. Sci. 101(3): 384-394.

Rolle, R., Satin, M., 2002. Basic requirements for the transfer of fermentation technologies to developing countries. Int. J. Food Microbiol. 75: 181–187.

Sathe, G.B. and Mandal, S., 2016. Fermented products of India and its implication: a review. Asian J. Dairy Food Res. 35: 1–9.

Sekar, S. and Mariappan, S., 2007. Usage of traditional fermented products by Indian rural folks and IPR. Indian J. Trad. Know. 6: 111–120.

Singh, R.K. and Srivastava, R.C., 2009. Biocultural knowledge and Adi community: conservation and sustainability in the biodiversity hotspot of Arunachal Pradesh. Curr. Sci. 96(7): 883–884.

Singh, R.K. and Sureja, A.K., 2006. Community knowledge and sustainable natural resources management: learning from the Monpa tribe of Arunachal Pradesh. J. Transdisciplinary res. Southern Africa. 2(1): 73–102.

Singh, R.K., 2004. Conserving diversity and culture: Pem Dolma. Honey Bee. 15(3): 12–13.

Tamang, J.P., Tamang, N., Thapa, S., Dewan, S., Tamang, B., Yonzan, H., Rai, A.P., Chettri, R., Chakrabarty, J. and Kharel, N., 2012. Microorganism and nutritive value of ethnic fermented foods and alcoholic beverages of North-East India. Indian J. Trad. Know. 11: 7–25.

Turner, N.J. and Berkes, F., 2006. Coming to understanding: developing conservation through incremental learning in the Pacific Northwest. Human Ecology. 34: 495–513.

Yumnam, Y.J., 2008. Rich biodiversity of northeast India needs conservation. Curr. Sci. 95(3): 297.

Impact of Climate Change on Hydrological Cycle, Ecosystem, Fisheries and Food Security, Pages: 439–458
Edited by: B. Madhusoodana Kurup, M.R. Boopendranath, M. Harikrishnan and A.V. Shibu

CHAPTER - 37

Utilization of Secondary Raw Materials from Fish for Developing High Value Novel Foods and Dietary Supplements

Suseela Mathew* and C.N. Ravishankar

Central Institute of Fisheries Technology, P.O. Matsyapuri, Cochin-682029, India
**E-mail: suseela1962@gmail.com*

Nutraceutical is a combination of two words, "nutrition" and "pharmaceutical," and the word nutraceutical was coined by Stephen L. DeFelice in 1989 (Wildman *et al.*, 2006). Nutraceuticals are food products of natural origin from both terrestrial and marine sources having healthcare importance. The word nutraceuticals comprise of variety of products derived from terrestrial and marine sources (isolated nutrients, dietary supplements, and genetically engineered designer foods, herbal products, processed foods, and Beverages). Recent report says that nutraceuticals provides a positive healthcare approach with tremendous therapeutic impacts on human body (Das *et al.*, 2012; Bagchi *et al.*, 2015). The nutraceutical industry has identified a wide range of phytochemicals described as phytoestrogens, terpenoids, limonoids, glucosinolates phytosterols, polyphenols, carotenoids, flavonoids, isoflavonoids, and anthocyanidins having therapeutic effects on human health as antioxidants, anti-inflammatory, antibacterial, anti-allergic, anti-fungal, chemopreventive, immunomodulatory etc., (Gupta and Prakash, 2014; Karwande and Borade, 2015).

Classification of Nutraceuticals

Based on the bio-functional properties of bioactive compounds from terrestrial and marine sources are classified in to following categories:

- Dietary Supplements
- Functional food
- Medicinal food

Dietary supplements

According to the Dietary Supplement Health and Education Act (DSHEA), 1994 in USA, dietary supplements are defined as products comprised of "dietary constituents" and orally administered to supplement the nutritional requirement of diet. The "Dietary constituents" refers to bioactive components comprising of amino acids, vitamins, minerals, fibres, important metabolites, and certain enzymes. The dietary supplements also include extracts available in tablets, capsules, powders, liquids, and in any other dosage form (Radhika *et al.*, 2011).

Functional foods

Functional foods are foods derived from natural origin enriched in nutrients and are being fortified with essential nutrients (Jones, 2002). As per the Health Canada, functional food defines a regular food with an ingredient having specific therapeutic effect along with nutritional value (Wildman *et al.*, 2006). Whereas in Japan, functional foods are assessed on the basis of three important standards: (i) functional foods must be derived from natural sources and consumed in their native state instead of processed in different dosage forms like tablet, capsule, or powder; (ii) consumed regularly as a part of daily diet; and (iii) exert a dual role in prevention and management of disease and contribute in biological processes (Arai, 1996).

Medicinal foods

Medical foods are foods that are specially formulated to be consumed internally under the supervision of a physician, which is intended for the dietary management of particular disease that has distinctive nutritional needs that cannot be met by normal diet alone. Dietary supplements and functional foods do not meet these criteria and are not classified as medical food (Radhika *et al.*, 2011).

Marine ecosystems consist of diverse group including fish, shellfishes, sponges, corals, sea anemones, hydrozoans, jellyfish, macroalgae, microalgae, bacteria, cyanobacteria etc. (Thorpe *et al.*, 2000). Because of these vast biodiversity, marine ecosystems are often categorized as one of the richest treasure houses of biomolecules. The biomolecules from marine ecosystems are known to have potential bioactive properties like antioxidant, antimicrobial, antiviral, antiparasitic, anti-inflammatory, antifibrotic, anticancer activities etc. By virtue of their bio-functional properties, it finds wide applications in many arenas, viz, pharmaceutical, nutraceutical, biomedical and cosmetic industries (Barrow and Shahidi, 2008; Shahidi and Zhong, 2008). Among the diverse marine organisms, seafood is considered inherently functional in nature because of its richness in vital nutrients such as proteins, vitamins, minerals, omega-3 fatty acids, etc. In the present scenario fish being nutritionally rich, plays a vital role in combating the hunger and malnutrition in the under developed as well as developing counties. Despite the concern over food insecurity is becoming so prominent in many parts of the world, till date there are no such proper action plans for the sustainable utilization of the food produced. Moreover, the quantum of food losses or waste generation is escalating at an alarming rate annually and its impact on both the environment, social, economic and health aspects are largely overlooked. Fish processing sector tends to generate huge quantity of waste which includes head, skin, trimmings, fins, frames and viscera (Dekkers *et al.*, 2011), from which the bioactive compounds can be extracted. In the recent past, the researchers across the globe has been showing interest towards proper utilization of marine products and waste to extract novel bio-molecules by adopting bio-processing technologies. The extracted bio-molecules which will be having bio-functional properties can be used as functional foods and nutraceuticals.

In this review, emphasis has been given to the bioactive compounds obtained from marine sources such as omega-3 rich fish oils, chitin, chitosan and their functional derivatives, minerals, chondroitin sulphate, hyaluronic acid, squalene, collagen, gelatin, vitamins, etc. The sources, extraction protocols, bioactivities and their applications have been discussed

Chitin and Chitosan

Chitin is natural biopolymer and it is a cationic amino polysaccharide composed of *N*-acetyl-d-glucosamine with β (1 → 4) glycosidic bonds between each monomer (Beaney *et al.* 2005). *N*-acetyl-d-glucosamine units are predominant polymorphic solid sate *viz.*, α, β, and γ chitin which differ in their degree of hydration, size of unit cell, and number of chitin chains per unit cell (Einbu, 2007). Among all the three polymorphic solid states, α-Chitin is the most abundant and is present in shellfish exoskeletons and fungal cell walls. In squid pens and diatoms β-Chitin is present, whereas γ-chitin is predominantly found in squid and cuttlefish stomach lining (Se-Kwon, 2010). The term chitosan is used when nitrogen content of chitin is more than 7% by weight or the degree of deacetylation is more than 60% (Peter *et al.* 1986; Gagne 1993). Chitosan is a biopolymer and it consists of d-glucosamine units obtained during the deacetylation of chitin by adopting hot alkali treatment (Tharanathan and Kittur, 2003). Similar to chitin, chitosan has also been described as one the most versatile biomaterial (Sandford, 1989). Chitin and chitosan are mainly present in many aquatic, terrestrial organisms and also found in some of the microorganisms (Tokura and Tamura, 2007). Bio-waste generated from aquatic and terrestrial source can be used as raw material for production of chitin and chitosan (Crestini *et al.*, 1996; Nwe and Stevens, 2008). As per the published reports, the yield of purified chitin and chitosan from the terrestrial source (insect and mushroom) is much lower than that of aquatic crustacean sources (Se-Kwon, 2010). Terrestrial sources (insect and mushroom), have a difficulty in the production process of chitin mainly due to the complex form, in insect chitin is complexed with melanin whereas in mushroom it is with glucan (Se-Kwon, 2010). Chitin was initially discovered in mushrooms by Henri Braconnot in France in the year

1811. Chitin was subjected to alkali treatment by Prof. C. Rouget in 1859, resulted to a compound which is unlike chitin, dissolves in acids. In the year 1894 the term "Chitosan" was coined by Hoppe-Seiler (McKay 1995). Initially chitin remains as an unutilized natural recourse for quite a long period, in the recent past appeal towards chitin and its derivatives namely chitosan and chito-oligosaccharides (COS) has increased due to their exceptional properties such as biodegradability, bio-renewability, biocompatibility, and physiological inertness (Varum *et al.*, 1991, Kumar, 2000, Beaney *et al.*, 2005, Kurita, 2006). Chitin is abundant in the marine organisms such as, lobster, crab, krill, cuttlefish, shrimp, and prawn. Bio-waste from the marine processing industries is the major resource used for production of chitin (Teng *et al.*, 2001).

The extraction of chitin form bio-waste comprises of thee-steps: deproteinization (DP), demineralization (DM), and decolorization (DC). Further, chitin has to undergo a deacetylation process to produce chitosan. Commonly, chemical and enzymatic methods are being used for production of chitin, chitosan and other derivatives. No and Meyers (1995) have well documented extraction procedure for the production of chitin from shellfish waste. Generally, first proteins are removed from bio-waste by treating with sodium hydroxide or potassium hydroxide solution at higher temperature. As per the published reports, best method for the deproteinization of crustacean waste was to treat with 1%–2% KOH at 90°C with a shell to alkali solution ratio of 1:20 (w/v). Normally, 1 h duration is required to extract 90% of the proteins and within a 2 h of duration most of the proteins can removed from the waste (Shahidi and Synowiecki, 1991). Demineralization process from the bio-waste / shell waste is done by adopting dilute acids extraction method (Sahidi and Synowiecki, 1991). Usually, solvent like ethanol, ether, acetone, sodium hypochlorite, hydrogen peroxide, or a combination of different solvents are being used at ambient temperature to decolorization (No and Meyers, 1995). Chitin has to undergo deacetylation (DA) process to become chitosan. Deacetylation process is done by using highly concentrated sodium or potassium hydroxide solutions (40%–50%) at high temperature (100°C) in an inert atmosphere (Peniston and Johnson, 1980, Johnson and Peniston 1982; Youling *et al.*, 2011). During deacetylation process higher temperature helps to reduce the molecular weight of the resultant chitosan (Lusena and Rose, 1953). After all reaction process, the product has to be filtered and washed with distilled water to get neutral pH. Further, the obtained product was dried in an oven for 24 h at 40 °C, then in a vacuum oven for 24 h at 40 °C (Youling *et al.*, 2011). Another method of extraction is by using enzymes, such as alcalase, chymotrypsin, papain, bromelain (Wang and Chio, 1998; He *et al.*, 2006; Wang *et al.* (2008). Enzyme extraction method seems to have high residual protein, reaction time and high cost. Hence, chemical extraction method was chosen over the enzyme extraction method (Percot *et al.*, 2003).

Chitin is known for its unique properties like, biodegradability, nontoxicity, physiological inertness, antibacterial properties, hydrophilicity, gel-forming properties and affinity for proteins. With these properties, it has found wide applications in the food industry, biosensors

(Krajewska, 2004), used in treatment of industrial pollutants and adsorbs silver thiosulfate complexes (Songkroah *et al.,* 2004) and actinides (Kosyakov *et al.*, 2002). Further, films and fibers (Austin and Brine, 1977; Hirano, 2001) can be can prepared from chitin and these films and fibers can be used in the field of medical and pharmaceutical as a wound dressing material and controlled delivery of drug (Hudson, 1998; Hudson, 1999; Rathke and Hudson, 1994; Yusof *et al.*, 2003;). In India, a few entrepreneurs are producing chitin and chitosan on a commercial scale under the technical guidance of the Central Institute of Fisheries Technology, Cochin. In-line with chitin, chitosan also finds extensive application in multidimensional sector, such as in food and nutrition, biotechnology, material science, drugs and pharmaceuticals, agriculture and environmental protection, dental and surgical appliances as a haemostatic agent, wound healing, biodegradable films as a replacement for artificial skins for removing toxic heavy metals, wine clarification, industrial effluent treatment, photography, cosmetic applications and textiles and as an immobilizing agent for enzymes and recently, it has role in gene therapy also (Se-Kwon, 2010). Chitosan is found to have properties like antimicrobial, antioxidant, texturizer, has applications as had it popular additives for various applications in food product development and preservation (Se-Kwon, 2010).

Glucosamine Hydrochloride

Glucosamine is chemically glucose in which a hydroxyl group on the second carbon atom is substituted with an amino group. Generally, glucosamine is produced from the crustacean's bio-waste (Xu and Wang 2004; Tahami, 1994). Glucosamine is part of the structural polysaccharides such as chitosan and chitin, which is present in the exoskeletons of crustacean and other arthropods. Even though, glucosamine was discovered long back, from the last two decades it has gained popular interest in the field of nutraceutical. Dietary supplementation of glucosamine in the form of glucosamine sulphate, glucosamine hydrochloride, or N-acetyl-glucosamine is proven to promising biomolecule for the treatment of osteoarthritis, knee pain, and back pain. (Houpt *et al.*, 1999; Luo *et al.*, 2005). Glucosamine is one of the most abundant monosaccharides and is produced commercially by the hydrolysis of coarsely ground chitosan with concentrated HCl (Sibi *et al.*, 2013). Glucosamine acts as a chosen substrate for the constitution of glycosaminoglycan chains. In addition, it plays a crucial role in production of aggrecan and proteoglycans which is important for attributing hydrophilicity to cartilage (Setnika *et al.*, 1991). Further, glucosamine known for its unique properties like anti-cancer, anti-inflammatory and antibacterial effects (Jung *et al.*, 2012; Oh *et al.*, 2007; Nagaoka *et al.*, 2011).

Chitin and Chitosan Oligomers

Owing to the higher molecular weight, high viscosity and water insolubility of chitin and chitosan, it often finds difficulties in its proper utilization process (Tao *et al.*, 2007). In the recent past, researchers have taken steps ahead to convert the chitin and Chitosan to their oligomers, which have attracted the scientific community as well as the commercial industry. Chitosan oligomers have low molecular weight, improved water solubility and some unique properties like antitumor activity, antifungal activity, antimicrobial activity and immunomodulatory effect (Hong *et al.*, 2002; Tsukada *et al.*, 1990; Suzuki and Fragrance. 1996; Suzuki *et al.*, 1992; Tokoro *et al.*, 1989; Yamada *et al.*, 1989; Kendra *et al.*, 1989; Uchid *et al.*, 1989). Chitin and chitosan oligomers can be obtained from two hydrolytic procedures namely, chemical hydrolysis (acid hydrolysis) and enzymatic hydrolysis (Jeon *et al.*, 2000). Chitin oligomers, which are available commercially, are prepared by adopting partial hydrolysis of chitin with concentrated HCl, followed by column chromatographic fractionation (Rupley, 1964). Traditional method for extraction of chitin oligomers are include, acid hydrolysis followed by neutralization, demineralization, charcoal-celite column fractionation, HPLC fractionation, and lyophilization (Takahashi *et al.*, 1995). Production of chitin oligomers by adopting combination method of a mild acid degradation and sonolysis was given by Takahashi *et al.*, (1995), wherein this method could be able to degrade the polymer without depending on the temperature and hydrolyze chitin by HCl under ultrasound irradiation. In the year 1957 for the first time chitosan oligomers produced by acid hydrolysis method (Horowitz *et al.*, 1957) which has lead to production of chitosan oligomers with lower degree of polymerization. Fluorohydrolysis of chitosan in anhydrous hydrogen fluoride (HF) leads to production of chitosan oligomers (Defaye *et al.* 1989). Another method for preparation of chitin and chitosan oligomers is by adopting enzyme hydrolysis method, wherein, enzymes such as chitinase, chitosanase, lysozyme, and cellulose are being used. This particular enzyme hydrolysis method was found to be good over the chemical hydrolysis method and this method produce oligomers with a high degree of polymerization (Uchida *et al.*, 1989). As chitin and chitosan oligomers have unique functional properties, found to have applications in various field such as, agricultural, biotechnological, food, nutritional, pharmaceutical, medical and other sectors (Jeon *et al.*, 2000).

Chondroitin Sulphate

Chondroitin sulphate (CS) consists of repeated disaccharide units of glucuronic acid (GlcA) and *N*-acetylgalactosamine (GalNAc) linked by β-(1→3) glycosidic bonds and sulfated in different carbon positions (CS no-sulfated is CS-O). Chondroitin sulpahte can be extracted from trachea of cow and ear and nose of pig, still shark cartilage found to be a good source of chondroitin sulpahte (Venugopal V., 2009). Normally, chondroitin sulphate is extracted by following various steps *viz.,* chemical hydrolysis of cartilage, breakdown of proteoglycan core, elimination of proteins, CS recovery and purification of CS. Chemical hydrolysis and breakdown of proteoglycans are done by alkali hydrolysis at high concentration of NaOH, Urea or guanidine HCL, after that precipitation of glycosaminoglycan is done by using cationic quaternary ammonium chemicals such as cetylpyridinium chloride, potassium thiocyanate, non-ionic detergents or alcoholic solutions

(Rodén *et al.*, 1972; Chascall *et al.*, 1994). Further, deproteinization is done by using trichloroacetic acid and purification of the extract is done with gel filtration / ion-exchange / size-exclusion chromatography (Patel *et al.*, 1980). In the recent past, various alternative extraction methods have been developed, the process includes digestion of cartilage and protein by enzymes, precipitations with alcoholic solutions, further, resuspension and neutralization of extract with salt solutions and final separation of chondroitin sulpfate is done by molecular-weight using ultrafiltration-diafiltration technologies (UF-DF) (Lignot *et al.*, 2003; Mollard *et al.*, 2005; Tadashi 2006 and Murado *et al.*, 2010). Chondroitin sulfate plays various roles in biological processes such as the function and elasticity of the articular cartilage, hemostasis, inflammation, cell development, cell adhesion, proliferation and differentiation by being an essential element of extracellular matrix of connective tissues (Schiraldi *et al.*, 2010).

Hyaluronic Acid

Hyaluronic acid can be obtained from the bio-waste like fish eyeball (Murado *et al.*, 2012) and Hyaluronic acid is also present in the cartilage matrix of fishes; but Hyaluronic acid content is very low and makes it economically not feasible for any industrial type of extraction (Imberty *et al.*, 2007). Further, Hyaluronic acid is also present in animal substrates like umbilical cord, rooster combs, synovial fluid from pig or bovine (Murado *et al.*, 2012; Nakano *et al.*, 1994; Cullis-Hill, 1989; Marcellin *et al.*, 2009). The conventional method for extraction of Hyaluronic acid from animal source includes preparation of material, washing the material with water or alcohol, followed by aqueous or organic solvent extraction, after that precipitation is done by adding cetylpiridinium chloride, followed by filtering and successive extraction with chloroform, finally centrifugation and purification is done by chromatography (Swann, 1968; Balazs, 1979). The above said extraction method was costly, laborious, time-consuming and lead to contaminated Hyaluronic acid (O'Regan *et al.*, 194). An alternative extraction method was developed, which includes deproteinization of bovine substrate with xylenesulphonate-Na (Radulescu *et al.*, 1997), followed by extraction with water, there after precipitation with ethanol and purification by DEAE-cellulose (Gao *et al.*, 1996) further, fractional precipitation using ethanol combined with enzymatic protein hydrolysis (Rodén *et al.*, 1972). Mainly because of its unique properties, Hyaluronic acid finds several biomedical applications *viz.* viscosupplementattion in osteoarthritis treatment, as aid in eye surgery and wound regeneration. Further, hyaluronic acid finds its applications in drug delivery, tissue engineering applications, gene delivery applications, targeted drug delivery, tumor treatment, environmental applications and sensors (Sudha and Rose, 2014).

Collagen, Gelatin and Collagen Peptides

Collagen and its partially denatured form, gelatin, is finding extensive applications in food, pharmaceutical, cosmetic, packaging and medical sectors owing to its diverse functional and biological properties. Currently, bovine and porcine skin, bone and cartilage are being utilized for the extraction of collagen and gelatin. However, the use of bovine and porcine collagen has created serious religious concerns as its usage doesn't satisfy the kosher (Judaism) and halal (Islam) requirements (Karim & Bhat, 2009). It has also raised serious health issues as there is a chance of disease transmission such as bovine spongiform encephalopathy from cattle and foot-and-mouth disease from pig and other prion mediated diseases (Wilesmith *et al.*, 1991, Ogawa *et al.*, 2004). Hence, researchers are in search of suitable alternative sources for collagen and gelatin extraction. Skin and scales which constitutes about 30% and 5% of the total seafood processing discards respectively are considered as the richest source for collagen and gelatin (Suresh and Prabhu, 2013; Pal *et al.*, 2015). Hence, seafood and seafood processing discards can be looked upon as alternative sources for mammalian collagen and gelatin and moreover, it fulfills the requirements of kosher and Halal products.

The collagen obtained from seafood processing discards is often categorized into three major types: fish collagens, collagens from invertebrate origin and marine mammals (Ehrlich, 2015). Out of the 29 genetically distinct collagens reported as of now, seafood collagen falls in the category of type I (Pal *et al.*, 2015; Benjakul *et al.*, 2012). Collagen derived from marine sources is finding wide applications in various sectors due to its biocompatibility, biodegradability, high cell adhesion properties and weak antigenicity (Pal *et al.*, 2015; Yamada *et al.*, 2014). Moreover, its acceptance as a functional food ingredient in beverages, dietary supplements, drinks and confectionery is growing remarkably over the years due to its nutritive properties (Antoniewski and Barringer, 2010; Bilek and Bayram, 2015).

Another major application of collagen is to act as a source for extraction of collagen hydrolysates, peptides, gelatin and gelatin peptides. Collagen peptides which remain inactive within the parent polypeptide sequences of collagen will become active after its release by processes such as gastrointestinal digestion, hydrolysis etc. (Kim and Wijesekara, 2010; Suresh and Prabhu, 2013; Liu *et al.*, 2015). The peptides thus produced will be having about 3-20 aminoacid residues and its bioactivities depend to a greater extent on the aminoacid composition (Menon and Lele., 2015). In general, the properties of the peptides will greatly depend on the extraction methods. Reports says that the seafood derived collagen peptides have numerous bioactivities such as antioxidant, antimicrobial, antihypertensive, metal chelating, tyrosinase inhibitory, immunomodulatory, neuroprotective, antifreeze, wound healing, cell-proliferation, activities (Liang *et al.*, 2014, Chi *et al.*, 2014; Zhuang *et al.*, 2009). It has been reported that the ingestion of collagen peptides can alleviate the pain in osteoarthritic patients. Collagencin, an antimicrobial peptide extracted from Atlantic mackerel byproducts was found to have antibacterial activity with its maximum effect against *Staphylococcus aureus* growth (Ennaas *et al.*, 2006). Gelatin, the denatured form of collagen, by virtue of its surface active properties finds extensive applications in the food, pharmaceutical and biomedical industries (Balti *et al.*, 2011). Gelatin obtained from marine sources depends to a greater extent on its thermal stability, rheological and gelling properties. Gelatin peptides are reported to antihypertensive, antioxidant properties. The major difference between fish and mammalian gelatin lies in the iminoacid composition, viz, proline and hydroxyproline. The lower content of iminoacids is attributed as one of the probable reason for its low gelling and melting temperature (Haug *et al.*, 2004). Gelatin obtained by pre-treatment of collagen with acid and alkali is referred to as Type-A and Type –B respectively. Though there are different methods, acid treatment remains as the most commonly used method for the extraction of gelatin from fishes.

Collagen and gelatin are noted mainly because of their gelling and surface properties which includes thickening, texturizing, film-forming, gel formation, water binding capacity, emulsion formation and stabilization, adhesion and cohesion etc. (Schrieber and Gareis, 2007). Though the gelation formation mechanism of both collagen and gelatin are of thermo-reversible nature, both happens in opposite directions, ie. lowering the temperature melts collagen gels whereas gelatin gels can be melted by increasing the temperature. Gelatin is mostly used in the preparation of desserts because of its unique melt-in-the-mouth property. It has been reported that by either increasing the gelatin concentration or by using a mixture of cold and warm water fish gelatin, fish skin gelatin can be used to make desserts that are having similar properties of that of the desserts prepared from pork skin gelatin. Both gelatin and collagen, when covalently cross-linked, form matrices and swells in the presence of aqueous solutions to form hydrogels. Being economical and biodegradable in nature, gelatin hydrogels finds numerous applications in the field of agriculture, medicine, pharmacy, and biodegradable food packaging.

Fish Lipids

The health benefits of long chain omega-3 polyunsaturated fatty acids (PUFA) are well demonstrated and documented by researchers worldwide (Connor, 2000). The major omega-3 PUFA, viz, eicosapentaenoic acid (EPA C20:5) and docosahexaenoic acid (DHA C22:6) are required by humans, but cannot be synthesized endogenously and hence considered as essential fatty acids. Therefore, the requirements for these fatty acids must be obtained from the diet. The intake of long chain omega-3 PUFA is promoted by many health organizations owing to the health benefits associated with it. An average intake of 0.2 g and 0.65g of EPA and DHA a day is recommended by the European Academy of Nutritional Sciences (EANS) and International Society for the Study of Fatty Acids and Lipids (ISSFAL) respectively (deDeckere *et al.*, 1998). Though the consumption of omega-3 fatty acids has been recommended by many of the health organizations, there should be an appropriate balance between the omega-6 to omega-3 fatty acids in the diet.

Among the different dietary sources, fish oil remains as the most excellent and economical source of omega-3 PUFA. Having a high content of fat soluble vitamins and lipids, especially EPA, cod liver has been exploited as an omega-3 PUFA source for development of nutraceuticals (Mondello *et al.*, 2006; Falch *et al.*, 2006). Consumption of fish oil in adequate quantities is reported to fight against various ailments such as cardiovascular diseases, cancer, hypertension, Alzheimer's disease, diabetes, arthritis, autoimmune disorders and to improve overall functioning of brain and retina (Simopoulos, 1991; de Deckere, 1999; Connor, 2000; Banning, 2005; Yokoyama *et al.*, 2007; Cole *et al.*, 2009 and Tur, Bibiloni, Sureda and Pons, 2012). Hence, fish oil is being used as a functional ingredient and is fortified into a wide variety of foods. However, due to the high degree of unsaturation, fish oil is highly susceptible to oxidation resulting in the formation of toxic hydroperoxides and other byproducts, off-flavors (Nawar 1996; Watkins and German 1998). Hence, fish oil is often encapsulated to retain its oxidative stability. Currently, fish oil powder is fortified into a wide range of food products such as bread, biscuits, cakes, bakery products, pasta etc. diet powder, fruit juice, milk powder, instant soups, infant formula, etc. (Trautwein., 2001).

Squalene

Squalene (2, 6, 10, 15, 19, 23-hexamethyl-2, 6, 10, 14, 18, 22- tetracosahexaene), a naturally occurring triterpenoid compound, is an intermediate in cholesterol synthesis. It is widely present in nature in many compounds such as wheat germ, rice bran, shark liver and olive oils and among all the sources identified, shark liver oil is considered to be the richest source accounting for about 40% of its weight. It has garnered considerable attention in the recent past because of its diverse bioactivities such as antioxidant, antilipidemic, membrane stabilizing, cardioprotective, chemopreventive, anti cancerous, antiaging properties etc (Ko *et al.*, 2002; Ivashkevich *et al.*, 1981; Passi *et al.*, 2002; Smith, 2000). It is also reported to protect human skin surface from oxidation owing to its reactive oxygen species scavenging activity (Kabuto *et al.*, 2013). Experimental studies have proven that, about 60–85% of the orally administered squalene is absorbed and distributed to various tissues (Strandberg, 1990; Miettinen, 1994; Gylling and Miettinen, 1994). So far, no report of toxicity is associated with higher squalene intake. Because of its diverse bioactivities, it finds wide applications in many sectors such as biomedical, cosmetic, drug delivery systems and even in food industries. Taking into consideration the diverse bioactivities, it can be said that daily dietary squalene supplementation can help in improving the overall health status of human. Moreover, when used as a functional food ingredient, it can facilitate several health benefits such as improved immune response, cholesterol lowering effect, antiaging properties.

Minerals

Marine organisms especially fish are considered as important source of minerals such as sodium, potassium, calcium, phosphorous and Magnesium. Fish bone which is often discarded after the removal of protein is an excellent source of calcium and hydroxy apatite. Being rich in minerals, fish bone powder can be fortified into several food products. However for fortification, the fish bone should be converted into an edible form by softening its structure by pre-treatment with how water or hot acetic acid or superheated steam. Calcium powder processed from the backbone of tuna is a potential nutraceutical. It can be used to combat calcium deficiency in children. Other than this, it can be even used in the fortification of fish products such as surimi (Shungan, 1996). Fortification of calcium in foods helps consumers in meeting the calcium requirements and may reduce the risk of osteoporosis.

Other than fish bone calcium, certain other minerals such as selenium, potassium, iodine, zinc, magnesium are more abundant in seafood than in meats. The higher intake of seafood diet will also ensure that adequate amount of iodine is obtained. Fish bone also serve as a source of hydroxyapatite [Ca10(PO4)6(OH)2], which finds extensive applications in biomedical and dental arenas as bone grafting material. The common bone substitution materials such as autografts, allografts and xenografts are mechanically unstable and incomptabile and hence doesn't provide a perfect healing effect. However, hydroxyapatite obtained from fishes is thermodynamically stable at physiological pH and plays an important role in bone bonding. Attempts have been made to isolate fish based hydroxyapatite and use as a promising alternative for synthetic hydroxyapatite (Ozawa and Suguru, 2002). Generally high temperature treatment is followed for hydroxyapatite extraction from fish bones (Choi *et al.*, 1999).

Conclusion

Nutraceuticals from fish and fishery waste have wide applications owing to the high nutritive value of fish, presence of polyunsaturated fatty acids, minerals, vitamins and all essential aminoacids. There is considerable scope for developing high value nutraceutical products from fishery waste whose applications range from cholesterol lowering, anti-obesity, anti-inflammatory, anti-arthritic and, in general, to improve the health of human beings. Incorporation of these nutraceuticals in common foods can be done so as to tackle the problem of malnutrition in the country and also to develop health promoting foods for the needy population. By developing high value nutraceuticals from fishery waste can solve the problem of tackling huge amount of waste generated during fish processing operations in the country.

References

Antoniewski, M.N. and Barringer, S.A., 2010. Meat shelf-life and extension using collagen/ gelatin coatings: a review. Critical Reviews in Food Science and Nutrition, 50(7): 644-653.

Austin, P.R., 1977. Brine J. Chitin films and fibers. USPatent 4,029,727; 1977.

Balazs, E.A., 1979. Ultrapure Hyaluronic Acid and the Use Thereof. U.S. Patent 4,141,973, 27 February 1979.

Balti, R., Jridi, M., Sila, A., Souissi, N., Nedjar-Arroume, N., Guillochon, D. and Nasri, M., 2011. Extraction and functional properties of gelatin from the skin of cuttlefish (Sepia officinalis) using smooth hound crude acid protease-aided process. Food Hydrocolloids 25(5): 943-950.

Banning, M., 2005. The role of omega-3-fatty acids in the prevention of cardiac events. British Journal of Nursing 25: 503-508.

Barrow, C. and Shahidi, F. (Eds.), 2007. Marine Nutraceuticals and Functional Foods. CRC Press, 506 p. ISBN 9780367202804

Beaney, P., Lizardi-Mendoza, J., and Healy, M., 2005. Comparison of chitins produced by chemical and bioprocessing methods. J. Chem. Technol. Biotechnol., 80: 145-150.

Beaney, P., Lizardi-Mendoza, J., and Healy, M., 2005. Comparison of chitins produced by chemical and bioprocessing methods. J. Chem. Technol. Biotechnol., 80: 145–150.

Benjakul, S., Nalinanon, S. and Shahidi, F., 2012. Fish collagen. In:.), Food Biochemistry and Food Processing, B.K. Simpson (Ed.), pp. 365–387, 2nd edn. Wiley-Blackwell, USA.

Bilek, S.E. and Bayram, S.K., 2015. Fruit juice drink production containing hydrolyzed collagen. Journal of Functional Foods 14: 562-569.

Chascall, V., Calabro, A., Midura, R.J and Yanagishita, M., 1994. Isolation and characterization of proteoglycans. In: Methods in Enzymology, Lennarz, W.J., Hart, G.W. (Eds.), pp. 390–417, Academic Press, San Diego, CA.

Chi, C.F., Cao, Z.H., Wang, B., Hu, F.Y., Li, Z.R. and Zhang, B., 2014. Antioxidant and functional properties of collagen hydrolysates from spanish mackerel skin as influenced by average molecular weight. Molecules 19(8): 11211-11230.

Choi, J.S., Lee, C.K., Jeon, Y.J., Byun, H.G. and Kim, S.K., 1999. Properties of the ceramic composites and glass ceramics prepared by using the natural hydroxyapatite derived from tuna bone. Journal of Korean Industrial and Engineering Chemistry 10: 394–399.

Cole, G.M., Ma, Q.L. and Frautschy, S.A., 2009. Omega-3 fatty acids and dementia. Prostaglandins, Leukotrienes and Essential Fatty Acids 81: 213-221.

Connor, W.E., 2000. Importance of n-3 fatty acids in health and disease. The American Journal of Clinical Nutrition, 71(Suppl.): 171S-175S.

Crestini, C., Kovac, B., and Giovannozzi-Sermanni, G., 1996. Production and isolation of chitosan by submerged and solid state fermentation from *Lentinus edodes*. Biotechnology and Bioengineering 50: 207–210

Cullis-Hill, D., 1989. Preparation of Hyaluronic Acid from Synovial Fluid. U.S. Patent 4,879,375, 7 November 1989.

de Deckere, E.A., 1999. Possible beneficial effects of fish and fish n-3 polyunsaturated fatty acids in breast and colorectal cancer. European Journal of Cancer Prevention 8: 213–221.

deDeckere E.A.M., Korver O., Verschuren P.M. and Katan M.B., 1998. Health aspects of fish and n-3 polyunsaturated fatty acids from plant and marine origin. Eur. J. Clin. Nutr., 52: 749-753.

Defaye, J., Gadelle, A. and Pedersen, C., 1989. In: Chitin and Chitosan (G. Skjak-Braek, T. Anthonsen, and P. Sandford, Eds.), pp. 415–429, Elsevier, London.

Ehrlich, H., 2015. Marine collagens. In: Biological Materials of Marine Origin, pp. 321-341, Springer Netherlands. ISBN 978-94-007-5730-1.

Einbu, A., 2007. Characterisation of chitin and a study of its acid-catalysed hydrolysis. Ph.D. Dissertation, Norwegian University of Science and Technology, Trondheim, Norway.

Ennaas, N., Hammami, R., Gomaa, A., Bédard, F., Biron, É., Subirade, M., Beaulieu, L. and Fliss, I., 2016. Collagencin, an antibacterial peptide from fish collagen: Activity, structure and interaction dynamics with membrane. Biochemical and Biophysical Research Communications, 473(2): 642-647.

Falch, E., Rustad, T., and Aursand, M., 2006. By-products from gadiform species as raw material for production of marine lipids as ingredients in food or feed. Process Biochemistry 41: 666-674.

Gao, Y.Q., Liu, J.H., Huo, X., Shan, Y.L., Xu, Z.X., 1996. The purification and identification of hyaluronic acid isolated from various tissues. Chin. Boichem. J., 12: 215–218.

Gylling, H. and Miettinen, T.A., 1994. Post absorptive metabolism of dietary squalene. Atherosclerosis 106(2): 169-78.

Haug, I.J., Draget, K.I. and Smidsrød, O., 2004. Physical and rheological properties of fish gelatin compared to mammalian gelatin. Food Hydrocolloids 18(2): 203-213.

He, H., Chen, X., Sun, C., Zhang, Y., and Gao, P. 2006. Preparation and functional evaluation of oligopeptide-enriched hydrolysate from shrimp (*Acetes chinensis)* treated with crude protease from *Bacillus* sp. SM98011. Bioresour. Technol., 97: 385–390.

Hirano, S., 2001. Wet-spinning and applications of functional fibers based on chitin and chitosan. In: Arguelles-Monal, W. (Ed.). Natural and synthetic Polymers: Challenges and Perspectives. Macromol Symp., 168., pp. 21–30, Wiley-VCH Verlag GmbH, Weinheim, Germany.

Hong, K.N., Na, Y.P., Shin, H.L., Samuel, P.M., 2002. Int. J. Food Microbiol., 74: 65-72.

Horowitz, S., Roseman, S. and Blumenthal, H.J. 1957. The Preparation of Glucosamine Oligosaccharides. I. Separation. J. Am. Chem. Soc., 79: 5046-5049.

Houpt, J.B., McMillan, R., Wein, C. and Paget-Dellio, S.D., 1999. Effect of Glucosamine hydrochloride in the treatment of pain of osteoarthritis of the knee. J. Rheumatol., 26: 2423-2430.

Hudson, S.M., 1998. Applications of chitin and chitosan as fiber and textile chemicals. In: Domard, A., Roberts, G.A.F., VaÚrum, K.M. (Eds.). Advances in Chitin Science, 2. pp. 590–599, Jacques Andre Publ., Lyon, France.

Hudson, S.M., 1999. The applications of chitin and chitosan to fiber and textile products. In: Chen, R.H. and Chen, H.C. (Eds.) Advances in Chitin Science, 3. pp. 80–87, National Taiwan Ocean University, Taiwan.

Imberty, A., Lortat-Jacobb, H. and Pérez, S., 2007. Structural view of glycosaminoglycan–protein interactions. Carbohydr. Res., 342: 430-439.

Ivashkevich, S.P., Apukhovskaia, L.I. and Vendt, V.P., 1981. Effect of sterols having different chemical structure and squalene on osmotic resistance of erythrocytes. Biokhimiia 46: 1420–1425.

Jeon, Y.J., Shahdi, F. and Kim, S.K., 2000. Preparation of chitin and chitosan oligomers and their application in physiological functional foods. Food Reviews International, 16: 159-176.

Johnson, E.L. and Peniston, Q.P. 1982. Utilization of shellfish waste for chitin and chitosan production. In: Chemistry and Biochemistry of Marine Food Products. Martin, G.J., Hebard, C.E. and Ward, W.R. (Eds.), pp. 514–522, AVI Publishing Co., Westport, CT.

Jung, C.W., Jo, J.R., Lee, S.H., Park, Y.K., Jung, N.K., Song, D.K., Bae, J., Nam, K.Y., Ha, J.S., Park, I.S. and Park, G.Y., 2012 . Anti-cancer properties of glucosamine-hydrochloride in YD-8 human oral cancer cells: Induction of the caspase-dependent apoptosis and down-regulation of HIF-1α. Toxicology in Vitro 26(1): 42-50.

Kabuto, H., Yamanushi, T.T., Janjua, N., Takayama, F. and Mankura, M., 2013. Effects of squalene/squalane on dopamine levels, antioxidant enzyme activity, and fatty acid composition in the striatum of Parkinson's disease mouse model. Journal of Oleo Science 62: 21–28.

Karim, A.A. and Bhat, R., 2009. Fish gelatin: properties, challenges, and prospects as an alternative to mammalian gelatins. Food Hydrocolloids 23(3): 563-576.

Kendra, D.F., Christian, D. and Hadwiger, L.A., 1989. *Fusarium solani*/pea interactions. Chitinase/β-glucanase digestion of sporelings and from fungal wall chitin actively inhibit fungal growth and enhance disease resistance. Physiol. Mol. Plant Path., 35: 215–230.

Kim, S.K. and Wijesekara, I., 2010. Development and biological activities of marine-derived bioactive peptides: A review. Journal of Functional Foods 2(1): 1-9.

Ko, T.F., Weng, T.M. and Chiou, R.Y., 2002. Squalene content and antioxidant activity of *Terminalia catappa* leaves and seeds. Journal of Agricultural and Food Chemistry 50: 5343-5348.

Ko, T.F., Weng, Y.M. and Chiou, R.Y.Y., 2002. Squalene content and antioxidant activity of *Terminalia catappa* leaves and seeds. J. Agric. Food Chem., 50: 5343-5348.

Kosyakov, V.N., Yakovlev, N.G. and Veleshko, I.E., 2002. Application of chitin-containing fiber material "mycoton" for actinide adsorption. J. Nucl. Sci. Technol. (Suppl. 3): 508–11.

Krajewska, B., 2004. Application of chitin- and chitosan-based materials for enzyme immobilizations: a review. Enzyme Microbiol. Technol., 35: 126–39.

Kumar, R. 2000. A review of chitin and chitosan applications. React Funct Polym., 46(1): 1–27.

Kurita, K. 2006. Chitin and chitosan: Functional biopolymers from marine crustaceans. Mar Biotechnol., 8: 203–226.

Liang, Q., Wang, L., He, Y., Wang, Z., Xu, J. and Ma, H., 2014. Hydrolysis kinetics and antioxidant activity of collagen under simulated gastrointestinal digestion. Journal of Functional Foods 11: 493-499.

Lignot, B., Lahogue, V. and Bourseau, P., 2003. Enzymatic extraction of chondroitin sulfate from skate cartilage and concentration-desalting by ultrafiltration. J. Biotechnol., 103: 281-284.

Liu, D., Nikoo, M., Boran, G., Zhou, P. and Regenstein, J.M., 2015. Collagen and gelatin. Annual Review of Food Science and Technology 6: 527-557.

Luo, J., Hu, Y.S., Wu, Y., and Fan, W.K., 2005. Effect of Glucosamine hydrochloride in ameliorating knee osteoarthritis. Chin. J. Clini. Rehabil., 9: 70-72.

Lusena, C.V. and Rose, R.C., 1953. Preparation and viscosity of chitosan. J. Fish. Res. Board Can. 10: 521–522.

Marcellin, E., Chen, W. and Nielsen, L.K., 2009. Microbial Hyaluronic acid biosynthesis. In: Microbial Production of Biopolymers and Polymer Precursors: Applications and Perspectives, Rehm, B.H.A. (Ed.), pp. 163-180, Caister Academic Press, Norwich, UK.

McKay, G., 1995. Use of Adsorbents for the Removal of Pollutants from Wastewaters. Taylor & Francis, Boca Raton, FL.

Menon, V.V. and Lele, S.S., 2015. Nutraceuticals and bioactive compounds from seafood processing waste. In: Handbook of Marine Biotechno*logy*, pp. 1405-1425, Springer, Berlin Heidelberg.

Miettinen, T.A. and Vanhanen, H., 1994. Serum concentration and metabolism of cholesterol during rapeseed oil and squalene feeding. Am. J. Clin. Nutr., 59: 356-63.

Mollard, L., Montillet, A., Horriere, C., Legrand, J and Nguyen, T.H., 2005. Method for obtaining avian biological products. U.S. Patent 6,844,424.

Mondello, L., Tranchida, P. Q., Dugo, P. and Dugo, G., 2006. Rapid, micro-scale preparation and very fast gas chromatographic separation of cod liver oil fatty acid methyl esters. Journal of Pharmaceutical and Biomedical Analysis 41: 1566-1570.

Murado, M.A., Fraguas, J., Montemayor, M.I., Vαzquez, J.A., and Gonzαlez, M.P., 2010. Preparation of highly purified Chondroitin sulphate from skate (*Raja clavata*) cartilage by-products. Process optimization including a new procedure of alkaline hydroalcoholic hydrolysis. Biochem. Eng. J., 49: 126-132.

Murado, M.A., Montemayor, M.I., Cabo, M.L., Vαzquez, J.A. and Gonzαlez, M.P., 2012. Optimization of extraction and purification process of hyaluronic acid from fish eyeball. Food Bioprod. Proc., 90: 491-498.

Nagaoka, I., Igarashi, M., Hua, J., Ju, Y., Yomogida, S. and Sakamoto, K., 2011. Recent aspects of the anti-inflammatory actions of glucosamine. Carbohydrate Polymers 84(2): 825-830.

Nakano, T., Nakano, K. and Sim, J.S., 1994. A simple rapid method to estimate hyaluronic acid concentrations in rooster comb and wattle using cellulose acetate electrophoresis. J. Agric. Food Chem., 42: 2766-2768.

No, H.K. and Meyers, S.P., 1995. Preparation and characterization of chitin and chitosan - A review. J Aquat Food Prod Technol., 4: 27–52.

Nwe, N. and Stevens, W.F., 2008. Production of chitin and chitosan and their applications in the medical and biological sector. In: Recent Research in Biomedical Aspects of Chitin and Chitosan, Tamura, H. (Ed.), pp. 161–176, Research Signpost, Trivandrum, Kerala, India.

O'Regan, M., Martini, I.,Crescenzi, F., de Luca, C. and Lansing, M., 1994. Molecular mechanisms and genetics of hyaluronan biosynthesis. Int. J. Biol. Macromol., 16: 283-286.

Oh, H.J., Lee, J.S., Song, D.K., Shin, D.H., Jang, B.C., Suh, S.I., Park, J.W., Suh, M.H. and Baek, W.K., 2007. D-glucosamine inhibits proliferation of human cancer cells through inhibition of p70S6K. Biochemical and Biophysical Research Communications 360(4): 840-845.

Ozawa, M. and Suguru, S., 2002. Microstructural development of natural hydroxyapatite originated from fish-bone waste through heat treatment. Journal of the American Ceramists Society 85: 1315-1317.

Pal, G.K., Nidheesh, T. and Suresh, P.V., 2015. Comparative study on characteristics and in vitro fibril formation ability of acid and pepsin soluble collagen from the skin of catla (Catla catla) and rohu (*Labeo rohita*). Food Research International 76: 804-812.

Passi, S., De Pita, O., Puddu, P. and Littarru, G.P., 2002. Lipophilic antioxidants in human sebum and aging. Free Radical Research 36: 471-477.

Patel, B., Ehrlich, J., Stivala, S.S., Singh, N.K., 1980. Comparative studies of mucopolysaccharides from marine animals. I. *Raja eglanteria* Bosc. J. Exp. Mar. Biol. Ecol., 46: 127–135.

Peniston, Q.P. and Johnson, E.L., 1980. Process for the manufacture of chitosan. Patent No. 4,195,175.

Percot, A., Viton, C., and Domard, A., 2003. Optimization of chitin extraction from shrimp shells. Biomacromolecules 4: 12–18.

Radulescu, G., Lupescu, I., Petrea, D.M. and Scurei, H., 1997. Hyaluronic acid extraction from vitreous fluid. Biotechnol. Lett., 2: 147-152.

Rathke, T.D. and Hudson, S.M., 1994. Review of chitin and chitosan as fiber and film formers. Rev. Macromol. Chem. Phys. C34: 375–437.

Rodén, L., Baker, J.R., Cifonelli, J.A and Mathews, M.B., 1972. Isolation and characterization of connective tissue polysaccharides. Methods Enzymol., 28: 73-140.

Rupley, J.A., 1964. The hydrolysis of chitin by concentrated hydrochloric acid, and the preparation of low-molecular-weight substrates for lysozyme. Biochim. Biophys. Acta., 83: 245-255. doi: 10.1016/0926-6526(64)90001-1.

Schiraldi, C., Cimini, D. and de Rosa, M., 2010. Production of chondroitin sulphate and chondroitin. Appl. Microbiol. Biotechnol., 87: 1209–1220.

Se-Kwon, K., 2010. Chitin, chitosan, oligosaccharides and their derivatives: Biological activities and applications; CRC Press-Taylor & Francis Group, Boca Raton.

Sen, D.P., 2005. Selected by-products from sea, in Advances in Fish Processing Technology, Allied Publisher, New Delhi. 616 p.

Setnika, R. Cereda, M.A. Pacini and L. Revel, 1991. Anti-reactive properties of Glucosamine sulfate, Arzneim.-Forsch/Drug Research 41: 157-161.

Shahidi, F. and Synowiecki, J., 1991. Isolation and characterization of nutrients and value-added products from snow crab (*Chionoecetes opilio*) and shrimp (*Pandalus borealis*) processing discards. J. Agric. Food Chem., 39: 1527–1532.

Shahidi, F. and Zhong, Y., 2008. Bioactive peptides. Journal of AOAC International, 91(4): 914-931.

Sibi, G., Dhananjaya, K., Ravikumar, K.R., Mallesha, H. and Venkatesha, R.T., 2013. Preparation of Glucosamine hydrochloride from crustacean shell waste and it's quantitation by RP-HPLC. American-Eurasian Journal of Scientific Research 8(2): 63-67.

Simopoulos, A.P., 1991. Omega-3 fatty acids in health and disease and in growth and development. American Journal of Clinical Nutrition 54: 438-447.

Smith, T.J., 2000. Squalene: potential chemopreventive agent. Expert Opinion on Investigational Drugs 9 (2000): 1841-1848.

Songkroah, C., Nakbanpote, W. and Thiravetyan, P., 2004. Recovery of silver–thiosulfate complexes with chitin. Process Biochem., 39: 1553–1559.

Strandberg, T.E., Tilvis, R.S. and Miettinen, T.A., 1990. Metabolic variables of cholesterol during squalene feeding in humans: comparison with cholestyramine treatment. J. Lipid Res., 31(9): 1637-1643.

Sudha, P.N. and Rose, M.H., 2014. Beneficial effects of hyaluronic acid. Adv. Food Nutr. Res., 72: 137-176.

Suresh, P.V. and Prabhu, G.N., 2013. Seafood. In: Valorization of Food Processing By-Products, M. Chandrasekaran (Ed.), pp. 685–736, Taylor & Francis, CRC Press, New York.

Suzuki S, Watanabe T, Mikami T, Matsumoto T, and Suzuki M. 1992. Immuo-enhancing effects of N-acetyl-chitohexanose. In: Advance in Chitin and Chitosan, Brine. C.J., Sandford, P.A. and Zikakis, J.P. (Eds.), pp. 96-105, Elsevier Applied Science, NY.

Suzuki, S., 1996. Studies on biological effects of water soluble lower homologous oligosaccharides of chitin and chitosan. Fragrance J., 15: 61-68.

Swann, D.A., 1968. Studies on hyaluronic acid: I. The preparation and properties of rooster comb hyaluronic acid. Biochim. Biophys. Acta 156: 17-30.

Tadashi, E., 2006. Sodium Chondroitin sulfate, Chondroitin-Sulfate-Containing Material and Processes for Producing the Same. U.S. Patent Appl. 20,060,014,256.

Tahami, M., 1994. Synthesis of chitosan and glucosamine from crustaceans wastes (Shrimp, Crab, Lobster). Iranian Fisheries Journal 3: 5-15.

Takahashi, Y., Miki, F., and Nagase, K., 1995. Effect of sonolysis on acid degradation of chitin to form oligosaccharides. Bull. Chem. Soc. Jpn., 68: 1851-1858.

Tao, S., Dongxiang, Z., Junlin, X. and Fang, M., 2007. Preparation of chitosan oligomers and their antioxidant activity. Eur. Food Res. Technol., 225: 451–456.

Teng, W.L., Khor, E., Tan, T.K., Lim, L.Y., and Tan, S.C., 2001. Concurrent production of chitin from shrimp shells and fungi. Carbohydr, Res., 332: 305-316.

Thorpe, J.P., Solé-Cava, A.M. and Watts, P.C., 2000. Exploited marine invertebrates: genetics and fisheries. Hydrobiologia 420(1): 65-184.

Tokoro, A., Kobayashi, M., Tatewaki, N., Suzuki, K., Okawa, Y., Mikami, T., Suzuki, S. and Suzuki, M., 1989. Microbiol. Immunol., 33(4): 357-367.

Tokura, S. and Tamura, H., 2007. Chitin and chitosan. In: Comprehensive Glycoscience, Kamerling, H., Boons, G., Lee, Y.C., Suzuki, A., Taniguchi, N. and Voragen, A.G.J. (Eds.), pp. 449–475. Elsevier Ltd., Amsterdam, The Netherlands.

Tsukada, K., Matsumoto, T., Aizawa, K., Tokoro, A., Naruse, R., Suzuki, S. and Suzuki, M., 1990. Jpn. J. Cancer Res., 81: 259.

Tur, J.A., Bibiloni, M.M., Sureda, A. and Pons, A., 2012. Dietary sources of omega-3 fatty acids: Public health risks and benefits. The British Journal of Nutrition 107(S2): 23-52.

Uchida, Y. Izume, M. and Ohtakara, A., 1989. In: Chitin and Chitosan, (G. Skjak-Braek, T. Anthonsen, and P. Sandford, Eds.), pp. 373–382, Elsevier, London.

Varum, K.M., Anthonsen, M.W., Grasdalen, H., and Smidsrod, O. 1991. Determination of the degree of *N*-acetylation and the distribution of *N*-acetyl groups in partially *N*-deacetylated chitins (chitosans) by high-field NMR spectroscopy. Carbohydr Res., 211: 17–23.

Wang, S.L. and Chio, S.H., 1998. Deproteinization of shrimp and crabshell with the protease of *Pseudomonas aeruginosa* K-187. Enzyme Microb. Technol., 22: 629–633.

Wang, S.L., Lin, H.T., Liang, T.W., Chen, Y.J., Yen, Y.H., and Guo, S.P., 2008. Reclamation of chitinous materials by bromelain for the preparation of antitumor and antifungal materials. Bioresour. Technol., 99: 4386–4393.

Wilesmith, J.W., Ryan, J.B. and Atkinson, M.J., 1991. Bovine spongiform encephalopathy: epidemiological studies on the origin. The Veterinary Record 128(9): 199-203.

Xu, Y.S. and Wang, Y.M., 2004. Preparation of D (+) glucosamine hydrochloride from crab shell, Chemistry Adhesion, 4 p.

Yamada, A., Shibuya, N., Kodama, O. and Akatsuka, T., 1993. Induction of Phytoalexin Formation in Suspension-cultured Rice Cells by N-Acetylchitooligosaccharides. Biosci. Biotech. Biochem., 57(3): 405-409.

Yamada, S., Yamamoto, K., Ikeda, T., Yanagiguchi, K. and Hayashi, Y., 2014. Potency of fish collagen as a scaffold for regenerative medicine. BioMed Research International, 2014. Article ID 302932. doi.org/10.1155/2014/302932.

Yokoyama, M., Origasa, H., Matsuzaki, M., Matsuzawa, Y., Saito, Y., Ishikawa, Y., Oikawa, S., Sasaki, J., Hishida, H., Itakura, H., Kita, T., Kitabatake, A., Nakaya, N., Sakata, T., Shimada, K. and Shirato, K., 2007. Effects of eicosapentaenoic acid on major coronary events in hypercholesterolaemic patients (JELIS): A randomized open-label, blinded endpoint analysis. Lancet 369(9567): 1090-1098.

Youling, Y., Betsy, M.C., Warren, O.H., and Joel, D.B., 2011. Deacetylation of chitosan: material characterization and in vitro evaluation via albumin adsorption and pre-osteoblastic cell cultures. Materials 4: 1399-1416.

Yusof, N.L., Wee, A., Lim, L.Y. and Khor, E., 2003. Flexible chitin films as potential wound-dressing materials: wound model studies. J. Biomed. Mater. Res. A, 66A: 224–32.

Zhuang, Y.L., Zhao, X. and Li, B.F., 2009. Optimization of antioxidant activity by response surface methodology in hydrolysates of jellyfish (*Rhopilema esculentum*) umbrella collagen. Journal of Zhejiang University Science B 10(8): 572-579.

Impact of Climate Change on Hydrological Cycle, Ecosystem, Fisheries and Food Security, Pages: 459–467
Edited by: B. Madhusoodana Kurup, M.R. Boopendranath, M. Harikrishnan and A.V. Shibu

CHAPTER - 38

Characterization of Collagen from the Skins of Three Different Commercially Important Fishery Resources

Moe Theingi Hlaing*[1], K. Elavarasan [2], U. Parvathy[2], C.G. Joshy[2] and George Ninan[2]

[1]*East Yangon University, Yangon, Myanmar*
[2]*ICAR-Central Institute of Fisheries Technology, Kochi-682029, India*
**E-mail: moetheingihlaing.dr@gmail.com*

Collagen is the most abundant biological macromolecules of extracellular matrix where it provides the major structural and mechanical support to tissues. Native collagens from different sources find applications in cosmetics, biomedical and pharmaceutical industries. Collagen's characteristics depend on the raw material and the extraction conditions, which subsequently determine its application. There is high demand in the food industry for collagen and gelatin on account of their high protein content and functional as well as bioactive properties (Nagai & Suzuki, 2000). In the biomedical and pharmaceutical fields, it is used as a vehicle for drugs, proteins and genes, as well as substitute for human skin, blood vessels and ligaments. Several studies have examined collagen in order to obtain bioactive compounds with antimicrobial, antioxidant and antihypertensive properties (Nagai *et al.*, 2004; George, 2009; Hema, 2015). The objectives of the study were to utilize the discarded skins from commercially important species along Kerala coast viz., cuttlefish, tilapia and lizard fish for the extraction of collagen followed by its characterization for exploring its application potentials.

Materials and Methods

The skins from cuttlefish, lizardfish and tilapia, collected from the industry were used for the study. The samples obtained were cleaned by scrubbing off the adhering flesh and other extraneous matter, cut into small pieces, and frozen stored at -20^{0}C till use. All the chemicals used for the study were of analytical grade purchased from Sigma as well as Himedia, India.

Skins of cuttlefish, lizard and tilapia fish were collected from local fish market, Kochi, Kerala, India. Skin was washed twice with cold tap water and scrapped with knife to remove flesh and scales. Further it was washed with cold demineralized water and cut into small pieces. Skin samples of approximately 100 g (wet weight basis) were soaked in 0.1 M NaOH (1:30 w/v) for 24 hours and retreated by soaking further in fresh 0.1M NaOH (1:30 w/v) solution for 6 hours.

Extraction of Acid Soluble Collagen

Acid soluble collagen was extracted according to the method of Bhuimbar *et al.* (2019) with slight modification. Pre-treated fish skin was soaked in 0.5 M acetic acid (1:25 w/v) for 48 hours. Entire process of extraction was performed at 4°C. After swelling, the skin was squeezed and viscous solution was fIltered through muslin cloth. The residue was again soaked with 0.5 M acetic acid (1:20 w/v). The extracts were filtered and both filtrates were mixed. This solution was centrifuged (10,000 ×g, 4°C) and the resultant supernatant was subjected to precipitation using 0.25M NaCl. Precipitate was collected by centrifugation at 10,000 ×g, 4°C, dissolved in 0.5M respective acid (1:5 w/v). The final solution was dialyzed against cold demineralized water for 192 hours to remove acetic acid completely using dialysis bag and lyophilized as ASC samples. For cuttlefish skin, prior soaking in 0.2M EDTA (1:20 w/v) was carried for 6 hours and properly grinded to become thick paste and then soaked with 0.5 M acetic acid (1:25 w/v).

Nutritional Profile

Proximate composition of tuna protein hydrolysates was estimated as per AOAC (2012). Protein content of the samples were estimated by kjelhdahl method. Amino acid profiling of the sample was done using HPLC (high-performance liquid chromatography) (Shimadzu Prominence, Japan) following the methodology of Ishida *et al.* (1981). A precolumn derivatization method was developed for the high-performance liquid chromatographic (HPLC) determination of aminoacids using o-phthalaldehyde (OPA) and 9-fluorenylmethyl-chloroformate (FMOC).

Yield and Colour

Yield of the sample was evaluated as the amount of collagen obtained upon drying to the initial amount of raw material (fish skin) used for extraction, expressed as percentage. Sample colour was measured using Hunter- Lab scan XE – Spectrocolorimeter (Color Flex,Hunter Associates Laboratory, Reston, USA.) at D-65 illuminant and 10° observer. Colour determinants were expressed in terms of L*(lightness), a*(+redness/-greenness), b*(+yellowness/-blueness).

UV-VIS Absorption Spectra

UV absorption spectrum of collagen samples was measured using a UV spectrophotometer. Samples were dissolved in 0.5 M acetic acid and UV spectra were measured between 190 and 600 nm at a scan speed of 2 nm/s with an interval of 1 nm.

SDS PAGE

The protein pattern was analyzed using sodium dodecyl sulfate poly-acrylamide gel electrophoresis (SDS-PAGE) according to the method of Laemmli (1970).

FTIR

FTIR spectrometer by KBr method using diffuse reflectance assembly was adopted to determine the spectra of the sample in the wavelength range of 4000 to 400 cm^{-1}. A known quantity of sample mixed with potassium bromide (1:100 (w/w)) was analysed for the reflectance spectra which were further transformed to transmission spectra.

Results and Discussion

Nutritional profile

The proximate composition of raw skin and fish skin collagen for the different species are given in Table 1. Collagen samples had low moisture contents (< 10%) while protein content ranged from about 73% to 89%. The amount of fat as well as ash were also low indicative of the efficacy of extraction process. Similar range were reported in previous studies as well (Kittiphattanabawon *et al.*, 2005; Hema *et al.*, 2013). Amino acid composition of collagen extracted from the skins of tilapia, cuttle fish and lizard fish is given in Table 2. Glycine was the most abundant amino acid in all three collagens from the fish skin and it ranged from 27-29%. Other dominant amino acids included glutamic acid, arginine, alanine, proline and aspartic acid. Previous studies also indicated similar amino acid profile for collagen from fish skin (Hema *et al.*, 2013; 2014).

Table 1: Proximate composition of skin and skin collagens

Parameters	Cuttle fish skin	Lizard fish skin	Tilapia skin	Cuttle fish skin collagen	Lizard fish ski collagen	Tilapia skin collagen
Moisture	70.85	65.45	64.61	9.46	7.48	6.43
Protein	24.04	29.84	28.54	89.39	72.65	88.95
Fat	0.76	1.04	1.58	0.32	0.43	0.47
Ash	2.746	2.09	3.71	0.43	0.4	0.75

Table 2: Amino acid composition of skin extracted collagens

Amino acids	Cuttlefish skin collagen	Lizardfish skin collagen	Tilapia collagen
Aspartic acid	8.158	5.954	6.596
Glutamic acid	11.243	9.707	10.897
Serine	4.562	3.156	4.160
Histidine	1.318	0.724	1.285
Glycine	26.842	29.029	28.093
Threonine	2.424	2.425	2.881
Arginine	9.381	8.275	9.686
Alanine	7.419	13.094	11.701
Tyrosine	1.419	0.539	0.635
Cysteine	1.971	1.373	2.693
Valine	2.162	2.248	0.204
Methionine	1.940	1.835	1.105
Phenylalanine	1.770	2.293	2.545
Isoleucine	2.451	0.990	1.342
Leucine	3.177	2.013	2.935
Lysine	2.576	5.877	6.825
Proline	11.231	10.472	6.428

Yield and Colour

Estimation of the collagen yields from the skin of the selected samples indicated maximum yield from the skin of Tilapia (16.06%) followed by collagen extracted from cuttle fish and lizard fish. There was not much significant difference in the yield of collagen extracted from cuttle fish (5.19%) and lizard fish (5.16%). Hema *et al.* (2013) reported ASC yields of 8.96%, 4.13% and 13.97% (wet weight basis) from skins of shark, Rohu and Tuna, respectively. Yields of collagens from different fish skins were previously reported (Jongjareonrak *et al.*, 2005; Huang *et al.*, 2011). The colour parameters of collagen extracted from different species are given in Table 3. Cuttlefish collagen had a pink colour on account of the pigments present in the raw material. However, collagen from lizard fish as well as tilapia had a lighter colour.

Table 3: Colour of skin extracted collagens

Collagen	L*	a*	b*
Cuttle fish	55.56 ± 0.99	4.79 ± 0.11	5.35 ± 0.12
Lizard fish	67.72 ± 0.04	0.59 ± 0.02	5.42 ± 0.17
Tilapia	66.16 ± 0.12	1.63 ± 0.02	6.61 ± 0.05

UV-VIS absorption spectra

The UV-VIS absorption spectra of skin collagens are given in Fig. 1. It was observed that absorbance peak was maximum at around 220 nm which was mainly due to the absorption by peptide bond. There was no absorption peak at 280 nm indicating the absence of aromatic amino acids and it is a typical characteristic of collagen. This is strengthened by the low value of amino acid tyrosine (Table 2) which has absorption maxima at 280 nm. Further, Kittiphattanabawon *et al.* (2010) stated the absence of absorption peak at 280 nm indicative of the collagen purity and efficacy of non-collagenous protein removal.

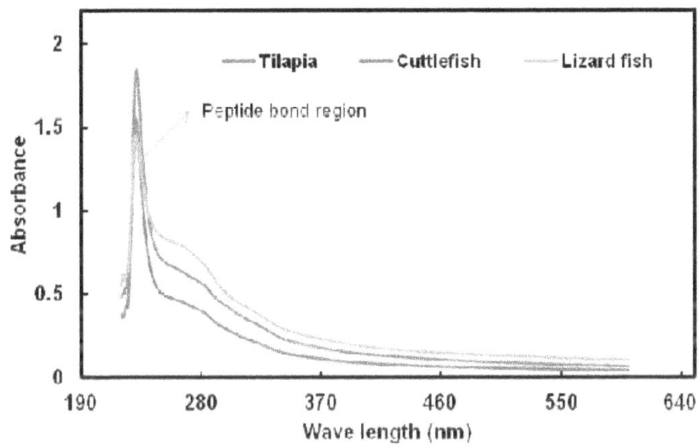

Fig. 1: UV-Vis absorption spectra of skin collagens

SDS PAGE

Fish skin collagen has typically two alpha 1, one alpha 2 chains, and a beta component structure that is recognized as type I collagen (Singh *et al.*, 2011). Type I collagen was characterized by two alpha chains of around 100 kDa and a beta component of about 200 kDa. In the present study, alpha 1 and alpha 2 bands were observed in all the three collagen samples. Alpha 2 band was found much lower to than alpha 1 while alpha 3 component was not detected which must possibility be due to co-migration with alpha 2. Beta components are indicative of the presence of dimers. Peptide bands below 97 kDa can be the product of collagen degradation during extraction. No major differences were observed on protein pattern of collagens extracted from three different species studied. The Molecular weight characterization of collagen isolates using SDS-PAGE is given in Fig 2.

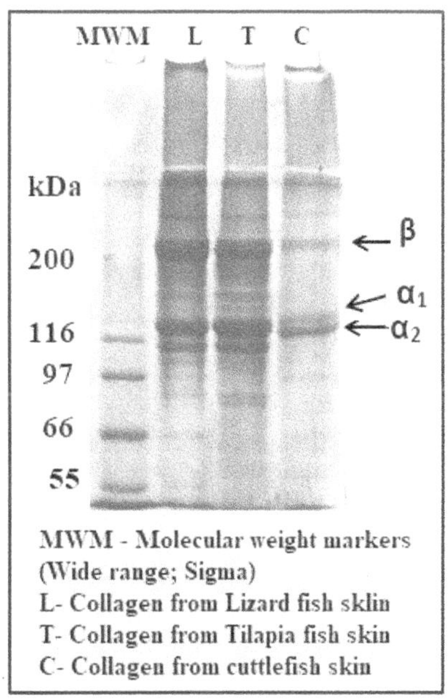

Fig. 2 Molecular weight characterization of skin collagens using SDS-PAGE

FTIR

The alterations in the secondary structure of proteins and polypeptides can be well revealed by Fourier Transform Infrared spectroscopy (Nikoo *et al.*, 2011). FTIR spectra of skin extracted ASC samples are depicted in the Fig. 3. The spectra for ASC were similar to those of previous studies from other fish species (Muyonga *et al.*, 2004; Singh *et al.*, 2011). Among these absorption bands, the Amide-I band between 1600 and 1700 cm-1 and Amide-II between 1565 and 1520 cm^{-1}, are the most useful peaks for infrared analysis of the secondary structure of proteins. The frequency range of 1660–1650 cm^{-1} is characteristic of α-helical structures (Hashim *et al.*, 2010) and the present study indicated the sample to have helical confirmation. The amide A band of ASC were found at a wave range of around 3300 cm^{-1}. Reports suggest that free N–H stretching vibration occurs in

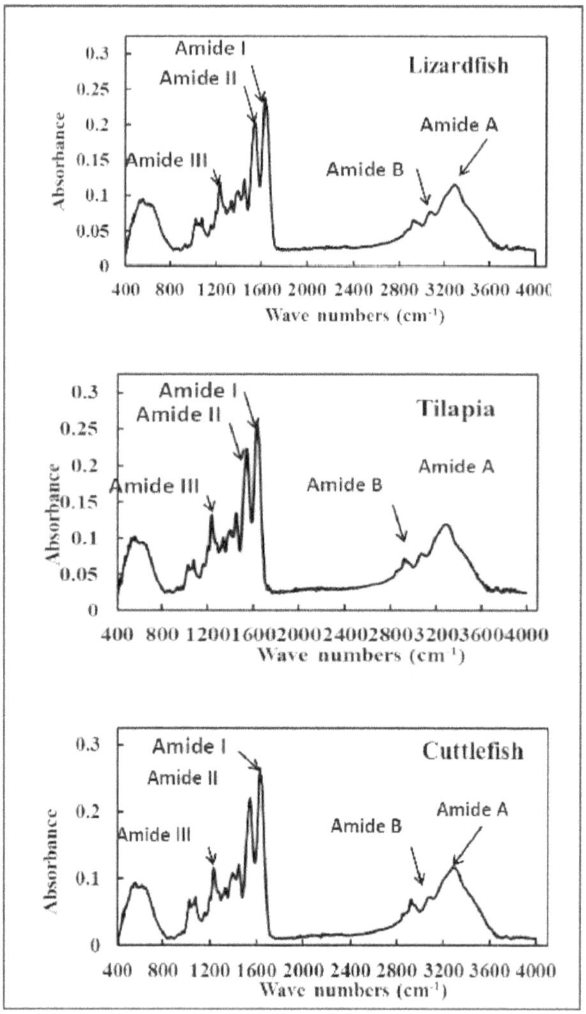

Fig. 3. FTIR pattern of skin collagens

the range of 3400–3440 cm^{-1} and when the NH group of a peptide is involved in a hydrogen bond, the position is shifted to lower frequencies, usually around 3300 cm^{-1} (Doyle *et al.*, 1975). Similar shifts were reported elsewhere (Singh *et al.*, 2011).

Conclusion

Present study explores the possibility of fish skin as a potential source of collagen. The yield was maximum for collagen extracted from the skin of tilapia. Nutritional evaluation indicated that collagen from all the three species skin were rich in protein with good profile of amino acids. From SDS-PAGE, amino-acid analysis, FTIR collagens extracted were identified as Type I. Further optimization studies can promise better yield and superior properties for collagen from these potential sources.

Acknowledgements

First author acknowledges ASEAN-India Research Training Fellowship (AIRTF) Scheme Under ASEAN – India Science & Technology Development Fund (AISDTF) Sanctioned by DST, Govt. of India for assistance in conduct of this research. Authors express their sincere thanks to Dr. Ravishankar C.N., Director, ICAR-CIFT, Kochi; Dr. Ashok Kumar, Head of Division, Fish Processing and all staffs of fish processing division for their support and encouragement.

References

AOAC, 2016. Official methods of analysis. 20th Edn., Association of Official Analytical Chemists, Washington DC, USA.

Bhuimbar, M.V., Bhagwat, P.K.and Dandge, P.B., 2019. Extraction and characterization of acid soluble collagen from fish waste: Development of collagen-chitosan blend as food packaging film. Journal of Environmental Chemical Engineering 7(2): 102983.

Doyle, B.B., Bendit, E.G. and Blout, E.R., 1975. Infrared spectroscopy of collagen and collagen like polypeptides. Biopolymers: Original Research on Biomolecules 14(5): 937-957.

Hashim, D.M., Man, Y.C., Norakasha, R., Shuhaimi, M., Salmah, Y. and Syahariza, Z.A., 2010. Potential use of Fourier Transform Infrared Spectroscopy for differentiation of bovine and porcine gelatins. Food Chemistry 118(3): 856-860.

Hema, G.S., 2015. Biochemical Characterization and Bio-evaluation of Collagen and Collagen Peptides Extracted and Purified from Fish Skin: in vitro and in vivo Studies on Antiarthritic and Wound Healing Properties, Ph.D. Thesis, Cochin University of Science and Technology, Cochin.

Hema, G.S., Shyni, K., Mathew, S., Anandan, R., Ninan, G. and Lakshmanan, P.T., 2013. A simple method for isolation of fish skin collagen-Biochemical characterization of skin collagen extracted from albacore tuna (*Thunnus alalunga*), dog shark (*Scoliodon sorrakowah*) and rohu (*Labeo rohita*). Annals of Biological Research 4(1): 271-278.

Huang, Y.R., Shiau, C.Y., Chen, H.H. and Huang, B.C., 2011. Isolation and characterization of acid and pepsin-solubilized collagens from the skin of balloon fish (*Diodon holocanthus*). Food Hydrocolloids 25(6): 1507-1513.

Ishida, Y., Fujita, T. and Asai, K., 1981. New detection and separation method for amino acids by high-performance liquid chromatography. Journal of Chromatography A, 204: 143-148.

Jafari, S.M., Assadpoor, E., He, Y. and Bhandari, B., 2008. Encapsulation efficiency of food flavours and oils during spray drying. Drying Technology 26(7): 816-835.

Jongjareonrak, A., Benjakul, S., Visessanguan, W., Nagai, T. and Tanaka, M., 2005. Isolation and characterisation of acid and pepsin-solubilised collagens from the skin of Brownstripe red snapper (*Lutjanus vitta*). Food Chemistry 93(3): 475-484.

Kittiphattanabawon, P., Benjakul, S., Visessanguan, W. and Shahidi, F., 2010. Isolation and characterization of collagen from the cartilages of brownbanded bamboo shark (*Chiloscyllium punctatum*) and blacktip shark (*Carcharhinus limbatus*). LWT-Food Science and Technology 43(5): 792-800.

Kittiphattanabawon, P., Benjakul, S., Visessanguan, W., Nagai, T. and Tanaka, M., 2005. Characterisation of acid-soluble collagen from skin and bone of bigeye snapper (*Priacanthus tayenus*). Food Chemistry 89(3): 363-372.

Laemmli, U.K., 1970. Cleavage of structural proteins during the assembly of the head of bacteriophage T4. Nature 227(5259): 680-685.

Nagai, T. and Suzuki, N., 2000. Isolation of collagen from fish waste material - skin, bone and fins. Food Chemistry 68(3): 277-281.

Nagai, T., Izumi, M. and Ishii, M., 2004. Fish scale collagen. Preparation and partial characterization. International Journal of Food Science and Technology 39(3): 239-244.

Nikoo, M., Xu, X., Benjakul, S., Xu, G., Ramirez-Suarez, J. C., Ehsani, A., Kasankala, L.M., Duan, X. and Abbas, S., 2011. Characterization of gelatin from the skin of farmed Amur sturgeon Acipenser schrenckii. International Aquatic Research (Islamic Azad University, Tonekabon Branch), 3(2): 135-145.

Ninan, G., 2012. Optimization of Process Parameters for The Extraction of Gelatin from the Skin of Freshwater Fish and the Evaluation of Physical and Chemical Characteristics. Ph.D. thesis, Cochin University of Science and Technology, Cochin.

Sathe, S.K. and Salunkhe, D.K., 1981. Functional properties of the great northern bean (*Phaseolus vulgaris* L.) proteins: emulsion, foaming, viscosity, and gelation properties. Journal of Food Science 46(1): 71-81.

Shahidi, F., Han, X.Q. and Synowiecki, J., 1995. Production and characteristics of protein hydrolysates from capelin (*Mallotus villosus*). Food Chemistry 53: 285-293.

Singh, P., Benjakul, S., Maqsood, S. and Kishimura, H., 2011. Isolation and characterisation of collagen extracted from the skin of striped catfish (*Pangasianodon hypophthalmus*). Food Chemistry 124(1): 97-105.

Impact of Climate Change on Hydrological Cycle, Ecosystem, Fisheries and Food Security, Pages: 469–476
Edited by: B. Madhusoodana Kurup, M.R. Boopendranath, M. Harikrishnan and A.V. Shibu

CHAPTER - 39

Consumer Preference for Value added Fish Products in Kochi (Kerala)

Mary Saniya Stephen[1*] and Dhanya Pulikkottil Rajan[2]

[1]*School of Industrial Fisheries, Cochin University of Science and Technology Lake-side campus, Fine arts Avenue, Cochin-682016, Kerala, India*
[2]*Department of Aquaculture, M.E.S. Asmabi College Padinjare Vemballur, Thrissur-680671 Kerala, India*
**E-mail: saniyastephen94@gmail.com>*

Fisheries sector has witnessed an impressive growth from a subsistence traditional activity to a well-developed commercial and diversified enterprise. It has been playing a pivotal role in the economic development by virtue of its potential contribution to employment generation, income augmentation, addressing food and nutritional security concerns and foreign exchange earnings. World fish production has increased immensely and the capture fishery has arrived at a state of over exploitation. An alarming increase in human population is resulting in widening the supply-demand gap, with the consequences of reduced availability, rising price and search for alternative resources to meet the gap (Clark, 1990).

Low priced fishes are either converted to fish meal or sold for curing. In general, deep-sea fishes may not be immediately acceptable to the consumers due to unfamiliarity in shape, size and colour of the new varieties. These low priced fishes are nutritionally and chemically in no way inferior to that of fishes of commercial importance. Therefore, collecting and processing meat of these fishes into diversified value-added products suiting to human consumption would result in effective utilization of the resource. Moreover, purchasing capacity of people is going to enhance in years to come and market prices are going to be high for fish (Salim, 2008).

Marketing of value-added fish products is completely different from traditional seafood trade. Market surveys, packaging and advertising are a few of the very important areas, which ultimately determine successful movement of new products. Most market channels currently used may not be suitable to trade value added fish products. A new appropriate channel would be supermarket chain; which want to procure directly from source of supply. Though market research has been conducted in the field of fresh and frozen fish, the domain of value-added fish and fish products remain less explored (Sabat *et al.*, 2008). According to a survey conducted by Corporate Catalyst India (CCI), India is the second largest producer of food and holds high potential in the global food market. On the whole, it is understood that the food processing industry is one of the largest in India presently growing at an average rate of 10% per annum (Ali *et al.*, 2009). Fish based snack products have not been able to capture the snack food market due to lack of innovative products keeping in pace with current consumer trend. Value addition for fish is an important strategy that will add economic value and possibly widen the market performance while reducing a problem of post harvest losses (Kyule *et al.*, 2014, Mohamad *et al.*, 2011). The popularity of fish products in the diet has increased due to the international recommendation to lower the total level of dietary fat (Akter *et al.*, 2013).

Earlier studies have employed structural equation modeling (Tudoran *et al.*, 2009), conjoint analysis (Harrison *et al.*, 2002; Boughanmi *et al.*, 2007; Hill *et al.*, 2008), sensory analysis, scoring method, quantitative descriptive analysis (Sveinsdottir *et al.*, 2009), conditional logit model (Jaffry *et al.*, 2004), ordered probit models (Brecard *et al.*, 2009) and cluster analysis (Arvanitoyannis *et al.*, 2004) to analyse the consumer preference and their willingness to pay for fish food products by providing information on health benefits, price, nutritive value and sensory quality of the products. Studies on consumers' attitudes, purchasing behavior and willingness to pay for safe and quality fishery products (Feng *et al.*, 2009) indicated that income, cultural differences, risk exposure to food safety, economy and trade are the forces that influence consumption of food products (Jensen, 2006), while product labeling (Jaffry *et al.*, 2004), eco-labelling, taste and price (Brecard *et al.*, 2009; Zhou *et al.*, 2010) were the determining attributes for consumer preferences.

Methodology

The present investigation on consumer preference for value added fish products was conducted in Kochi which is the largest city and the largest urban agglomeration in Kerala. Conjoint analysis was employed to analyse the data on consumer preferences. Conjoint analysis, conceptualised by Luce and Tukey (1964) and further developed by Green and Rao (1971) for marketing research, employs a decompositional method to estimate the structure of consumer preferences and consumer utility values of different attributes of a product or service. Consumer preference for the value added fish products like fish cutlet, fish roll and fish samosa developed under the project in comparison to meat and vegetable based snacks available in the snack food market was evaluated. The attributes selected for this study was: Ingredient, Price, Perceived quality and Cooking method at various levels, which are described in Table 1. Given the four attributes with their respective levels, a full factorial design requires 54 hypothetical combinations (multiplying each attribute level). In order to reduce the number of product profiles, which are manageable by the respondents, an orthogonal design was generated with 9 cards or combinations using SPSS software. Orthogonal design generated for data collection using SPSS software is given in Table 2.

The data collection was carried out from a simple random sample of 100 respondents around Kochi, using the orthogonal design. Those persons who are in-charge of food product purchases for their household or who have a say in the purchase activity were contacted for this study. The orthogonal design act as different combinations and these combinations are put as a questionnaire in front of the respondents to rank them based on their liking or utility they obtain from these combinations. Prior to the questionnaire an introduction about various value added fish products like fish cutlet, fish roll, fish samosa were given. It was noticeable that many respondents were unaware about some of the value added fish products compared to vegetable and meat products. The combinations were represented as Card 1 to Card 9 for convenience. This technique gives the consumer preference for value added fish products with respect to price, quality and cooking method.

Table 1: Label and attribute levels selected for the conjoint study

Attribute	What consumers perceive through this attribute	Attribute label	Attribute level
Ingredient	Taste	Snack based on	Meat; Fish; Vegetable.
Price	Economy	Price	Rs.10; Rs.15; Rs. 20.
Quality	Food safety	Perceived Quality	Yes/No.
Cooking method	Health benefit	Type of the product	Fried and crispy; Sandwich; Stuffed.

Table 2: Orthogonal design

Combination/Card	Ingredient	Price	Quality	Cooking method
1	Meat	Rs.15	Yes	Sandwich
2	Fish	Rs.10	Yes	Fried and crispy
3	Vegetable	Rs.15	No	Fried and crispy
4	Meat	Rs. 20	Yes	Fried and crispy
5	Meat	Rs.10	No	Stuffed
6	Fish	Rs. 20	No	Sandwich
7	Fish	Rs.15	Yes	Stuffed
8	Vegetable	Rs. 10	Yes	Sandwich
9	Vegetable	Rs. 20	Yes	Stuffed

Results and Discussion

Women were mainly responsible for making food purchase at the household. Most of the respondents (46%) belonged to the income group Rs.15,000 to Rs. 30,000 per month and only 17% of the respondents belonged to the higher income group. About 78% of the households were nuclear families as is the trend nowadays. The result shows that the highest utility score was achieved for vegetable based products. The consumer could compromise for 10% to 15% increase in price compared to the standard prices for snack items, as is shown for the high utility value for Rs.15 of price attribute. The highest utility score was achieved for vegetable based products (utility value=0.220), and a negative utility for fish based products (utility value=-0.103) (Fig. 1). The consumer could compromise for increase in price compared to the standard prices for snack items, as is shown for the high utility value for Rs. 15 of price attribute (utility value=0.316). But when price increases to Rs.20 a negative preference is found (Fig. 2). With respect to taste, fried snack items are highly preferred as indicated by the part-worth utility (utility value=0.203) corresponding to the level of method attribute. The customers didn't prefer sandwich products (Fig. 3). The utility values for the quality attribute gives an indication that the consumer is conscious about ensuring good quality of the product (utility value=0.330) (Fig. 4). The factors influencing the consumers among the investigated attributes were found to be in the following order (Fig. 5): Price>Quality>Cooking method>Ingredients. The highest utility score was achieved for vegetable based products (utility value=0.220), and negative utility was for fish based products (utility value = -0.103). Geethalakshmi *et al.* (2013) reported that the highest utility score was achieved for fish based products in Palakkad, Kerala, which doesn't coincide with the present study reports. The consumer could compromise for increase in price compared to the standard prices for snack items, as is shown for the high utility value for Rs. 15 of price attribute (utility value = 0.316).

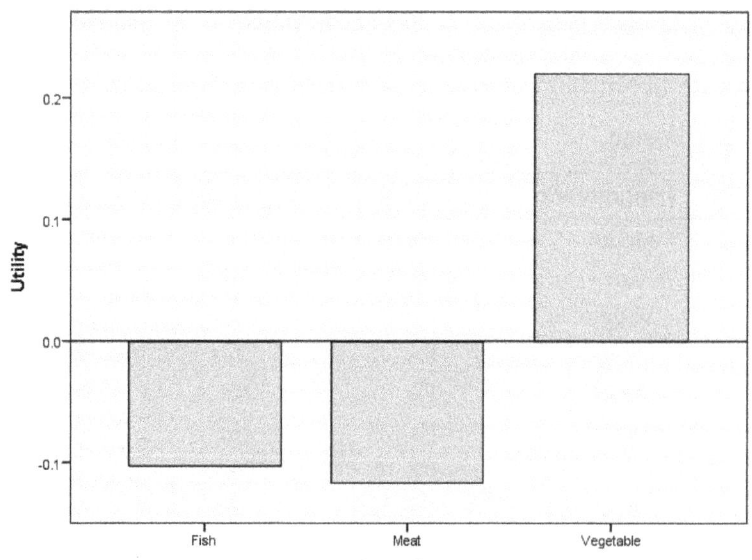

Fig. 1. The customer utility with respect to fish, meat and vegetable

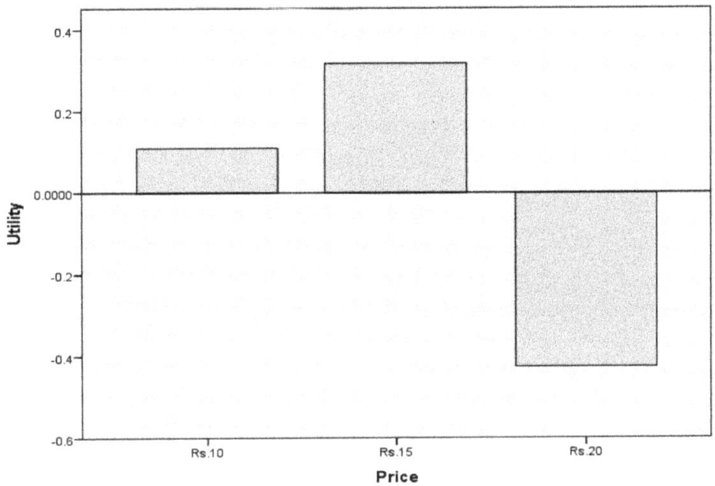

Fig. 2. Customer utility with respect to price

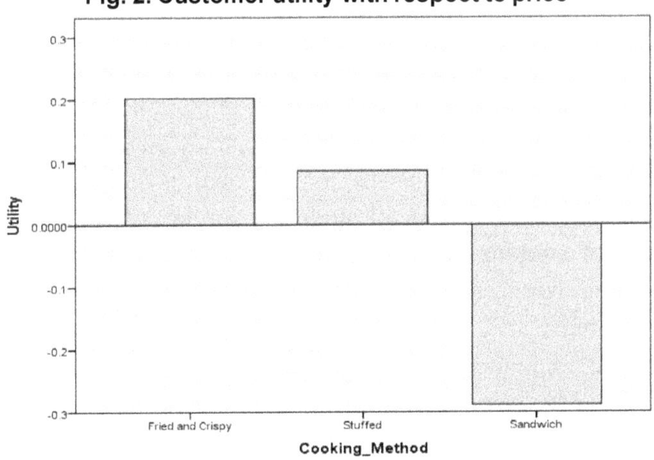

Fig. 3. Customer utility with respect to cooking method

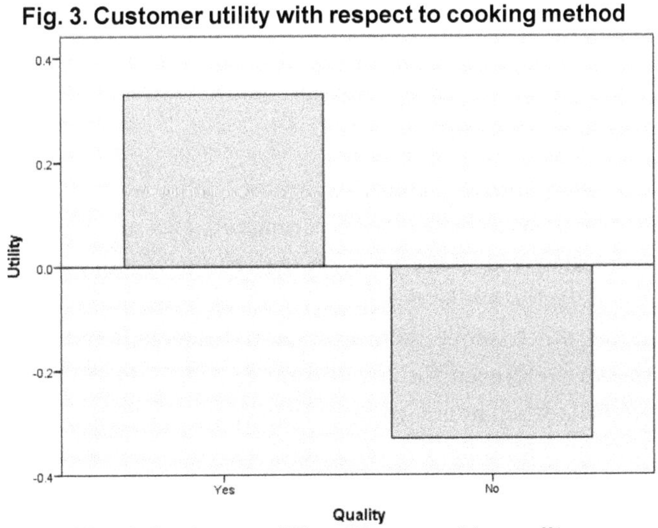

Fig. 4. Customer utility with respect to quality

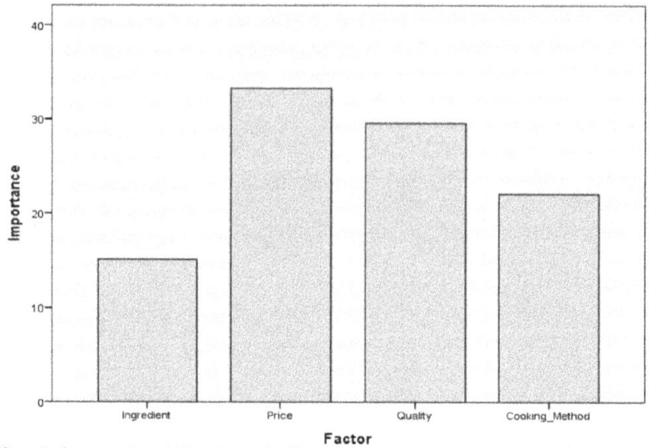

Fig. 5. Important factors influencing consumer preference

The study revealed that price, nutritional value and quality were the most important factors for guiding the fish purchasing behaviour of consumers. Debnath *et al.* (2011) during the study on consumer behaviour at organized fish retail outlet found that quality, nutritional value and price of products were the most important factors for guiding the fish purchasing behaviour of consumers. The result of the study concludes that the respondents were highly health conscious, and insisted on the quality of products. The preference structure for the method of cooking in all models is similar, with the fried method as the most preferred way of cooking, and the sandwich as the least preferred, contributing negatively to overall preferences. This low preference for microwaving is not consistent with the increased demand for convenience demonstrated by consumers in most western societies but may be explained by the perceived negative health consideration of microwaving expressed by some consumers in the sample, and the strong fishy smell linked to this cooking method. Present study has indicated that price is the important factor for consumer preference of value added fish products and ingredient is the least important factor.

Conclusion

During the study, it was noticeable that many respondents were unaware about various value added fish products available in the hypermarkets and supermarkets around Kochi. This has lead to decrease in the utility value. A majority of respondents regard nutritional value as an important factor while purchasing fish and fish products. The response of consumers has shown much scope for research and development on fish products as consumers have shown preference for crispy form of the products. The results show that value added fish products when hygienically prepared, and promoted with quality label can fetch higher prices compared to the conventional meat or vegetable based products in the domestic food market, given the obvious health benefits of fish. The consumer could compromise for 10% to 15% increase in price compared to the standard prices for snack items, as is shown for the high utility value for Rs.15 of price attribute. The results of the conjoint analysis show that the highest utility score was achieved for vegetable based products however there is higher scope for value added fish products due to its nutritional quality. Willingness to pay more for a good quality snack item was also observed. Growing income levels, changing lifestyles and promotional campaigns of the companies aiming to capture India's middle and upper middle class consumers would lead to dramatic explosion of processed food market in the coming years. Tastes and consumer habits change over time and as lifestyles evolve. It is therefore necessary, for producers to understand their market, so as to be able to adapt their products to the changing needs and expectations of consumers, and therefore to derive optimum value from their production.

References

Akter, M., Islamil, S.N., Reza, M.S., Shikha, F.H., and Kamal, M., 2013. Quality evaluation of fish ball prepared from frozen stored striped African catfish (*Pangasianodon hypophthalmus*). Journal of Agroforestry, Environment 7(1): 7-10.

Ali, J., Singh, S.P. and Ekanem, E., 2009. Efficiency and productivity changes in the Indian food processing industry: Determinants and policy implications. Int. Food Agribus. Man. Rev., 12(1): 43-66.

Arvanitoyannis, I.S., Krystallis, A., Panagiotaki, P. and Theodorou, A.J., 2004. A marketing survey on Greek consumers' attitudes towards fish. Aquaculture International, 12: 259–279

Boughanmi, H., Al-Musalami, J., Al-Oufi, H. and Zaibet, L. 2007. Estimating consumer preferences for value added fish products in Oman. J. Food Prod. Market., 13(2): 47-68.

Brecard, D., Hlaimi, B., Lucas, S., Perraudeau, Y. and Salladarre, F., 2009. Determinants of demand for green products: An application to eco-label demand for fish in Europe. Ecol. Econ., 69: 115-125.

Clark, C.W., 1990. Mathematical Bioeconomics: The Optimal Management of Renewable Resources, pp. 7-14, John Wiley and Sons. Inc. New York.

Feng, W., Zhang, J., Weisong, M., Zetian, F. and Xiaoshuan, Z., 2009. Consumers' perception towards quality and safety of fishery products. Food Control 20(10): 918-922

Geethalakshmi, V., Ashaletha, S., Daniel Raj and Nasser, M., 2013. Consumer preference and willingness to pay for value added fish products in Palakkad, Kerala. Indian Journal of Fisheries 60: 67-71.

Green, P.E. and Rao, V.R., 1971. Conjoint measurement for quantifying judgemental data. Journal of Marketing Research 8(3): 355-363.

Harrison, R.W, Stringer, T. and Prinyawiwatkul, W., 2002. Analysis of consumer preferences for value added seafood products derived from craw fish pp:157-170.

Hill, J.I., Nelson, R.G., Woods-Williams, K.L., Weese, S.J. and Whitis, G.N., 2008. Conjoint analysis of breaded catfish nuggets: Consumer preferences for price, product color, cooking method and country of origin. Paper presented at Southern Agricultural Economics Association Annual Meeting, February 2-6, 2008, USA.

Jensen, H.H., 2006. Changes in seafood consumer preference patterns and associated changes in risk exposure. Mar. Pollut. Bull. 53(10-12): 591-598.

Kyule, D.N., Yongo, E., Opiyo, M.A., Obiero, K., Munguti, J.M., and Charo-Karisa, H., 2014. Fish product development and market trials of fish and fish products in Kenya: a case study of Kirinyaga and Meru Counties. Livestock Research for Rural Development 26 (6): 1-9.

Mohamad, R., Ahmad, M. F., Abidin, A.Z.Z., and Mohamad, R., 2011. Consumers' acceptance and purchase intent of fresh water fish products developed by MARDI. Economic and Technology Management Review, 6: 77-82.

Sabat, S., Arpita Sharma and Shyam S. Salim, 2008. Consumption pattern and consumer preference for value-added fish and fish Products in north zone of India. J. Indian Fish Assoc., 35: 19-27

Sveinsdottir, K. Martinsdottir, E., Green-Peterson, D., Hyldig, G. Schelvis, R. and Delahunty, C., 2009. Sensory characteristics of different cod products related to consumer preferences and attitudes. Food Quality and Preference 20(2): 120-132.

Tudoran, A., Olsen, S.O. and Dopico, D.C., 2009. The effect of health benefit information on consumers health value, attitudes and intentions. Appetite 52(3): 568-579.

Impact of Climate Change on Hydrological Cycle, Ecosystem, Fisheries and Food Security, Pages: 477–486
Edited by: B. Madhusoodana Kurup, M.R. Boopendranath, M. Harikrishnan and A.V. Shibu

CHAPTER - 40

Chemical and Biophysical Properties of Gelatin Composites from Cephalopods for the Improvement of Textural Quality in Sausages

Roopa Rajan, L. Aranganathan and S.R. Radhika Rajasree*

Department of Fish Processing Technology
Kerala University of Fisheries and Ocean Studies, Panangad, Kochi-682506, Kerala, India
**E-mail: radhikarajasree@kufos.ac.in*

Gelatin is an important biopolymer, finding application in the food, pharmaceutical, biomedical and photographic industries as well as diverse technical uses. The recent improvements in fish processing technology enable development in the utilization of fish processing by-products for various applications. Gelatin by definition known as a product obtained by the partial hydrolysis of collagen derived from the skin, white connective tissue and bones of animals. As gelatins from fish may provide an alternative source that is acceptable for kosher (Jewish) and halal (Muslim) products, , the present study envisaged to explore the Octopus resources viz., *Cistopus indicus* (*C. indicus*) and *Amphioctopus marginatus* (*A. marginatus*) as a source for extraction and evaluation of its functional properties of blended with fish sausages.

Materials and Methods

Samples were collected from both east and west coasts of India viz., Thirupalakudi, Ramanathapuram district, Tamil Nadu and Thoppumpady, Kerala, India for *C. indicus* and *A. marginatus* respectively. Samples were transferred to laboratory and preserved at 4^0C prior to use. Skin separated from both species and washed with cold tap water. Cleaned skins were then placed in polyethylene bags and stored at -20°C until use. The storage time was less than one week.

Extraction of Gelatin

Frozen skins were cut into smaller pieces and then mixed with 0.05 M NaOH for 1h at 4°C with continuous stirring at a sample in alkaline solution ratio of 1:5 (w/v), to remove non-collagenous proteins following Jridi *et al.* (2013). The solution was changed every 20 minutes followed by the alkaline treated skins were washed with cold distilled water until neutral pH of wash water was obtained. The alkaline-treated skin was washed and then soaked in 100mM glycine-HCL buffer (pH-2.0) with a solid/solvent ratio 1:5 (w/v) for 10 h at 4°C with continuous stirring.The pH of the mixtures was then raised to 7.0 using 10 M NaOH and stirred gently for 1 h at 4°C to extract the gelatin. Treated skin mixtures were centrifuged to remove insoluble material. Finally, the supernatant were freeze-dried and obtained the dry gelatin powder.

Yield and proximate analysis of gelatin

The yield of gelatin obtained was calculated by the following;

$$\text{Yield (\%)} = \frac{\text{Weight of dried gelatin (g)}}{\text{Weight of wet skin (g)}} \times 100$$

Moisture, protein, lipid and carbohydrate contents of gelatin were determined according to the AOAC (2000).

Colour

Colour measurement of gelatin was made by using a Lab Scan XE spectrophotometer (Hunter Lab scan XE, USA) and the tristimulus L*, a*, b* measurement mode was used. The samples were filled into clear petri dish and readings were taken. The procedure was performed in triplicate for each sample (Kumar, *et al.*, 2012).

Amino acid Analysis by High Performance Liquid Chromatography

The amino acids profile was analyzed by using high performance liquid chromatography (Lachrome Hitachi HPLC). 20 ìl of the sample was injected manually in HPLC equipped with DENALI C18 5 MICRO MM 4.6 mm x 150 mm (cat no. DEN-5C18-15046) and UV-vis detector was used. The mobile phase was a gradient prepared from two solutions A and B. Solution A= 20: 80 acetonitrile: 25 Mm potassium Phosphate, pH 3.3 and solution B = 80: 20 CAN: 25 Mm potassium Phosphate, pH 3.3. Sample of 5 µl of amino acids standards mixture was added. The flow rate was 1 ml min^{-1} and the detection wavelength is at 254 nm. Gradient was 0 to 75% B over 15 min. The composition of amino acids was identified with standard amino acids.

UV-Visible spectrophotometer

The gelatin solution (0.5%) (w/v) was heated at 60°C for 30 minutes and cooled to ambient temperature (25°C). The solution was filtered using whatmann no.4 filter paper.The UV absorption spectra recorded in the wavelength range of 200-800nm and plotted to get absorption spectra (Chandra *et al.*, 2015).

Fourier Transform Infrared Spectroscopy (FT-IR)

Gelatin samples were subjected to FT-IR analysis using Bruker Model EQUINOX 55 FTIR spectrometer (Bruker, Ettlingen, Germany) equipped with a deuterated L-alanine triglycine sulphate (DLATGS) detector. The horizontal attenuated total reflectance (HATR) accessory was mounted in the sample compartment. The internal reflection crystal (Pike Technologies, Madison, WI, USA), made of Zinc selenide, had a 45° angle of incidence of the IR beam. Spectra were acquired in the IR range of 4000-400 cm^{-1} (mid-IR region) at 25°C (Nagarajan *et al.*, 2012). Dried gelatin (4 mg) was mixed with KBr and analyzed in the range of 4000 cm^{-1} to 400 cm^{-1} using FT-IR instrument at 25°C. Automatic signals were collected in 32 scans at a resolution of 4 cm-^{1}, against a background spectrum recorded from the clean and empty cell at 25°C. Analysis was carried out using the OPUS 3.0 data collection software programme.

X-Ray diffraction (XRD)

X-ray diffraction patterns of gelatin were obtained using X-ray diffractometer in the angular range of 550 (2q) with Nickel filtered Cu K α radiation (1 ¼ 0.154 nm) at a voltage of 40 kV and current of 40Ma (Nuamsrinuan *et al.*, 2015).

Water holding capacity (WHC)

Water-holding capacity was measured by adding distilled water (50 ml) to 0.5 g of gelatin in a centrifuge tube and held at room temperature for 1 h. The gelatin solutions were mixed for 5 sec every 15 min and centrifuged at 450 g for 20 min. The upper phase was removed and the centrifuge tube was drained for 30 min on a filter paper after tilting to 45° angle. Water-holding capacity was calculated as the weight of the contents of the tube after draining divided by the weight of the dried gelatin, and expressed as the weight % of dried gelatin.

Viscosity measurement

Viscosity of gelatin sample was determined according to the method of (Cho *et al.*, 2005). Gelatin solutions (10 g/100 ml) were prepared. Viscosity was determined using a Brookfield digital viscometer (Model LV-DV-II, Brookfield Engineering; MA, USA) equipped with No.18 spindle (Model LV) at 60 rpm at 40 ±10°C.

Determination of Gelling and Melting Temperature

10 ml of gelatin sample was taken onto test tubes kept in deep freezer (-20°C).The gelling temperature was noted. The frozen gelatin was kept in water bath at 40°C and the time period corresponding to the melting temperature was recorded. (Anchana Devi *et al.*, 2016).

Preparation of sausage

Fish sausage was made with the following ingredients: 60% Fish by-products, 27% water (ice and cold-water), 8% edible modified starch, 2%NaCl, 0.05%carrageenan, 0.045% ascorbic acid and 2% vegetable oil.(Souissi *et al.*, 2016).

Textural analysis of sausage

TPA was performed as described by Dolores Romero de Avila *et al.*, 2014 with slight modifications. Briefly, three cylinders of 1.5 cm height and 2 cm diameter were prepared from each sample. A double compression cycle test was performed up to 30% compression of the original portion height with 1.27 cm diameter flat-faced Centurion XVI version: 16.2.04 using ANOVA analysis. All experiments were done in triplicate and differences were considered significant at $p < 0.05$. (Souissi *et al.*, 2016).

Results and Discussion

Yield and protein estimation of gelatin

The yield of gelatin extracted from Octopus skin which was observed as 4% for *C. Indicus* and 6% for *A. Marginatus*, respectively. The difference in the yield obtained from different species may depend on age of the animal, proximate composition, collagen content, method of extraction and species type has an influence on the yield percentage (Songchotikunpan *et al.*, 2008). The extracted dried gelatins had high protein content (Table 1). Similar results were obtained *croaker fish (*Muyonga *et al., 2004),* cuttle fish *(*Balti *et al., 2011)* and giant squid (*D. gigas*) (Uriarte-Montoya *et al., 2011*). Colour of gelatin from Octopus skin was expressed in terms of L*, a*and b*. Lightness and yellowness was more compared to redness in the octopus species which might be due to molecular weight that influences the colour of the skin. (Jridi *et al.*, 2013).

Table 1: Proximate composition of gelatin

pH	6.5
Moisture (%)	9.77
Protein (%)	88.98
Lipid (%)	0.92
Ash (%)	0.085
*L**	34.23±0.025
*a**	0.1575±0.042
*b**	4.145±0.031

Amino Acid Analysis

The amino acid composition affects the gelatin's physical and chemical properties. The amino acid composition was expressed as residues per 1000 amino acid residues. C.indicus shown higher amounts of glycine, proline and arginine. The composition of gelatin is similar to that of the collagen from which it was prepared (Fig. 1) (Haffidz *et al.*, 2011). Gelatins have the same proportion of alanine, glutamic acid, cysteine, isoleucine, tyrosine, phenylalanine, lysine and arginine. In spite of differences in amino acid compositions, all types of gelatins have almost constant hydroxyl amino acid content. All amino acids commonly found in proteins occur in gelatin with the probable exception of tryptophan and cystine, which might have resulted due to keratins and other protein contaminants.

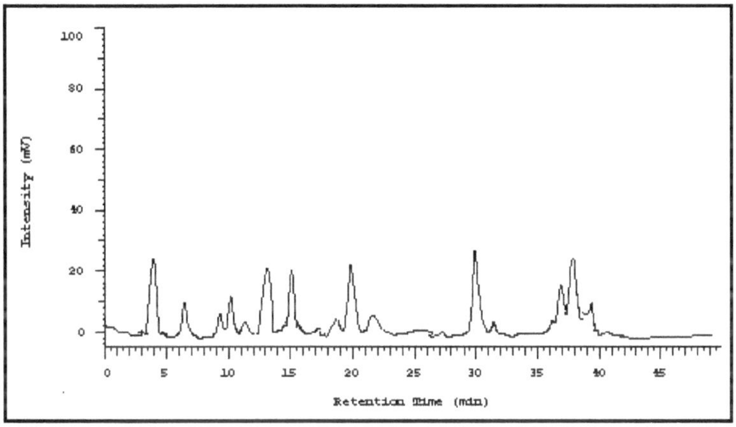

Fig. 1. Amino acid profile of gelatin

UV-Visible Spectroscopy

The UV-visible absorption spectra of gelatin solution are depicted in the Fig. 2 and 3. Absorption peak spotted at 210 nm for *C. Indicus* and 193nm for *A. marginatus,* indicated the presence of peptide bonds in the polypeptide chains of gelatin.

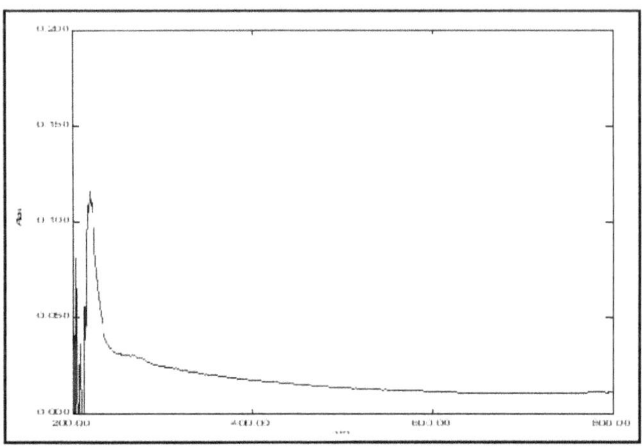

Fig. 2. UV absorption spectrum of gela tin from *Cistopus indicus*

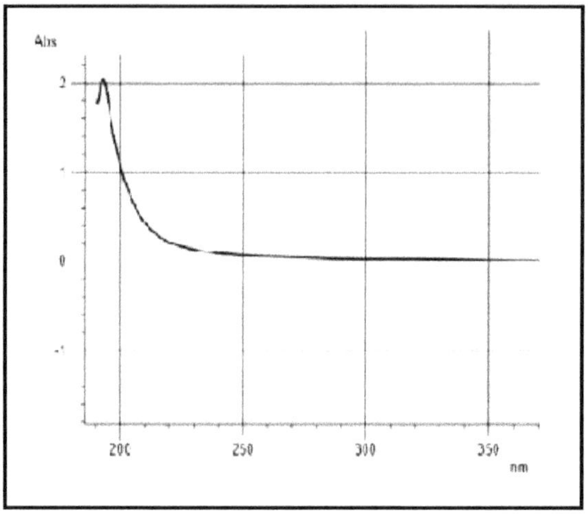

Fig. 3. UV absorption spectrum of gelatin from *Amphioctopus marginatus*

Fourier Transform Infrared Spectroscopy (FT-IR)

FT-IR analysis of gelatin in Fig. 4 and 5 showed presence of different functional groups. The peak at 1043.49cm^{-1} has been assigned to C-O, glycosidic bond. Other peaks located at 3103.46 OH group, 2605.83 carboxylic acid, 1598.99 aramatics, 1124.50cm^{-1} carbohydrate, 1598.99 amide II, the presence of amide bands in the gelatin was in agreement with the results.

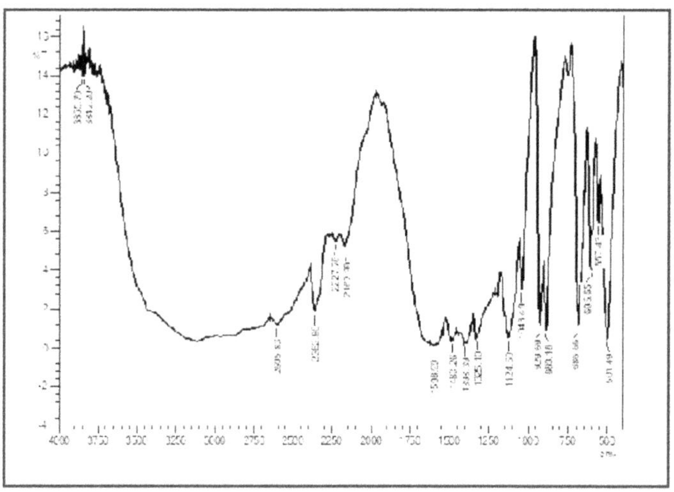

Fig. 4 FT-IR spectra of gelatin from *Cistopus indicus*

X-ray Diffraction

Collagen has a uniform triple-helical conformation with semi-crystalline structure, which is determined by its repeating (Gly-XY) sequence pattern and high imino acid content.

Polypeptides with these sequence features also adopt this conformation (Brodsky *et al.*, 2004) generally. X-ray diffraction is used to investigate the collagen fibril distribution and orientation (Bigi *et al.*, 2001). XRD pattern of gelatin showed (Fig. 6) diffraction peaks at 20° and 21° that was identified as characteristic peaks assigned to the triple-helical crystalline structure of collagen and gelatin.

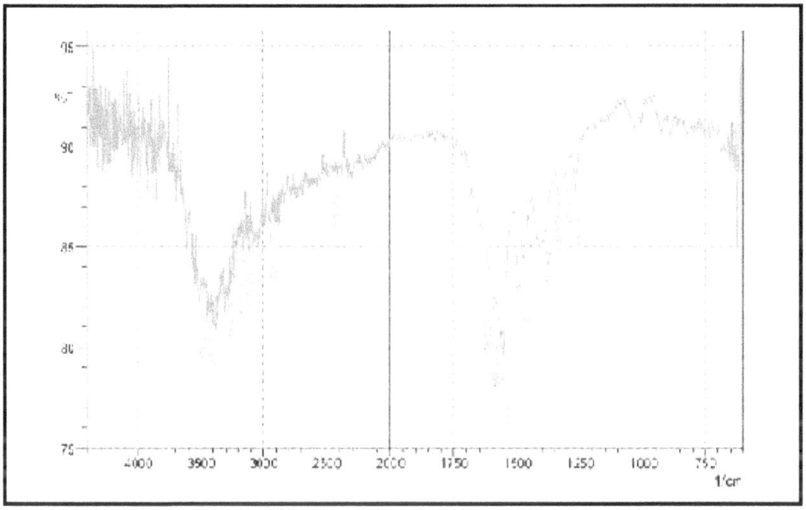

Fig. 5 FT-IR spectra of gelatin from *Amphioctopus marginatus*

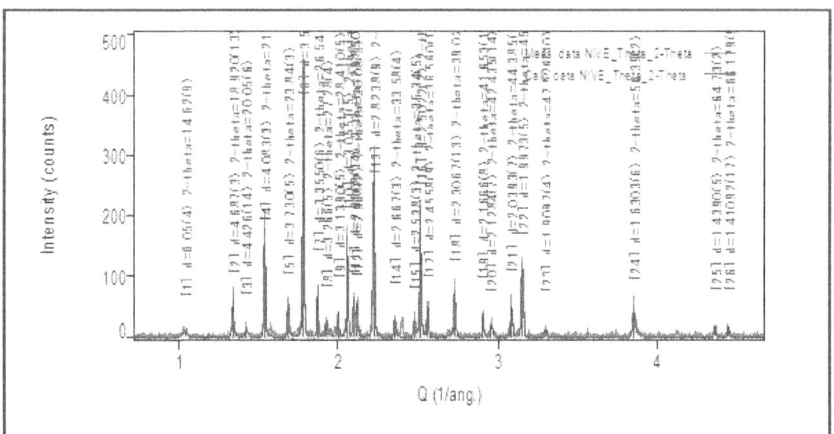

Fig. 6. X-ray diffraction analysis of gelatin from *Amphioctopus marginatus*

Water Holding Capacity

Water holding capacity of extracted gelatin was calculated to be 1.56% which means that it had a capacity to retain 0.015gm of water per gram. The water holding capacity of gelatin may be attributed to its microstructure, specifically to it primary chemical structure, hydrophilic/hydrophobic balance and particle size.

Viscosity

The viscosity of the gelatin solution increased significantly with increasing concentration and with decreasing temperature (Table 2). This abrupt increase was probably due to reorganization of the freely moving gelatin coils into a region of entangled chains, as well as to intermolecular hydrodynamic interactions between gelatin molecules and the solvent along with few additional factors like temperature, concentration and pH which might have influenced the viscosity of solution.

Table 2: Viscosity measurement of gelatin

CP	RPM	Temp. (°C)	% Dark force
2.00	15	30.8	1.0
1.62	50	30.9	2.7
1.59	100	30.9	5.3
1.56	150	30.9	7.8
1.56	200	30.9	10.4

Gelling and melting temperature

The extracted gelatin has higher gelling temperature compared with commercial gelatin might be due to the presence of impurities. The melting temperature of the extracted gelatin was found to be 45°C at temperature -20°C the gelling time was found to be 30 minutes.

Textural analysis of sausage

In a meat system, texture profile analysis is used to evaluate not only the consumer desirability of the products but also the structural integrity of the protein matrix. The hardness of the different formulations prepared, increased slightly with an increase in gelatin which is proportional to the amount of the gelatin added (Table 3). This increase in hardness between sausages can be explained by the protein network formed by gelatin, starch and other components of the octopus especially proteins.

Table 3: Textural analysis of sausage

Gelatin %	Hardness N	Adhesive-ness, Ns	Springiness 10^{-2} (m)	Cohesive-ness10^{-2} (m)	Gummi-ness	Chewi-ness	Resilience 10^{-2} (m)
0	2337.247	-0.090	0.891	0.686	1602.671	1427.731	0.412
0.25	1831.589	-	0.923	0.674	1234.193	1138.193	0.366
0.50	1491.790	-1.325	0.919	0.631	941.161	864.941	0.377
0.75	2156.512	-5.479	0.905	0.720	1552.477	1404.882	0.469
1	1672.407	-4.884	0.915	0.687	1149.494	1052.353	0.390

Conclusion

Fish sausages were prepared from fish by-products, without or with the addition of gelatin at different levels. The addition of gelatin to fish sausage resulted in a significant improvement of the physicochemical properties. Further, the addition of gelatin increased protein content and the moisture and reduced cooking loss. Indeed, addition of gelatin yielded products with enhanced textural parameters and the most appreciated sausage was with 1% gelatin. Fish sausages can be conserved for 15 days without alteration especially those supplemented with gelatin. Further studies are required for the commercial production of gelatin from marine sources for food applications.

References

Anchana Devi, Kamatchi, P., Leela, K., 2016. Extraction, characterization and application of gelatin from Carcharhinus amblyrhyncho and Sphyraena barracuda. IOSR J. Biotechnol. Biochem. 2(6): 40-49.

AOAC, 2000. Official Methods of Analysis. Association of Official Analytical Chemists International. Maryland, USA.

Balti, R., Jridi, M., Sila, A., Souissi, N., Nedjar-Arroume, N., Guillochon, D. and Nasri, M., 2011. Extraction and functional properties of gelatin from the skin of cuttlefish (*Sepia officinalis*) using smooth hound crude acid protease-aided process. Food Hydrocoll. 25(5): 943-950. doi:10.1016/j.foodhyd.2010.09.005.

Bigi, A., Burghammer, M., Falconi, R., Koch, M. H. ., Panzavolta, S. and Riekel, C., 2001. Twisted plywood pattern of collagen fibrils in teleost scales: An X-ray Diffraction Investigation. J. Struct. Biol., 136(2): 137–143. doi:10.1006/jsbi.2001.4426.

Brodsky, B., Eikenberry, E. F., Belbruno, K. C., Sterling, K., 1982. Variations in collagen fibril structure in tendons. Biopolymers. 21(5), 935–951. http://doi:10.1002/bip.360210507.

Chandra, M.V. and Shamasundar, B.A., 2015. Rheological properties of gelatin prepared from the swim bladders of freshwater fish Catla catla. Food Hydrocoll. 48: 47–54. doi:10.1016/j.foodhyd.2015.01.022.

Cho, S.M., Gu, Y.S. and Kim, S.B., 2005. Extracting optimization and physical properties of yellowfin tuna (*Thunnus albacares*) skin gelatin compared to mammalian gelatins. Food Hydrocoll. 19(2): 221–229. doi:10.1016/j.foodhyd.2004.05.005.

Dolores Romcro de Avila, M., Isabel Cambero, M., Ordóñez, J. A., de la Hoz, L., Herrero, A.M., 2014. Rheological behaviour of commercial cooked meat products evaluated by tensile test and texture profile analysis (TPA). Meat Sci. 98(2): 310-315. doi:10.1016/j.meatsci.2014.05.003.

Jridi, M., Nasri, R., Lassoued, I., Souissi, N., Mbarek, A., Barkia, A., Nasri, M., 2013. Chemical and biophysical properties of gelatins extracted from alkali-pretreated skin of cuttlefish (*Sepia officinalis*) using pepsin. Int. Food Res. J. 54(2): 1680–1687. doi:10.1016/ j.foodres.2013.09.026.

Kumar, J.J., Devivaraprasad Reddy, A., Sunita Maharia., Hanjabam Mandakini Devi., Vidya Sagar Reddy, G., Venkateshwarlu, G., 2012. Characterization of fish gelatin from Blackspotted Croaker (Protonibea diacanthus). Arch. Appl. Sci. Res. 4 (3): 1353-1358.

Nagarajan, M., Benjakul, S., Prodpran, T., Songtipya, P. and Kishimura, H., 2012. Characteristics and functional properties of gelatin from splendid squid (*Loligo formosana*) skin as affected by extraction temperatures. Food Hydrocoll. 29(2): 389–397. doi:10.1016/j.foodhyd.2012.04.001.

Nuamsrinuan, N., Suttisiri, N., Hoonnivathana, E., Limsuwan, P. and Naemchanthara, K., 2015. Characterization and Gel Properties of Gelatin Extracted from Waste Fish Scales. Appl. Mech. Mater. 804: 187–190. doi.org/10.4028/www.scientific.net/amm.804.187.

Raja Mohd Hafidz, R. N., Yaakob, C. M., Amin, I. Noorfaizan, A., 2011. Chemical and functional properties of bovine and porcine skin gelatin. Int. Food. Res. J., 18: 787-791.

Songchotikunpan, P., Tattiyakul, J. and Supaphol, P., 2008. Extraction and electrospinning of gelatin from fish skin. Int. J. Bio. Macro. 42(3): 247–255. http://doi:10.1016/j.ijbiomac.2007.11.005.

Souissi, N., Jridi, M., Nasri, R., Ben Slama, R., Njeh, M., Nasri, M., 2016. Effects of the edible cuttlefish gelatin on textural, sensorial and physicochemical quality of octopus sausage. LWT. 65: 18–24. doi:10.1016/j.lwt.2015.07.051.

Uriarte-Montoya, M. H., Santacruz-Ortega, H., Cinco-Moroyoqui, F.J., Rouzaud-Sαndez, O., Plascencia-Jatomea, M., Ezquerra-Brauer, J.M., 2011. Giant squid skin gelatin: Chemical composition and biophysical characterization. Int. Food. Res. J. 44(10): 3243–3249. doi:10.1016/j.foodres.2011.08.018.

Impact of Climate Change on Hydrological Cycle, Ecosystem, Fisheries and Food Security, Pages: 487–496
Edited by: B. Madhusoodana Kurup, M.R. Boopendranath, M. Harikrishnan and A.V. Shibu

CHAPTER - 41

Quality Evaluation of Ready to Cook, Pressurized Indian White Prawns (*Fenneropenaeus indicus*)

Ginson Joseph[1*] and J. Bindu[2]

[1]*St. Albert's College, Kochi-682018, Kerala, India*
[2]*Fish Processing Division, Central Institute of Fisheries Technology, Kochi-682029, Kerala, India*
**E-mail: jinsonjosephif@gmail.com*

Indian white prawn is one of the major commercial seafood in the world, which has high value and demand in domestic and international markets. In India, frozen shrimp is the major item in the export basket, in terms of quantity and value, accounting for a share of 34% in quantity and 67% of the total export revenue during 2017-18, by exporting a total quantity of 3,57,528 MT, worth US $3703 million. Since 2010, India has assumed a position of greater influence in the global shrimp trade, and it has well capitalized. However even today, country's shrimp export industry is facing several constraints including rejection of consignments (33 rejections in 2015), mainly due to the incidence of pathogens, inability to preserve seafood with natural characteristics, lack of novel technologies for the generation of value added products, high operational costs, etc. This situation demands the need to strictly adhere to the food safety and other standards in seafood products particularly in major export markets, predominantly those relating to regulatory and customer requirements. In lieu of these issues, Govt. of India has adopted stringent measures in addressing technical and quality issues in fish / shrimp processing, reinforcing the structure and modus operandi of the supply chain, and has given more focus to export promotion through product diversification and value addition. Presently large number of seafood processing industries are moving towards new product development and innovative propositions through new processing methods, to cater healthy and safe, next generation convenience food products.

One of the most successful developments in this area is the introduction of High Pressure Processing (HPP), an innovative non-thermal food processing technology, which is successfully implemented by all types of food industries worldwide. The technology has the potential for meeting the demands of the 21st century consumers (convenience foods, higher sensorial and nutritional quality, additive free/natural, functional products, etc.). Thermal methods, traditionally used in the food industry for food preservation, carry disadvantages like vitamin destruction or flavour changes that can be avoided with HPP. The technology also features environmental friendly processes, zero waste production, low operational costs and optimum energy consumption. HPP acts instantaneously and uniformly through a mass of food independent of size, shape and food composition. In addition to food preservation, this technology is also applicable to develop various value added food products (Hayashi, 1990). Microbial flora of seafood could be minimized through this technology and increased the quality of foods with minimum changes in its organoleptic properties (Hoover, 1993). Globally, consumers increasingly demand for convenience foods with highest quality in terms of natural flavor and taste, and which are free from additives and preservatives. This demand has triggered the need for the adoption of non thermal approaches to food processing, of which HPP has proven to be very valuable. So this research work was to investigate the combined effect of high pressure processing and marination over the biochemical and microbiological quality indices and shelf life of the peeled and undeveined Indian white prawn during chilled storage.

Materials and Methods

Raw material

Raw material was collected from Fort Cochin, India and brought to the laboratory in chilled condition. Prawns were washed with potable water and de-shelled manually. The prawn meat was then marinated with a mixture of condiments like chilli powder (1.5% wt), salt (1.25% wt) and turmeric (0.2 % wt) and kept at refrigerated temperature for 30 min. 100 g of marinated sample was vacuum (Sevana Quick Seal Machine, India) packed in EVOH (Ethylene-vinyl alcohol copolymers) packs. After packing the samples were divided into four batches of which one was kept as control and the remaining were subjected to pressurization.

High pressure treatment of samples

Pressurization of the samples was carried out in a high pressure processing equipment (Stansted Fluid Power Ltd, Stansted, Essex, and UKModel No FPG71009/2C). The pressure vessel was of 2 L capacity, 30% monopropylene glycol in water was used as the pressurization fluid. The temperature of high pressure transmission fluid inside the pressure chamber during pressurization was monitored through a K-type thermocouple. Three different pressures 100, 150 and 200 MPa with a pressurization time of 5 min at a temperature of 25 °C were subjected to the different lots of the samples. After depressurization the samples were stored immediately in ice and the samples were drawn periodically for biochemical and microbiological analysis.

Biochemical and microbial analysis

The following biochemical aspects were studied. pH was monitored by using a glass electrode digital pH meter (Cyberscan 510, Eutech Instruments, Singapore) (APHA, 1998). Micro diffusion method has been applied to measure TVB-N 11. (Conway, 1962). TBA value of the sample was determined as described by Tarladgis *et al.* (1960) and FFA by AOCS method (1989). Bacterial load of the samples was counted as per Ryser and Schuman (2013). Sensory analysis was done by a 12 non trained taste panelist using a 9-point hedonic scale (Meilgaard *et al.*, 1999). Two-way analysis of variance (ANOVA) was conducted to find the direct and interaction effect of pressure and storage days on the dependent quality indices. Sampling was done in triplicate and mean values were taken.

Results and Discussion

Changes in pH

Variations in pH of both pressurized and unpressurized samples during storage were depicted in Fig. 1. The initial values for control, 100, 150 and 200 MPa treated sample were 6.95, 6.8, 6.9 and 7.08, respectively. Shamshad *et al.*, (1990) made an attempt to evaluate the changes in pH of untreated fresh prawns stored at temperatures between 0 to 35 °C and they found an increase of pH from 7 to 8.25 on 16th day of storage.

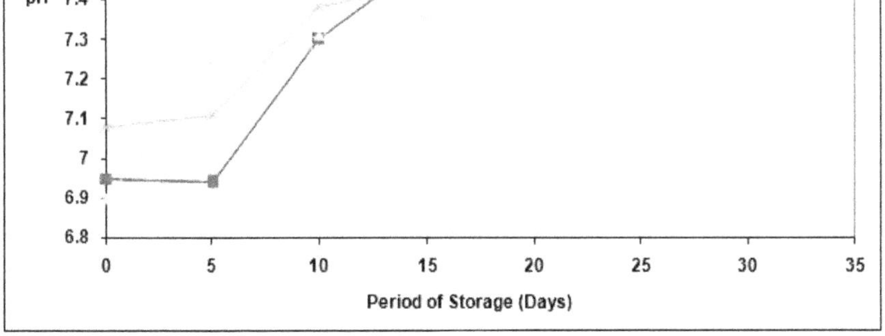

Fig. 1. Changes in pH of pressurized and unpressurized samples during chilled storage

Increasing pH value in all the samples was observed during storage. There is a significant difference of pH content except during 20th and 25th day of storage and 30th and 35th day of storage ($p<0.05$), respectively. The unpressurized sample was rejected after 20th day of storage. This increase of pH of the sample was mainly due to the production of different alkaline volatile compounds (Erkan *et al.*, 2011).

Changes in TVB-N

TVB-N content of pressurized and unpressurized samples was presented in Fig. 2. According to Velankar & Govindan (1958) TVB-N is used as quality index for ice stored shrimps. The initial values of TVB-N for pressurized and 100, 150 and 200 MPa pressure treated samples were 12.00, 11.45, 10.25 and 10.65 mg N_2 $100g^{-1}$, respectively. Statistical analysis revealed that there was no significant difference on the TVB-N values of 100 and 200 MPa treated sample ($p>0.05$), whereas control and 150 MPa treated samples indicated heterogeneous TVB-N content ($p<0.05$). An enhancing trend of TVB-N content was noticed in control and pressurized samples during chill storage. Similar findings have been reported by Bindu *et al.* (2013) during ice storage of pressurized shrimps. Acceptability score of TVB-N content is 30–35 mg N_2 $100g^{-1}$ (Connell, 1995). The unpressurized sample reached the point of rejection after 20 days of storage when the values reached 37.21 mg N_2 $100g^{-1}$. For 100 and 150 MPa pressure treated prawn the values were 27.48 and 25.87 mg N_2 $100g^{-1}$, respectively which have reached a margin of rejection; whereas, for 200 MPa sample showed a content of 28.75 mg N_2 $100g^{-1}$ on 30th day of storage. Higher TVB-N content in the samples might be due to the activity of the endogenous enzymes (Botta *et al.*, 1984) and also the formation of metabolic compounds by the action of microbes.

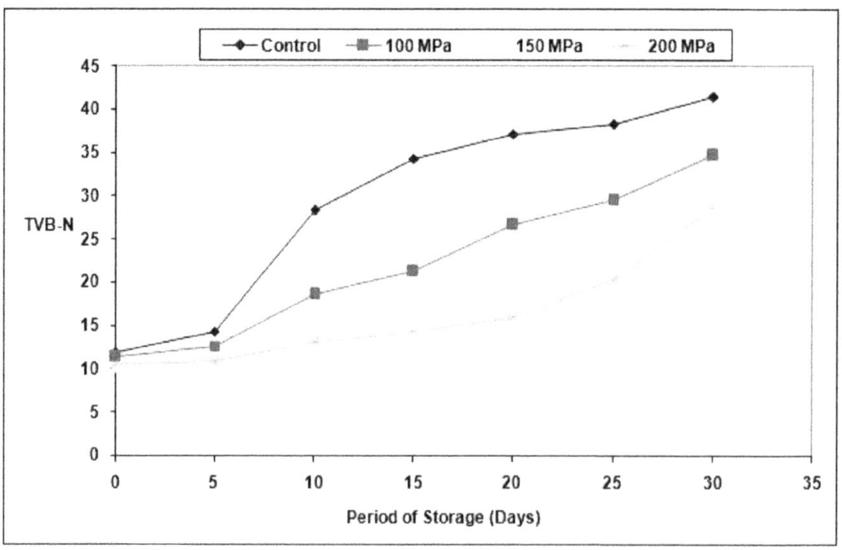

Fig. 2. Changes in TVB-N of pressurized and unpressurized samples during chilled storage

Changes in FFA

The free fatty acid content in pressurized and unpressurized samples showed lower values immediately after high pressure treatment (Fig. 3). Heterogenous FFA content was observed in unpressurized and 200 MPa treated sample ($p<0.05$); whereas on significant difference on FFA content was noticed in 100 and 150 MPa treated samples ($p>0.05$). Initial FFA value was 10.1 mg % oleic acid for unpressurized sample, which was decreased to 8.1, 8.2 and 7.6 mg % oleic acid after pressure treatment of 100, 150 and 200 MPa, respectively. Reduction of FFA content in pressurized sample might be due to the antioxidant effect of the spices in the samples. Period of storage has significant effect on FFA values of all sample ($p<0.05$) and it enhanced during storage. Similar findings were noticed in pressurized carp fillets (Sequeira-Munoz *et al.*, 2006). This could be due to the exposure of the cellular lipid deposition by the breakdown of the cell during pressurization and storage. FFA content in 100, 150 and 200 MPa treated samples were 16.3, 17.4 and 15.88 mg % oleic acid respectively at the end of storage.

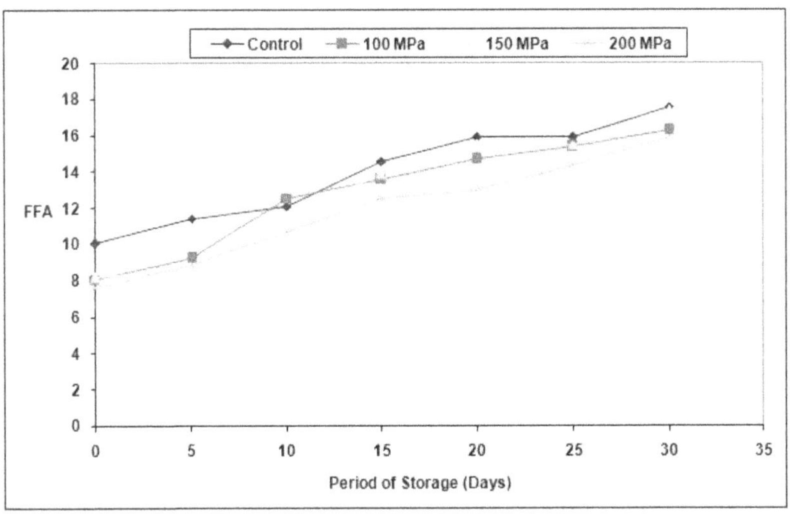

Fig. 3. Changes in FFA of pressurized and unpressurized samples during chilled storage

Changes in TBA

TBA content of the samples were significantly enhanced after pressurization ($p<0.05$) (Fig. 4). Available citations reveal that liberated metal ions from haem protein by the action of pressure causes enhanced lipid oxidation (Cheah and Ledward, 1995). However, combined effect of pressurization and antioxidant properties of chilly and turmeric powder could be the reason for the low levels of TBA in pressurized samples. In addition to this, the changes occurring in the cell membrane during pressure treatment may also make the fish muscle less susceptible to oxidation even when haeme proteins were released (Hultin *et al.*, 1982). However, gradual enhancement of TBA content of the sample has been increased during storage. This is in agreement with Marshall *et al.* (2005). There was no significant difference of TBA content in all sample in each days of storage ($p<0.05$). The level of acceptability TBA content is 1-2 mg malonaldehyde kg^{-1} (Goulas and Kontominas, 2007). TBA content was in the acceptable limit even after 30 days of storage this might be due to the low fat content in prawns. Since prawns have a low fat content when compared to other fish species the TBA levels did not reach the upper exceeding limit during the 30 days of storage.

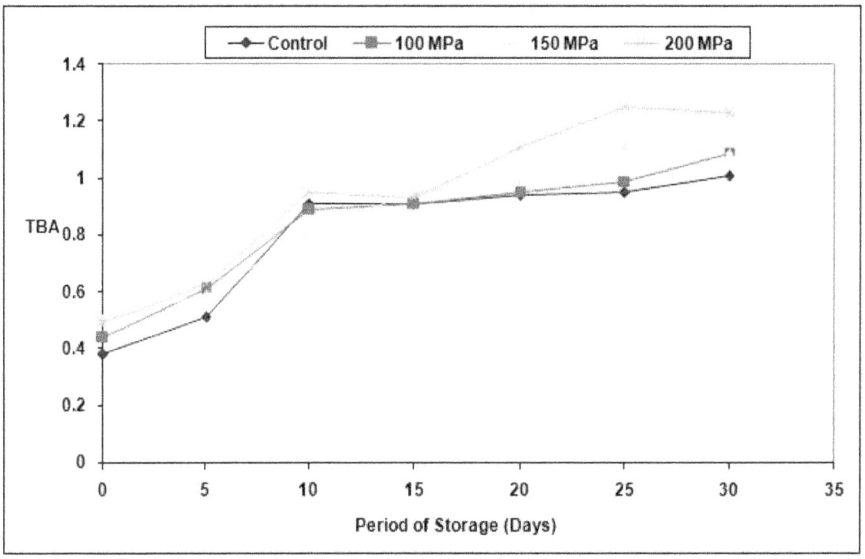

Fig. 4. Changes in TBA of pressurized and unpressurized samples during chilled storage

Changes in total plate count

Total bacterial load of unpressurized sample was 4.8 Log_{10} CFU/g. Average bacterial load of prawns in India was 10^4 -10^7 (Jacob *et al.*, 1962). Significant reduction of TPC was noticed after HP-treatment ($p<0.05$) (Fig. 5). Bacterial load of 200, 150 and 100 MPa treated samples was 3.8, 4.0 and 4.2 Log_{10} CFU/g, respectively. Less bacterial count in pressurized sample could be due to impermeable cell membranes (Houjaij, 2000). HP-treatment also denatures enzymes and nucleic acids necessary for biological activities and reproduction of bacterial cell. Cell division of bacteria was retarded with the application of pressure (Zobell, 1964) and intracellular vacuoles collapse at pressures level less than 88 psig (Walsby, 1972). Total plate count of all samples enhanced with period of storage and heterogonous bacterial load in the samples for each day of storage ($p<0.05$). Similar findings was observed in pressurized abalone stored at chilled condition (Briones *et al.*, 2010). As per the limit of ICMSF (1986) unpressurized sample, 100, 150 and 200 MPa samples were rejected on 20, 22, 28 and 30th day of storage. Similar trend has been observed by Montero *et al.* (2001).

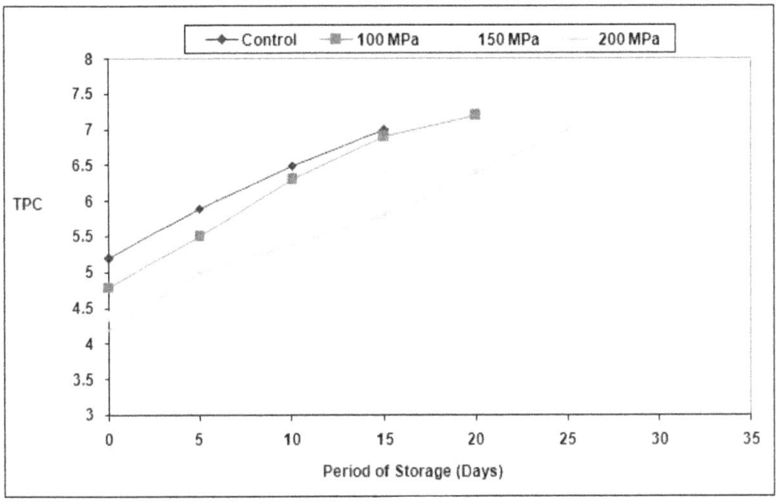

Fig. 5. Changes in TPC of pressurized and unpressurized samples during chilled storage.

Changes in overall acceptability score

Sensory methods are universally applied for the estimation of quality of food in the markets, retail shops and table of the consumers. Sensory evaluation is one of the oldest and most widespread means of evaluating the acceptability and edibility which can be defined as the scientific discipline used to evoke, measure, analyze and interpret the characteristics of food as perceived by the senses of sight, smell, taste and touch. Variations in overall acceptability scores of pressurized and unpressurized samples during storage are presented in Fig. 6. Heterogeneous overall acceptability score was showed

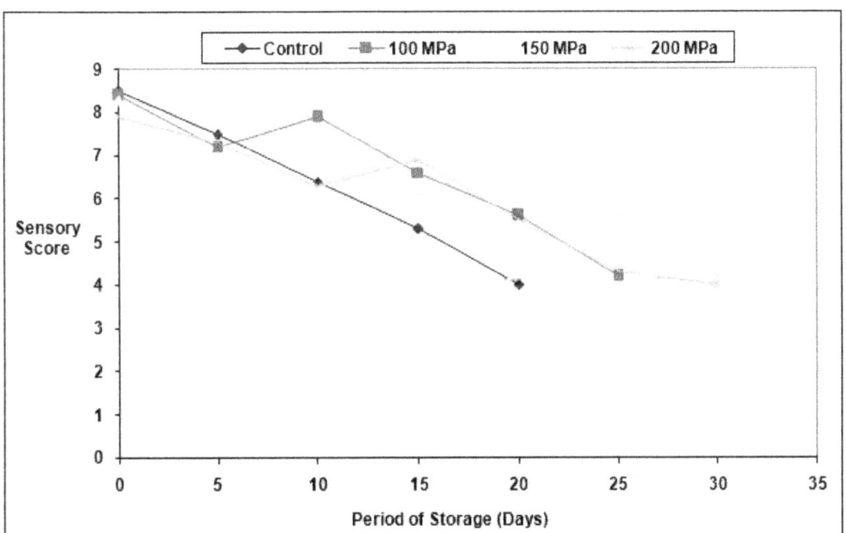

Fig. 6. Changes in Overall acceptability score of pressurized and unpressurized samples during chilled storage

in all samples after pressurization ($p<0.05$). Unpressurized sample showed a faster rejection rate compared to HP-treated samples. Overall acceptability the samples was reduced with the period of storage ($p<0.05$). In the present study the unpressurized samples showed a shelf life of 20 days and then it rejected due to off taste and flavour. Erkan *et al.* (2011) found 14 days shelf life extension in pressurized cold smoked salmon than control samples stored at 2 °C.

Conclusion

This study demonstrated that pressurization techniques delays biochemical and microbiological spoilage and extend the shelf life of ready to cook marinated prawn during iced storage. Biochemical and microbiological quality indices such as pH, TVB-N, FFA, TBA, TPC and sensory score of the samples had shown significant changes with pressure and storage periods. Unpressurized control sample ie., prawns treated only with spice condiments had an extended shelf life, which further improved by the application of pressurization. Pressurization improves the impregnation of condiments into the prawns. Hence it can be concluded that the microbial quality of the samples improvised by the effect of pressurization and marination. Hence, this food processing technology enhances flavours and quality of the seafood by impregnation of different types of spices and flavors.

Acknowledgements

The authors acknowledge the financial assistance provided by the National Agricultural InnovationProject (NAIP), Indian Council of Agricultural Research [grant number: NAIP/C4/C-30027/2008-09] for carrying out this work.

References

AOCS, 1989. Official methods and recommended practices of American Oil Chemists Society-(5th Edn.), AOCS, Champaign, USA.

APHA, 1998. Standard Methods for Examination of Water and Water Waste. 20th Edn., American Public Health Association, American Water Works Association, Water Environmental Federation, Washington, D.C.

Bindu, J., Ginson, J., Kamalakanth, C.K., Asha, K.K. and Srinivasa Gopal, T.K., 2013. Physico-chemical changes in high pressure treated Indian white prawn (*Fenneropenaeus indicus*) during chill storage. Food Sci & Emer. Techn. 17: 37–42.

Botta, J.R., Lauder, J.T. and Jewer, M.A., 1984. Effect of methodology on total volatile basic nitrogen (TVB-N) determination as an index of quality of fresh Atlantic cod (*Gadus morhua*). J. of Food Sci. 49(3): 734–736.

Briones, L.S., Reyes, J.E., Tabilo-Munizaga, G.E. and Pérez-Won, M.O., 2010. Microbial shelf-life extension of chilled Coho salmon (*Oncorhynchus kisutch*) and abalone (*Haliotis rufescens*) by high hydrostatic pressure treatment. Food Cont., 21: 1530-1535.

Burke, R.M. and Monahan, 2003. The tenderization of shin beef using a citrus juice marinate. Meat Sci., 63: 161-168.

Cheah, P.B. and Ledward, D.A., 1995. High-pressure effects on lipid oxidation. J. American Oil Chemists' Soci. 72(9):1059–63.

Connell, J. J., 1995. Control of Fish Quality (4th edn.), Fishing News Books Limited., London.

Conway, E.J., 1962. Microdiffusion Analysis and Volumetric Error (5th edn.). Parch Goskey and Sockwood, London.

Erkan, N., Ürtener G., Alpas, H., Selçuk, A., Ozden, O. and Buzrul S., 2011. The effect of different high pressure conditions on the quality and shelf life of cold smoked fish. Food Sci. and Emer. Techn. 12(2): 104–110.

Goulas, A.E. and Kontominas, M.G. 2007. Combined effect of light salting, modified atmosphere packaging and oregano essential oil on the shelf-life of sea bream (*Sparus aurata*): biochemical and sensory attributes. Food Chem., 100: 287-296.

Hayashi, R., 1990. Application of high pressure to processing and preservation: philosophy and development. In: W.E.L. Spiess, and H. Schubert (Eds.), Engineering and Food, pp 815-826, Elsevier Applied Science, London.

Hoover, D.G., 1993. Pressure effects on biological systems. Food Tech. 47: 150–155.

Houjaij, N., 2000. The application of high pressure treatment and its effect on the quality attributes. M.Sc Thesis, pp. 14-28, Department of Food Science and Agricultural Chemistry, Macdonald Campus of McGill University, Montreal, Quebec, Canada.

Hultin, H.O., Mc-Donald, R.E. and Kelleher, S.D. 1982. Lipid oxidation in fish muscle microsomes. In: R.E. Martin., G.J. Flick, C.E. Hebard, and D.R. Ward. (Eds.). Chemistry and Biochemistry of Marine Food Products, AVI Publishing Co., Westport, CT.

ICMSF, 1986. Microorganisms in foods 2. Sampling for food microbiological analysis: Principle and Specific Applications, 2nd Edn., University of Toronto, Canada.

Ignacio, O., Alicia, R., Gipsy, T.M., Mario, P.W. and Santiago, P.A., 2010. Effect of hydrostatic high-pressure treatment on proteins, lipids and nucleotides in chilled farmed salmon (*Oncorhynchus kisutch*) muscle. Euro. Food Res. Tech. 230: 925–934.

Jacob, S.S., Iyer, K.M., Nair, M.R. and Pillai, V.K., 1962. Quality studies on round, headless and peeled and deveined prawns held in ice storage. Indian J. Fish., 9(2): 97.

Marshall, M.R., Kristinsson, H. and Balaban, M.O. 2005. Effect of high pressure treatment on Omega-3 fatty acids in fish muscle. Report, University of Florida.

Meilgaard, M., Civille, G.V. and Carr, B.T., 1999. Sensory evaluation techniques (3rd Edn.). CRC Press, Boca Raton.

Montero, P., Lopez-Caballero, M.E. and Perez-Mateos, M., 2001. The effect of inhibitors and high pressure treatment to prevent melanosis and microbial growth on chilled prawns (*Penaeus japonicus*). J. Food Sci. 66(8): 1201-1206.

Ryser, E.T. and Schuman, J., 2013. Aerobic plate count. In: Compendium of Method for the Microbiological Examination of Foods. American Public Health Association, Washington, D.C, USA. doi: 10.2105/MBEF.0222.013.

Sequeira-Munoz, A., Chevalier, D., LeBail, A., Ramaswamy, H.S., Simpson, B.K. 2006. Physicochemical changes induced in carp (*Cyprinus carpio*) fillets by high pressure processing at low temperature. Food Sci. Emer. Techn. 7: 13–18.

Tarladgis, B.G., Watts, B.M. and Yonathan, M., 1960. Distillation method for the determination of malonaldehyde in rancid foods. J. American Oil Che. Soc. 37(1): 44–48.

Velankar, N.K. and Govindan, T.K., 1958. Preliminary study of the distribution of non-protein nitrogen in some marine fishes and invertebrates, Proceedings of the Indian Academy of Science, Vol. 1+7B: 202-209.

Walsby, A.E., 1972. Gas-filled structures providing buoyancy in photosynthetic organisms. In: M.A., Sleigh and A.G., Macdonald (Eds.). The Effects of Pressure on Living Organisms, Academic Press, New York.

Zobell, C.E. 1964. Hydrostatic pressure as a factor affecting the activities of marine microbes. In: Recent Researches in the Fields of Hydrosphere, Atmosphere, and Nuclear Geochemistry, Maruzen, Tokyo.

Impact of Climate Change on Hydrological Cycle, Ecosystem, Fisheries and Food Security, Pages: 497-508
Edited by: B. Madhusoodana Kurup, M.R. Boopendranath, M. Harikrishnan and A.V. Shibu

CHAPTER - 42

Biochemical Characteristics of Pepsins from two Cichlids, *Etroplus suratensis* and *Oreochromis mossambicus*

H.S. Hari Sankar[1&2*], Babu Philip[1] and M. Harikrishnan[2]

[1]*School of Industrial Fisheries,*
Cochin University of Science and Technology, Cochin-682016, Kerala, India
[2]*Department of Marine Biology, Microbiology and Biochemistry,*
Cochin University of Science and Technology, Cochin-682016, Kerala, India
**E-mail: harisankarhs@gmail.com*

Acid proteases are enzymes that control various key processes in many organisms. Acidic proteases like cathepsins D and E were stored in lyzosomes within the cell whereas others like pepsin, cyprosin and endothiapepsin are form mammals, plants and fungal origins respectively were secreted out. Among them, pepsins constitute a widely distributed enzyme super family that performs protein degradation in highly acidic conditions (Davies, 1990). In the history of evolution, pepsin is considered as the primary enzyme in the aspartic protease family. It has been identified in all vertebrates without exception and it is the principal proteolytic enzyme of vertebrate gastric fluid. The chief cells present in the fundic glands of stomach secretes pepsin as an inactive enzyme (zymogen or pro-enzyme) called pepsinogen. In fishes, two categories of zymogens designated as pepsinogen A and C were identified (Feng *et al.*, 2008; Tanji *et al.*, 2009). These inactive pro-proteins are automatically converted to active enzyme pepsin at below pH 6.0 in the stomach. In fishes, two categories of pepsin forms, I and II were identified. Species similar to sardine (*Sardinops melanostica*) and capelin (*Mallotus villosus*) posses Pepsin I (Noda and Murakami, 1981; Gildberg, 1983) and species like cod (*Gadus morhua*) and salmon (*Oncorhynchus keta*) have pepsin II in abundance (Gildberg, 1988; Sanchez-Chiang *et al.*, 1987). Castillo-Yanez (2004) isolated an aspartic protease similar to pepsin II class from Monterey sardine viscera. Commercially, pepsin is prepared from swine stomachs. In the leather industry, before tanning, fur and residual tissue from animal hides were eliminated by use of crude pepsin (Choudhary *et al.*, 2004). Pethick and Money Ltd.

(2002) patented a method to recover silver in photographic films by using pepsin to digest gelatine coat embedded with silver compounds for photo reaction. Fish wastes have extensively been exploited as ingredients in commercial feeds and also organic fertilizers (Gildberg and Raa, 1977; El-Beltagy *et al.*, 2004). Extraction of enzymes like trypsin, chymotrypsin, elastase and valuables such as collagen, ù-3 fatty acids from fish viscera (Yoshida *et al.*, 1999; Genicot *et al.*, 1996; Klomklao *et al.*, 2006a; 2006b; Castillo-Yαñez *et al.*, 2006; Gildberg and Overbo, 1990; Nalinanon *et al.*, 2007) were reported. However, conventional fish waste disposal techniques are inadequate for effective recovery of valuable ingredients from fish wastes.

Pepsin is one of the plenteous among these valuable products that can be recovered and purified from fish viscera and which be used in industrial processes like collagen extraction (Benjakul *et al.*, 2009), gelatin extraction (Nalinanon *et al.*, 2008) and cheese making (Aehle, 2007). Even though fish pepsin based cheese manufacturing is not commercialized, use of pepsins from cod and tuna were demonstrated in production of cheese (Han, 1993; Tavares *et al.*, 1997; Gildberg, 1992; Aehle, 2007). Biochemical characterization of *E. suratensis* and *O. mossambicus* pepsins is one of the objectives of this present work. Denaturation of proteins will lead to impaired biological activity and kinetics of pepsin denaturation due to temperature is another objective. pH affects enzyme stability there by its performance. This study is designed to account the change in pepsin stability during variation in pH.

Materials and Methods

Isolation of pepsin

Live, mature *E. suratensis* and *O. mossambicus* of comparable size (10-12cm) were purchased from the Fisheries station, Kerala University of Fisheries and Ocean Studies, Puthuvyppu. The fishes were acclimated in laboratory condition for seven days before the experimental analysis. A commercial, pelleted fish feed (Godrej Agrovet P Ltd) was given *ad libitum.* Before sampling, the fishes were subjected to 12 h starvation and sacrificed by providing a cold shock. The fishes were dissected instantly and tissues homogenate was prepared as illustrated in Hari Sankar (2015). The homogenate was then centrifuged at 4°C at 10,000*g* for 10 min. The protein content of homogenate was estimated by Lowry *et al* (1951).

Pepsin specific activity

The specific activity was determined by Anson's method (1938) by using haemoglobin as substrate. The optical density after pepsin hydrolysis reaction was analysed at 280nm. A standard tyrosine curve was prepared and the specific enzyme activity was expressed in Anson Unit (mMole of Tyr/ min/ mg protein).

Biochemical characterization of pepsin

In order to obtain optimum temperature for pepsin hydrolysis, the reactants were incubated at an interval of 5^0C temperatures from 20^0C to 80^0C. Curve depicting specific activity against temperature was plotted for determining the optimum temperature. Similarly, by using different non- reactive buffer systems commonly utilized to attain specific pH values. A plot of activity against pH to obtain the optimum pH was drawn. A series of haemoglobin concentrations ranging from 0.5% to 3% with pH 2 was used to analyze the optimum substrate concentration and an activity against concentration graph was constructed.

Thermal stability of pepsin

Tissue homogenates were kept at different temperatures and from the incubated enzyme, samples were taken at time intervals of 0, 5, 10...and 60 minutes and analyzed the specific activity for understanding the thermal inactivation kinetics of peptic hydrolysis.

pH Stability

Non-reacting buffers of pH 1 to 7 were mixed with equal volumes of tissue homogenate and subsequent incubation for an hour prior to specific activity was estimated. The stability of enzymes at various pH was expressed as relative activity.

Results and Discussion

The specific activity of pepsin extracted from *E. suratensis* varied from 0.43±0.29 Anson's unit at pH 1.0 to to a maximum of 1.39±0.03 at pH 4.50 (Fig.1.). The specific activity recorded at different pH levels differed significantly ($F= 8.01$, $p<0.01$). Mean specific activity was perceptibly higher at pH 4.0, 4.5 and 5.0. In *O. mossambicus*, the specific activity values recorded at pH values from 1.0 to 5.0 were significantly different ($F=5.44$, $p<0.01$). High mean specific activity values were recorded at pH 2.0 to 4.0 and the highest activity was recorded at a pH of 2.0 (3.40±0.26) (Fig. 2.). Fig. 3 and Fig. 4 revealed the effect of haemoglobin substrate concentration on pepsin from *O. mossambicus* and *E. suratensis* respectively. The pepsin activity curves attained substrate saturation at different haemoglobin concentration, specifically, *O. mossambicus* at 1.5% and *E. suratensis* at 2%. Relatively higher activity was shown at pH 2-3 and over 50% of the activity was lost pH higher than 4. The temperature optimum of *O. mossambicus* pepsin was 40°C (Fig. 5). Up to 45°C it was stable and subsequent increase in temperature decreased the enzyme activity and above 55°C the enzyme was not stable (Fig. 6). The maximum *E. suratensis* pepsin activity was obtained over temperature 35 to 45°C (Fig. 7). Similar to *O. mossambicus*, *E. suratensis* pepsin was stable up to 45^0C and it is not stable beyond 50°C. The activity of pepsin lost on further increase in temperature (Fig. 8.). At pH 4, the tilapia pepsin showed its maximum stability and the stability has been decreased to 80% in pH 2.5 and 5-5.5 (Fig. 9.). It was exhibited a decreased stability at higher pH values. Fig.10 depicts the pH stability of *E.* suratensis pepsin. The figures 6, 8, 9 and 10 were adopted from Hari Sankar (2015).

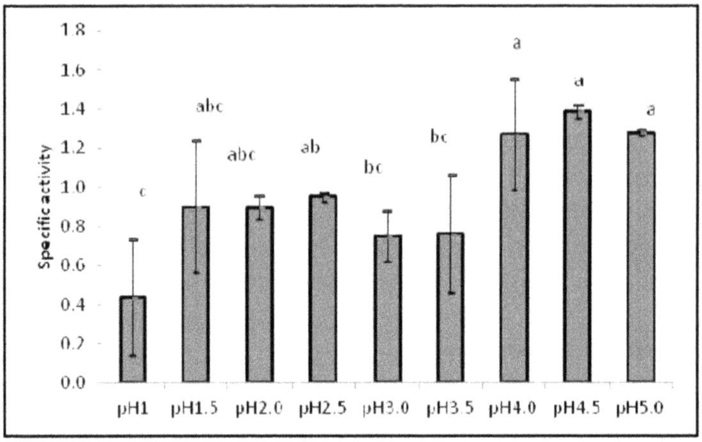

Fig. 1. Effect of pH on specific activity of pepsin isolated from *E. suratensis*

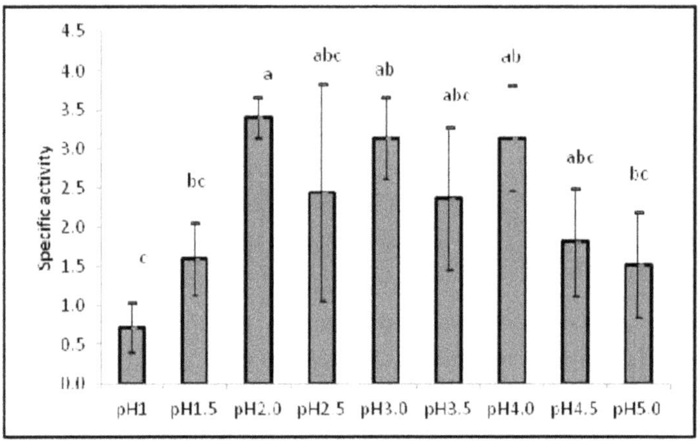

Fig. 2. Effect of pH on specific activity of pepsin from *O. mossambicus*

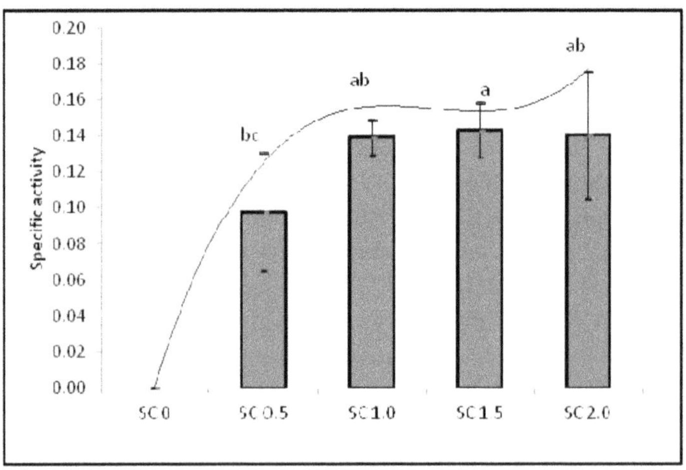

Fig. 3. Effect of substrate concentration on specific activity of pepsin separated from *O. mossambicus*

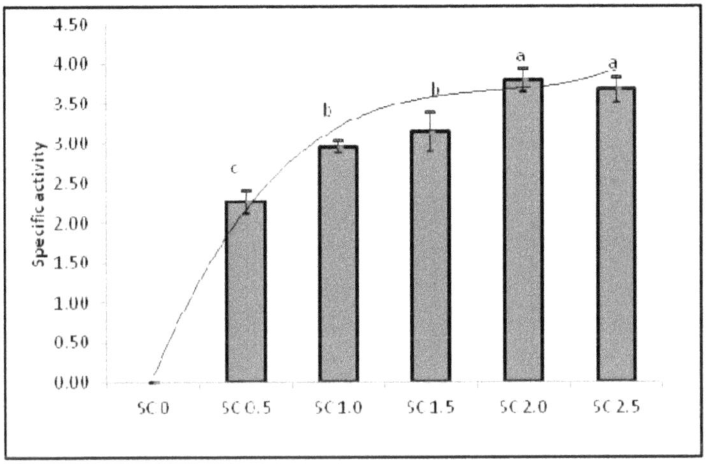

Fig. 4. Effect of substrate concentration on specific activity of pepsin obtained from *E. suratensis*

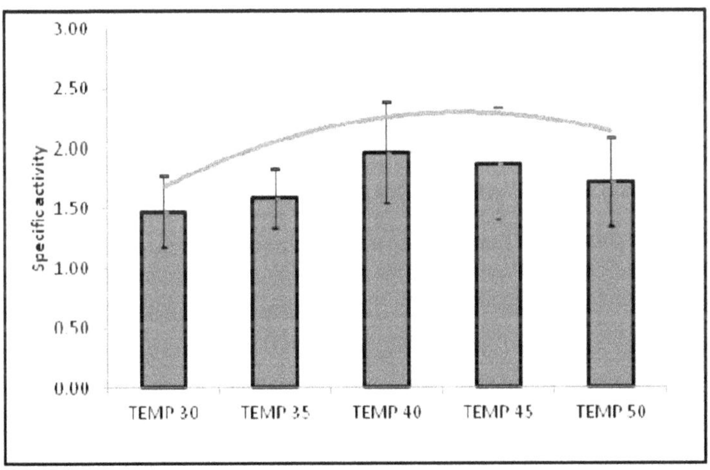

Fig. 5. Effect of temperature on specific activity of pepsin from *O. Mossambicus*

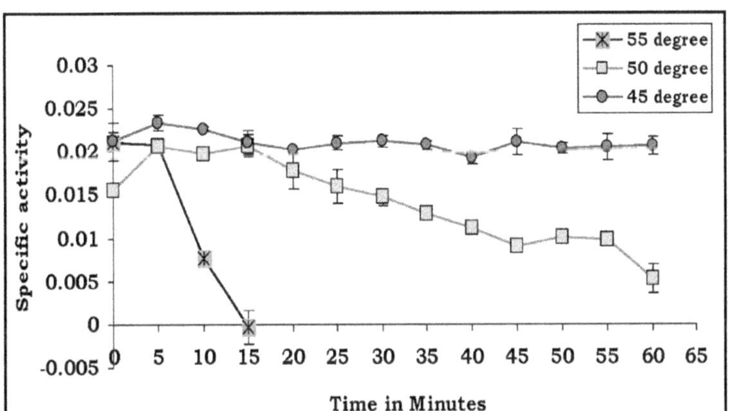

Fig. 6. Thermal inactivation kinetics of pepsin from *O. mossambicus* (Source: Hari Sankar, 2015).

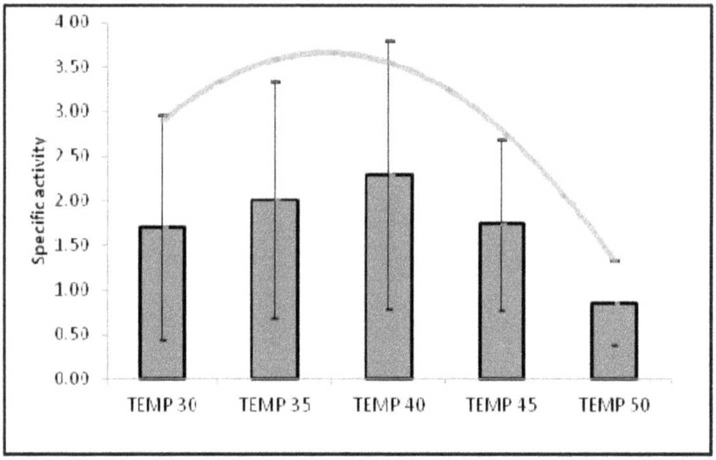

Fig. 7. Effect of temperature on specific activity of pepsin extracted from *E. suratensis* (Source: Hari Sankar, 2015).

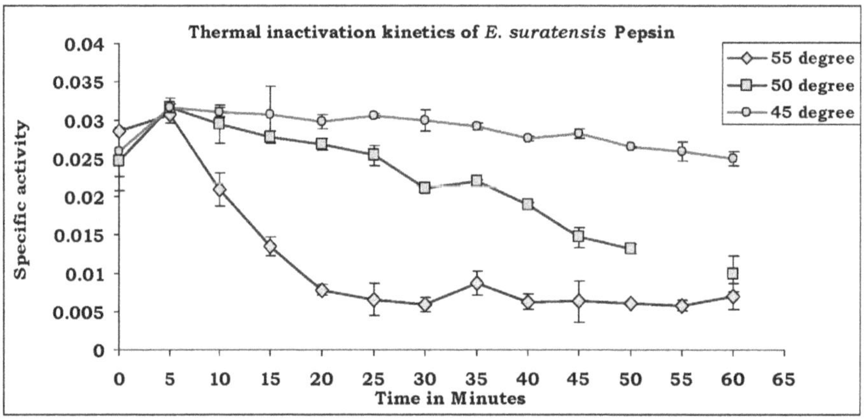

Fig. 8. Thermal inactivation kinetics of pepsin of pepsin extracted from *E. suratensis* (Source: Hari Sankar, 2015).

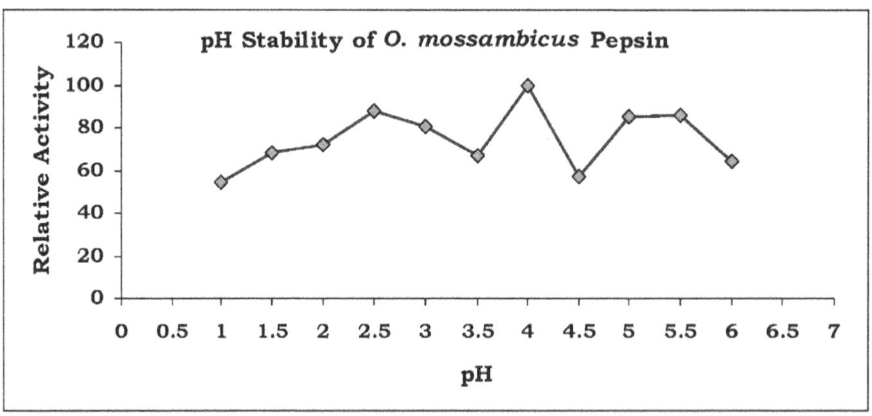

Fig. 9. pH stability of pepsin extracted from *O. mossambicus* (Source: Hari Sankar, 2015).

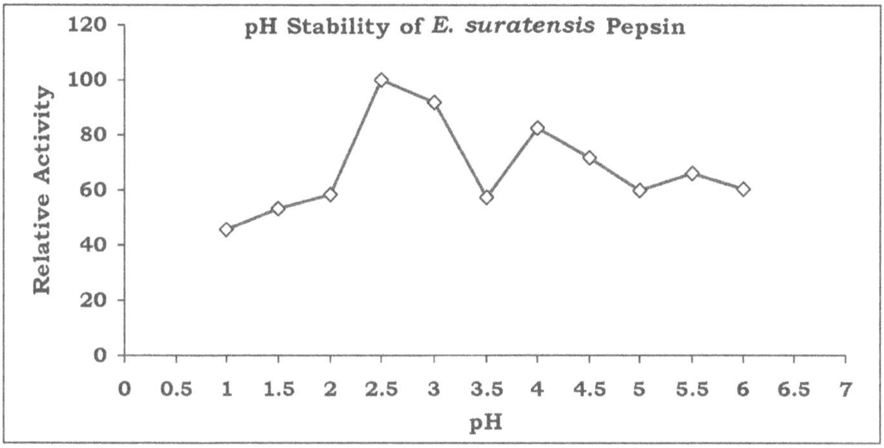

Fig. 10. pH stability of pepsin extracted from *E. suratensis* (Source: Hari Sankar, 2015).

Acid proteases have gained attention of many researches due to variety of functions, from digestion to disease transmission, performed by them. Various infectious disease caused by organisms like human immune deficiency virus (HIV) utilises acid proteinases for their propagation and researchers are focusing on activity, kinetics and inhibitors of such proteases nowadays (Campos, 2003). Recently several investigations were designed to describe digestive protease activities and to discover practical applications for digestive enzymes (Jiang *et al.*, 1991; Garcýa-Carreno, 1992; Haard, 1992). Studies on protease activity of various marine and fresh water organisms have been employed to facilitate cost effective feeds for better aquaculture practice of such species through conscientious awareness about their digestive capabilities (Clark *et al.*, 1985; Alarcon *et al.*, 1998). The identification of feed components (especially the proteins with essential amino acids) with comparatively higher digestibility can be made easier by this practice (Lan and Pan, 1993, Eshel *et al.*, 1993). Pepsin, an aspartic acid protease, belongs to the largest family of acid proteases.

pH is known to have ample influence on the specific activity of fish pepsin. The pH ranges of optimum enzyme activity and better enzyme stability are very important for enzymes like pepsin while considering its possibility of using as an industrial enzyme. The presence of two optimal pHs in the crude tissue homogenates (Fig. 1 and Fig. 2) clearly indicates the occurrence of two distinct pepsins, Pepsin I and Pepsin II. Bjelland *et al.* (1988) identified the existence of Pepsin I (more active in relatively weak acidic condition) and II (active in strong acidic environment). Mammalian pepsin highly active at pH of 1.5 –2.0 however, many pepsins from fish viscera revealed optimum activity at less acidic conditions (Gildberg, 1988; Gildberg, 2004). Pepsin A1 and pepsin A2 isoenzymes from rock cod *Trematomusbernacchii* displayed their optimum at pH 2.5 and 2.0 respectively and above these optimal pH, a slow decline of their relative activity was observed (Brier *et al.*, 2007). Nalinanon *et al.* (2008) identified only one type of pepsinogen and pepsin in Albacore tuna and skip jack tuna (Nalinanon *et al.*, 2009) in contrast to two isoforms of pepsins from Tongol tuna.

Acidic protease from the stomach of Nile tilapia (*Tilapia nilotica*) displayed an optimal pH at 2.5 and better stability around pH 2 to 6 (El-Beltagy *et al.*, 2004). Pepsin from Pacific yellowfin tuna *Thunnus albacares* has an optimum pH of 2.5 and differ in their crystalline state, specificity, stability in alkali like properties from other pepsins (Norris and Mathies, 1953; Northrop, 1931). Two to three major pepsins with pH optima around 2 and 4 was identified in many fishes (Gildberg, 1983). Acid proteases from fish consist of cathepsin D, an intracellular aspartic proteases that plays a significant function in yolk protein digestion during development of eggs (Romano *et al.*, 2004). Arowana fish (*Scleropages formosus*) pepsin have pH optima 1.5–2.0 (Natalia *et al.*, 2004).

Another factor that influences the activity of fish pepsin is temperature. Depending on the species and nature of habitat temperature for optimum enzyme activity varies (Gildberg, 1988; Pavlisko *et al.*, 1997; Shahidi and Janak Kamil, 2001). In contrast with high values of optimum temperature for pepsins from warm water habitats, a lower temperature optimum has been displayed by fishes from colder waters (Bjelland *et al.*, 1988; Noda and Murakami, 1981; Squires *et al.*, 1986; Sanchez-Chiang *et al.*, 1987). Cold water fish enxymes have a low Arrhenius activation energy as a proof of their low optimal temperature and high heat labiality in contrast with warm water counter parts (Simpson and Haard, 1987). The scheme of temperature inactivation for *O. mossambicus* and *E. suratensis* pepsins exhibited similar characteristics. Both pepsin activities were quite stable up to 45°C and get inactivated very quickly at 55°C (Fig. 7 and Fig. 8). The sharp decrease of thermal stability was attributed to the obliteration of tertiary structure of enzymes and denaturation of pepsin (Nalinanon *et al.*, 2009).

Conclusion

The results can be reviewed as the pepsin from *E. suratensis* was active with maximum velocity over a broad range of optimum pH and temperature when compared to that of *O. mossambicus*. Pepsins from these fishes displayed similar thermal stability. The gastric pepsin from both fishes exhibited similar temperature stability. The *O. mossambicus* pepsin is able to produce higher specific activity under a variety of physiochemical conditions.

References

Aehle, W., 2007. Enzymes in industry: production and applications (3rd Edn.), pp 136-137, Wiley-VCH, Weinheim, Germany.

Alarcon, F.J., Dýaz, M., Moyano, F.J.and Abellan, E., 1998. Characterization and functional properties of digestive proteases in two sparids; gilthead seabream (*Sparus aurata*) and common dentex (*Dentex dentex*). Fish Physiol Biochem., 19: 257–267. doi: 10.1023/ A:1007717708491.

Anson, M., 1938. The Estimation of Pepsin, Trypsin, Papain and Cathepsin with Hemoglobin, J. Gen. Physiol. 22: 79-89.

Benjakul, S., Klomklao, S. and Simpson, B., 2009. Enzymes in fish processing, In: Enzymes in Food Technology (Whitehurst, R.J., Oort, M.V., Eds.), pp. 211-225, John Wiley & Sons, Chichester, UK.

Bjelland, S., Gildberg, A. and Volden, G., 1988. Degradation of human epidermal keratin by fish pepsin. Arch. Dermatol. Res., 280(2): 119-123. doi:10.1007/BF00417716

Brier, S., Maria, G., Carginale, V., Capasso, A., Wu, Y., Taylor, R.M., Borotto, N.B., Capasso, C. And Engen, J.R.., 2007. Purification and characterization of pepsins A1 and A2 from the Antarctic rock cod *Trematomus bernacchii*. FEBS J. 274(23): 6152–6166. doi:10.1111/j.1742-4658.2007.06136.x

Campos, L., Sancho, J., 2003. The active site of pepsin is formed in the intermediate conformation dominant at mildly acidic pH. FEBS Lett. 538(1), 89-95.

Castillo-Yañez, F.J., Pacheco-Aguilar, R. and Garcia-Carreño, F.L., 2006. Purification and biochemical characterization of chymotrypsin from the viscera of Monterey sardine (*Sardinops sagax caeruleus*). Food Chem., 99: 252-259.

Castillo-Yañez, F.J., Pacheco-Aguilar, R., Garcia-Carreño, F.L., Toro, M.D.L.A.N., 2004. Characterization of acidic proteolytic enzymes from monterey sardine (*Sardinops sagax caerulea*) viscera. Food Chem. 85(3): 343-350. doi:10.1016/j.foodchem. 2003.07.008

Choudhary, R.B., Jana, A.K. and Jha, M.K., 2004. Enzyme technology applications in leather processing. Indian J. Chem. Technol., 11: 659-671.

Clark, J., MacDonald, N.L. and Stark, J.R., 1985. Metabolism in marine flatfish-III. Measurement of elastase activity in the digestive tract of dover sole (*Solea solea* L.). Comp. Biochem. Physiol. B. 81(3): 695-700.

Davies, D.R., 1990. The structure and function of the aspartic proteinases. *Ann. Rev.* Biophys. and Biophys. Chem. 19: 189-215.

El-Beltagy, A.E., El-Adawy, T.A., Rahma, E.H. and El-Bedawey, A.A., 2004. Purification and characterization of an acidic protease from the viscera of bolti fish (*Tilapia nilotica*). Food Chem. 86(1): 33-39. doi: 10.1016/j.foodchem.2003.08.009

Eshel, A., Lindner, P., Smirnoff, P., Newton, S. and Harpaz, S., 1993. Comparative study of proteolytic enzymes in the digestive tracts of the European sea bass and hybrid striped bass reared in freshwater. Comp. Biochem. Physiol. A. 106: 627-634.

Feng, S., Li, W. and Lin, H., 2008. Characterization and expression of the pepsinogen C gene and determination of pepsin-like enzyme activity from orange-spotted grouper (*Epinephelus coioides*). Comp. Biochem. Physiol. B 149(2): 275-284. doi:10.1016/j.cbpb.2007.09.017

Garcia-Carreno, F.L., 1992. Protease inhibition in theory and practice. Biotechnol. Educ. 3: 145-150.

Genicot, S., Rentier-Delrue, F., Edward, D., VanBeeumen, J. and Gerday, C., 1996. Trypsin and trypsinogen from an Antarctic fish: molecular basis of cold adaptation. Biochim. Biophys. Acta. 1298(1): 45-57. doi: 10.1016/S0167-4838(96)00095-7

Gildberg, A. and Overbo, K., 1990. Purification and characterization of pancreatic elastase from Atlantic cod (*Gadus morhua*). Comp. Biochem. Physiol. B, 97(4): 775-782. doi: 10.1016/0305-0491(90)90122-A.

Gildberg, A., 1983. Purification and characterization of pepsins from the arctic fish capelin. Comp. Biochem. Physiol. A 75(3): 337-342. doi:10.1016/0300-9629(83)90090-7

Gildberg, A., 1988. Aspartic proteinases in fishes and aquatic invertebrates. Comp. Biochem. Physiol. B, 91: 425-435.

Gildberg, A., Raa, J., 1977. Properties of a propionic acid/formic acid preserved silage of cod viscera. J. Sci. Food. Agric., 28(7): 647-653. doi: 10.1002/jsfa.2740280713

Gildberg, A.R., 2004. Enzymes and bioactive peptides from fish waste related to fish silage, fish feed and fish sauce production. J. Aquat. Food Prod. Technol. 13(2): 3-11.

Haard, N.F., 1992. A review of proteolytic enzymes from marine organisms and their application in the food industry. J. Aquat. Food Prod. Technol., 1: 17-36. doi:10.1300/J030v01n01_05

Hari Sankar, H.S., 2015. The Major Digestive Enzymes in *Etroplus suratensis* and *Oreochromis mossambicus* - Distribution and Characteristics. Ph.D. Thesis, Cochin University of Science and Technology, Cochin.

Jiang, S., Moody, M. and Chen, H., 1991. Purification and characterization of proteases from digestive tract of grass shrimp (*Penaeus monodon*). *J. Food Sci.*, 56: 322-326.

Klomklao, S., Benjakul, S., Visessanguan, W., Kishimura, H. and Simpson, B.K., 2006a. Purification and characterization of trypsin from the spleen of tongol tuna (*Thunnus tonggol*). J. Agr. Food Chem., 54(15): 5617–5622. doi: 10.1021/jf060699d.

Klomklao, S., Benjakul, S., Visessanguan, W., Kishimura, H., Simpson, B.K. and Saeki, H. 2006b. Trypsins from yellowfin tuna (*Thunnus albacares*) spleen: Purification and characterization. Comp. Biochem. Physiol. B, 144(1): 47-56.

Lan, C.C. and Pan, B.S., 1993. In vitro digestibility simulating the proteolysis of feed protein in the midgut gland of grass shrimp (*Penaeus monodon*). Aquaculture 109: 59-70.

Lowry, O.H., Rosebrough, N.J., Fan, A.L. and Randall, R.J., 1951. Protein measurement with Folin phenol reagent. J. Biol. Chem., 193(1): 256–275.

Nalinanon, S., Benjakul, S., Visessanguan, W. and Kishimura, H., 2007. Use of pepsin for collagen extraction from the skin of bigeye snapper (*Priacanthus tayenus*). Food Chem., 104(2): 593-601. doi: 10.1016/j.foodchem.2006.12.035.

Nalinanon, S., Benjakul, S., Visessanguan, W. and Kishimura, H., 2008. Improvement of gelatin extraction from bigeye snapper skin using pepsin-aided process in combination with protease inhibitor. Food Hydrocoll., 22(4): 615-622. doi: 10.1016/j.foodhyd. 2007.01.012.

Nalinanon, S., Benjakul, S., Visessanguan, W. and Kishimura, H., 2009. Partitioning of protease from stomach of albacore tuna (*Thunnus alalunga*) by aqueous two-phase systems. Process Biochem., 44: 471-476. DOI:10.1016/j.procbio.2008.12.018.

Natalia, Y., Hashim, R., Ali, A., Chong, A., 2004. Characterization of digestive enzymes in a carnivorous ornamental fish, the Asian bony tongue *Scleropages formosus* (Osteoglossidae). Aquaculture. 233, 305–320.

Noda, M. and Murakami, K., 1981. Studies on proteinases from the digestive organs of sardine. II. purification and characterization of two acid proteinases from the stomach. Biochim. Biophys. Acta., 658(1): 27-34. doi: 10.1016/0005-2744(81)90246-1.

Norris, E.R. and Mathies, J.C., 1953. Preparation, properties and crystallization of tuna pepsin. J. Biol. Chem., 204: 673-680.

Northrop, J.H., 1931. Crystalline pepsin III. preparation of active crystalline pepsin from inactive denatured pepsin. J. Gen. Physiol., 14(6): 713-724.

Pavlisko, A., Rial, A., Vecchi, S. and Coppes, Z., 1997. Properties of pepsin and trypsin isolated from the digestive tract of *Parona signata* "Palometa". J. Food Biochem., 21(3): 289-308. doi: 10.1111/j.1745-4514.1997.tb00210.x.

Pethick and Money Ltd., 2002. Method for recovering silver from waste photographic film and paper. US Patent. US6335042.

Romano, M., Rosanova, P., Anteo, C. and Limatola, E., 2004. Vertebrate yolk proteins: a review. Mol. Reprod. Dev., 69(1): 109-116.

Sanchez-Chiang, L., Cisternas, E. and Ponce, O., 1987. Partial purification of pepsins of adult and juvenile salmon fish (*Oncorhynchus keta*). Effect of NaCl on proteolytic activities. Comp. Biochem. Physiol. B, 87: 793-797.

Shahidi, F. and Janak Kamil, Y.V.A., 2001. Enzymes from fish and aquatic invertebrates and their application in the food industry. Trends Food Sci. Technol., 12: 435-464.

Simpson, B.K. and Haard, N.F., 1987. Cold-adapted enzymes from fish. In: Food Biotechnology (Knorr, D.W., Ed.), pp. 495-527, Marcel Dekker, Inc., New York.

Squires, E.J., Haard, N.F. and Feltham, L.A., 1986. Gastric proteases of the Greenland cod (*Gadus ogac*) I. Isolation and kinetic properties. Biochem. Cell Biol., 64: 205-214.

Tanji, M., Yakabe, E., Kubota, K., Kageyama, T., Ichinose, M., Miki, K., Ito, H. and Takahashi, K., 2009. Structural and phylogenetic comparison of three pepsinogens from Pacific bluefin tuna: molecular evolution of fish pepsinogens. Comp. Biochem. Physiol. B, 152(1): 9-19. doi: 10.1016/j.cbpb.2008.08.006.

Yoshida, H., Terashima, M. and Takahashi, Y., 1999. Production of organic acids and amino acids from fish meat by sub critical water hydrolysis. Biotechnol. Progr., 15(6): 1090-1094.

VII

Climate Change and Socio-economic Scenarios

Impact of Climate Change on Hydrological Cycle, Ecosystem, Fisheries and Food Security, Pages: 511–537
Edited by: B. Madhusoodana Kurup, M.R. Boopendranath, M. Harikrishnan and A.V. Shibu

CHAPTER - 43

Geoinformatics for Conservation Planning in Coastal Fisheries: Monitoring Hilsa (Hamilton-Buchanan, 1822) through Remote Sensing in Coastal Bangladesh

Md. Abu Syed[1*], M. Tanvir Hasan[1] and Ahmed Tahmid Raihan[2]

[1]*Nansen-Bangladesh International Centre for Coastal, Ocean and Climate Studies (NABIC) at Bangladesh Centre for Advance Studies (BCAS), House-10, Road-16 A, Gulshan-1, Dhaka-1212, Bangladesh*
[2]*Hochschule fur Technik Stuttgart, Schellingstraße 24, 70174 Stuttgart, Germany*
**E-mail: abu.syed@bcas.net; mabusyed@gmail.com*

Among five varieties of *Tenualosa* spp. (*T. toli*, *T. macrura*, *T. thibaudeaui*, *T. reevesii* and *T. ilisha*), Hilsa shad, *Tenualosa ilisha* is considered the major fishery component of Ganga-Brahmaputra-Meghna river system, due to high consumer preference (Alam *et al.*, 2012). About 50-60% of global Hilsa catch is recorded from Bangladesh (Ahsan *et al.*, 2014) and the contribution of Hilsa to GDP of Bangladesh is 1% (Mohammed, 2014). *Tenualosa ilisha* (Hamilton) fishery covers 10% of Bangladesh's total fish production (BBS, 2014). It is a diadromous fish species and considered the largest single fishery for both marine and inland waters of Bangladesh (Amin *et al.*, 2002). Hilsa became economically one of the most important fish of Bangladesh because of its contribution towards national GDP and export business. Hilsa also has socio-cultural significance because of its integration with the culture in the region over the last millennia. People of Bangladesh have been preparing a number of delightful and mouthwatering dishes with Hilsa in religious (e.g., Eid, Puja), cultural (e.g., Bengali new year) and in marriage ceremonies (Jahan *et al.*, 2015). Sometimes, Hilsa become interesting point of discussion among policy makers in Bangladesh and India because of its popularity among population of West Bengal and Bihar in India. However, this important fishery declined in the late 1990s that led the Government to formulate the Hilsa Fishery Management Action Plan

(HFMAP) and started its implementation from 2005 (DoF, 2019). Since then, Hilsa production increased at a rate of 5% per year till 2015 (Rahman *et al.*, 2018). To improve the annual incremental production further, Department of Fisheries (DoF) and WorldFish have jointly been implementing "Enhanced Coastal Fisheries in Bangladesh (ECOFISH-Bangladesh)", a USAID supported project (2014-2019). The authors from Bangladesh Centre for Advanced Studies (BCAS) had been engaged in this ecosystem assessment works. Authors supported the DoF and local communities to establish a science-based "adaptive co-management" that focuses on the brood hilsa protection, juvenile conservation, illegal fishing, monitoring and evaluation of conservation efforts and overall ecosystem resilience involving all stakeholders.

During their lifecycle, Hilsa shad occurs in a diverse range of aquatic ecosystems including marine, estuarine and riverine (Hossain *et al.*, 2014). Rahman and Cowx (2006) found that the lunar cycle has great influences in the life span of Hilsa, which includes migration, spawning, and feeding. As the species becomes active in terms of feeding, migration, spawning and swimming during full moon and new moon, i.e., spring tide, the catch rates by gill nets around this time increases (Rahman, 2001).

The indigenous or local knowledge acquired over the time by the community/individuals in this particular ecosystem plays an important role in detection and catching of any specific fish (here it is Hilsa) abundance in any specific part of rivers and estuaries around the world (Aswani and Lauer, 2014). However, Department of Fisheries (2008) reported that the inland catch of Hilsa is decreasing year to year, while the marine catch is increasing. Recent news reports and authors observation showed increased catch in inland water following various conservation efforts. Economic value of riverine Hilsa is always higher than the value in marine because its tastes change when it enters in the fresh water. Increased sedimentation and increased fishing density are the main reasons behind the decrease of catch over the time in fresh water (Jahan *et al.*, 2015). Several studies have concluded that an emerging human population and increasing demand for aquatic protein has resulted in overfishing of Jatka (young Hilsa length <25 cm) and adult Hilsa in gill net fishery (Amin *et al.*, 2008; 2002; Rahman *et al.*, 2013). The Global Environmental Facility (GEF) study concluded that, the Hilsa catch could be increased by 45% by imposing restrictions on Jatka fishing at four major fish landing sites (Ali, 2013). In 2003, following a further decline in the annual landings of Hilsa, the Government of Bangladesh established Hilsa Fisheries Management Action Plan (HFMAP). The main objectives of HFMAP were sustainable conservation of Hilsa through establishment of five sanctuaries, establishment of four nationally important spawning grounds, implementation of a nationwide 15-day fishing ban in October only in spawning grounds, enforcement of the 'Protection and Conservation of Fish Act (1950)' and offering compensation to fishers affected by the ban in the form of rice and/or support in developing alternative income-generating activities.

The Hilsa supports an estimated 0.45 million full-time and 2.5 million part-time fishers. In the face of over-exploitation, the exploitation rate increased from 0.33 to 0.66 between 1990 and 2002. This has led the Bangladeshi government to declare five sites in the fishing grounds as 'Hilsa sanctuaries', where fishing is banned during the breeding season. To compensate for lost earnings, the government provides 'affected' fisher communities, which total 224,102 households in 2014-15, with food and alternative income-generating activities. The objectives of Hilsha fishery conservation of the Department of Fisheries (DoF), Bangladesh are to: (i) increase Hilsa production by saving Jatka and brood Hilsa, (ii) enhance and strengthen Hilsa conservation activities and effective management of sanctuary, and (iii) create mass awareness about the importance of Jatka conservation. Compensation appears to have improved livelihoods, but may not have incentivized compliance with regulations. Fishing bans may have helped increase Hilsa abundance, but strong spatial variation in how fishers view the fishery and its regulations, combined with remote sensing data, suggest declining habitat suitability which may be masking the benefits in some areas.

The project, 'Investing in Hilsa fishery as an Economic Infrastructure for Bangladesh' encompassed activities designed to address key areas of work such as improved science based fisheries management decision making, strengthened fisheries adaptive co-management, and enhanced socio-ecological resilience of target communities. The objectives of the present case study presented here had been (i) to identify different boats for Hilsa fishing during ban and non-ban periods from RADAR (RAdio Detection And Ranging) satellite imageries; and (ii) to evaluate the Hilsa fish habitat by analyzing the satellite imagery with the corresponding measured ecological parameters.

Satellite Remote Sensing in Monitoring Coastal Fishery Habitats

Remotely sensed imageries data of both active (Radarsat-2) and passive (Landsat-8) were used in this study to monitor the Hilsa habitat in the south-central to southwest coastal area of Bangladesh. The detail materials and methods used in this study are summarized in following sub-sections:

Study area

The study area has been selected south-central to south-western coastal zone of Bangladesh, where Hilsa fishing is predominant. Important considerations in selecting the remotely sensed microwave RADAR, Radarsat-2 data were based on: (i) night time scanning by the sensor (considering fishing during night time); (ii) Hilsa sanctuaries; and (iii) Jatka sanctuaries. To serve the purpose, Radarsat-2, SLC (Single Look Complex) images (5 meter resolution) of ban period (April 22, 2015; Time- 11:56:06 pm, GMT+6) and non-ban period (July 3, 2015; Time-11:56:08 pm, GMT+6) were procured and analyzed. The Radarsat image of ban period covered an area of 16,434 km^2 and another one of non-ban period covered 17,248 km^2. The Radarsat images of two time periods covered different geographical areas. However, for change detection and comparison of two time periods a common overlap area by both the images was selected for advanced analyses. This common overlap area covered 12,676 km^2 which is hereafter referred to as defined study area of boat counts (Fig. 1). This enabled the research team to comparing the boats during the ban and non-ban periods. The study area of boat counts is situated between 23°3'24" - 21°56'11" N latitude and 89°31'44" - 90°30'45" E longitude. The area includes 10 major rivers in 7 districts of south-central and south-western coast of Bangladesh. The major rivers are Tentulia, Arial Khan, Bishkhali, Baleshwar, Kali Ganga, Sandha, Bighai, Galachipa, Payra and Kalabadar in Bagerhat, Jhalkathi, Bhola, Patuakhali, Pirojpur, Barguna and Barishal districts.

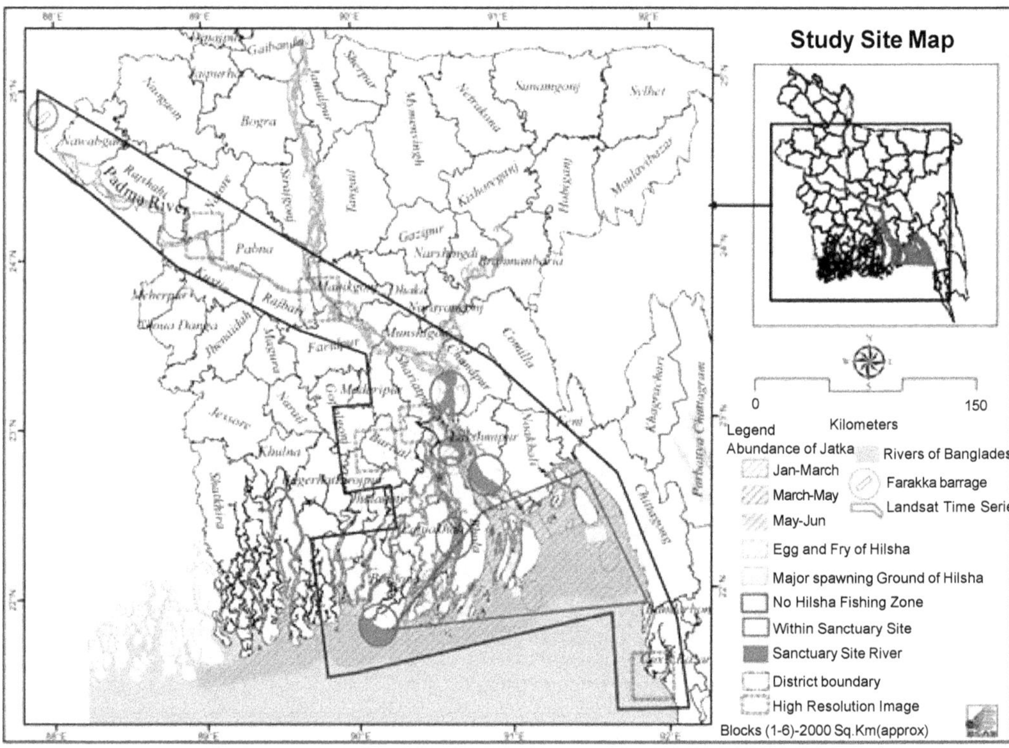

Fig. 1. Map shows study area covered by optical Landsat 8 and RADARSAT 2 data; and different sanctuaries and control sites of the study

Normally two types of fishing boats are used for catching Hilsa fish in Bangladesh: small and large boats. Small boats are normally 4-7 m long and 1-2 m wide made of wood with a low-power engine that are mainly engaged in catching Hilsa fish in the rivers. Large boats are normally 10-20 m long and 4-10 m wide made of also wood with high powered engine. Large boats need to get a permit for fishing in the deep sea.

RADAR image preprocessing

RADAR stands for RAdio Detection And Ranging. Radar is an active sensor system, transmitting a signal of electromagnetic energy, illuminating the terrain, and recording or measuring the response returned from the target or surface features Radar system operates in the microwave region of the electromagnetic spectrum. A radar system operates in a specific wavelength or frequency and records the signal response from the ground or target at single specific wavelength. A radar image is a representation of the amount of energy backscattered from a surface (Fig. 2). On the contrary optical multi spectral scanner (MSS) interacts with target in the shorter wavelength region and in a range of wavelength, which are called 'bands'. A radar image is functions of wavelength, polarization (Fig. 3) and swath width. However in addition to those system parameters determine the resolution and radiometric quality of images. Based on system parameters radar systems can be different types. In the early stage of radar system development real aperture radar (RAR) and side looking airborne radar (SLAR) systems were developed. These systems produce echoes whose phase is random from pulse to pulse. So, it appears enhanced resolution by processing the SLAR or RAR images. In the process of improved radar system development, the latest synthetic aperture radar (SAR) systems were developed. In case of RAR azimuth resolution is set by the width of the azimuth antenna pattern. On the other hand, SAR preserves the inter-pulse phase structure of the received Fig. 2: Concept of energy interactions in an imaging radar/SAR system; Fig. 3: Concept of imaging radar/SAR system signals. The SAR image processor mainly depends on this coherent phase information to enhance azimuth resolution. This process of enhancing azimuth resolution is called focusing. As a result, in case of SAR azimuth, resolution is much smaller than antenna pattern width. Since the launch of Radarsat-1, maritime surveillance has been a key application of the data from the satellite, and this trend has continued with Radarsat-2. Large areas of interest coupled with the limited accessibility of the maritime environment lend itself to the use of space borne radar. The two key maritime surveillance applications for Radarsat-2 are ship detection and oil slick detection. Ship detection is primarily focused on illegal dumping and illegal fishing, and oil slick detection is primarily focused on the detection of natural oil seeps, illegal dumping or accidental discharge of oil from offshore drilling platforms.

In general, a large incidence angle is preferred for ship detection and HH polarization is the preferred choice for ship and boat detection. Given these conflicting demands, selection of the best imaging geometry was problematic for end users that were interested in both applications for a single area of interest (AOI). Radarsat-2 introduced additional polarization modes. The dual-polarized modes, which balance spatial coverage and higher resolution for detection, have effectively eliminated the constraints for ship, boat and oil spill detection inherent with Radarsat-1. Radar imagery is used for a wide range of marine, wetland ecosystem application, such as: ecosystem assessment, monitoring and surveillance. Launched in 2007 Radarsat-2 improves ship detection with its Ultra-Fine beam mode (3 m spatial resolution) and offers the potential for ship classification. The most important characteristics and advantages of radar remote sensing are summarized as follow (Hoffer, *et al.*, 1994):

- Active system;
- Side looking imaging system;
- All-weather capability (penetration through cloud, fog, atmospheric dust);
- Relative high spatial resolution;
- Can penetrate surface features including vegetation and water; and
- High variability in configuration (wavelength, polarization and depression angle).

Boat detection using Radarsat-2 depends on three main parameters: wind, ship properties, and radar characteristics (MDA 2005, 2007, 2011; Vachon and Tunaley, 2007; Pichel *et al.*, 2004; Crips, 2004; Greidanus *et al.*, 2004). For a given ship, detection improves as the wind speed decreases. The improved detection for lower wind speeds is due to the reduction of ocean-surface roughness and the corresponding reduction in the radar backscatter, leading to higher signal-to-clutter ratio (SCR) and better detection. As the wind speed increases, the ocean-surface backscatter increases which reduces the signal-to-clutter ration and impedes detection (Skrunes *et al.*, 2014). The detection also

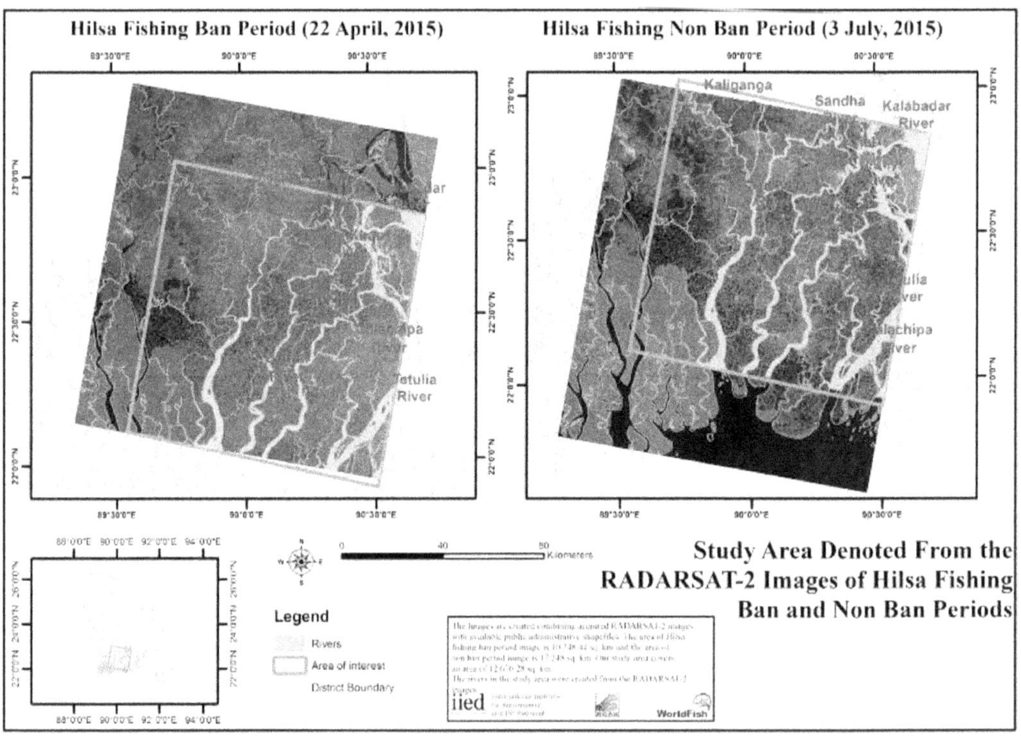

Fig. 2. Study area denoted from Radarsat-2 images of Hilsa fishing ban and non-ban periods

depends on the wind speed relative to the radar look-direction. Detection is better when the wind is blowing perpendicular to the radar look-direction and worse when the wind is blowing toward the radar look-direction.

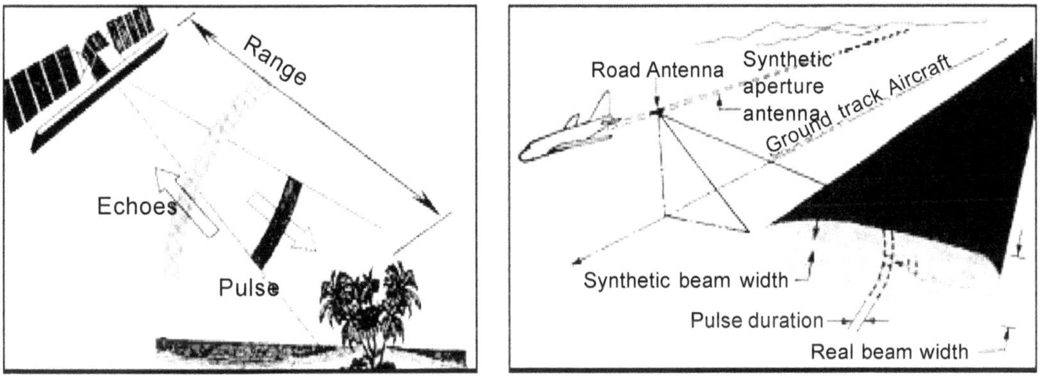

Fig. 3. Concept of energy interactions in an imaging radar/SAR system (L) and Concept of imaging radar/SAR system

With respect to ship properties, ship detection depends on the ship length, the ship's primary construction material and the ship's orientation relative to the radar look-direction. The radar return is proportional to the ship's length, so the longer the ship, the higher the probability of detection. Detection also depends on the ship's primary construction material. Metal produces a stronger radar return than wood or fiberglass. As a guideline, the minimum detectable ship length is roughly one-half the radar resolution for metal ships and approximately equal to the ship length for a non-metallic ships. MDA (2011) verified the relationship between radar resolution and detection of metallic ships or wooden boats.

The three radar characteristics that affect detection are resolution, incidence angle, and polarization (Choudhury *et al.*, 1994; MDA 2005). Ship or boat detection depends on boat or ship length and width, so for a constant boat or ship length, the detection improves as the radar resolution improves. The minimum detectable ship length decreases as the radar resolution improves. Ship or boat detection also improves with increasing incidence angle (MDA 2005; Crisp 2004). The radar backscatter is largely invariant with incidence angle, but the backscatter from the ocean decreases significantly with incidence angle (Cloude and Pottier 1996). Therefore, at a small incidence angle, the SCR is smaller than the SCR at a larger incidence angle. Detection also depends on the radar polarization. In general, the use of co-polarized HH delivers better results than VV since the ocean-surface backscatter from HH is less than VV. The use of HH provides good performance at large incidence angle and for low wind speeds, but at small incidence angles and high wind speeds, the use of HH is problematic. To mitigate these issues, cross-polarization (e.g. HV) is used. The HV sea-surface backscatter from the sea's surface is largely invariant with incidence angle and is considerably weaker than the co-polarized backscatter (MDA, 2011).

Remotely sensed images in raw format contain no reference to the geographic location of the data. To integrate these data with other data in GIS, it is required to correct and adapt them geometrically (Hurni, 1997). Information extracted from remotely sensed images is to be integrated in the form of map data in GIS or presented to the users in a map. Remotely sensed images are transformed to the scale and projection of map of the same area (Mather, 1999). This transformation of scale and projection properties of remotely sensed data called geometric correction. Geometric correction of remotely sensed images is done to make the data compatible for the following purposes: (i) transform an image to match a map projection; (ii) locate points of interests on map and images: (iii) bring adjacent images to registration; (iii) overlay temporal sequences of images of the same area; (iv) integrate images from different sensors, and (v) GIS operations.

The different distortions of the image geometry are not realized in certain sequence, but happen all together and therefore, cannot be corrected step wise (Hurni, 1997). Thus the corrections of all distortions were done at once with the execution of affine transformation, which combine all the separate distortions. Affine transformation can be presented, following Hurni (1997). The overall accuracy of the transformation is indicated by the average of the errors in the reference points, which is commonly called root mean square error (RMSE) or sigma. Over all positional error has been kept to a limit by maintaining sigma value below 0.2. All transformations were done using and Erdas Imagine.

Characteristically, radar images are different from the optical images. Radar images are not made of smooth gray scale as in the optical images it contains a mixture of light and dark pixels. Active radar sensor gives a burst of coherent radiation, which returns from the target features. Like light from a laser, the waves emitted by active sensors travel in phase and interact minimally on their way to the target features. Following the interactions with the target features, waves no longer remain in phase. Different distances they travel from target features, or single versus multiple bounces cause speckle. Once out of phase, radar waves can interact to produce light and dark pixels is called speckle noise (ERDAS, 2014). Speckle noise is fundamental to any active sensor like radar or laser caused by interference of waves returned from an ensemble of independent scatters, due to a surface roughness on the scale of wavelength (Ulaby, *et al.*, 1981). Radiometric distribution of the reflectivity from distributed scene at a single point of observation is attributed by speckle (Raney, 1998). Speckle noise appears to be the main problem in radar data interpretation, processing and classification. This is why prior to the utilization of radar data speckle noise must be reduced. However, speckle reduction techniques used to reduce speckle noise will also produce changes in the image information.

There are different methods for speckle reduction. One way to produce multi-look images during radar data formation and the other way to smooth the data with a smoothing filter that is appropriate for specific application. Adaptive speckle reduction filters promise better results that compared to conventional linear filters. Because linear filters are constrained by the principle of conservation of confusion that, what is gained in image variance is lost in image resolution (Raney, 1998). Prior to application of speckle reduction filter, the objectives are clearly identified. To perform quantitative image analysis following speckle reduction, implicit requirements on the radiometric transformations must be fulfilled (Raney, 1998). On the contrary, with the objective of simply deriving an image that looks better, less stringent method could be applied. Still it is necessary to compare the performance of various speckle reduction schemes. Quantitative review of adaptive filters summarized that in their original form the adaptive filters (Raney, 1998) preserve the value of homogeneous areas within ~0.1 dB. These filters work by deriving estimates of the variation of image brightness over a small sub area of the image and replace the observed values by a new value computed from the local statistics of the image. From the above ideas the following speckle reduction method was adopted.

Radar image post-processing

Post-processing steps include firstly, rectifying image for identification of boats. For that, setting up criteria of pixel values is needed. Then classifying and counting different types of boats follow next. After the enhancement of Radarsat-2, SLC images, boats were identified through careful visual interpolation. Though, noises are inherent to coherent images, but vessels and boats are often visible in Radarsat SAR images as bright point targets against the clutter background (Vachon, *et al.*, 1999; Iehara *et al.*, 2001). Some moving fishing vessels are clearly detected as white lines through the rivers (Fig. 4).

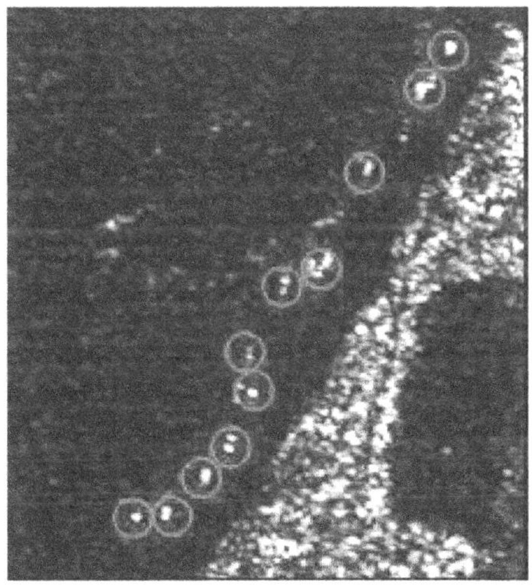

Fig. 4: Visible boats (red circled) in Baleshwar River (Dated 22 April, 2015; period: ban)

In midst of highly sediment enriched rivers and canals, which is indicated by noise in the images even after advanced image rectification, boats are not easily detectable (Staples & Rodrigues 2013). To detect boats in such regions, we have fitted point-target return shapes to the image at locations where a pixel value within a given threshold (3000-10000) (Fig. 5). The returns were assumed to be circularly symmetrical and with a Gaussian, i.e., exp. (r^2/a), profile having variable width a. This type of thresholding is global thresholding. It stores the intensities of the pixels in an array. The threshold is calculated by using total mean and variance. Based on this threshold value each pixel is set to either 0 or 1, i.e., background or foreground. Thus here the change of image takes place only once. The following formulas are used to calculate the total mean and variance (Devi, 2006). The pixels are divided into 2 classes, C1 with gray levels [1, ..., t] and C2 with gray levels [t+1,... ,L].

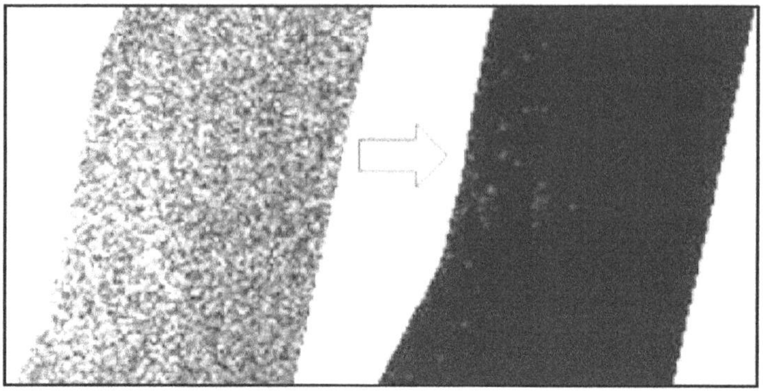

Fig. 5: Detected boats (red dots) in noisy regions through pixel threshold

The probability distribution for the two classes is:

$$C_1: p_1/w_1(t), \ldots, p_t/w_1(t) \, and$$
$$C_2: p_{t+1}/w_2(t), \ldots, p_L/w_2(t)$$

$$Where, w_1(t) = \sum_{i-l}^{t} p_i$$

and

$$w_2(t) = \sum_{i=t+l}^{L} p_i$$

Also, the means for the two classes are:

$$\mu_1 = \sum_{i=l} ip_i / w_1(t)$$
$$\mu_2 = \sum_{i=t+l} ip_i / w_2(t)$$

Using Discriminant Analysis, Otsu (1979) defined the between-class variance of the threshold image as:

$$\sigma_B^2 = w_1(\mu_1 - \mu_T)^2 + w_2(\mu_2 - \mu_T)^2$$

After detection of the boats, automated shapefiles with the same threshold are generated. Then they are cross checked with manually generated shapefiles depending on the pixels. Then Geodatabase of ten detected major rivers from the image is updated with frequency of boats in both ban and non-ban periods. Average number of boats per ten square kilometers are also detected in each river because of low number of boats (i.e. less than one), intensity of boats per ten square kilometers are determined. The procedure was shown in the flowchart below (Fig. 6):

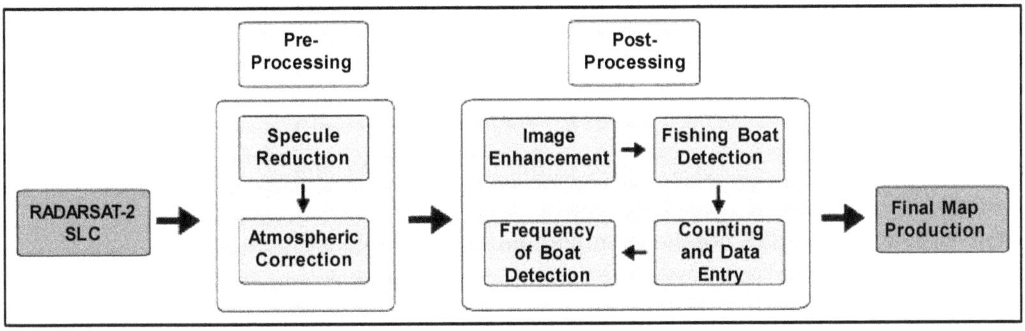

Fig. 6. Procedure of fishing boat detection in the Area of Interest

To figure out the frequency of boats in any particular region of the area of interest, multi distance spatial cluster analysis through 'Ripley's K function' is done. Mathematically, the Multi-Distance Spatial Cluster Analysis tool uses a common transformation of Ripley's k-function where the expected result with a random set of points is equal to the input distance (Haase 1995, Greidanus *et al.*, 2004). The transformation L(d) is shown below:

$$L(d) = \sqrt{\frac{A\sum_{i=1}^{N}\sum_{j=1, j\neq i}^{N} k_{i,j}}{\pi N(N-1)}}$$

Where, A is area, N is the number of points, d is the distance and k (i, j) is the weight, which (if there is no boundary correction) is 1 when the distance between i and j is less than or equal to d and 0 when the distance between i and j is greater than d. When edge correction is applied, the weight of k (i, j) is modified slightly.

Monitoring ecological parameters

A spatio-temporal variability of ecological parameters which are related to Hilsa fish habitat was evaluated using the ecological parameters derived from Landsat-8 data. Ecological parameters were analyzed at the laboratories from the collected water samples and in-situ measurements by different sensors at 12 sites in 4 locations near Chandpur (lower Meghna river), Daulatkhan (Shahbazpur channel), Lalmohan (Tentulia river), and Patuakhali (Andharmanik river) (Hasan et al 2016). Water samples for all physical, chemical, and biological parameters were collected monthly from Chandpur site and bi-monthly from other two sites. Physical parameters included air and water temperature, transparency, total dissolved solids (TDS), total suspended solids (TSS), and conductivity, while hydrological parameters were water current velocity and depth. Dissolved oxygen (DO), pH, total alkalinity, salinity, nitrate-nitrogen (NO_3-N), nitrite-nitrogen (NO_2-N), ammonia-nitrogen (NH_3-N), phosphate-phosphorus (PO_4-P) and chlorophyll-a were measured as chemical parameters. In-situ temperature was measured by HQ 40D sensor and water transparency was measured using a Secchi disc of 30 cm dia. In-situ pH, conductivity, DO, NO_3-N, NO_2-N, NH_3-N, PO_4-P and salinity were measured using HQ 40D multi-parameter and Hach Kit (DR 4000) equipped with spectrophotometer (DR 6000). Total alkalinity was determined in the laboratory by methyl orange indicator and standard EDTA solution by titrimetric method (APHA-AWWA-WEF, 1999). Chlorophyll-a was estimated using DR 6000 spectrophotometer at 664 and 750 nm wavelengths (Felip & Catalan 2000). The water samples were collected at each site from a vertical depth up to 2 m. More details of in-situ measurements and laboratory analyses can be found in Wahab *et al.* (2014).

Landsat imagery

Analyzing the Gonadosomatic Index (GSI) of Hilsa fish from the study area, four spawning times without cloud cover i.e., January, March, May and November were selected for analyses of Landsat images. Landsat-8 images of November 6, 2013; January 25, 2014; March 14, 2014; and May 17, 2014 of a scene with path 137 and row 44 were downloaded from the U.S. Geological Survey's Landsat archives (http://earthexplorer.usgs.gov) for this study corresponding to the bi-monthly observed ecological parameters of river water. The scene covers 9 sites at 3 locations out of 12 sites at 4 locations of monitoring.

Landsat-8 satellite launched on February 11, 2013 consists of two science instruments: the Operational Land Imager (OLI) and the Thermal Infrared Sensor (TIRS). These two sensors provide seasonal coverage of the global landmass at a spatial resolution of 30 m (visible bands 1-4, NIR band 5, and SWIR bands 6-7); 100 m (thermal bands 10-11); and 15 m (panchromatic band 8). The OLI collects data for two new bands, a coastal band (band 1) and a cirrus band (band 9), as well as the heritage (old bands also implemented in previous Landsat satellites) Landsat multispectral bands. Additionally, the bandwidth has been refined for six of the heritage bands. The TIRS carries two additional thermal infrared bands (bands 10 and 11). TIRS collects data for two more narrow spectral bands in the thermal region formerly covered by one wide spectral band on Landsats 4-7. The OLI and TIRS data are registered together to create radiometrically, geometrically, and terrain-corrected 12-bit data products. The Landsat-8 scene size is 185-km-cross-track-by-180-km-along-track. The nominal spacecraft altitude is 705 km. Cartographic accuracy is 12 m or better including compensation for terrain effects in the level-1 data products (NASA, 2016).

Mapping Ecological Parameters

The ecological parameters were first evaluated with each band of Landsat images by linear regression between each ecological parameter values and the digital numbers of all bands, band ratios, and indices such as normalized difference vegetation index (NDVI) and normalized difference water index (NDWI) of each Landsat image at 9 sites. NDVI and NDWI (Mcfeeters, 1996; Rogers and Kearney, 2003) are widely used indices for detecting vegetation and water in the landmass; however, in this study we selected these two indices along with the bands and band ratios to evaluate any relation with the measured ecological parameters with them. The equations of the indices are as follows:

$$\text{NDVI} = \frac{B_{NIR} - B_{\text{Red}}}{B_{NIR} - B_{\text{Red}}} \quad \text{... (1)}$$

$$\text{NDWI} = \frac{B_{\text{Red}} - B_{SWIR}}{B_{\text{Red}} - B_{SWIR}} \quad \text{... (2)}$$

Where B_{NIR} is the pixel value (digital number) of near-infrared band (band 5 in Landsat-8), B_{Red} is pixel value of red band (band 4 in Landsat-8), and B_{SWIR} is the pixel value of a shortwave-infrared band (here taken band 7 in Landsat-8). Although some non-linear regressions gave better results than the linear, however, linear was selected as preferred regression to minimize the related errors propagated to the extrapolated values. After the evaluation of linear regressions, the band or band ratio or any index with the highest value of regression was selected and converted into the corresponding parameters. The negative or any abnormal lower or upper values than the measured values were converted into zero or excluded in the final maps. The cloud affected areas were excluded from the maps.

Performance of Landsat-8 images for coastal ecological parameters

The cluster analysis encompasses a number of algorithms and methods to identify structures within the data as homogeneous group of cases that has been applied to wide range of research problems (Anderberg, 2014). The hierarchical clustering method has been performed to produce much more stable cluster boundaries as well as to identify a reasonable optimal number of classes (Bladon *et al.*, 2016) in this study we applied the within cluster sum of squares error (SSE) metric (Zhang, *et al.*, 2016) and elbow method (Kashani and Graettinger, 2015) to set optimal number of cluster in the dataset. The hierarchical agglomerative clustering algorithm helps in mapping the number of cluster formed groups of class or factors of classification through a dendrogram. The study dendogram approach of hierarchical agglomerative clustering method applied in Bangladesh South West coastal region found 3 number of clusters whereas more than 86% variability was explained with less SSE (<0.23) in original dataset. The three clusters are named as high for >0.7, medium for 0.4-0.7, low for <0.4. Within the above mentioned classes the correlation of determination are graphically presented through pixel color and dendogram approach. Euclidian distance and Wards method have been applied to generate the dendogram which represents the homogeneous group of data structure. The computation was done using R statistical programming language.

Results and Discussion

Analysis of Radarsat-2 images

From the data gathered following the methods described earlier, maps of boat frequency, density and cluster during both ban and non-ban periods could be produced. Difference maps of the two periods were also produced. Densities were calculated number of boats /10 km^2. The unit of 10 km^2 area was taken because of number of boats found to be less than one per square kilometer. This doesn't give us a clear overview of boat density in the study area.

Boat density and distribution pattern and seasonal variation in different rivers of sanctuary and non-sanctuary areas in the study area are represented in Fig. 7 and 8. These also postulate that boat density or number of boats in ban period is quite less compared to that found during non-ban period. This could be attributed to the positive impacts or response to the enforcement of ban in the area and vigilance. The primary target was to monitor the sanctuary area in the image which had been Tentulia River. In Tentulia River, during ban period, a total of 67 boats were found which meant 4 boats/ 10 sq. km. It is clear from the analysis; fishing activity goes on during ban period in sanctuary area. Areas other than sanctuary area, considered as 'control area' were free from fishing ban except fishing of Jatka for a period of eight months (November 1, 2013 to June 30, 2014). In the control areas, Jatka fishing ban was outdated in the second image of non-ban period. As there are no other restrictions than Jatka fishing ban in control areas, it is assumed that, fishermen of those areas are more flexible about their

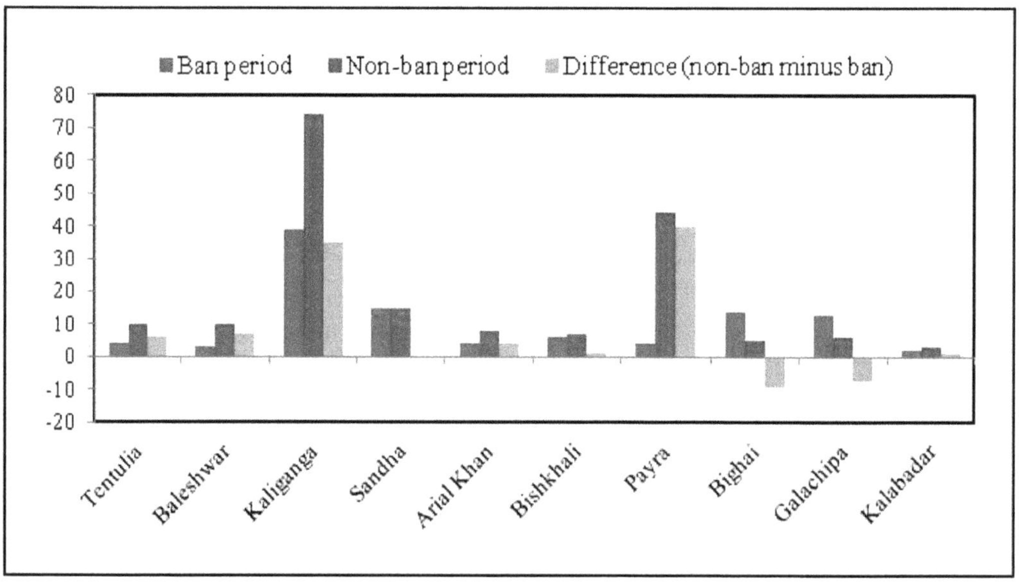

Fig. 7. Boat density and distribution pattern during ban and non-ban period as mapped by Radarsat-2 images

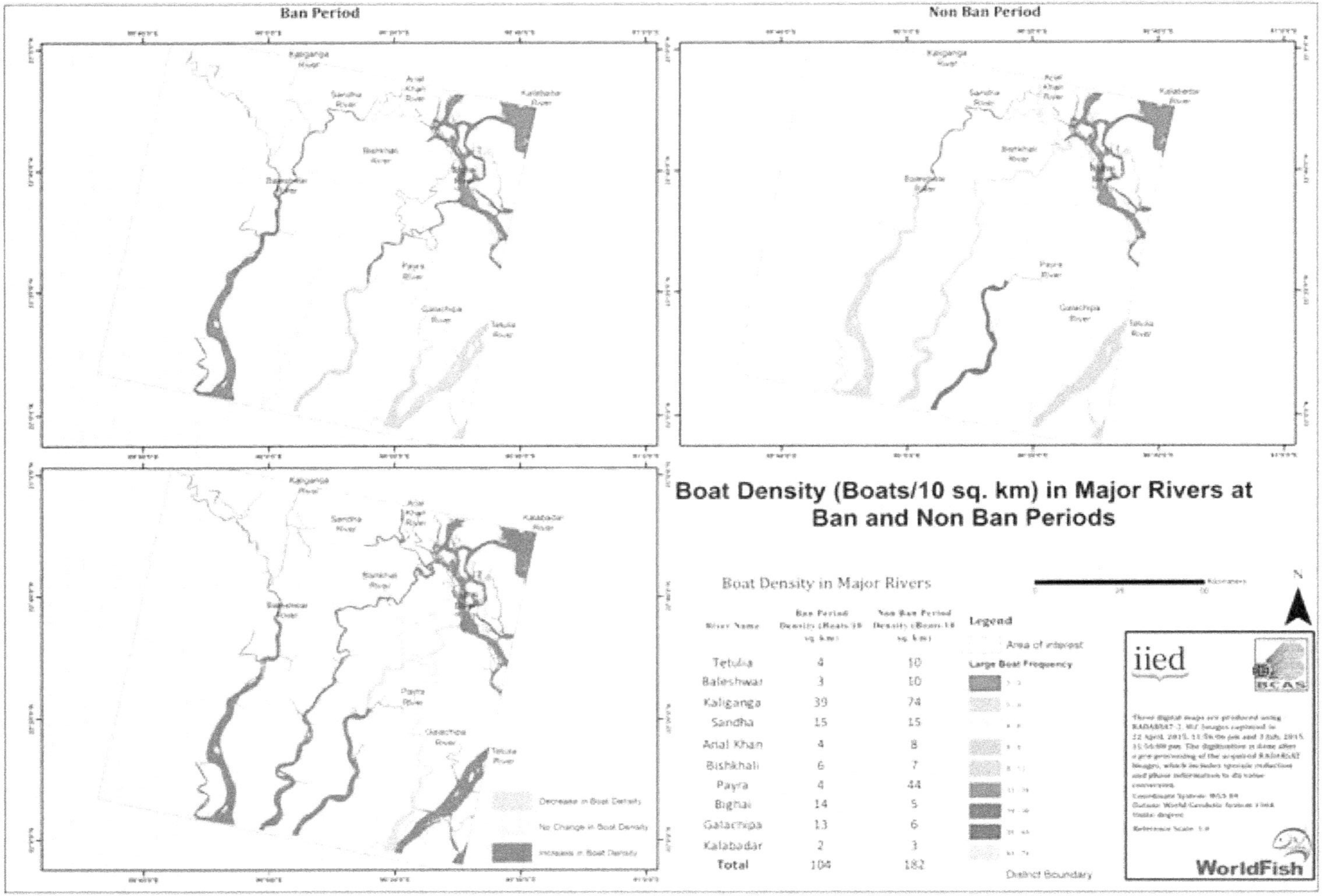

Fig. 8: Boat density (boats/10 sq. km.) in major rivers during both ban and non-ban periods

fishing time. This is why during ban period, less boats in control areas and the number increases during the Hilsa fishing season i.e. non-ban period were observed. Most of the rivers showed this pattern except two rivers Bighai and Galacipa where more number of boats found during ban period then non-ban period. Hence, it is important to understand why this has been happening in these two rivers and is there anything DoF or other stakeholders can do (more engagement with fishers, check whether incentive packages are reaching fishers of these two rivers or not? If not, what DoF can do in this to improve the situation there?

Fishing boat clusters of both ban and non-ban periods were determined through spatial statistics and visualized in the map below (Bladon, 2016). The map shows boat clusters in different regions of rivers in both sanctuary and control areas, which can be helpful to determine the habitats or routes of Hilsa in the rivers.

It was revealed by the analysis that during non-ban period, the river Kaliganga has the highest number of fishing boats which is 111 boats in total. Throughout the ban period, less fishing boats are identified in the sanctuary region i.e., Tentulia river. A total of 67 fishing boats were identified in Tentulia River. But, considering our focused sanctuary region of 217 km2 (app.), which included some portion of Kalabadar River alongside with Tentulia River, holds a total of 66 boats during ban period and 382 boats during non-ban period (Fig. 9). On the other hand the control area of 818 sq. km (app.) holds total of 764 fishing boats (Bladon et al 2016). It was revealed that comparison of density sanctuary area holds double density of boats (18 boats/10sq. km) of the control area (9 boats/10 sq. km.) during non-ban period. Analyzing the boat cluster image (Bladon et al 2016), Bighai River works as the fishing hotspot during ban period as a lot of boats cluster in areas of Bighai River during that time. And during non-ban period, in portions of Kalabadar and Tentulia Rivers i.e. portions of sanctuary areas found to fishing hotspot. But, lower portion of Tentulia River, which is considered sanctuary area, holds a moderate amount of boats (15-25 boats cluster) during both ban and non-ban periods (Fig. 9).

According to Hossain *et al.* (2014), the breeding phase of Hilsa is coordinated with new and full moon of monsoon wet season (March-September), which is considered with high turbidity, resilient fresh water flow, low salinity (0-5ppt.) and enormous tidal disparity. Again, according to BOBP (1987), the major spawning periods for Hilsa are October-November, June-July and January-March. These significant numbers of spawning periods are because of the suitability of coastal, estuarine and river waters of Bangladesh to Brood Hilsa (Shafi *et al.*, 1978; Quddus *et al.*, 1984).

Though, there is a noticeable boat density in the sanctuary region during non-ban period, but noticeable amount of boats exist during the ban period too. The assumed observation from this output and a number of previous news reports is, most of the boats during ban-period were fishing for 'Jatka' to meet the unfulfilled demand of Hilsa fish in Bangladesh. When juvenile fishes (10-26 cm) begin to migrate towards the brackish

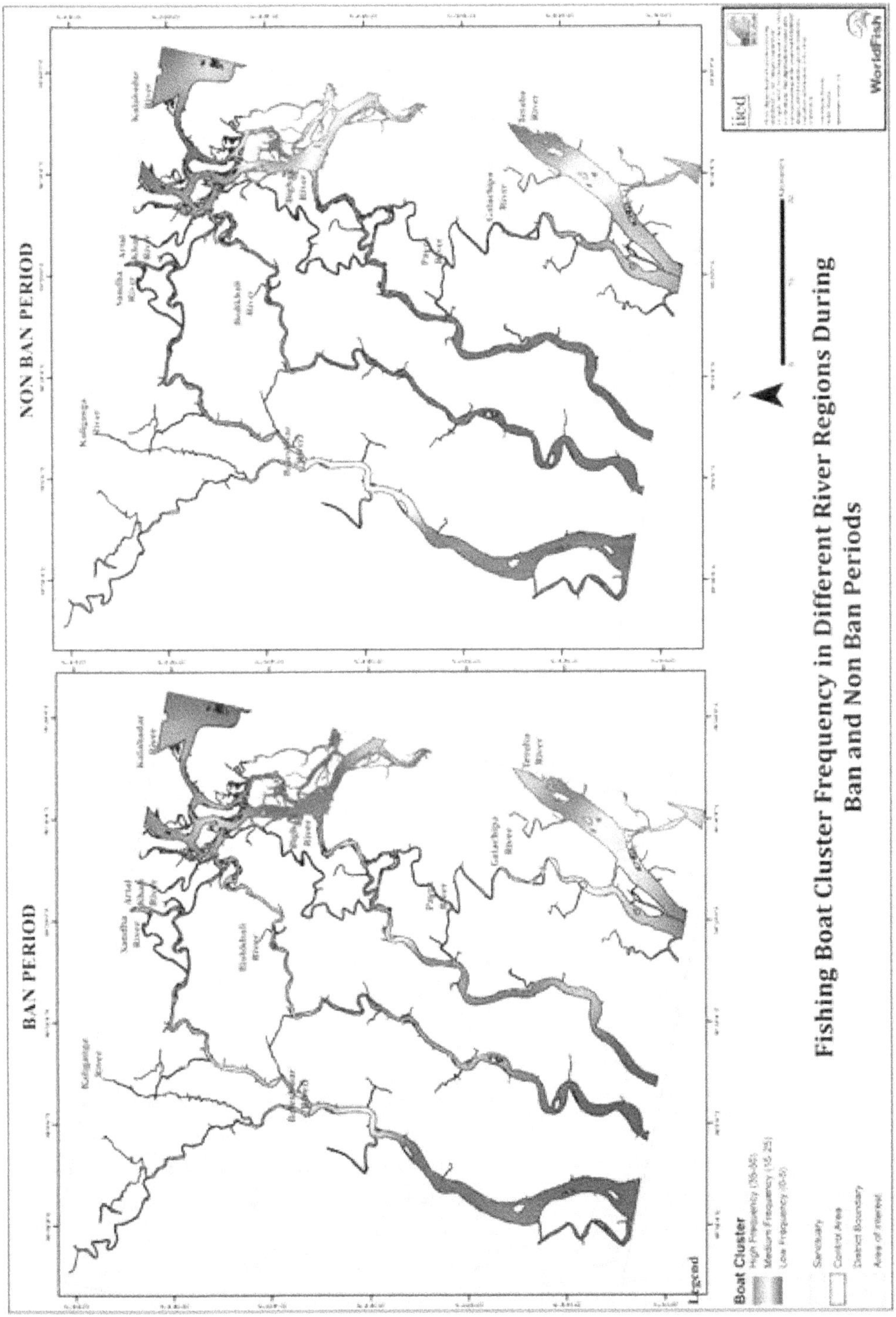

Fig.9: Fishing boat cluster frequency during ban and non-ban periods

water, they get caught in the illegal synthetic nylon fiber net and according to the findings of the research, in time of Hilsa fishing ban. This activity is contributing to the drastic degradation in Hilsa biodiversity in coastal region of Bangladesh. But, after enforcement of the ban period, the illegal fishing has decreased in comparison with number of fishing boats in non-ban period.

There are some limitations and uncertainties of extracting small fishing boats from the RADAR images. It was observed that the small boats can be detected if the digital number (DN) of Radarsat-2 image are in the threshold values >3000. However, as the small boats are normally 5-7 m long and 1-2 m width, the spatial resolution of 5 m of Radarsat-2 images are still low for detecting these boats. So, there might be uncertainties of detecting small boats from the images. However, possibilities of other types of boats were considered carefully. Furthermore, the common area of both Radarsat-2 images covered only 45% of Tentulia River (217 km^2) which is one of the declared major sanctuary areas.

Analysis of Landsat-8 images

Bladon *et al.* (2016) carried out linear regression between the observed values of ecological parameters and Landsat-8 individual band values, band ratios and indices of November 6, 2013; January 25; March 14; and May 17 of 2014, respectively. The green marked with bold font coefficient of determinants (R^2) were found the highest among the R^2 values between each parameter and each band, band ratio and indices. After evaluating R^2 values of all the images, 11 parameters were selected for mapping spatial variation of the ecological parameters to evaluate the Hilsa habitat and abundance. The selected parameters include chlorophyll-a, dissolved oxygen, electrical conductivity, salinity, total dissolved solid, total suspended solid, phosphate-phosphorus (PO_4-P), water current velocity, water transparency, water temperature and pH of upper 2 m of the river water of each date. The observed ecological parameter values of corresponding image dates were interpolated linearly from the monthly (3 sites of Chandpur) and bi-monthly (6 sites of Bhola) observed values. Air temperature, alkalinity, NO_3-N, NO_2-N, NH_3-N, and water depth were not selected because of not high R^2 values at all dates or not relevance with the Hilsa habitat. The R^2 values were ranging from 0.32 to 0.99.

Hierarchical clustering represents the values of linear regression between two variables that are closely packed together, expanding clusters in any direction where there are nearby points, thus dealing with different shapes of clusters. Coefficient of determination (R^2) of linear regression between 11 ecological parameters and 21 different Landsat-8 band values, band ratios, and indices of November 6, 2013, January 25, 2014, March 14, 2014, May 17, 2014 have been presented through color maps of matrices (Figure 24) whereas a single pixel represents the R^2 (0 to 1) of respective variables. Based on the R^2 between ecological parameters at 9 sites and band values, band ratios, and indices

(NDVI and NDWI) are agglomerated using dendrogram approach for defining four time period (Bladon *et al.*, 2016) within these matrix plots the white and red pixels indicate the highest and the lowest R^2 values, respectively. In the mentioned four time periods R^2 represent the certain band are more sensitive to represent certain ecological parameters. In addition to investigate present status of TDS, TSS, EC, salinity, DO, pH, and water transparency of respective river using most of all bands and ratios might be useful as well. Eleven ecological parameters are truly represented except some cases through all bands and band rations in March 2014 and May 2014 in a nutshell, above mentioned cluster analysis suggested as well that the band 1 (coastal/aerosol), 5 (NIR), 6, 7 (SWIR) and their ratios performed very well for evaluating the spatio-temporal dynamics of ecological parameters which enhance the Hilsa habitat (Bladon *et al.*, 2016).

Spatio-temporal variability of ecological parameters

This section summarizes the spatio-temporal variability of 11 selected (high R^2) ecological parameters for mapping spatial variation of the ecological parameters in 4 time period (November 2013, January, March and May, 2014) to evaluate the Hilsa habitat and abundance in the coastal rivers of Bangladesh.

Chlorophyll-a

Hilsa habitat occurs in water areas where phytoplankton density is high. Felip and Catalan (2000) found a non-linear relationship (power-law) between Chlorophyll-a content and total phytoplankton bio-volume in water. Recent gut analysis of Tenualosa Ilisha by Hasan *et al.* (2016) found various phytoplankton species as major food ingredients in different experimental Hilsa fishes (Hasan et. al., 2016). The analysis has revealed that phytoplankton density is high around March in Padma and Meghna rivers (15-20 mg/l). In Meghna River, Chlorophyll-a density remains moderate around May. The concentration of Chlorophyll-a is medium in Meghna and Tentulia rivers in January (Shafi *et al.*, 1978). In November, the concentration is very low in the river systems. So, it can be stated that high Hilsa habitat should occur during March, May and possibility of moderate Hilsa occurrence prevails in January. Landsat 8 images were found to be very useful in visualizing spatio-temporal variation of Chlorophyll-a and dissolved oxygen in mg/l (ppm) in coastal rivers of Bangladesh for study periods (a) November 6, 2013; (b) January 22, 2014; (c) March 14, 2014; and (d) May 17, 2014.

Suitable dissolved oxygen condition for Hilsa abundance is above 5 mg/L (Mojumdar, 1939). Regarding the image analyses, most suitable dissolved oxygen environments for Hilsa abundance were November, January and March. While dissolved oxygen concentration remains the lowest during May. In middle portion of lower Meghna river, dissolved oxygen level goes up to 10 mg/l during January. The highest observed dissolved oxygen concentration is up to scattered around portions of Meghna and during May, dissolved oxygen content is usually low according to the analyses (Bladon *et al.*, 2016).

Electrical conductivity

Conductivity is a measurement of how well an aqueous solution can carry an electrical current. It is commonly used to determine the levels of impurities in the water, typically the more impurities in the water, the higher the conductivity value will be. Conductivity is directly proportionate to salinity (Rhoades *et al.*, 1999). It was observed that November and January are not suitable for Hilsa stock in our study areas because of high electric conductivity. On the other hand, during March and May, when conductivity is relatively low, a suitable environment for Hilsa stock is created.

Salinity

Most suitable salinity range for occurrence of Hilsa stock in river regions is 0 - 2 ppt (Jhingran and Natarajan, 1969). November and March are the most suitable for Hilsa habitat as salinity level remains in the expected range during the time. While May is moderately suitable for Hilsa occurrence (2 - 5 ppt) and January is least suitable regarding salinity level of the rivers in study area.

Total dissolved solids (TDS)

From the image analyses, it was observed that the whole study region remains high in TDS during November, which makes it less suitable for occurrence of Hilsa stock. November is just after the monsoon period when >80% of total annual rainfall occurs and river discharge is quite high with very high sediment load. The TDS levels in January, March and May were lower than in November, especially in upper Meghna river which is suitable for occurrence of Hilsa stock in January, March and May.

Total suspended solids (TSS)

As TSS can be observed in huge amount in Shabazpur channel in November, January and March, it may create blockage in the migratory route of Hilsa (Hasan *et al.*, 2015). However, Padma and lower Meghna river contained quite low TSS during that period. Only in May, there was opportunity of Hilsa stock occurrence in lower Meghna and high possibility of occurrence in Padma and lower Meghna river during March and November due to low concentration of TSS in the regions during the season.

Phosphate-phosphorus (PO_4-P)

According to Bhatnagar *et al.* (2004), at least 1 ppm of PO_4-P is good for plankton production. But, the highest amount of PO_4-P found in our study area has been 0.4 ppm in upper portion of Meghna River during December and in lower portion during May, indicating that the study area is low in PO_4-P.

Water current velocity

Most suitable water current condition for Hilsa is between 0.1 and 0.15 m s^{-1} and moderately suitable condition is 0.05 - 0.2 m s^{-1} (Nair, 1958). The Landsat 8 image analyses showed that the most suitable current condition occurs during January in very small portions of Meghna River and in surrounding small rivers and canals of the study area. Most unsuitable condition of current in study area occurs during May (0.35 - 0.6 m s^{-1}). In November and March, the condition remains somewhat suitable in some parts of the study area as the range varies from 0.03 - 1 m s^{-1} in the whole area.

Water transparency

According to Ahmed (1993), Chlorophyll-a and water transparency are inversely related. Maximum water transparency (80 cm) was recorded during November, January and March while the minimum transparency level was same in all the observation months. The minimal upper limit of transparency (40 cm) was found during May in almost the whole area of interest.

Water temperature

Most apposite temperature for Hilsa occurrence varies between 20 and 25°C (Jafri, 1988). From the analyses, we can state the most suitable time for Hilsa stock occurrence is during January while the temperature range becomes moderate during November (19 – 28 °C). Considering aspect of temperature, March is moderately suitable for Hilsa stock occurrence while May is the least suitable.

pH

Suitable pH range for occurrence of Hilsa habitat is 6 - 7.5 (Ghosh and Nagpal, 1968) which makes January the most suitable month for Hilsa stock occurrence according to the image analyses. The range becomes moderate (5 - 9) from most suitable in November and May; while almost the whole study area remains unsuitable for Hilsa stock during March. The above spatio-temporal dynamics of ecological parameters shows that Hilsa abundance increases from the coastal areas to the inland rivers from November-January to March-May. Because there is a 'nutrition front' which encroaches from the sea to the inland rivers during the dry season (November – May). Hilsa swims to the inland water with the nutrition front against the upstream freshwater flow, and the more Hilsa swims against the freshwater flow, the more it remains tasty and for this reason, 'Padma's Hilsa' is renowned in the region.

Discussion

The 11 habitat parameters used were Chlorophyll-a, dissolved oxygen, electrical conductivity, salinity, total dissolved solids, total suspended solids, phosphate-phosphorus (PO_4-P), water current velocity, water transparency, water temperature and pH of upper 2 m of water in the study area. The sensitivity analysis using dendrogram revealed how different bands of Landsat 8 images during four time periods representing above parameters. The band and band ratio for instance B7, B7/B9, B1, B9, B2 found to show good performance in monitoring water transparency. To demonstrate water temperature the band of B7 and B7/B9 performance are significant as well as to monitor the water current the use of B10, B6, B11 are very effective in January 2014 (Bladon *et al.*, 2016). Eleven ecological parameters are truly represented except in some cases, through all bands and band ratios in March 2014 and May 2014. The cluster analysis using dendrogram showed that the band 1 (coastal/aerosol); 5 (NIR); 6, 7 (SWIR) and their ratios performed very well for evaluating the spatio-temporal dynamics of ecological parameters which enhance the Hilsa habitat.

The boat density during the ban period may be attributed to the unauthorized Hilsha or 'Jatka' fishing. This indiscriminate 'Jatka' fishing had been contributing to the drastic degradation in Hilsa population and catch in coastal region of Bangladesh. However, recent conservation efforts with enforcement of the ban period and sanctuaries, the illegal fishing has decreased in comparison with number of fishing boats in non-ban period. The RADAR data could detect the small boats if the digital number (DN) of Radarsat-2 image are in the threshold values >3000. However, the spatial resolution of 5 m of Radarsat-2 images are still low to detect the small boats of sizes 5-7 m long and 1-2 m width.

The resultant maps are found to be instantly informative and provided very clear visual indication of which areas are most important as Hilsa habitat and which areas are frequented by the fishers. The maps also provided quantitative data, which used to analyze the relative impact of different Hilsa conservation and management to ensure that conservation goals may be met at minimum negative impact to the habitats and the industry.

Conclusion

Remotely sensed Radarsat-2 RADAR and Landsat-8 optical data were found to be capable of identifying different boats for Hilsa fishing during ban and non-ban periods and spatio-temporal dynamics of ecological parameters in coastal rivers. The following major findings can be summarized from the remotely sensed data analyses:

- High (5-m spatial) resolution Radarsat-2 images could be successfully used for identifying, mapping and monitoring boats for fishing during all season as it can penetrate clouds which visible sensors cannot. Microwave RADAR data are collected mainly during mid-night (12 am), hence, it found to be very useful in monitoring when people normally go for fishing during ban period. These RADAR images can be used for monitoring fishing boats during ban and non-ban periods.

- The freely available Landsat-8 optical images performed well in mapping ecological parameters and showed significant relationships between image and the observed ecological parameters those linked to Hilsa habitat. Hence, Landsat-8 imagery may be used for monitoring Hilsa abundance areas mainly during cloud-free dry period.
- It was revealed that sanctuary area holds double density of boats (18 boats/10 km^2) of the control area (9 boats/10 km^2) during non-ban period. Analyzing the boat cluster image, it may be said, Bighai River works as the fishing hotspot during ban period as a lot of boats cluster in areas of Bighai River during that time. And during non-ban period, in portions of Kalabadar and Tentulia Rivers i.e., portions of sanctuary were found to be fishing hotspots. However, downstream portion of Tentulia River, which is sanctuary area, holds a moderate 15-25 boats cluster during both ban and non-ban periods.
- There had been still a large number of small fishing boats (67; 4 boats/10 km^2) catch Hilsa illegally on 22 April 2015 during ban period in Tentulia river although larger number of boats (151; 10 boats/10 km^2) were present on 3 July 2016 during non-ban period.
- Spatio-temporal dynamics of eleven ecological parameters could be analyzed with the satisfactory high coefficient of determinants (R^2) (0.33 – 0.99) between observed ecological parameters and the band values, band ratios or indices of NDVI and NDWI.
- The near and short-wave infrared bands of Landsat 8 (bands 5, 6, and 7) and the new coastal/aerosol band (band 1) of Landsat 8 imagery performed very well for assessing the spatio-temporal dynamics of ecological parameters which improve the Hilsa habitat.
- From the spatio-temporal dynamics of ecological parameters (November - May), it can be summarized that Hilsa abundance may increase from the coastal areas to the inland rivers from November-January to March-May. There is a 'nutrition front' which arrives from the seashore to the inland rivers during the dry season (November - May). Hilsa swims to the inland water with this nutrition front against the upstream freshwater flow and get caught in a massive number. The exiting ban period is from March to end of April. So, in line with the nutrition front and the boat counts, re-adjustment for Hilsa fishing ban may be required.
- A new legislative structure may be time specific instead of area-time specific e.g., ban period for spawning may be readjusted during 15 days of each lunar calendar month (7 days before and 7 days after the full moon day) in all rivers including small and narrow rivers like Payra and Kaliganga which become fishing hotspots. So, ban should be implemented in all rivers during spawning period of Hilsa, as nutrient front proceeds to those inland rivers too.

Acknowledgements

The authors acknowledge the financial support of USAID. Special thanks are given to the EcoFishBD project team of the WorldFish for their active support in the implementation of the research and to the colleagues of Department of Fisheries, Bangladesh for their continuous inspiration, sincere interest and support towards the implementation of the research activities.

References

Amin, S.M.N., Rahman, M.A., Mazid, M.A., Haldar, G.C. and Milton, D.A. 2008. Catch per unit effort, exploitation level and production of Hilsa Shad in Bangladesh waters. Asian Fish. Sci., 21:175-187

Amin, S.N., Rahman, M.A., Haldar, G.C., Mazid, M.A. and Milton, D. 2002. Population dynamics and stock assessment of Hilsa shad, *Tenualosa ilisha* in Bangladesh. Asian Fish. Sci. 15(2): 123-128.

Anderberg, M. R. 2014. Cluster Analysis for Applications: Probability and Mathematical Statistics: A Series of Monographs And Textbooks. Vol. 19. Academic Press, 2014, 376 pp.

APHA-AWWA-WEF 1999. Standard Methods for the Examination of Water and Wastewater. American Public Health Association, Water Environment Federation. Retrieved July 24, 2016, from www.mwa.co.th/download/file_upload/SMWW_1000-3000.pdf

BBS (Bangladesh Bureau of Statistics) 2014. Statistical Year Book Bangladesh, 2014. Edn. 34, 154.

Bladon, A., Syed, A., Hassan, S.M.T., Raihan, A.T., Uddin, N., Ali, L., Ali, S., Hussein, B., Mohammed, E.Y., Porras, I., Steele, P. 2016. Finding evidence for the impact of incentive-based hilsa fishery management in Bangladesh: combining theory testing and remote sensing methods. IIED Working Paper. IIED, London. http://pubs.iied.org/16625IIED ISBN 978-1-78431-424-8

BOBP (Bay of Bengal Programme), 1987. Hilsa Investigation in Bangladesh. UNDP, Colombo, 113.

Choudhury, B.J., Sippel, S.J., Hamilton, S.K., Melak, J.M. 1994. Determination of inundation area in the Amazon River floodplain using the SMMR 37 GHz polarization difference. Remote Sensing and Environment, 48(1): 70-76.

Cloude, S.R., and Pottier, E. 1996. A review of target decomposition theorems in radar polarimetry. IEEE Transactions on Geoscience and Remote Sensing 34(2): 498–518.

Crisp, D.J., 2004. The state-of-the art in ship detection in Synthetic Aperture Radar Imagery. Research Report DSTO-RR-0272, Defense Science and Technology Organisation, Australia.

Devi, H.K.A., 2006. Thresholding: A pixel-level image processing methodology preprocessing technique for an OCR system for the Brahmi Script. Ancient Asia 1(1): 161-165.

DoF (Department of Fisheries), 2019. Fish Week Compendium 2019 (Bangladesh: Department of Fisheries, Ministry of Fisheries & Livestock, Bangladesh), 160 pp.

ERDAS, 2014. Chapter-8: Radar concepts. ERDAS Field Guide-2014. Publisher: ERDAS Inc., Atlanta, Georgia: 307-342.

Felip, M. and Catalan, J., 2000. The relationship between phytoplankton bio-volume and chlorophyll in a deep oligotrophic lake: decoupling in their spatial and temporal maxima. Journal of Plankton Research, 22(1): 91-106.

Ghosh, A.H. and Nagpal, T.D., 1968. On the winter breeding of *Hilsa ilisha* (Hamilton) in the Ganga River system. Proceedings of Indo-Pacific Fisheries Council, 13(11): 132–142.

Greidanus, H., Clayton, P., Indregard, M., Staples, G., Suzuki, N., Vachon, P., Wackerman, C., Tennvassas, T., Mallorqui, J., Kourti, N., Ringrose, R. and Melief, H., 2004. Benchmarking operational SAR ship detection. In: Proc. International Geosci. Remote Sens. Symposium (IGARSS), 20-24, 2004, Anchorage, Alaska.

Haase, P., 1995. Spatial pattern analysis in ecology based on Ripley's K function: Introduction and methods of edge correction. Journal of vegetation science, 6(4), 575-582.

Hamilton-Buchanan, F. 1822. An account of the fishes found in river Ganges and its branches. Edinburgh and London, 405 pp.

Hasan, K.M.M., Ahmed, Z.F., Wahab, M.A. and Mohammed, E.Y., 2016. Food and feeding ecology of Hilsa (*Tenualosa ilisha*) in Bangladesh's Meghna River basin. IIED, 7-11.

Hoffer, R.M., (1994). Challenges in developing and applying remote sensing to ecosystem management. Remote Sensing and GIS in Ecosystem Management, Editor: V. Alaric Sample, Publisher: Island Press, Washington D.C., Covelo, California, 25-40.

Hossain, M. S., Sarker, S., Chowdhury, S. R., & Sharifuzzaman, S. M. (2014). Discovering spawning ground of Hilsa shad (*Tenualosa ilisha*) in the coastal waters of Bangladesh. Ecological Modeling, 282, 59-68.

Hurni, H., 1997. Concepts of sustainable land management. ITC Journal, 3(4): 210-215.

Hussain, M., 2016. Fisheries Statistics In Bangladesh: Issues, Challenges And Plans. Asia and Pacific Commission on Agricultural Statistics (from February 15 to February 19, 2016), Bhutan.

Iehara, M., Ouchi, K., Takami, I., Morimura, K. and Kumano, S., 2001. Detection of ships using cross-correlation of split-look SAR images. IGARSS 2001. Scanning the Present and Resolving the Future. Proceedings. IEEE 2001 International Geoscience and Remote Sensing Symposium (Cat. No.01CH37217). doi:10.1109/igarss.2001.977078

Jafri, S.I.H., 1988. Biology and fishery of river shad (palla)-a review. Pakistan Journal of Agricultural Research, 9(2): 252-263.

Jahan, I., Ahsan, D. and Farque, M.H., 2015. Fishers' local knowledge on impact of climate change and anthropogenic interferences on Hilsa fishery in South Asia: evidence from Bangladesh. Environment, Development and Sustainability 19: 461-478

Jhingran, V.G. and Natarajan, A.V., 1969. A study of the fisheries and fish populations of the Chilka Lake during the period 1957-65. J. Inland Fish. Soc. India, 1: 49-126.

Kashani, A.G. and Graettinger, A.J. 2015. Cluster-based roof covering damage detection in ground-based LIDAR data. Automation in Construction, 58: 19-27.

Mather, P.M. and Koch, M., 1999. Preprocessing of remotely sensed data. Computer processing of remotely sensed images - An introduction, John Wiley & Sons Ltd, The Atrium, Southern Gate, Chichester, West Sussex, PO19 8SQ, UK.

Mcfeeters, S.K., 1996. The use of the Normalized Difference Water Index (NDWI) in the delineation of open water features. International Journal of Remote Sensing, 17(7): 1425–1432. http://doi.org/10.1080/01431169608948714

MDA, 2005. Ship Detection Performance with RADARSAT-1 and ENVISAT, MDA Technical paper 01-3356-4, 2005.

MDA, 2007. Polar Epsilon: Near Rear Time Ship Detection, POL-RP-52-5672, 2007.

MDA, 2011. Marine Environmental Surveillance Improvements with RADARSAT-2, GSI 191, June 2011.

Mojumdar, C.H., 1939. Foreshore fishing in the eastern parts of the Bay of Bengal. Sci. Cult, 5(4): 211-219.

Nair, P.V., 1958. Seasonal changes in the gonads of *Hilsa ilisha* (Hamilton). Philipp. J. Sci, 87(3): 255-276.

NASA, 2016. Landsat 8. Retrieved July 25, 2016, from http://landsat.gsfc.nasa.gov/?p=3186

Otsu, N., 1979. A threshold selection method from gray-level histograms. IEEE Trans. Sys. Man. Cyber. 9(1): 62–66

Quddus, M.M., Shimizu, A.M.and Nose, Y., 1984. Comparison of age and growth of two types of *Hilsa ilisha* in Bangladesh waters. Bulletin of Japanese Society of Scientific Fisheries, 50 (1): 51–57

Rahman, M., Rahman, B., Hasan, S., Flura, A.T. and Haidar, M., 2013. Impact of eleven days fishing ban in the major spawning grounds of hilsa (Tenualosa ilisha) on its breeding success. Bangladesh Research publications Journal 9: 116–122.

Rahman, M.J. 2001. Population biology and management of Hilsa shad (*Tenualosa ilisha*) in Bangladesh, PhD thesis, University of Hull.

Rahman, M.J. and Cowx, I.G., 2006. Lunar Periodicity in Growth Increment formation in otoloiths of Hilsha shad Tenusa Ilisha-Clupiedae in Bangladesh Waters, Fisheries Sci. 81: 342-344

Rahman, M.J., Wahab, M.A. Nahiduzzaman, M., Haque, A.B.M. and Cohen, P., 2018. Hilsa fishery management in Bangladesh, IOP Conf. Series: Earth and Environmental Science 414 (2020) 012018 IOP Publishing doi:10.1088/1755-1315/414/1/012018

Raney, R.K., 1998. The delay/Doppler radar altimeter. IEEE Transactions of Geoscience and Remote Sensing, 36(5): 1578-1588.

Shafi, M., Quddus, M.M.A., Islam, N., 1978. Maturation and spawning of Hilsa Ilisha (Hamilton-Buchanan) of the river Meghna. Dacca University Studies-B, 26 (2), 63–71.

Skrunes, S. Brekke, C. and Eltoft, T., 2014. Characterization of Marine Surface Slicks by Radarsat-2 Multi-Polarization Features, IEEE Transactions on Geoscience and Remote Sensing,

Staples, G. and Rodrigues, D.F., 2013. Maritime environmental surveillance with RADARSAT-2 Anais XVI Simpósio Brasileiro de Sensoriamento Remoto - SBSR, Foz do Iguaçu, PR, Brasil, 13 a 18 de abril de 2013, INPE

Ulaby, F.T., Moore, R.K. and Fung, A.K., 1981. Microwave Remote Sensing: Microwave Remote Sensing Fundamentals and Radiometry. Vol. 1., Addison-Wesley Publishing Company, Advanced Book Program/World Science Division.

Vachon, P. and Tunaley, J., 2007. RADARSAT-2 Maritime Surveillance Mode Compatibility Study, DRDC Ottawa ECR-2005-054, 2005.

Vachon, P.W., Campbell, J.W.M., Bjerkelund, C.A., Dobson, F.W. and Rey, M.T., 1999. Ship detection by the RADARSAT SAR: Validation of detection model predictions. Canadian Journal of Remote Sensing, 23(1): 48-59.

Wahab, M.A., Ahmed, Z.F. and Hasan, K.M.M. 2014. Darwin initiative annual report on economic incentives to conserve Hilsa fish (*Tenualosa ilisha*) in Bangladesh (from September 2013 to August 2014). BAU-IIED-BCAS, Dhaka, Bangladesh.

Zhang, Y., Moges, S. and Block, P., 2016. Optimal Cluster Analysis for Objective Regionalization of Seasonal Precipitation in Regions of High Spatial-Temporal Variability: Application to Western Ethiopia. Journal of Climate, 29: 3697-3717

Impact of Climate Change on Hydrological Cycle, Ecosystem, Fisheries and Food Security, Pages: 539–546
Edited by: B. Madhusoodana Kurup, M.R. Boopendranath, M. Harikrishnan and A.V. Shibu

CHAPTER - 44

Implications and Ramifications of Extreme Weather on the Livelihood and Economy of Aquafarmers in Central Kerala

Patralika Mukhopadhyay*, Krishna Pankaj, R. Shruthy, A.V. Shibu, and M. Harikrishnan

School of Industrial Fisheries, Cochin University of Science and Technology, Kochi-682 016, Kerala, India
**E-mail: patralika2710@gmail.com*

Climate change is presently acknowledged as a significant threat to the biodiversity. It is considered as a man-made global environmental challenge (Anon, 2010). International endeavours to brace climate change began with the adoption of the United Nations Framework Convention on Climate Change in 1992. The importance and significance of the vulnerability of natural and human systems to climatic changes and adaptation to such changes is increasingly being realized (Anon, 2010). Consequently, there is now a growing recognition of the vulnerability and impacts of climate change on the key sectors of economic development (Anon, 2010). Most of the cities in the world are now the hotspots of risk from extreme weather events and the level of risk in such cities are likely to increase due to a combination of development and rising intensities of extreme weather events (Re, 2004; Hanson *et al.*, 2011; Ranger *et al.*, 2011).

Climate change has caused extreme impacts in India. Rise in average global temperatures has been attributed to lack of rainfall for prolonged periods followed by sudden bout of excessive rainfall, causing extreme weather events, particularly floods, which result in loss of lives, shelter, agricultural yields leading to huge revenue losses. The Southwest monsoon in Kerala rendered 30 percent excess rainfall in the state of Kerala. All 44 rivers in the state were overflowing and water from 35 of its 39 dams had to be released leading to extreme conditions of flood in the state. (Anon, 2018). This unprecedented deluge unleashed by excess rainfall, overflowing rivers and brimming dams had a grievous effect on the fish farmers of Kerala. The river Periyar, the longest river in the state (PWD, 1974; CESS, 1984) supports a vast irrigation system and is also the lifeline of Central Kerala. The presence of rich freshwater source, supported profitable fresh water fish farming in these regions. Ornamental fish farming is also getting popular in this area since the demand for the same is steadily growing in the state (Velayudhan, 2005). However, fish farming is facing difficulties because of the climate change and extreme weather conditions. An attempt is made in this study to appraise the damages caused by recent deluge and extreme weather on fish farming activities and economic loss for the fish farmers in the area under study.

Materials and Methods

Three farms located at Perumbavur, located at the bank of river Periyar in the district of Ernakulam and the second study area Pizhala, located in the Kadamakkudi panchayat, Kerala, situated on the Northern side of Kochi and Southern side of Varapuzha which is surrounded by the river Periyar were selected for the study. The study includes collection of onsite data using both qualitative and quantitative methods (Hartley, 1997). The survey was conducted during the period from December 2018 to April 2019. Finfish cage culture is one of the prime sources of income for fish farmers in this area. Impact of deluge was severe on cage culture units located at Pizhala and hence it is also brought under the survey to assess the aftermath of deluge on cage farming operations.

The data collection was carried out in the selected farms of the Perumbavur and Pizhala area through direct interviews with the fish farmers of the respective areas by using structured questionnaires. The questionnaire used composed of three sections. Section 1 included questions related to general information on the farming site such as experience in fish farming, total number of ponds, total area of ponds, average water depth, nature of ownership, type of culture practices, types of species being cultured, stocking density, harvesting time and yield. Section 2 included questions related to the consequences of the flood that took place (date of occurrence, how the farm was affected, flood stage water level, submergence span of farm, distance of the farm to the nearest water body, number and type of species lost during the flood, quantum of loss incurred due to flood, effect of flood on the fishes, nature of damage on the ponds, details of equipments or infrastructure or machineries lost and cost of the total loss incurred. Section 3 included questions related to post flood management and mitigation measures adopted by farmers and authorities.

Results and Discussion

Freshwater Fish Farms

Among the three farms surveyed in Perumbavur , farm A, started in 2008 is one of the largest fish farms located near to Keezhillam canal where both edible and ornamental species are cultured in massive scale. Arowana (*Scleropages formosus)*, Giant gourami (*Osphronemus goramy*), Alligator gar (*Atractosteus spatula*), Koi carp (*Cyprinus carpio*), Tilapia (*Oreochromis niloticus*) Gold fish (*Carassius auratus*), aquarium shark (*Epalzeorhynchos bicolor*), Angel fish (*Pterophyllum scalare*), Guppy (*Poecilia reticulata*), Molly (*Poecilia sphenops*), Platyfish (*Xiphophorus maculatus*), Sword tail (*Xiphophorus helleri*), Oscar (*Astronotus ocellatus*), Flower horn (*Vieja synspilus*), Piranha (*Pygocentrus nattereri*), Indian major carps (*Catla catla, Labeo rohita, Cirrhinus mrigala*) and two exotic carps (*Hypophthalmichthys molitrix*; *Ctenopharyngodon idella*) are mainly cultured in this farm. The farm consists of a total area of 2.6 hectares with indoor and outdoor farming units (Table 1).

Table 1: Total number, total area, average area of ponds and average water depth in selected farms

	Farm 1	Farm 2	Farm 3
Total No. of Ponds/Tanks	20	55	15
Total area of ponds (hectare)	2.6	1.2	2.02
Average area of ponds (hectare)	2.02	1	1.6
Average water depth (m) - Rainy Season	4	3	4
Average water depth (m) - Dry Season	3	2	3

Outdoor farm comprises of 20 ponds and the indoor farming unit works with funding of FIRMA and NFDB. In indoor farm, rearing of angel fish (*Pterophyllum scalare*), Guppy (*Poecilia reticulata*), Platyfish (*Xiphophorus maculatus)*, Koi carp (*Cyprinus carpio*) were done and in outdoor, a wide variety of fishes such as Arowana (*Scleropages formosus)*, Tilapia (*Oreochromis niloticus*), Goldfish (*Carassius auratus*), Indian major carps (*Catla catla, Labeo rohita, Cirrhinus mrigala*), Alligator gar (*Atractosteus spatula*), Giant gourami (*Osphronemus goramy*), Piranha (*Pygocentrus nattereri*), Arapaima (*Arapaima gigas*), Malabar snakehead (*Channa diplogramma*), Aquarium shark (*Epalzeorhynchos bicolor)*, and exotic carps (*Hypophthalmichthys molitrix*; *Ctenopharyngodon idella)* were stocked. The average pond size and depth of the pond are 2.02 ha and 3.5 m, respectively. Polyculture was carried out in the farm based on the feeding habits of the fishes. Multiple stocking twice in the year was adopted. The rate of per hectare cost of fish production in this farm is given in Table 2. A yield of 5000 kg/ha is obtained with a stocking density of 7000 fingerlings per hectare. The farmer maintained the physico-chemical parameters of the ponds within the optimum limits and the fishes are sampled periodically to check on their health, growth, presence of parasites or any kind of bacterial diseases.

Table 2: Per hectare cost of fish production in selected farms (A,B,C) at Perumbavur area.

	Farm A	Farm B	Farm C	Farm A	Farm B	Farm C
	Cost, Rs.	%	Cost, Rs.	%	Cost, Rs.	%
Human labour	80,000	4.14	68,000	4.29	75,000	5.40
Fingerlings	5,00,000	25.90	4,50,000	28.40	350,000	25.58
Feed	480000	24.80	3,85,000	24.30	4,50,000	32.89
Fertilizer	90,000	4.66	80,500	5.08	70,000	5.10
Cowdung and manure	90,000	4.66	65,000	4.10	50,000	3.65
Lime and disinfectants	85,000	4.40	58,500	3.69	48,000	3.50
Miscellaneous	50,000	2.59	40,000	2.52	25,000	1.82
Interest on operating cost	55000	2.84	35,900	2.26	20,000	1.46
Total variable costs	1,43,0000	–	1,18,2900	–	10,88,000	–
Land use cost	1,50,000	7.70	1,00,000	6.31	80,000	5.84
Depreciation of farm implements	3,50,000	18.10	3,00,000	18.90	2,00,000	14.60
Total fixed costs	5,00,000	–	4,00,000	–	2,80,000	–
Total costs	19,30,000	100.00	15,82,900	100.00	13,68,000	100.00

The farm B, located near Keezhilam canal in Perumbavur region is under lease ownership. The owner of the farm gives rights to the lessee to use his property in exchange of periodical rental payments. The total extent of pond is about 1.2 ha and the average pond depth is found to be 3.5 m (Table 1). The fish farming season starts with stocking the ponds during January and harvested during June and July. A large variety of edible fish, ornamental fishes and aquatic plants are cultivated in this unit. The farmer stocks edible species viz., Indian major carps (*Catla catla*, *Labeo rohita*, *Cirrhinus mrigala)*, one exotic carp (*Hypophthalmichthys molitrix)*, Pangasius (*Pangasianodon hypophthalmus*) and Tilapia (*Oreochromis niloticus*) in 60% of total farm area. Polyculture is mainly practised here which improves water quality and create better balance in the pond ecosystem thereby enhances fish production from the farm. Besides, monoculture of Pangasius species is also undertaken in this farm which yields considerable income to the farmer. The Pangasius *(Pangasianodon hypophthalmus)* stocked at the rate of 10-20 fish/m^2 yields up to 2 ton/ha/year. Usually this species is stocked at high densities and grown for around 6 -8 months before reaching its harvest weight of around 1 kg. The major varieties of ornamental species stocked in this farm includes, fighter fish (*Betta splendens*), angel fish (*Pterophyllum scalare*), guppy (*Poecilia reticulata*), gold fish (*Carassius auratus*), southern platy fish (*Xiphophorus maculatus*), green sword tail (*Xiphophorus helleri*), tiger barb (*Puntigrus tetrazona*) and sucker fish (*Hypostomus plecostomus*). Arrowana (*Scleropages formosus*), one of the sought after ornamental fish species, fetching very high unit price in ornamental fish markets was also cultured in this farm. The physico-chemical parameters of the ponds were kept within the optimum limits and the fishes are sampled periodically to check on their health, growth, presence of parasites or any kind of diseases. The farmer maintains wholesale and retail outlets attached to the farm.

The third farm C, established in 2011, located in the banks of Keezhillam canal in Perumbavur region is under individual ownership. Indian Major Carps (*Catla catla, Labeo rohita, Cirrhinus mrigala*), 3 exotic carps (*Hypophthalmichthys molitrix, Ctenopharyngodon idella, Cyprinus carpio*), Tilapia (*Oreochromis niloticus*) and ornamental fishes like Guppy (*Poecilia reticulata*), Molly (*Poecilia sphenops*), Angel fish (*Pterophyllum scalare*) and Gold fish (*Carassius auratus*) were cultivated in this farm. The farm yields high profit on both ornamental and edible fish culture. The average pond size here was found to be 1.6 ha and average pond depth was 3.5 m (Table 1). The seasonal fish farming starts from April to December and was harvested from December to January. The per hectare cost of fish production in this farm is given in Table 2. The overall fish production of this farm is about 4800 kg/ha/year with a stocking density of 6500/ha/year. Water quality parameters and the health of the fishes are periodically monitored.

Cage culture

The cage culture farms located at Pizhala was also subjected to survey. Different cage materials were used for cage farms like High Density Polyethylene (HDPE), Galvanised iron, and PVC pipes. The details of dimension and stocking density of cages ranged between 2x2x1.5 m^3 to 8x4x2 m^3 for 200 to 1000 fishes, respectively. The average investment details of cage culture units in Pizhala are given in Table 3. Introduction of low cost cages and availability of good quality seeds of finfishes facilitated popularisation of cage culture in Pizhala area (Aswathy and Imelda, 2018).The average stocking density depends on the carrying capacity of the cages and the feeding habits of the cultured species. The selection of species is based on several criteria like growth rate, food conversion ability, feeding habit, availability of quality seeds, market demand and disease resistance capability. Composite culture of Sea bass (*Lates calcarifer*) and Pearl spot (*Etroplus suratensis*) are commonly practiced in cage culture units at Pizhala. About 75 to 78 % survival rate was recorded in cages. The average period of farming duration was 6 to 8 months and the yield per cage ranges between 650 and 835kg. The composition of Seabass and Pearl spot per cage was 85 and 15%, respectively. In order to avoid net fouling and occurrence of diseases, the cages are kept under constant monitoring and webbing panels were changed periodically.

Table 3: Average investment details of cage culture units in Pizhala (in Rs.)

Particulars	Size 8x4x2m^3	Size 6x3x1.5m^3	Size 2.5x2.5x1.5m^3
Fixed investment	26,467	20,500	12,500
License charges	1,500	1,500	1,500
Labour charges	36,000	30,000	18,000
Input cost – Fish fingerlings	40,500	32,000	24,000
Input cost – Fish Feed	1,45,500	1,25,000	1,10,000
Miscellaneous expenses	10,000	8,000	5,000
Operational cost	2,33,500	1,96,500	1,58,500
Total cost	2,59,967	2,17,000	1,71,000
Gross revenue	4,04,500	3,20,000	2,70,000
Net profit	**1,44,533**	**1,03,000**	**99,000**

Implications of Flood

From the study, it was found that during the period of excess rainfall from June to August, the water depth in the ponds increased by 33.33% in farm A. The canal water has overflowed and reached the farm by 2nd week of August. 10 earthen ponds were completely submerged in water. Fish stocked worth Rs.5 lakhs were washed away by flood. These farms incurred loss of infrastructure, physical facilities, equipments, utensils (air compressors, water pump, motor, cast nets), fish seed fertilizers, etc. The flooding in Farm A caused an overall economic loss to the tune of 20 lakhs. Farm B, also suffered similar consequences due to flood. The fish feeds and the fertilizer bags were washed away in the flood waters. 50% of the Arowanas which were the chief profit making species of that farm, were swept away. The estimated economic loss due to flooding was around 20 lakhs. Flooding caused significant damages to the Farm C as well. Flooding caused an increase in water level by 35% along with damages to pond dykes, fish seed, fish feed, fertilizers and also the marketable sized fishes. The total loss to the farm was around Rs. 25 lakhs.

Cage fish farmers of Pizhala have suffered a major setback with the recent flooding of the rivers and water bodies around Ernakulam. Around 200 fish cages located near to Pizhala island suffered substantial economic loss. About 68% of cages were washed away in flood waters and fully damaged. 14% of cages were partially damaged and only 8% of cages survived after the floods. Economic loss due to loss of infrastructure was estimated as rupees Rs. 2.96 crores. The estimated economic loss of the stocked species of Asian seabass (*Lates calcarifer*) and Pearl spot (*Etroplus suratensis*), in Pizhala area was about Rs. 2.82 crores.

Post-Flood Management of Farms

Almost all the farmers have constructed pond dykes to a significant level to prevent such losses in future. Some of the farms have not yet recovered from the damages caused due the floods. In Farm A, the farmers have constructed fencing structures around the ponds to prevent the escape of larger fishes. Ponds needed to be deepened by removing the silts and debris. The earthen bunds also have been reconstructed in all the farms. Increasing the height of pond embankments and construction of containment structures will prevent the escape of farm fish to open waters (Casimiro *et al.*, 2018). Possibility of relocating the ponds from the flood channels and excavating new channels to divert flood water could also be explored. The escape of exotic species viz., Arapaima and Alligator Gar can cause threat to aquatic biodiversity during post-flood period, which could be controlled either by keeping wall of webbing along the embankments or fixing barriers in outlets.

Farmers involved in cage culture operations are now adopting safety measures to protect the cages from climate related eventualities. In order to save the cages from the flood water current, the newly built cages were erected by leaving 1 m gap from bottom of water bodies, which will allow free flow of sediments debris below bottom of cage. Additional mooring facilities are also provided to prevent the erosion of cages. Farmers undergo training to empower themselves to address climate change extremities so that such losses incurred can be reduced in future. The farmers need financial support and due compensation during the time of natural disasters and catastrophes, to revive the farming operation.

Conclusion

In many parts of the world, during the recent past, natural and human-induced disasters have become more frequent and increasingly destructive. Fishers and aqua-farmers who depend on water resources for their livelihoods have been seriously affected by loss of life and property. In this context, fishers and farmers should be equipped with preparedness programmes before the disaster strikes and there should be empowerment programme for the fish farming communities to adapt to the climate change situations. Institutional support and financial assistance should also extend to fish farmers and cage culture units affected by climate related hazards. Timely loss assessment and release of necessary grants will boost the morale of the farmers and contribute towards better productivity from the inland water sources.

Acknowledgement

The authors would like to express profound gratitude and deep sense of indebtedness to the farmers of all the farms for providing appropriate information required for the study.

References

Anon, 2010. Climate Change and India: A 4x4 Assessment. Government of India. 164 p.

Anon, 2011. Production, Accessibility, Marketing and Consumption Patterns of Freshwater Aquaculture Products in Asia: A Cross-Country Comparison. FAO, Rome. 275 p.

Anon, 2018. The Deluge. Editorial, The Indian Express, August 17, 2018.

Aswathy, N. and Imelda Joseph, 2018. Economic viability of cage farming of Asian seabass in the coastal waters of Kerala. International Journal of Fisheries and Aquatic Studies 6(5): 368-371

Carter, T.R., Jones, R.N., Lu, X., Bhadwal, S., Conde, C., Mearns, L.O., O'Neill, B.C., Rounsevell, M.D.A. and Zurek, M.B., 2007. New assessment methods and the characterisation of future conditions. In: Parry, M.L., Canziani, O.F., Palutikof, J.P., van der Linden, P.J. and Hanson, C.E. (Eds.) Climate Change 2007: Impacts, Adaptation and Vulnerability. Contribution of Working Group II to the Fourth Assessment Report of The Intergovernmental Panel on Climate Change, pp. 133–171, Cambridge University Press, Cambridge,.

Casimiro, A.C.R., Garcia, D.A.Z., Vidotto-Magnoni, A.P., Britton, J.R., Agostinho, A.A., De Almeida, F.S., Orsi, M.L., 2018. Escapes of non-native fish from flooded aquaculture facilities: the case of Paranapanema River, southern Brazil. Zoologia 35: 1-6. https://doi.org/10.3897/zoologia.35.e14638

CESS, 1984. Resource Atlas of Kerala. Centre for Earth Science Studies, Thiruvananthapuram, 39 p.

Hanson, S., Nicholls, R., Ranger, N., Hallegatte, S., Corfee-Morlot, J,, Herweijer, C. And Chateau, J., 2011. A global ranking of port cities with high exposure to climate extremes. Climate Change 104: 89-111. doi: 10.1007/s10584-010-9977-4.

Hartley., J.F., Case study research 1997. In: Cassell, C. and Symon, G. (Eds.), Essential Guide to Qualitative Methods in Organizational Research, pp. 323-333, Sage Publications, London.

PWD, 1974. Water Resources of Kerala. Public Works Department, Government of Kerala, Thiruvananthapuram, 125 p.

Ranger, N., Hallegatte, S., Bhattacharya, S., Bachu, M., Priya, S., Dhore, K., Rafique, F., Mathur, P.,· Naville, N., Henriet, F., Herweijer,C., Pohit, S. and Corfee-Morlot, J., 2011. An assessment of the potential impact of climate change on flood risk in Mumbai. Springer. Climatic Change 104: 139-167. doi 10.1007/s10584-010-9979-2.

Re, M., 2004. Megacities - megarisks: trends and challenges for insurance and risk management. http://www.munichre.com/publications/302-04271_en.pdf.

Velayudhan, T.D., 2005. Development of ornamental fish culture in Kerala - In relation to the upliftment of rural women. Fishing Chimes, 25: 25-27.

Impact of Climate Change on Hydrological Cycle, Ecosystem, Fisheries and Food Security, Pages: 547–556
Edited by: B. Madhusoodana Kurup, M.R. Boopendranath, M. Harikrishnan and A.V. Shibu

CHAPTER - 45

Socio-Economics Profile, Environmental and Climate Change Impacts of Fishermen Community at Pulicat Lake

N. Thirunavukkarasu* and R. Sumithraa

Department of Advanced Zoology & Biotechnology
Dr. Ambedkar Government Arts College (Autonomous)
Vyasarpadi, Chennai-600039, Tamil Nadu, India
**E-mail: marinethiru@gmail.com*

The small scale fisheries sector is mostly the livelihood occupation of the group of population within the extreme poverty that leads to a serious social and economic issues. Due to lack of knowledge in the socio-economic conditions of fishers and fishing communities, leads to poor planning and implementation of various fisheries management programs. Fishing is a highly unorganized and unregulated activity which provides employment and livelihood to millions of fisherfolk either directly or indirectly. Small scale fishing communities are the poorest and most disadvantaged population in the world (Bene *et al.,* 2000). Fishermen even comparatively less income than others in low skill jobs (Scott, 1954). Poverty is one of the major threats to the fishermen communities and is still an unsolved social issue in many of the developing countries, particularly in Asia and Africa. Kurien (2005) state that the global level about 5.8 million small scale fishers earn less than US$1 per day and continue to be in a state of poverty. Fernando (1981) suggested the improving the socio-economic conditions of fishermen at all fishing centres, sufficient cold storage and ice plant facilities must be provided to prevent fish from going waste. The scanty income and the numerous demands upon it, keep the fishermen in a permanent state of bondage to the moneylenders. Sathiadas and Panikkar (1991) have collected information relating to house pattern, ownership of means of production, employment status, occupation, annual income, indebtedness, marketing problems *etc.* Jayaselvi (2016) studied the economic and health status of fishermen in Tiruchendur,

Tamilnadu. Sudharsanam (2005) investigated the risk factors for sick children in a fishermen community on Pondicherry. Bhuyan and Islam (2016) analysed the socio- economic status of the fishing community of the Maghna river adjacent to Narsinghi district at Bangladesh. Jayanthi *et al.* (2016) focused on poverty and income equality of the fishermen community of at Nagapattinam District, Tamilnadu. Mahender *et al.* (2016) carried out the survey of the socio-economics of fishermen community of Chenugonipally Pedda Cheruvu Gadwal, Mahabubanagar District, Telengana. Jasna and Palai (2016) were studied the socio-economics of fishermen population at Chombala Harbour, Kerala. Salim *et al.* (2017) have studied the appraisal of socio economics status of fishers among different sectors of Kerala. The present study focused on the socio-economic profile of Pulicat lake villages.

Climate change is a phenomenon where distribution of weather pattern changes spatially and temporally, modifying the distribution and productivity of marine and fresh water species and impacts the sustainability of fisheries, eventually on livelihood communities that depend on fisheries. Changing seawater temperature and current flows affects marine fish stocks. Warming of water may impact fish diversity, distribution and abundance of phenology. Acidification of water, sea level rise, storms, floods and drought will severely affect the production of fishes. Coastal habitats and resources are likely to be impacted through warming sea temperatures, extreme of nutrient enrichment (eutrophication) and invasive species. Even a small difference of 10°C in seawater may affect their distribution and life processes. Temperature affects other ocean parameters such as salinity, pH and dissolved oxygen. All these factors have a synergistic effect on the biota ranging from microscopic plankton to large fishes. Physiological functions like growth, reproduction, spawning and recruitment are directly influenced by change of temperature. Primary and secondary productions, fish distribution (changes in migration patterns), fish abundance and phenology affected change of temperature. The water temperatures were uniformly high all over the lake 30°C with incredibly high salinities of 70 to 80 ppt (Sanjeeva Raj, 2010). Climate change has its own specific impacts on the biodiversity. Natural calamities by cyclones, grazing and browsing are alarming in causing damage to the flora and fauna of the Pulicat lake. This study focused on environmental and climate change impacts on fishermen community.

Materials and Methods

This study is based on the primary data, the data was collected through 400 personal interviews from various fishermen from the sample villages. The socio-economic parameters were age, gender, education, marital status, family, religion, capital source, housing, income, savings, government schemes and health care awareness analyzed. The factors for the environmental and climate change issues were collected from the fishermen villages.

Results and Discussion

The finding of the current study is revealed that 10.5% of the respondents were of age below 25 whereas 63% of the respondents were between the age group 25 and 50. The respondents of age group between 50 and 65 were 20.5% and people of age above 65 responded were only 6% (Fig. 1). The Fig. 2 revealed that the percentage of male respondents was 76.7% whereas the percentage of female respondents was 23.3%. The Fig.3 represented the educational status of the respondents, in that illiterate rate of the respondents were 39%, 49% of the respondents completed their matriculation education, 10% of the respondents completed their undergraduate whereas only 2% of the respondents completed their postgraduate. From Fig. 4 which is showed that 84% of the respondents were married whereas 16% of the respondents were unmarried. The Fig.5 indicated that out of total respondents, 88% of the population were hindus, 6.7% were christians and 5.3% were muslims. From Fig.6 indicated that the respondents consist of 80.7% of nuclear family whereas 19.3% of the population was joint family. The Fig. 7 revealed that 80.7% of the respondents had been using their own capital for boats whereas 19.3% of the respondents had been borrowing capital for boats for fishing. The data revealed that out of total respondents 1.7% of the respondents were living in hut, 11.5% respondents were living under their own concrete roof house whereas 86.8% were living under government built concrete roof. None of the respondents were living in tiled house (Fig. 8). The Fig. 9 revealed that 38% of the respondents were earning between 2000 and 3500, 38.2% respondents were earning between 3500 and 5000 and 23.8% respondents were earning above 5000 rupees as their monthly income. The Fig.10 stated that 64.7% of the respondents were into savings for their capital as well as for their livelihood but 35.3% of the respondents were not into savings. The Fig. 11 showed the types of boats, out of total respondents 99.8% of the respondents were using normal boats for fishing whereas 0.2% were using mechanized boats for fishing. 71.7% of the respondents got benefits from various government schemes whereas 28.3% were not benefited from the government schemes (Fig. 12). Out of total respondents from the village it shows that 53.7% of the respondents were aware of healthcare schemes and its various aspects whereas 46.3% were not fully aware of the healthcare schemes of the government (Fig. 13).

Fishing still remains a hazardous and stressful occupations its present difficult physical conditions, dislocation, isolation and less than ideal personal habits. Fishermen have prolonged hours of continuous work, which are found to be correlated with high cigarette and alcohol consumption. Diet lacks fruits and vegetables and meals are eaten at very erratic intervals. Fishermen are prone to excess ultraviolet radiation due to constant exposure to sun. Statistical studies in the past have shown that fishermen are prone to develop skin and lip cancer. Fishermen have lower socio-economic status and their illiteracy adds to their poor oral hygiene, which may influence general and oral health. The access of fishermen to dental services is very limited, making regular check-ups and treatment of caries difficult. Fishermen in the coastal area should be provided with proper sanitation and medical facilities, drinking water and electricity.

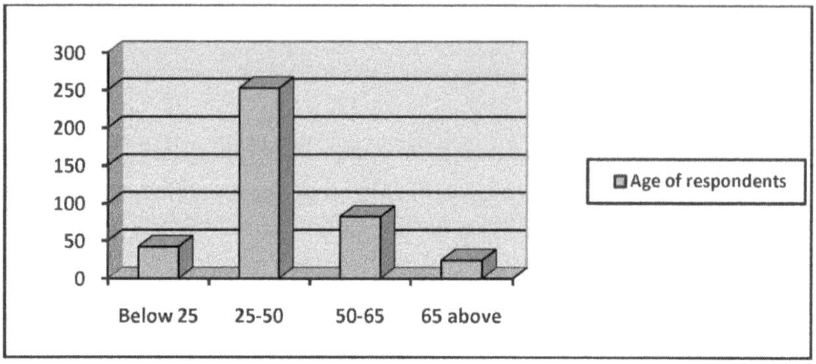

Fig. 1. Age of respondents

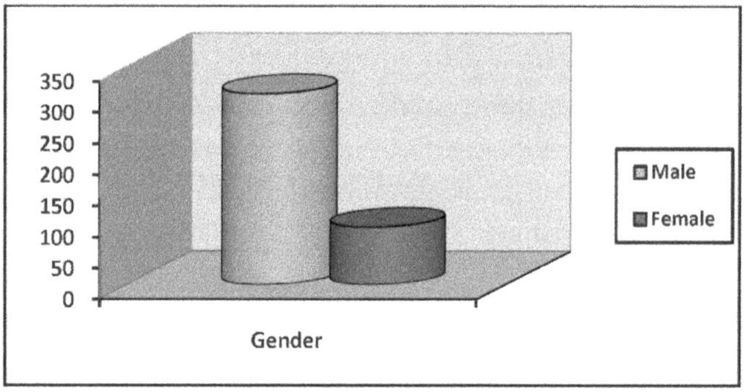

Fig. 2. Gender status

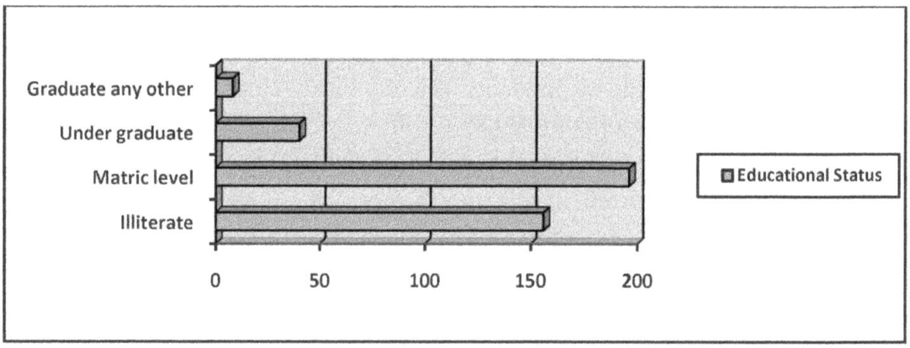

Fig. 3. Educational status

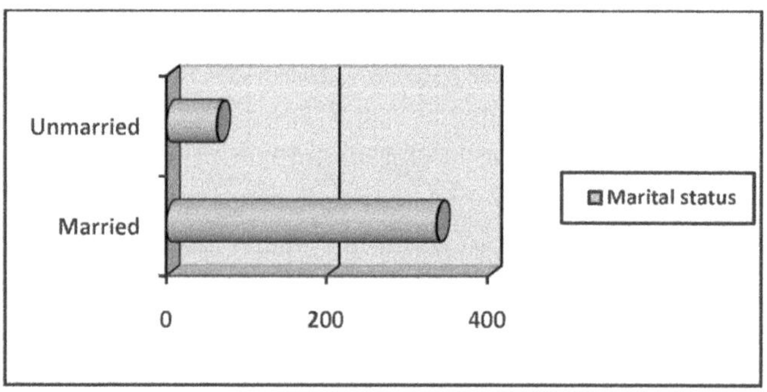

Fig. 4. Marital status

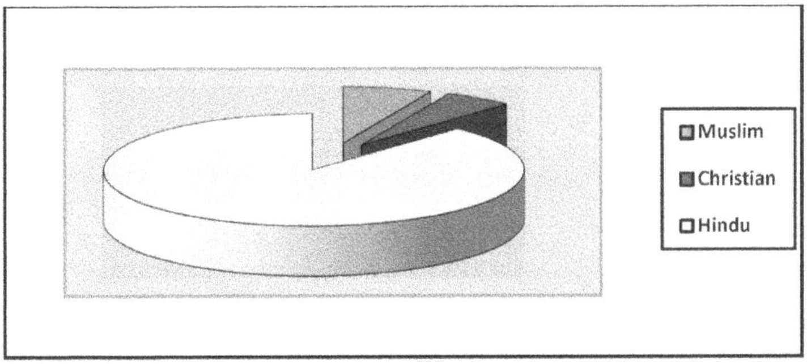

Fig. 5. Religion distribution

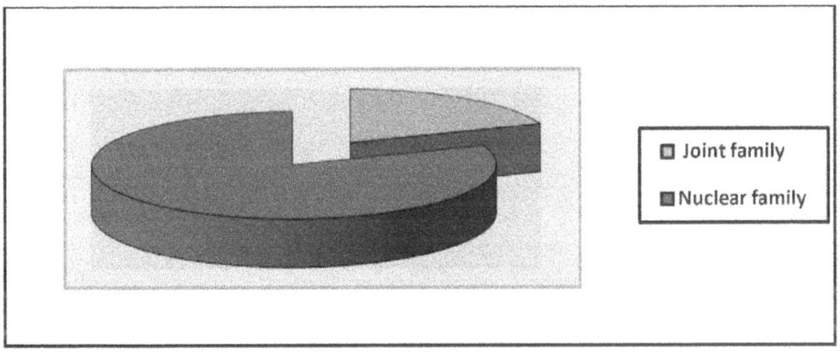

Fig. 6. Type of family

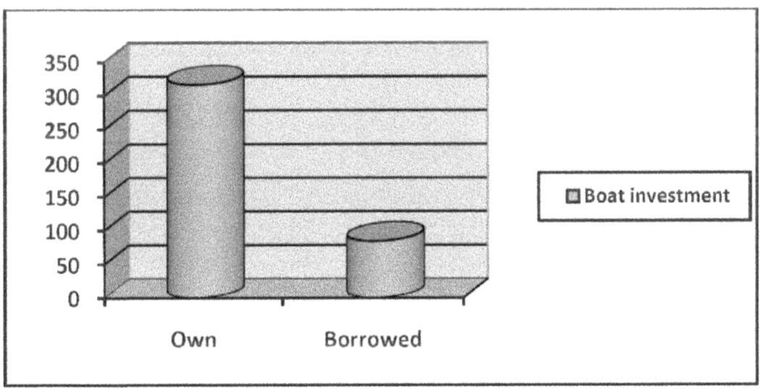

Fig. 7. Capital source

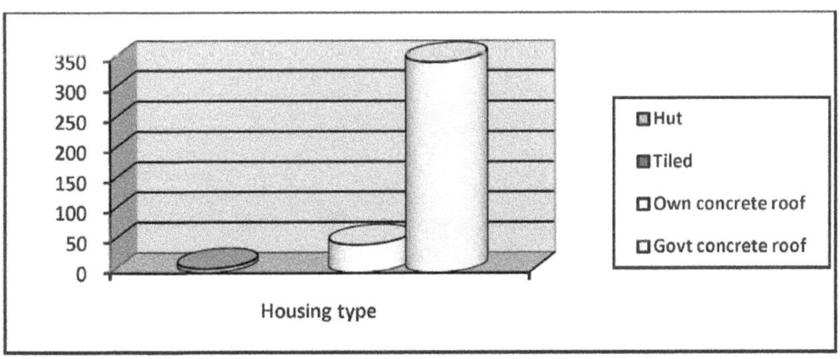

Fig. 8. Type of housing

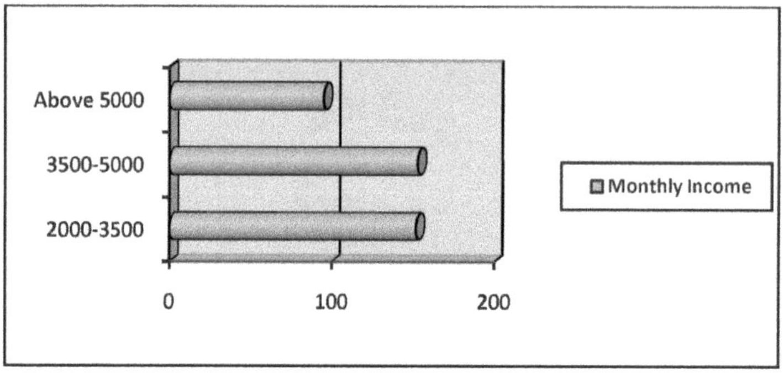

Fig. 9. Monthly income (Rs.)

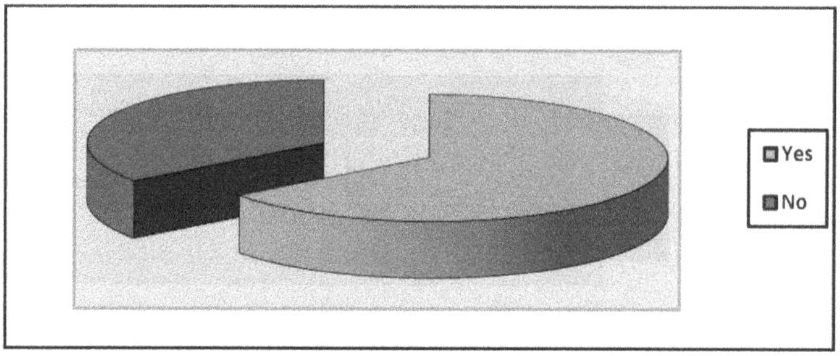

Fig.10. Fishermen savings

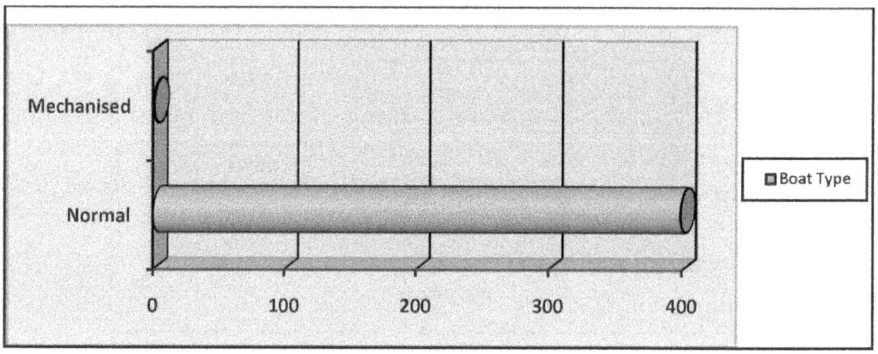

Fig.11. Types of boats

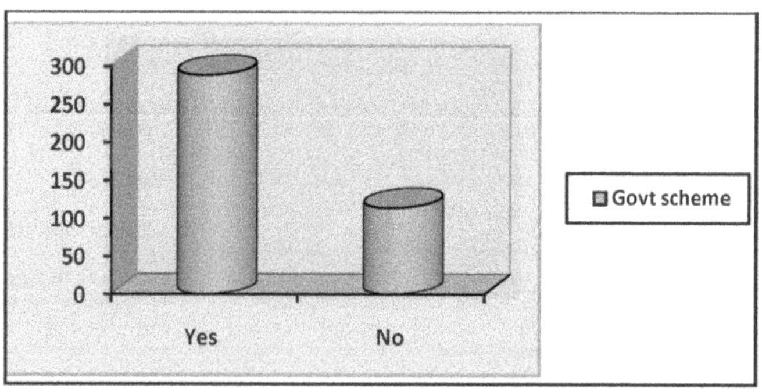

Fig.12. Benefit from government scheme

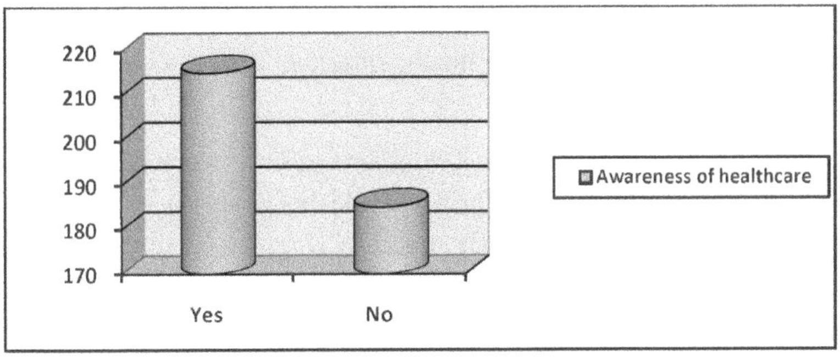

Fig. 13. Awareness of healthcare

Ecological crisis which includes drastic topographical and hydrological changes has brought several marine species to decline in numbers / not seen any more in the lake. Comparing with earlier records, diversity in the lake has been diminished due to the recent threats faced by the Pulicat lake such as siltation, lake mouth-closure, shrinkage and pollution. Siltation of the lake, reducing the water holding capacity so that the water in the lake does not last long enough for continued feeding and breeding of birds. Sand bar formation at the mouth of the lagoon threatens the complete ecological balance of the area. Wetlands are important repositories of biodiversity, some have become important sanctuaries for birds. The main reasons for these changes are depth reduction, reduction in freshwater flow and changes in water quality. It has been getting silted up at the rate of about one meter, per century. It buries all the benthic (bottom) habitats, flora and sessile fauna, resulting in a large scale or mass mortality of biodiversity. The lake mouth is now becoming shallower than before and sometimes it is completely closed. This greatly affects the fishery of Pulicat lake by blocking the movement aquatic organisms between the lake and the sea. The port construction diverts the currents to settle its sand particles in the lake mouth. The flora and fauna of the lake ecosystem are changing both quantitatively and qualitatively. The important migratory waterfowls are pelicans, herons, egrets, storks, flamingos, ducks, shorebirds, gulls and terns. Pulicat is the third most important wetland for migratory bird population along the eastern seaboard of India and during the spring and autumn migratory seasons (Jacobson and Sanjeeva Raj, 2009). The reduction of migratory birds are due to depletion in the feeding ground and the impact of global warming on the seasonal changes.

Introduction of fibre boats to the fishermen community which in turn to overfishing, juveniles catching and untargeted fishing, hence the wealth has been depleted. The reproductive cycles of the fin- and shellfishes will get affected by nonylphenol compounds emission through the debris of fibre boats which mimic the reproductive steroids (Gokulakrishnan *et al.*, 2014). Construction of bridge connecting from Pulicat town to Lighthouse, Kuppam and other surrounding villages have had its toll on destroying seagrass beds and its associated organisms. Mining of shells were very large in remote areas of the lake, thus causing depletion in biological diversity. Unethical practice of fishing through

"Padi-valai", fishing net with very fine mesh has affected the aquatic resources. Edible oyster of the lake were lost now due to heavy siltation. This also causes rapid shrinking water spread area of the lagoon.

Assessment of the biodiversity of a particular region is very important to formulate appropriate conservation strategies. Priority should be given to in-situ conservation by protecting the natural habitats. People's Biodiversity Register (PBR) should maintained by village panchayat, is a record of traditional knowledge and practices of sustainable use of local bioresources and conservation biodiversity. They may form an appropriate instrument for designing conservation efforts. This beautiful "lagoon" is faced with the surmounting challenges due to pollution, topographical and hydrological changes which may compromise the beneficial biodiverse ecological service of the Pulicat lake. Ensuring sustainability in fish production is crucial as it provides scope for enhancing self-employment, income generation and subsequently the quality of life.

Pulicat lake faces several anthropogenic, industrial and environmental issues threatening out only the livelihoods of fisher folk. Major threats of the lake are pollution from sewage, pesticides, industrial wastes, oil spills from boats carries Arani and Kalangi rivers draining into the lake, petrochemical complex, power plant and the satellite port of Ennore creek. The 630 MW North Central Thermal Power Station (NCTPS), the Ennore satellite port project and a Petrochemical park are major threats to the lake ecosystem. NCTPS draws 44 lakh litres of freshwater from the Ennore creek, lets out hot coolant water into the Buckingham canal and discharges toxic fly ash in the form of slurry into the lake. This has been increase of temperature $5^{O}C$ at the outlets. Despite precipitatous and chemical filters the fly ash contains toxic elements such as arsenic, cadmium, mercury, lead, manganese, fluorine and beryllium. Fly ash is especially harmful as it can be inhaled and leads to permanent respiratory disorder, dermatitis, asthma, bronchitis and cancer. The silica in fly ash could cause silicosis.

The emerging problem is the tourism, tourists who arrive are taken to *Casuarina* plantation to spent time. They carry much garbage and drop them. The plastic bags and other non-degradable materials have been collecting for a long time. Tourism industry in Pulicat lake gives much caution of concern towards ecological disturbance. Promoting eco-tourism on this fascinating Pulicat lake, would provide alternate livelihoods for traditional fishermen through punting tourist boats, setting up restaurants, souvenir sale, renting up camping sites and tents and through helping as guides for bird-watchers and tourists.

Increasing human population and industrialization during the last few decades have, however, led to an increasing pressure on this Pulicat ecosystem. Pulicat lake has been supporting rich and diverse fisheries from which a large population of the rural poor eked a living. Unfortunately, during the last 30 years, the lake has increasingly become a victim of diverse environmental tragedies, including over fishing, habitat fragmentation and pollution from chemical fertilisers, pesticides and thermal power. Prawn farms not only drain the little water left over during the summer, but also discharge untreated effluents from the farms into the lake. Consequently, the benthic habitats as well as their benthic communities of organisms get degraded and depleted, as a result of which the natural food-chains and food-webs of wetland birds have been seriously affected.

Conclusion

Fishermen have lower socio economic status and their illiteracy adds to their poor oral hygiene, which may influence general and oral health. Majority of the fishermen are stuggling for their welfare in the socio economic and social security. Currently the fishermen have poor economic and health status. There is an urgent need of government's attention in this regard to make them aware of importance of their health. The government can provide sufficient financial facilities at low rate of interest to the fishermen, make available health care facilities at cheaper cost and conduct awareness camp of health care services and health insurance. Evolving a technically sound and comprehensive approach, and people's participation in conserving this waterlogged wealth is vital. The wider and the deeper lake mouth is and the longer it opens into the sea, during the year, the more will be the vital inputs in the sea, like nutrients, oxygen, plankton and fish-feed, which are the very basis life and food chains in this lake. Protection and conservation of biodiversity is the urgent need of the hour in order to maintain the balance of nature and support the availability of natural resources for future generations. Unless the conservation strategies and crash programme for the sustainable development of Pulicat lake is evolved and implemented immediately, this lake will face imminent danger of sequential degradation into small pools, salt marshes, mud flats and ultimately a sure "ecological death".

Acknowledgements

The authors are thankful to Tamil Nadu State Council for Science and Technology (TNSCST), Chennai, India for the financial support.

References

Bene, C., Mindjimba, K., Belal, E. and Jolley,T., 2000. Evaluationg livelihood atrategies and the role of inland fisheries in rural development and poverty alleviation. The Case of the Yaere floodplain in North Cameroon. IIFET Proceedings, 15p.

Buyan, S., and Islam, S., 2016. Present status of socio-economic conditions of the fishing community of the Meghna River adjacent to Narsingdi District, Bangladesh. Journal of Fisheries and Livestock Production. 4: 192. doi:10.4172/2332-2608.1000192.

Fernando, A., 1981. Community development and Infrastructure facilities for improving the socio-economic conditions of fishermen, CMFRI Bulletin 30: 45-55.

Gokulakrishnan, S., Madhu Magesh, K., Premjothi, R. and Moses Inbaraj. R., 2014. Current status of macro fauna and its threat in Pulicat lake ecosystem. Ecol. Env. & Cons. 20 (Spl. Issue. Coast. Res. Manag.): 145 – 152.

Jacobson, O.W. and Sanjeeva Raj, P.J., 2009. Birds of Pulicat Lake – A Checklist. Pulicat Lake Bird Lovers' Society, 59 p.

Jasna, P.T. and Sanghamitra Palai, 2016. Socio-economic conditions of fishermen population with special reference to Chombala Harbour, Kerala. Indian Journal of Economics and Development 4(4): 1-8.

Jayaselvi, S. 2016. An economic and health status of fishermen in Tiruchendur. Shanlex International Journal of Arts, Science and Humanities 4 (1): 35-52.

Jeyanthi, P., Balasubramaniam, S. and Charles Jeeva, 2016. Socio-economic status of fishermen operating boats in Nagapattinam District, Tamil Nadu: a special focus on poverty and income inequality, Indian J. Fish. 63(2): 151-153.

Kurien, J. 2005. Responsible fish trade and food security. FAO Fisheries Technical Paper. No. 456, FAO, Rome: 102 p.

Mahender, J., Rajashekar, A.V. and Laxmappa, B., 2016. Study on socio-economic status of fishermen community of Chenugonipall Pedda Cheruvu Gadwal, Mahabubnagar Dist. Telangana. International Journal of Zoology 6(2): 10-13.

Salim, S.S., Narayanakumar, R., Sathiadas, R. and Manjusha, R., 2017. Appraisal of the socio-economic status of fishers among the different sectors in Kerala, South-West coast of India. Indian. J. Fish., 64 (1): 66-71.

Sanjeeva Raj, P.J., 2010. Management of Pulicat Lake. Wildlife Management Division, Andhra Pradesh, India., 42 p.

Sathiadas, R. and Panikar, K.K.P., 1991. Socio-economics of traditional fishermen in Tirunelveli coast, Tamilnadu. Journal of the Marine Biological Association of India, 33 (1&2): 175-181.

Scott, H.G., 1954. The economic theory of a common-property resource, the fishery. J. Polit. Econ., 62: 124-142.

Sudharsanam, M.B., 2006. Risk factors for sick children in a fishermen community in Pondicherry. Indian Journal of Community Medicine 31(4): 308-309.

Author Index

J

K

L

M

N

P

R

S

T

U

V

Z

Subject Index

D

E

F

G

H

I

J

K

L

M

N

O

P

Q

R

For Product Safety Concerns and Information please contact our EU
representative GPSR@taylorandfrancis.com
Taylor & Francis Verlag GmbH, Kaufingerstraße 24, 80331 München, Germany

www.ingramcontent.com/pod-product-compliance
Ingram Content Group UK Ltd.
Pitfield, Milton Keynes, MK11 3LW, UK
UKHW052224190926
475702UK00007B/1185
* 9 7 8 1 0 3 2 2 9 0 4 3 0 *